CW01333881

AN ENCYCLOPAEDIA OF PLANTS in MYTH, LEGEND, MAGIC and LORE

AN ENCYCLOPAEDIA OF PLANTS in MYTH, LEGEND, MAGIC and LORE

STUART PHILLIPS

ROBERT HALE · LONDON

First published in Great Britain 2012

ISBN 978-0-7090-9150-9

Robert Hale Limited
Clerkenwell House
Clerkenwell Green
London EC1R 0HT

www.halebooks.com

A catalogue record for this book is available from the British Library

2 4 6 8 10 9 7 5 3 1

Illustrations by Jill Glover
Designed and typeset by Paul Saunders
Printed by Imprint Digital Limited, India

This book is dedicated to the first professional gardener
I ever met, and the one who taught me the most,

CLARENCE WILFRED PHILLIPS,

my father

	flaw / goal	conflict
bk1	attachment / hatred	ego
bk2	dissociation / identity	self
bk3	forgetfulness / memory	mind
bk4	social influence / incapacity	power
bk5	relaxed / stressed	hormones
bk6	agressive / submissive	gender
bk7	knowledge / ignorance	love
bk7/8	fantasy / reality	perception

CONTENTS

PREFACE

No natural exaltation in the sky,
No scape of nature, no distemper'd day,
No common wind, no customed event,
But they will pluck away its natural cause,
And call them meteors, prodigies, and signs,
Abortive, presage, and tongues of heaven,
Plainly denouncing vengeance.

– William Shakespeare, *King John*

Plants have always held something of a fascination for me. This comes not, as many might think, from the love of their cultivation, but rather from the way in which they are central to so many aspects of our lives. It started, I believe, with English literature classes at school and the imagery of Shakespeare, Pope, Marlowe, Donne, Webster and others; the mixture of once-held beliefs and classical references that laced throughout their poetical and dramatic works.

When I began my training as a gardener it was impressed upon me that I needed to know much more about plants than their names. I suspect that the tutor had in mind that I should learn about their countries of origin, their methods of propagation and their cultivation requirements. However, I also learned about the symbolic meanings of plants and their significance in mythology and folklore. This came in part from trying to learn botanical names, and discovering that many of the 'Latin' names were derived from Greek, and some from classical Greek myths. It was about then that I discovered Gertrude Jekyll.

In a collection of articles by Gertrude Jekyll I found a short piece she had written on the subject of the common names, or folk names, of plants. The one that hooked my attention was that for *Sempervivum tectorum*, the common houseleek. This plant has the wonderful alternative common name of 'welcome-home-husband-however-drunk-you-be'. How does a plant ever get a name like that? Having found one bizarre name I began to discover others of which I had previously been blissfully ignorant. Some of these names are now disappearing as a single, standard, common name is adopted. This is due, in part, to the use in magazines, and on television and radio, of one common name. In consequence some of the more interesting names are lost, and it is also becoming more difficult to track down plants referred to by writers in the sixteenth and seventeenth centuries who used these older common names.

As I have collected these snippets of information I have discovered that there are other people equally interested in them for all sorts of different reasons, not the least of which is the general desire for knowledge. Somewhere between these covers is the answer to such burning questions as 'Why do the Welsh fans brandish leeks at the Millennium Stadium?' 'What have the holly and the ivy got to do with Christmas?' and 'What is wassailing?' There are also some details regarding the use of plants in magical rites. These are included because many of our 'old wives' tales' actually relate to magical practices. Most would be classed as simple spells, such as those to discover whom a maid would marry, or cures for warts. Others are there to provide protection against the Devil and the forces of evil. It is important to remember that our folklore includes numerous stories of imps, elves, faeries, brownies and boggarts. Belief in these pre-dates the acceptance of Christianity across the country. Plants were widely used to give protection from their mischief or malice and to gain their favour. Inevitably the use of plants as protective amulets was later applied to ward off the dark forces spoken of in the Christian tradition.

The entries in this book are not all British plants, or even those found growing within the UK. I make no apology for this, as there is ever more interest in plants from elsewhere around the world, and an increasing desire to cultivate exotic species in our homes.

INTRODUCTION

Folklore is a mirror held up to the tangle of beliefs and practices of people long ago: so it often appears like a collection of isolated legends and superstitions.

Folklore is neither fixed nor finite but rather changes and develops with each passing era. Rarely is anything actually lost, rather it is adapted, expanded and absorbed into the new system of ideas and beliefs which are imposed on top of what has gone before. In consequence, looking at any aspect of folklore is rather like looking at a web where each strand is connected to all the others and there is layer after layer working out from a central point. Our folklore is a hotchpotch of myths and legends, traditional practices and old wives' tales. All of which is served up with a liberal dressing of superstition, a sprinkling of exaggeration and a seasoning of a few grains of truth.

Plants have been key to the survival of mankind since the dawn of creation. We have been reliant on the plant kingdom for our food, shelter, clothing, medicines and even the very air we breathe. It is hardly surprising, therefore, that plants have been enmeshed within this tangle of tales and beliefs.

> To primitive people who thus symbolized natural phenomena, vegetable life was, in a manner, glorified, because it sustained all other life. The tree supplied lumber, fuel, house, thatch, cordage, boats, shields and tools as well as fruit and medicine.
>
> – Charles M. Skinner *Myths and Legends of Flowers, Trees, Fruits and Plants*

When, often as a result of exploration or invasion, new plants were introduced, the folk stories and traditions associated with them were also introduced. Sometimes the stories might travel without the plant and so became applied to a completely different species. Consequently, in times when people might live their whole lives without venturing beyond the boundaries of their own village community, very similar beliefs could be found associated with unrelated plants which might have only the slightest similarity in appearance. Further confusion is added because several plants might have the same common name and so the traditional beliefs about one plant can suddenly be found being applied to another. This is particularly worrying where the plant has found use in folk medicine and herbal remedies.

The entries in this book have been arranged in alphabetical order according to the botanical names, although each individual entry is still headed with the common name. This is done to allow for comparison to be made between related species. It is for this same reason that the botanical family is also identified in each case. Common names are given as, unless the reader is a keen horticulturist or a botanist, this is the name most likely to be recognized. However, as common names may vary from place to place various common names by which the plant is, or has been, known are listed. Many of the more entertaining common names are being lost as a result of standardization in the various magazines and television programmes that deal with plants.

Where applicable the planetary influence, element and 'gender' of each plant are given. These tend to relate to the plants used in magic and in astrological herbalism. Seventeenth-century botanists and herbalists, such as Nicholas Culpeper (1616–1654), believed that plants and ailments were governed by the constellations. Likewise, different planets governed various parts of the human body. Thus, the sun governed the heart, circulation and spinal column. The moon reigned over growth, fertility, breasts, stomach, womb and menstrual flow. In fact all body fluids were, like the tides, controlled by the phases of the moon. Mercury controlled the central nervous system together with those organs involved in speech, hearing and respiration. The complexion, genitalia and hidden workings of the body's cells were all under the domain of Venus. Mars ruled libido, muscles and body vitality. Saturn presided over ageing, teeth, bones and hardening processes. The liver, abdomen, spleen and kidneys were under the influence of Jupiter, as was digestion. It should be remembered that the followers of this particular system were unaware of any other planets beyond these seven. They would prescribe their remedies with careful note as to the astrological influence on the plant. This was the right belief for its time, as there was a widespread acceptance of astrology as the key to gaining an understanding of the universe.

This is not the only system of herbalism referred to in the text, as there is also frequent mention of the Doctrine of Signatures. This dates from the sixteenth century and so pre-dates Culpeper. There are allusions to a system of this basis in the writings of Galen (AD 131–200) but it did not really appear as a complete philosophy until the publication of *Signatura Rerum* (*The Signature Of All Things*) by Jacob Boehme in the first part of the seventeenth century. This Doctrine of Signatures was not formulated as a guide for the practitioners of medicine but was a spiritual concept founded on the theory that God marked everything in creation with a sign. This sign indicated the true purpose of the object or organism. It therefore followed that the merciful Almighty, having created all ailments, had created an appropriate treatment if only we could read its 'signature'. Basically, things were taken as treatments for what they looked like. Hence *Pulmonaria*, having a speckled leaf, was said to look like a lung and so could be used to treat illnesses of the lungs and respiratory system. From this the plant gains the common name of lungwort. This was the system propounded by Theophratus Bombast von Hohenheim (1493–1541) who is probably better known by the Latinized name of Paraceleus.

His ideas were seized upon by Giambattista Porta for his *Phytognomonica* which was published in Naples in 1588. In its simplest sense Porta suggested that long-lived plants could be used to lengthen life expectancy but short-lived plants would shorten life. In Britain this system was promoted by William Coles (*Art of Simpling*, 1656), who informed us that it was sin and Satan that were the cause of all mankind's ailments but that God had made herbs for man to use and 'given them particular signatures whereby man may read even in legible characters the use of them'.

It is difficult to know how far the Doctrine of Signatures was accepted by the educated classes and it was already being dismissed by writers such as Dodoens (1517–85) before the end of the sixteenth century. However, it can still be found mentioned in works well into the eighteenth century.

Both of the above systems of herbalism are reliant to some extent on 'sympathetic magic', the idea that things can be influenced by items that look like them or 'correspond' to them. This is the same belief that underlies the use of voodoo dolls in Tahiti or of mandrake pappettes in Europe. It can be seen in the suggestion that orchid bulbs, tomatoes and other things that vaguely resemble testicles might prove to be aphrodisiacs.

Another belief which was once common was that our characters were formed as a result of the relative levels of the four humours or fluxes of the body: blood, phlegm, yellow bile and black bile. Blood gave a sanguine character, phlegm a phlegmatic character, yellow bile a choleric character and black bile a melancholic character. Some of the entries in the text suggest that a plant relieves melancholia or lifts the spirits. Rather than simply treating the symptoms of what we now call depression, the plant was believed to drive out the 'black bile' that causes the condition.

This collection was never conceived as a herbal, but inevitably various references to herbal remedies have been included. These are, for the most part, folk remedies rather than homeopathic herbal treatments. In any event, great caution must always be exercised when using plants, or plant extracts, as medicines, as many can prove to be extremely poisonous. They can be as likely to kill as to cure. In like manner, reference is occasionally made to the hallucinogenic properties of certain plants. Whilst this use may be reported it is not a recommendation, as many of the plants that have been use in that way in the past are extremely toxic.

Various plants have, at some point, become closely associated with particular deities. Where these can be identified they have been noted. This may be useful in tracing stories relating to the name given to the plant or in looking at the way in which a particular species has been identified with different religious figures. In this way it is possible to trace ideas which, having been applied to one goddess, might then be applied to another and finally to the Virgin Mary. It is often interesting to see the development of beliefs around a plant as it becomes 'Christianized'. Similarly as the various festivals and religious feasts were absorbed into the Christian calendar, so the plants and flowers used in decorations at that time were given a Christian significance.

The use of plants in symbolism dates from time immemorial and is still relevant today. Most of us would know the olive branch as a symbol of peace and will give red roses as tokens of love. Some of these traditional meanings are now being forgotten. For example how many people would instantly know that a picture of an ear of corn near to a fall of water was emblematic of plenty? Some knowledge of this imagery can be helpful in getting under the skin of poetry, prose and plays. Shakespeare and the other writers of his time made great use of such images to colour their work. The audiences of the day would probably have had a better grasp of the significance of the plant alluded to than a modern audience.

Many modern readers will also probably not be aware of the key dates in the Christian and pagan calendars. Whilst we remember Christmas, Easter and Halloween most people are unaware of Samhain, Beltane, All Soul's Night and others. Festivals such as May Day and Midsummer have lost their historical significance and the traditional activities once associated with these feasts have all but gone. There were, of course, rites and spells linked to these various festivities, many of which required particular plants, just as we associate mistletoe and holly with Christmas even now.

From this framework of classical symbolism grew the language of flowers. A full index of plants and their respective meanings has been included in the appendices at the back of the text, but where appropriate the meanings of individual plants have been included under that entry. The language of flowers came to its height in the Victorian age but, as will be seen in the full list, became somewhat unwieldy and cumbersome to use. If, however, you are in any way doubtful of the power of plants as an emblem you need only watch the Six Nations rugby tournament. The English rose, Scottish thistle, Welsh leek and Irish shamrock will all be very much in evidence.

The vast majority of the information included in the book has been gleaned from printed material rather than from oral tradition, which may only serve to compound the errors of others and to introduce a few new ones for the future. This need not be a problem in itself, as folklore, by its very nature, has evolved and developed over the passage of years. For the greater part of our history the development of folk traditions has depended on the spoken word and the fallible memories of people. The diversity of our folklore is due to the alterations and adaptations that have occurred as a result of this elaborate game of Chinese whispers. Even the classical tales of myth and legend have not been exempt from being twisted and changed in their retelling. Consequently there are several versions of the same story, all of which are equally 'correct'.

I hope that readers will find the contents of this book entertaining and useful. What started as a personal collection by a self-confessed 'plantaholic' has grown out of all proportion. It is far from complete, and the greatest problem in preparing it for publication has been re-editing having found yet another juicy snippet of information that ought to be included. A compilation of this nature is unlikely ever to be absolutely definitive and so, as the frequency at which new information appears has become significantly less marked, it was felt that the 'story so far' could safely be offered for publication.

Stuart R. Phillips

Acknowledgements

This book would never have been completed but for the help, encouragement and enthusiasm of many other people, some of whom deserve individual mention for their particular assistance.

Thanks must be given to the library staff of Reaseheath College in Cheshire and Moulton College, Northampton, particularly Peter Dale, Janet Jones and Gary Meades.

Although a number of people helped with the research, the following have particularly generously given of their time, skill and knowledge: Jackie Bagnall (Reaseheath College), Louise (Glade) Bustard (Glasgow Botanic Gardens), Dr Cutler FLS (Royal Botanic Gardens, Kew), Ian Phillips (brother and book buyer) and Maria Paz Tilley.

Over the book's several years gestation period there have been people who have read bits and piece, then corrected them, made suggestions and generally kept the whole thing moving. Amongst the foremost of these are: Tania Glover, Roger Clarke, Caroline Barton, Pauline (Polly) Smith, Nick Warliker and Dave Watkins.

To these people and all the others who have helped I extend my unreserved thanks.

DICTIONARY OF PLANTS IN FOLKLORE, MYTHOLOGY AND MAGIC

FIR

Botanical Name: *Abies* spp.
Family: Pinaceae
Associated with: Adonis; Artemis; Attis; Bacchus; Cybele; Diana; Dionysus; Druantia; Erigone; Hathor; Io; Neptune; Osiris; Pan
Meanings: Elevation; time

Although the common name of fir correctly belongs to the genus of plants *Abies* it has frequently been applied to almost any coniferous tree species. Amongst the foremost of the plants with which the folklore of this plant overlaps are the pine, spruce, larch and Christmas tree.

The lightweight nature of the timber of this group of trees led to its use in the construction of triremes, ancient ships that had three ranks of oars. Along with the cypress and the yew, the fir was commonly used for funeral pyres in ancient times and so, like these other two species, is linked with the great circle of death and rebirth.

In Russia, in a village near to Dorpat, whenever rain was needed three men would climb the fir trees in an old, sacred, grove. One of them would drum on an old kettle to imitate the sound of thunder whilst another would knock firebrands together to make sparks symbolizing the lightning. The third, the rainmaker, would drip twigs into a container of water and sprinkle the droplets all around, mimicking the raindrops. This was, presumably supposed to encourage the rain to follow their example and is clearly a form of sympathetic magic.

In the interpretation of dreams, seeing yourself in a forest of fir trees is supposed to be a sign of suffering, either current or impending.

See also *Pinus* (pine), Christmas tree (*Picea abies*), Scots pine (*Pinus sylvestris*) and Larch (*Larix decidua*).

INDIAN LICORICE

Botanical Name: *Abrus pecatorius*
Family: Leguminosae
Common Names: Crab's eyes; goonteh; gunga; jequirity beans; prayer beads; rati; rosary pea; wild liquorice

The attractive red and white seeds produced by the Indian liquorice have been used in the manufacture of rosaries. They are unsuitable for necklaces as, like the seed of most legumes, they are very poisonous.

In the 1890s it was claimed that, from a close observation of changes in this plant's growth patterns, it was possible to predict extremes of weather, such as typhoons and hurricanes. It was further claimed that major movements in the earth's crust, including earthquakes and volcanic eruptions could also be predicted. The plant was said to be especially attuned to fluctuations in the earth's magnetic field, which resulted in the changes in its growth. These claims were eventually brought to the attention of the director of the Royal Botanic Gardens at Kew, who was asked to investigate them. A full study was made, and the claims proved to be totally false. However, in spite of this, the assertion that this plant foretells storms and natural disasters is still repeated in books and magazine articles from time to time.

ACACIA

Botanical Name: *Acacia* spp.
Family: Leguminosae
Common Names: Altar of the tabernacle; crown of thorns; wattle
Planet: Sun
Element: Air
Associated with: Allat; Al-Uzzah; Astarte; Diana; Ishtar; Osiris; Ra
Meanings: Friendship; platonic love; 'ours is a chaste love' (white); elegance; 'you possess a queen's majesty' (pink); secret love; 'let us disclose our hearts to no one' (yellow)

This is a group of attractive flowering trees and shrubs that come, for the most part, from the warmer regions of the world, typically Australia and Africa. The 'mulberry' mentioned in the Bible (1 Chronicles 14: 15) is, most probably, a species of acacia. They should not be confused with *Robinia pseudoacacia*, the false acacia or locust tree.

Myth and Legend

Amongst the Arab races the acacia was considered sacred to the love goddess Al-Uzzah, whose symbol is a cluster of acacia trees. Indeed, it was thought to be so sacred that if anyone broke so much as a twig from the tree they would die within a year. Umbrella acacias were known as the 'daughters of Allat'. Allat forms a third of the triple goddess along with Al-Uzzah and Manat.

In the myth of Osiris, the acacia is identified by some with the tree that became the resting place for the sarcophagus containing the body of the god. Osiris has, therefore, been called variously 'the one in the tree', 'lord of the acacia tree' and 'the one in the acacia'. Later in the same story, after the body of Osiris had been cut to pieces by his brother Set, we are informed that Isis, the wife of Osiris, and Horus, his son, sailed around the Nile delta in an ark made of acacia wood searching for the various body parts.

The ark of Isis is not the only one reputed to have been made of acacia wood; those of Noah, Deucalion, Xisuthros and Parnapishtim are all said to have been constructed of this timber. In Greek myth Danae, and her son Perseus, were set adrift in an acacia wood ark by her father, Acrisius.

Magic and Lore

Acacia was one of the four trees of the gods in ancient China and is generally considered a protective plant. A sprig placed over a bed wards off evil. It is considered sacred by Buddhists and Hindus, and in some eastern countries it is worn in the turban as a protective amulet. It may be this association, or its use as symbolic of eternal life, that led to its use as a symbol in Freemasonry.

Acacia pyracantha, the golden wattle, is the national flower of Australia.

MIMOSA

Botanical Name: *Acacia dealbata*
Family: Leguminosae
Gender: Feminine
Planet: Saturn
Element: Water
Meanings: Sensitive love; courtesy; 'You are too brusque with my tender feelings'

This plant should not be confused with the 'sensitive plant', *Mimosa pudica*.

In common with other acacias, mimosa is alleged to have protective qualities and can be used to remove curses or spells cast against you. To do this it is necessary only to bathe in water containing an infusion of mimosa, or to sponge the cleansing infusion over your body. This will also act as a protection against future troubles. To purify buildings, or areas of land, mimosa must be scattered around the whole site.

In addition to its protective powers mimosa can encourage prophetic dreams; it need only be placed beneath the pillow before going to bed. There are those who say that, like lilac and hawthorn, mimosa should never be taken into the house, as it will ultimately bring bad luck.

GUM ARABIC

Botanical Name: *Acacia senegal*
Family: Leguminosae
Common Names: Cape gum; Gum arabic tree; Egyptian thorn
Gender: Masculine
Planet: Saturn
Element: Air
Dieties: Astarte; Diana; Ishtar; Osiris; Ra

The timber of this tree is used in the construction of temples in India, and for fuelling sacred fires. When burnt with sandalwood it is supposed to stimulate the psychic powers. It can be used in magic rites to bring platonic affection, and in spells to attract money.

Gum arabic is added to incense or placed on charcoal because, on being burnt, it wards off negativity and evil, thus purifying the whole area.

Most texts state that Kordofan gum and Senegal gum are derived from *Acacia senegal*.

RED ACACIA

Botanical Name: *Acacia seyal*
Family: Leguminosae
Common Name: Shittim wood

The shittah tree is mentioned only once in the Bible ('I will plant in the wilderness … the shittah tree', Isaiah 41), but its wood is referred to many times as shittim, the plural of *shittah* in Hebrew. It has been speculated that Moses would naturally use shittim wood when he constructed the Ark of the Covenant and the Tabernacle; however, no one can be sure which species of acacia was actually meant.

The wood is soft, coarse-grained and easy to work, and polishes well, but it is susceptible to insect attack and quickly discolours easily with mould. The ancient Egyptians used the timber for making coffins. During the Exodus the Israelites absorbed many Egyptian beliefs and customs. The Egyptians had long held the acacia as a symbol of eternal life and it is easy to see how the traditional Egyptian use of the timber could have been taken on by the Israelites.

BEAR'S BREECHES

Botanical Names: *Acanthus spinosus*; *Acanthus mollis*
Family: Acanthaceae
Common Name: Brank ursine
Diety: Acantha
Meanings: Artifice; love of fine arts

The motif of acanthus leaves appears atop columns from the Roman and Greek periods. It is said that the used of this design came about after the young daughter of Callimachos died whilst he was employed designing pillars in Corinth. He is said to have sent a basket of flowers to her grave but to prevent them being blown away by the wind he placed a tile over the top. When next he visited her grave an acanthus plant had grown up around the basket and where the leaves reached to the top of the tile the leaf margins bent back around its corners. Being struck by the beauty of the leaf against the form of the tile Callimachos introduced the motif of the leaf into his designs, curling the edges of the leaf motif around the tops of his columns.

MAPLE

Botanical Name: *Acer* spp.
Family: Aceraceae
Gender: Masculine
Planet: Jupiter
Element: Air
Meaning: Reserve

The maples are predominantly a group of trees, although there are a few shrubby species. The name *Acer*, meaning sharp, refers either to the angular nature of the leaves, or to the use to which the timber was once put. The hardness of the wood made it ideal for making spear shafts. The timber is greatly valued for magical purposes and has been used for making magic wands. The leaves are used in love spells and rituals to attract money.

According to Alsatian folklore, bringing maple branches into the house would ensure protection against bats, which would then not dare to enter. It also ensured that nesting storks were safe against disturbance and even protected the chicks from being killed whilst still in the shell.

Amongst the learned rustic weather forecasters the turning of maple leaves, so that the undersides are seen, is a sign that rain is to be expected.

Medicinal

Passing a young child through the branches of a maple was traditionally thought to encourage good health and a long life for the child. This has obvious links to the sympathetic magic use of Ash saplings in a cure for ruptures.

FIELD MAPLE

Botanical Name: *Acer campestre*
Family: Aceraceae
Common Names: Common maple; dog maple; oak; oak apple

Confusingly, the field maple was once occasionally known as 'oak', or sometimes as 'oak apple', in various parts of Britain. This seems to originate in the tradition of wearing pieces of oak on 29 May, Oak Apple Day. The penalty for failing to display the oak was to be pinched or stung with nettles. It would seem that in some communities wearing a sprig of field maple was accepted as a reasonable substitute.

The foliage of the field maple would, of course, be much easier to reach than that of an oak, as it is a relatively small tree. The timber of this tree is, likewise, compact and fine grained. It can, sometimes, be very beautifully marked which led to its use in furniture-making and the manufacture of violin cases.

SYCAMORE

Botanical Name: *Acer pseudoplatanus*
Family: Aceraceae
Common Names: Faddy tree; great maple; grief tree; may tree; peweep tree; plane tree (Scot.); Scottish maple; seggy; segumber; share; tulip tree; whistle tree; whistle wood. (Seeds may be called: chats; horseshoes; knives and forks; locks and keys.)
Meanings: Curiosity; 'I wish to know more about you'; reserve; woodland beauty

Sycamore is supposed to be a tree of protection and favours. It survives in many situations that other tree find inhospitable. It is not the sycamore referred to in the Bible, which is a species of fig, *Ficus sycamorus.*

Myth and Legend

Irish legend tells of a great sycamore that grew at Clonenagh, County Laois, which was associated with St Fintan (died AD 603). Tradition has it that it grew on the site of his well, but that the well was desecrated by livestock being allowed to drink from it. The consequence of this desecration was that the tree was miraculously transported to a new site at Cromogue, some 3 miles away, which was the site of another church dedicated to the saint. It was reported that water could be found in a cavity in the tree even in the driest weather. This was thought to come from a natural spring immediately beneath the tree roots. The water from this 'well' could be used as a panacea to treat all manner of ailments and afflictions.

Magic and Lore

In Scotland the feudal lairds chose sycamores as dool or joug trees, that is to say they were selected to act as gallows, from which erring peasants would be hanged as an example to others. It is from this usage that the sycamore became known as the grief tree, and it is thus usually considered to be a tree of bad luck in Scotland.

In rural areas it is said that when the bark of a sycamore remains smooth and white through the autumn it is a clear indication that the winter will be mild.

Traditionally, to dream of sycamore is an omen of jealousy, if the dreamer is a married person. If they are single then such a dream indicates that they will soon be married.

YARROW

Botanical Name: *Achillea millefolium*
Family: Compositae
Common Names: Achillea; angel flower; arrowroot; bad man's plaything; bloodwort; bunch o' daisies; camel; cammock; carpenter's weed; death flower; Devil's nettle; Devil's plaything; Devil's rattle; dog daisy; eerie; fever plant; field hop; *gearwe*; goose tongue; green arrow; hemming and sowing; *herbe aux charpentiers*; *herbe militaris*; hundred leaved grass; knight's milfoil; knyghten; lady's mantle; melancholy; milfoil; militaris; military herb; moleery-fea; mother-die; mother of thousands; noble yarrow; nose bleed; old man's mustard; old man's pepper; sanguinary; seven year's love; snake's grass; sneezewort; soldier's woundwort; staunch griss; staunch weed; sweet nuts; tansy; thousand-leaved clover; thousand seal; thousand weed; traveller's ease; wild pepper; woundwort; yarrel; yallow; yarra-grass; yarroway; yerw
Gender: Feminine
Planet: Venus
Element: Water
Deity: Achilles
Meaning: War

Yarrow is the birthday flower of 16 January and is supposed to be symbolic of heart-ache and cure. The common name seems to be derived from the old English 'gearwe', which referred to the many indentations of the leaf.

Over the centuries yarrow has been used to make snuff and, in the Middle Ages, combined to form gruit, a mixture of yarrow with other herbs to flavour beers before the use of hops. Gruit might be made up from a selection of ingredients including yarrow, juniper berries, sweet gale (*Myrica gale)*, mugwort (*Artemisia vulgaris*), heather (*Calluna vulgaris*), Labrador tea (*Rhododendron tormentosum*), ginger (*Zingiber officinale*), cinnamon (*Cinnamonum zeylanicum*), caraway seeds (*Carum carvi*), aniseed (*Pimpinella anisum*), nutmeg (*Myristica fragrans*) and even hops (*Humulus lupulus*). Some of these plants were known as having mild narcotic properties and others known for their preservative qualities. The actual mixture of gruit would vary in order to give differing flavours to the beers. The use of gruit died out as hops became the standard flavouring for beers.

Myth and Legend

The botanical name comes from Achilles, the hero of Greek mythology. The legend tells that Thetis, the mother of Achilles, dipped her son into the magical waters of the River Styx and as a result he was rendered invulnerable. His only weak spot was that point of his foot by which his mother had held him, literally Achilles' heel. Achilles was eventually slain, during the Trojan wars, by an arrow fired at him by Paris but directed by the god Apollo. This struck him in the heel and led to his death.

Achilles was a pupil of Chiron, a centaur, who taught the art of healing with plants. When the Greeks invaded Troy, at the start of the Trojan wars, Telephus, king of the Myceans and the son-in-law of King Priam, tried to prevent them from landing. The god Bacchus intervened causing Telephus to stumble over a vine, and Achilles wounded him with a spear. The wound failed to heal, and Telephus consulted the oracle for guidance. He was told that Achillea would heal him. Rather than seeking out the plant he went to Achilles, who agreed to heal him in return for being guided into Troy. The Greek chieftain scraped rust from the spear used to wound Telephus and from these scrapings sprang up the milfoil. A wider told legend tells that Achilles used the milfoil to treat the wounds of his soldiers at the siege of Troy.

Magic and Lore

Like many other white-flowered plants, it is thought to be unlucky to pick yarrow and take it into the house, as it would cause sickness or the death of the collector's mother. Occasionally it is seen as a protective herb, and strewing it on floors was supposed to be a way of keeping witches away.

Originally yarrow was seen as a plant dedicated to the Devil and had names such as 'Devil's plaything' or 'Devil's rattle'. It was because of this association with the Devil that, in the seventeenth century, a witch was tried for using yarrow in her incantations. However, if suitably collected and treated it could be used in divination and to ward off evil. The tradition in Sussex was that if yarrow was to be used it must be gathered from the grave of a young man. Elsewhere in Britain this is refined still further with the direction that it should be collected by the light of a full moon at midnight, and that the person who gathered it should never have previously seen the churchyard from whence it comes. It might be worn as a protective amulet, or carried in the hand to give courage and prevent fear.

In France and Ireland yarrow is dedicated to St John, and on St John's Eve (23 June) it would be hung up in the house to dispel illnesses. In the USA, Native Americans burnt yarrow in order to drive away evil spirits. In China the stems of one yarrow species have been a favourite source of stalks to be used in I Ching. The species used produces straight stems up to 900 mm long. In the Orkneys there is a belief that the consumption of a tea made from yarrow will dispel melancholia, and elsewhere it is said that this will heighten the psychic powers and that holding the yarrow up to the eyes gives the power of second sight. Witches are supposed to have used the plant in potions to grant strength and courage, see the future or to inspire love.

As with many other herbs there are rituals and incantations to be observed when gathering yarrow. The following is a traditional Gaelic incantation, translated by Kenneth Jackson:

> I will pick the smooth yarrow that my figure may be sweeter, that my lips may be warmer, that my voice may be gladder. May my voice be like a sunbeam, may my lips be like the juice of the strawberry.
>
> May I be an island in the sea, may I be a hill in the land, may I be a star in the dark time, may I be a staff to the weak one: I shall wound every man, no man shall hurt me.

Unlikely as it seems the yarrow plant is also used in love spells. Simply carrying the plant should attract those to you that you would most like to see. If you seek to learn whom you will marry then the yarrow must be sown into a red flannel and this placed under your pillow before you go to bed. If you say this rhyme before you go to sleep you can be sure of dreaming of your future spouse.

> Thou pretty herb of Venus' tree,
> Thy true name is Yarrow,
> Now who my bosom friend must be,
> Pray tell Thou me to-morrow.
>
> – Halliwell, *Popular Rhymes*

An alternative to this is to place the flowers of the yarrow under the pillow saying:

> Yarrow, sweet yarrow, the first that I have found,
> In the name of Jesus Christ, I pluck it from the ground,
> As Jesus loved sweet Mary, and took her for his dear,
> So in a dream this night I hope my true love will appear.

A more complicated variation on this states that a young woman must gather the yarrow at midnight, from the grave of a young man whilst saying:

> Yarrow, yarrow, I seek the yarrow,
> And now thee I have found,
> I pray to the good Lord Jesus,
> As I pluck thee from the ground.

It should then be taken home and put into the right stocking which should then be tied to the left leg. The young woman had to get into bed backwards whilst saying, 'Goodnight to thee yarrow' three times. After that she had to repeat, again three times:

> Goodnight pretty yarrow,
> I pray thee sweet yarrow,
> Tell me by the morrow,
> Who shall my true love be.

She was then sure to dream of her future spouse.

Alternatively a girl might go out on May Eve and remove nine leaves from a yarrow plant whilst saying:

Yarrow for yarrow, if yarrow you be,
By this time tomorrow my true love to see;
The colour of his hair, the clothes he doth wear,
The first words he will speak when he comes to court me.

The nine leaves should then be put under her pillow to ensure that she would dream of her future love.

More simply, in Irish tradition, she might cut a sod containing yarrow and put a piece beneath her pillow to be sure of dreaming of her future husband; the size of the piece of turf is not specified. It is usually stated that she must not speak from the time of cutting the sod until the time that she falls asleep. A variation of this recommends that, to encourage someone to fall in love with you, you must 'sow' an ounce of yarrow collected from a dead man's grave and put it beneath the pillow of the person whose love you seek.

Other suggestions for identifying a future husband are to cut across the stem, when she will find inside the initial of her true love, or to throw yarrow onto the fire, when an image of her future husband will appear in the flames.

These practices of conjuring up images of a future lover are not restricted to women. In the Hebrides there is a sad tale of a young man who is hopelessly in love with a girl, but is heartbroken when she marries another. He uses yarrow to enable him to conjure up an image of her. As he pines for his lost love he fades away until his father ends up carrying all that remains of him around in fishing creel.

The ferny leaves of the yarrow were once used in wedding flowers and the bridal bouquet, as a result of which it gained the name of 'Venus tree'. One superstition claims that including it in the wedding bouquet, hanging a bunch of the flowers over the wedding bed, or eating a little of it at the wedding dinner will ensure that the couple's love lasts for at least seven years.

You can also check that your lover still cares for you by inserting a plug of yarrow leaves up your nose then reciting the rhyme:

Yarroway, Yarroway, you wear a white blow,
If my love love me, my nose will bleed now.
If my love don't love me it 'on't bleed a drop,
If my love do love me, 'twill bleed every drop.

There are several variations of this rhyme.

If a young woman went into a moonlit field where yarrow was growing and, with her eyes tight shut, picked some with the dew still on it, she could find out whether she would become popular with the opposite sex. If the pieces she had picked were still damp the following morning it was a sure sign that young men would soon start to take an interest in her.

Yarrow is not only linked to attracting members of the opposite sex. A quite different tradition says that if you smear your hands with yarrow, and plunge them into the river, it will attract fish to you.

To dream of picking yarrow, for use in herbal medications, is an omen that the dreamer can expect to hear good news in the near future.

Medicinal

The name 'nose bleed' may be a reference to a use of the plant as a herbal remedy or from its use in divination. In *Gerard's Herbal* a nosebleed was identified as a cure for headaches, as the head pain must be due to too much blood pressure in the head: 'the leaves being put into the nose do cause it to bleed, and easeth the pain of the megrin.' Smelling yarrow flowers was, on the other hand, said to be a cure for nosebleeds.

Milfoil has, of course, for centuries been recommended for the treatment of bruises, cuts and wounds, an infusion of the leaves being applied externally, hence the references to it as the military plant. It is also called 'carpenter's weed' as it was used to treat chisel and saw cuts.

Yarrow has also been dried and powdered to be used as snuff and it is from this that names such as 'old man's pepper', and 'old man's mustard' arise. It is more likely, however, that is was a close relative of yarrow, *Achillea ptarmicta,* which was commonly used as *ptarmica* comes from the Greek meaning sneeze making. It has also been claimed that yarrow, applied to the scalp as an infusion, is a sure way of preventing baldness, but it is unable to remedy hair loss once it has started. This may have resulted from confusion between yarrow and *Artemisia.*

Yarrow was said to drive toxins from the body and so was recommended as a treatment for snakebites, amongst other problems. It has been used to treat toothaches, given as a cure for colds and has even been taken as a substitute for quinine in the treatment of fevers. A decoction of yarrow might be prescribed as a remedy for piles. Yarrow tea is said to be an aid to sleep.

In general terms it was thought that adding a little yarrow to other herbs being used would intensify their effect.

In *A Garden Of Herbs*, Eleanour Sinclair Rohde says:

> Just because it is the custom, we make use of all the showier 'herbs', which now fill our kitchen gardens, not only because they are pleasant, but also because of their health-giving properties; but why neglect the older herbs: sage, thyme, yarrow, wild strawberry leaves, violet and primrose leaves, angelica, balm, rosemary, fennel, agrimony, borage, betony, cowslip flowers and leaves, elder, tansy and many others? The old herbal teas are wonderful tonics, and some of them, balm tea, for instance are delicious.

MONKSHOOD

Botanical Name: *Aconitum napellus*
Family: Ranunculaceae
Common Names: Aconite; auld wives' hood; bear's foot; birds of paradise; blue rocket; captain-over-the-garden; chariot and horses; common monkshood; cupid's car; doves-in-the-ark; dumbledore's delight; English monkshood; friar's cap; granny's nightcap; hecates; helmet flower; lady Lavinia's dove carriage; leopard's bane; Luckie's mutch; Noah's ark; Odin's helm; official aconite; old woman in her bed; old woman in her bed with her shoes on; old woman's nightcap; soldier's cap; storm hat; Thor's hat; thung; Turk's cap; Tyr's helm; Venus' chariot drawn by doves; wolfsbane; wolf's hat.
Gender: Feminine
Planet: Saturn
Element: Water
Deities: Hecate; Hercules; Minerva; Saturn
Meanings: Chivalry; crime; knight-errantry; misanthropy; 'Your attentions are unwelcome'

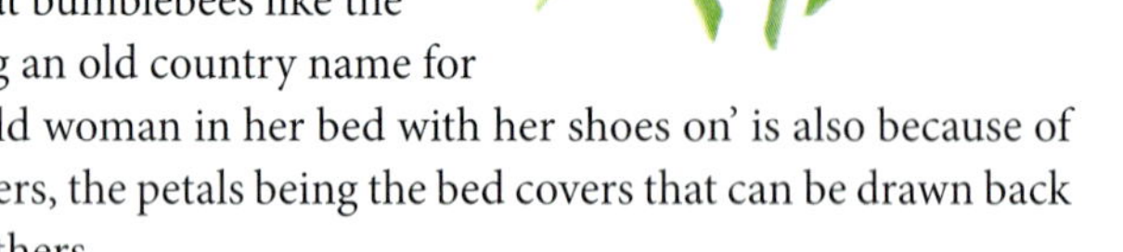

Many of the common names for this plant are references to the flower shape's resemblance to a hood or hat. The name 'dumbledore's delight' is not a reference to Harry Potter's mentor but rather refers to the fact that bumblebees like the flowers,'dumbledore' being an old country name for a bumblebee. The name 'old woman in her bed with her shoes on' is also because of the appearance of the flowers, the petals being the bed covers that can be drawn back to reveal her shoes, the anthers.

Myth and Legend

The whole of this plant is poisonous and there are tales of whole villages being wiped out due to poisoning of their water supply with *Aconitum*. It is claimed, in myth, that it grew from saliva that fell from the foaming mouth of Cerberus, the huge, three-headed dog that guarded the entrance to the 'infernal regions'. Hercules dragged Cerberus out onto the earth, at the command of Eurysteus, as one of his twelve great tasks.

According to Pliny the botanical name '*Aconitum*' is derived from the city of Aconis, a port on the Black Sea. Other sources suggest that it is derived from *akon*, a dart, as barbarous races used its poison to tip their arrows. *Napellus* is derived from the Latin to mean 'little turnip'. The common name 'wolfsbane' is said, by some, to come from hungry wolves digging up the roots when seeking food in winter, and then dying as a result. Alternatively it is suggested that arrows were dipped in the sap, or that the sap was added to bait, in order to kill wolves. Legends about the plant are conflicting. Some report that it can be used as a sure protection against werewolves and vampires. Other writers say that an injured werewolf would seek it out as a cure for wounds.

In mythology it was the witch goddess Hecate who was first to use monkshood. Elsewhere we are informed that Medea, the daughter of King Aeetes, used the toxic plant to prepare a poisoned cup of wine for Theseus. However Aegeus, the father of Theseus, interceded when he discerned his identity. In another tale, Arachne hanged herself after challenging the goddess Athene to a spinning contest and losing. The goddess sprinkled the juice of monkshood over the corpse and transformed Arachne into a spider.

Magic and Lore

Witches still supposedly use monkshood as an ingredient of their flying ointment (its use is said to give the sensation of floating off the ground). In a mixture with nightshade and henbane it is alleged to give the user the power to fly. All of these plants are, of course, extremely poisonous. Witches are also said to have used small quantities, mixed with other herbs, to enable them to make contact with the spirits of the dead. Furthermore monkshood is supposed to convey the power of invisibility. In order for those involved in sorcery to render themselves invisible a monkshood seed must be wrapped in the skin of a lizard. The bearer may then become invisible at will.

Wolfsbane has been ascribed with magical powers relating to werewolves and other lycanthropes. Variously it is said either to repel them or to make a person into a werewolf if it is worn, smelled, or eaten. Confusingly, it is also said to kill werewolves if they wear, smell, or eat the plant.

When trying to communicate with 'the other side', witches might use small quantities of monkshood, mixed with other herbs, to induce a trance.

CALAMUS ROOT

Botanical Name: *Acorus calamus*
Family: Araceae
Common Names: Flag root; gladdon; myrtle flag; sweet flag; sweet rush; sweet-scented rush
Gender: Feminine
Planet: Moon
Element: Water
Meanings: Fitness; resignation

Calamus is a popular waterside or bog garden plant; indeed some would say that it is indispensable in such situations as mosquitoes are supposed to dislike it and are said never to be found where calamus is growing.

The foliage and roots have a pleasant odour when crushed, and this led to it being a favourite 'strewing herb'. It would be thrown onto stone floors so that the fragrance released when it was walked on hid any foul smells in the air.

Magic and Lore

There are also magical uses for this plant. It is a plant of good luck, and will bring good fortune to the gardener who grows it. Small pieces secreted in the corners of the kitchen will ensure that all who live there never suffer poverty and hunger. The dried, powdered root can be added to other preparations, making spells stronger and binding them more firmly to their target. In addition powdered calamus can be added to an incense used to encourage healing. A necklace made up from the strung seeds of the plant can be worn to achieve the same end.

HERB CHRISTOPHER

Botanical Name: *Acteae spicata*
Family: Ranunculaceae
Common Names: Baneberry; bugbane; toad root
Planet: Saturn

Myth and Legend

Herb Christopher is known in Norway as *troldbaer* – troll berry – because the fruits are poisonous. Elsewhere it has been called *hexenkraut* – witch's plant – and *teufelsbeer* – Devil's berry. By the fourteenth century the plant had gained salvation, leaving its sinister background behind, to become *herba christofori* – St Christopher's herb. This may be because of the position in which the flowers are supported, being said to resemble the saint supporting the infant Christ on his shoulders. In this respect the plant and the saint are well matched.

In the Golden Legend, a compilation of tales by Archbishop Jacobus de Voragine, first published in 1480 (translated 1483), Christopher was a Canaanite soldier named Reprobus, sometimes described as the rejected one, a giant ogre or troll with a face like a dog, whose quest was to serve the mightiest ruler. He had many adventures, including bearing a small, but very heavy, child across a river. This child is later revealed to be Jesus Christ. It is from this that Christopher became the patron saint of travellers, and his name translates as 'bearer of the anointed one'.

Magic and Lore

This poisonous plant, used by witches and sorcerers, is marked as having become sacred because a cross or star is revealed if you cut through the root. The name 'toad root' derives from the fact that toads seem to be attracted by the scent of its roots.

MAIDENHAIR

Botanical Name: *Adiantum capillus-veneris*; *Adiantum pedatum*
Family: Pteridaceae
Common Names: *Capillaire commun*; *or de montpellier*; hair of Venus
Gender: Feminine
Planet: Venus
Element: Water
Deity: Venus
Meaning: Discretion

Magic and Lore

Adiantum is the 'true' maidenhair, and takes its name from the Greek *adiantos*, meaning 'unwetted'. The leaflets repel moisture so that they stay dry even when the frond is dipped in water. This makes the preparation of the plant for magical uses quite difficult. To be effective the foliage must be immersed in water for a short time. It can then be worn, or simply kept in the bedroom, in order to endow grace, love and beauty on the owner.

Medicinal

Maidenhair fern could be used to treat baldness and other problems of a more medical nature. John Gerard, in his herbal of 1597, says of maidenhair, 'It consumeth and wasteth away the King's Evil and other hard swellings, and it maketh the haire of the head or beard to grow that is fallen and pulled off.'

ADONIS FLOWER

Botanical Names: *Adonis annua*; *Adonis autumnalis*; *Adonis vernalis*
Family: Ranunculaceae
Common Names: False hellebore; Jack-in-the-green; love-lies-bleeding; pheasant's eye; red chamomile; red mathes; red Morocco: rose-a-rubie; sweet vernal
Deity: Adonis
Meanings: Painful recollections: sorrowful remembrance

One of the difficulties in identifying plants by their common name comes about because the same name is applied to different plants. The adonis flower, for some people, is the common field poppy (*Papaver rhoeas*), for others it refers to the anemones (the names *Adonis*, *Anemone* and *Pulsatilla* were regularly interchanged in the past, all three plants being in the same family). Indeed it would be appropriate as a name for those harbingers of spring, as the name comes from Adonis, the Greek god of plants. In Greek myth Adonis, like Persephone, disappeared into the earth in the late autumn to reappear in the following spring.

Myth and Legend

In France this plant is called *goute-de-sang*, as it is said to have sprung up from the blood of Adonis, who was gored to death whilst out hunting. Adonis was wooed unsuccessfully by the goddess Aphrodite who, in turn, was loved by Ares the god of war. It was Ares, in the form of a wild boar, who attacked Adonis and killed him. Aphrodite, viewing the form of the dead Adonis, magically caused the plant to grow from his blood. At Byblos there was a local belief that the waters turned red every year with the blood of this hunter, at the time when the rains washed the colour out of the soil into the river. A festival to Adonis was held annually at Athens, Alexandria and Byblos, for which festivals 'Adonis Gardens' were prepared. The gardens were pots planted with lettuce, dill and fennel. After the festival the pots were thrown away and, consequently, short-lived things may be referred to as 'Adonis gardens'. See also wood anemone *(Anemone nemerosa)*.

HORSE CHESTNUT

Botanical Name: *Aesculus hippocastanum*
Family: Hippocastanaceae
Common Names: Buckeye; conker tree
Gender: Masculine
Planet: Jupiter
Element: Fire
Meaning: Luxuriance

Although many people believe that the horse chestnut is a British native plant it was actually introduced from the Balkans, at the end of the sixteenth century. It rose in popularity throughout the seventeenth and eighteenth centuries, becoming widespread across the whole of Britain. Poets and artists used the horse chestnut as a symbol of the arrival of summer. John Evelyn wrote of it, 'All over England it lights up its Christmas candles in May.'

Myth and Legend

In a somewhat less than straightforward way the humble conker may also be credited with having played a part in the formation of the State of Israel. Following the outbreak of the First World War there was a shortage, in Britain, of acetone, a chemical required in the pharmaceutical, explosives and rubber industries. Chaim Weizman, a Jewish professor of chemistry, found that it could be extracted from the fermented pulp of conkers. Although there was soon an annual gathering of the necessary horse chestnut seeds, it was clear that the crop could not be easily expanded and the supply would be insufficient to meet the demand. Weizman took up the challenge and soon discovered alternative sources, finding that acetone could be extracted from the pulp of rice and maize.

In 1918, following the war, Palestine became vitally important to Europe, as it was the gateway to the Suez Canal. The debt owed to Weizman was partly responsible for the Balfour Declaration. The most significant clause of this declaration by the Earl of Balfour, the British Prime Minister, was a guarantee that the Jews would have a permanent home in Palestine including the town of Nazareth and the city of Jerusalem. A provisional government was set up in 1948, and a fully recognized government established the following year, with Weizman as the Prime Minister.

Magic and Lore

The power of a conker can be put to good use in a simple money spell. If you wish to attract money you can carry a conker, wrapped in a bank note, in a sachet or pouch. More generally, the carrying of a conker should bring the bearer good fortune and keeping one in the house will attract luck to the home.

The game of conkers was developed long after the introduction of the horse chestnut into Britain. It is rather unclear whether the word 'conker' is a corruption of 'conqueror' or derived from the call made before the start of a game:

> Obly, Oblionker, my best conker,
> Obli, Obli Oh! My first go.

Medicinal

In folk medicine carrying of a conker, the seed of the horse chestnut tree, can bring numerous benefits. It prevents the bearer suffering from rheumatism, arthritis, backache, or colds. However, in some parts of the country it is said only to be effective if it has been stolen or begged from someone else. If that was not enough, it will also 'draw out' any aches and pains, and ensure success in any activities undertaken. Carrying three conkers also prevents giddiness.

GRAINS OF PARADISE

Botanical Name: *Aframomum melequeta*
Family: Zingiberaceae
Common Names: African pepper; Guinea grains; mallaquetta pepper; melequeta
Gender: Masculine
Planet: Mars
Element: Fire

The grains of paradise plant is used to make wishes. If you want your wish to come true you must take a handful of the plant's leaves then, starting in the north, toss a few towards each of the points of the compass ending in the west. As you turn clockwise throwing the leaves, make your wish.

Perhaps it is not surprising, considering what some people would wish for, that grains of paradise might be added to sachets and used in spells to gain love, luck, lust and lucre.

BUCHU

Botanical Names: *Agathosma betulina*; *Agathosma crenulata*
(Synonym: *Barosma betulina*)
Family: Rutaceae
Common Names: Bookoo; bucoo; buku; oval buchu; short buchu
Gender: Feminine
Planet: Moon
Element: Water

Buchu is used by those wishing to see into the future. Drinking a tea made from this plant is said to heighten the psychic powers, enabling one to tell fortunes and make prophecies. When it is added to incense and burnt in a bedroom, it will bring on prophetic dreams.

AGAVE

Botanical Names: *Agave americana*; *Agave lechuguilla*
Family: Agavaceae
Common Names: American aloe; maguey; century plant; sisal
Gender: Masculine
Planet: Mars
Element: Fire
Deities: Agave; Mary

Although reasonably attractive the agave is not a particularly outstanding plant, until it sets up its huge flowering spike, when it becomes quite striking. Its spiky nature has led the Mexican Indians, in the region to which the plant is native, to plant it as a protective hedge around their homes. The plant, however, has more uses than merely keeping out wild beasts. Fibres from the leaves are used to make ropes and the sap used to make tequila and mescal.

Agave gained the name 'century plant' because of the long period before it comes into flower. In appropriate conditions, however, it will flower in much less than one hundred years, and it will also produce a great deal of vegetative growth.

AGRIMONY

Botanical Name: *Agrimonia eupatoria*
Family: Rosaceae
Common Names: Aaron's rod; church steeples; clot bur; cockburr; cockleburr; common agrimony; egremoine; faerie's wand; *garclive*; golden rod; harvest lice; hemp agrimony; lemonade; lemon flower; money-in-both-pockets; philanthropos; rat's tails; salt and pepper; sticklewort; tea plant

Gender: Masculine
Planet: Jupiter (in Cancer)
Element: Air
Meanings: Gratitude; 'Please accept my thanks for your token'

The name *Agrimonia* comes from the Greek word *agremone*, which is applied to plants used in the medicinal treatment of eye conditions. The species name *eupatoria* commemorates King Mithridates Eupator, who was a renowned maker of herbal remedies.

Magic and Lore

Agrimony has been used as a protective plant in magical spells. It is added to sachets to give protection to the bearer as it 'banishes negative energies and evil spirits'. In older writings it is reported as warding off goblins and all evil influences. It will repel spells cast against the bearer, returning them onto the person who cast them, and is a sure antidote to poisoning of all types. It is also reported that the skilled practitioner can use agrimony to detect the presence of witches.

Medicinal

There has been a long history of agrimony having been used to purify blood. Anglo-Saxon herbalists, who called it *garclive*, did not suggest that it should be used for eye infections, but that it was a treatment for warts, wounds and snakebites. It was one of the fifty-seven herbs used to make the holy salve that was supposed to give protection from goblins, evil and poisoning.

In the Middle Ages it was prescribed to aid sleep, simply being placed beneath the patient's head. All the time the plant stays there the patient will sleep.

> If it be leyd under mann's heed,
> He shall sleepyn as he were dead;
> He shal never drede ne wakyn
> Till fro under his heed it be takyn.

In *A Modern Herbal* (1710), William Salmon recommends the use of agrimony to stop bleeding. The preparation he suggests is blood curdling. It required one ounce of powdered agrimony, mixed with half an ounce each of catchu, powdered toads and dried man's blood and would be prescribed for both internal and external wounds. Parkinson follows the thinking of the Anglo-Saxons and in his herbal of 1640 he recommends the taking of a decoction of the plant, 'made with wine and drunke', as a treatment for the stings and bites of serpents. He adds that where it is applied externally it will draw out splinters and thorns.

> Next these here Egremony is,
> That helps the serpents' biting
> – *Muses' Elysium*

Agrimony has a long history of use as a spring tonic, to purify the blood, and in the time of Chaucer its name appears in the form of 'egrimoyne' when it was used, with mugwort and vinegar, for 'a bad back' and 'alle woundes'.

HOLLYHOCK

Botanical Names: *Alcea rosea*; *Alcea* spp.
Family: Malvaceae
Common Names: *Malva benedicta*; *caulis santi*
Element: Water
Deity: Althea
Meanings: Ambition; Fecundity; female ambition (white); fruitfulness; truthfulness; 'You inspire me to achieve great things'

Myth and Legend

According to legend, at midsummer an island would appear where the River Severn meets the River Wye on the border between Gloucestershire and Wales. It was a paradise filled with trees, flowers and songbirds, in the midst of which was a castle inhabited by faerie folk. Mortals were allowed to visit this faerie land via a tunnel beneath the river and, although the faeries were never to be seen, invisible hands provided food and drink to all who wanted it and charming music filled the air. There was but one rule to be observed by all who visited the faerie island, and that was that nothing should be removed.

One year, however, a little girl wanted to take home a bunch of flowers that she had picked on the island. Despite her mother forbidding her to do so the girl hid one flower in her pocket and took it with her when she returned through the tunnel to go home. As soon as she left the tunnel she was turned into a hollyhock, the pink of her pinafore dress becoming its pink flowers, and the faerie island disappeared, never to be seen again. It is said that faeries love hollyhocks, especially pink ones.

Magic and Lore

It is a plant of midsummer and the yellow hollyhock was commonly used as a church decoration on 29 June as a commemoration for the death of St John the Baptist. As a guide to gardeners it is usually claimed that

Where hollyhocks grow
Beans won't go.

Medicinal

The common name may be derived from *holy hocc,* meaning 'a blessed herb' because of its healing properties. A variation of this suggestion says that the plant was used to treat horses, the leaves being placed against their heels, hence 'hock'. It may also have been assumed that the plant originated from the Holy Land (although it was actually introduced from China), which explained the use of 'holy'.

LADY'S MANTLE

Botanical Names: *Alchemilla vulgaris*; *Alchemilla mollis*
Family: Rosaceae
Common Names: Bear's foot; dew's cup; duck's foot; elf-shot; lamb's foot; nine hooks; *leontopodium*; lion's foot; stellaria.
Gender: Feminine
Planet: Venus
Element: Water
Deity: Mary
Meaning: Fashion

The common name of 'lady's mantle' (in the German form of *frauenmantle*) first appears in the work of the sixteenth-century botanist Jerome Bock (usually known by the latinized form of his name, Tragus). This name refers to the plant being dedicated to the Virgin Mary. The leaf edges are said to resemble the folds of a cape. Elsewhere it was referred to as *leontopdium*, or lion's foot, probably due to the spreading roots.

It gains its botanical name from the Arabic word *Alkmelych*, referring to its supposed ability to work wonders. Alchemilla was literally the alchemists' plant; however, there is some disagreement as to whether it was the plant or dew collected from the foliage that was used. In many respects this is immaterial as, even if it was dew, the plant would have imparted its power into the droplets of water caught in the wrinkles on its leaves.

Medicinal

The dew off the leaves was considered by some to be an aphrodisiac, and to enhance the fertility of those who drank it. It is perhaps because this dew was caught in the wrinkles of the plant leaves that it was recommended as an anti-wrinkle lotion. Pillows may be stuffed with the flowers for those suffering insomnia, to induce sleep. A decoction of the plant, drunk for twenty days, was recommended as a treatment for deep wounds. The seventeenth-century herbalist Nicholas Culpeper claimed that women with sagging breasts should smear them with the juice of the plant in order to make them more firm and rounded.

TRUE UNICORN ROOT

Botanical Name: *Aletris farinosa*
Family: Liliaceae
Common Names: Ague grass; ague root; aloe root; bettie grass; bitter root; black root; blazing star; colic root; crow corn; devil's bit; star grass; star root; starwort; unicorn root

True unicorn root is another herb used in protective magic, the powdered root being sprinkled about a house to guard everyone in the household from evil. Individuals need only carry a small sachet of the powdered herb to protect themselves. In addition to warding off evil forces this herb can be used to break spells and curses cast against you.

ALKANET

Botanical Name: *Alkanna tintoria*
Family: Boraginaceae
Common Names: Anchusa; bugloss; dyer's bugloss; orchanet; Spanish bugloss
Gender: Feminine
Planet: Venus
Element: Water
Meanings: Falsehood; mendacity

Anchusa, an alternative name for alkanet, is derived from the Greek *anchousa*, meaning 'paint', as a red dye can be prepared from the roots.

In magic the plant might be burnt in order to drive off all negativity from an area, or to attract prosperity to those living in the area.

Medicinal

Dioscorides recommends the use of alkanet for the treatment of snakebites. He adds that if the root of the plant is chewed, and spittle directed into the snake's mouth, it will immediately be killed.

The common alkanet found in Britain is *Alkanna officinalis*. The specific epithet *officinalis* indicates that it has been used as a medicinal herb.

ONION

Botanical Name: *Allium cepa*
Family: Liliaceae
Common Names: Oingnum; onyoun; shallot (a sub group of the onion); unyoun; yn-leac
Gender: Masculine
Planet: Mars
Element: Fire
Deities: Hecate; Isis

Magic and Lore

The onion was once much more highly thought of than it is today. In many parts of the world it was believed that members of the onion family would give great strength and pain resistance to all who ate them. Some authors go so far as to say that onions were venerated in ancient Egypt and invoked during the taking of oaths. The tradition that rubbing onions on the skin prevented pain led to their use by schoolboys due to receive the cane as a form of corporal punishment. In addition to there being no sensation of pain it was said that it would cause the cane to split when the first blow landed.

A particular form of divination, called cromniomancy, uses onions. Those faced with having to make a particularly difficult decision should scratch each of the options onto an onion, the onions being put into a dark place and left to sprout. Whichever is the first to shoot will indicate the solution. In a similar way a young woman might choose between her suitors. She would mark the name of a suitor on each onion and then place the onions on the church altar on Christmas Day and leave them there to sprout. Once again, the first to shoot will indicate which man she should choose.

On St Thomas's Day (21 December), in Derbyshire, girls would peel a large red onion and then insert nine pins into it, one in the centre and the others radially. As the pins were inserted they would recite the following rhyme:

Good St Thomas, do me right,
Send me my true love tonight,
In his clothes and his array,
Which he doth wear every day,
That I may see him in the face,
And in my arms may him embrace.

The onion would then be put under the pillow before retiring to bed. This would ensure that they dreamed of their future husband.

Another variation on this says that 'St Thomas onions' should be peeled, wrapped in a clean handkerchief and placed beneath the young woman's head when she retires to bed. Furthermore, she must wear a clean nightdress and the bedding (and even the room) must be clean. The woman should then spread her arms wide and say:

St Thomas pray do me right,
And let my true love come tonight,
That I may behold his face,
And him in my kind arms embrace.

She is assured of beholding her future husband in her dreams and to be sure it is him she should try to get hold of him. If he gives her the slip, she must try again.

Onions can be used to induce other types of prophetic dreams. Placing one beneath your pillow, when you retire to bed may not only cause you to dream of a future lover but also help you to see the answer to a problem that has been troubling you. Where onions appear in a dream it portends the need for hard work, a loss of property and many worries. A dream of fried onions means that a friend is ill but will recover. In his book *Folklore of Plants* Thistleton-Dyer says:

To dream of eating onions means
Much strife in thy domestic scenes,
Secrets found out or else betrayed,
And many falsehoods made and said.

It may be because of their shape that onions were once recommended as an aphrodisiac. Eating them was supposed to induce venery, or magical lust. Burning onion skins in the hearth was said to attract wealth, whereas dropping them on the ground would dispel any good luck and drive away all prosperity.

There is even a place in the countryman's weather forecasting kit for an onion. According to the old rhyme:

Onion skin very thin,
Mild winter coming in.
Onion skin thick and tough,
Coming winter cold and rough.

Another important attribute of onions is their power as a protective herb. Sticking a small onion with black-headed pins can form a protective amulet. When this is left on a windowsill it will prevent any evil from entering the house. Growing onions in the garden will also ward off evil forces, and the dried flowers, when used in a decorative display, retain some protective powers. The protective powers of the onion may be the reason that it was often thought lucky to carry a small onion in your pocket, particularly as it protected against evil and illnesses.

The cleansing powers of onions are used in magical rite; the knives and swords used in ceremonies can be purified by rubbing them with the cut surface of a fresh onion. Shallots are also used in purification rites, and may be added to bath water in order to 'cure' those who suffer from continual bad luck. Carrying an onion is claimed to be a way of ensuring that venomous creatures will never bite you. Snakes especially were believed to dislike onions.

There are several superstitions associated with the planting of onions. Folklore suggests that the best time for planting is on 21 March, and if planted during the rising of one of the zodiacal signs then the crop will not be 'long in the ground'. If you miss a row when planting onions it is an indication that someone in the household will die within the year. All the sets must be upright, of course, or else they will grow downwards through the earth.

Medicinal

In addition to being a powerful protection against the forces of evil, onions are widely recommended as folk remedies to treat various ailments. A traditional English saying claims:

> If an onion is eaten each morning before breakfast,
> All the doctors might ride one horse.

Simply hanging a string of onions in a house will prevent infectious diseases. The disease enters the onions, leaving the members of the household safe from harm. One house on the Rows in Chester, built in 1652, is marked with the inscription 'God's Providence is Mine Inheritance'. It is said that, in the seventeenth century, the plague stopped there as the family hung strings of onions at their door. Unfortunately, the plague was recorded in Chester in 1605 and in 1647–48, which makes the claim more than a little suspect. More recently, in the foot and mouth outbreak of 1968, one Cheshire farm escaped despite the disease affecting other farms all around. The farmer attributed this to his wife having placed onions all along the windowsills and doorways of the cattle sheds.

Halved or quartered onions placed around the house will also absorb diseases and any negative vibrations. For this reason it is particularly important to inspect onions closely before cooking with them. Any that appear unhealthy must be discarded. To cure an ailment the infected area must be rubbed with the cut edge of an onion, whilst visualizing the illness entering the bulb. The onion should then be burned, or smashed and the pieces buried, to destroy the infection. Chilblains, for example, could be treated by rubbing with salt and an onion. To remove warts it is sufficient merely to rub them with an onion, which should then be tossed over the right shoulder, and you must walk away without a backward glance.

Baldness was treated by rubbing the scalp with onion and honey, to stimulate fresh hair growth. Toothache could be eased, according to a Lincolnshire tradition, by wrapping an onion skin, like a thimble, around the patient's big toe. One remedy for earache was to hold a roasted onion against the affected ear. Hanging a large red onion on the bedhead will prevent illness, or aid the recuperation of those already ill. In *The Englishman's Doctor*, 1608, Sir John Harrington suggests:

> If your hound by hap should bite his maister,
> With honey, rew and onyons make a plaister.

In Australia boiled onions were used to treat worms in children. Boiled or roasted onions might be given as a cold cure and cooked or raw onions recommended for the treatment of rheumatism. Raw onion was supposed to purify the blood. An onion poultice could be applied to the throat, chest and soles of the feet to treat croup in young children.

Finally, it is said that onions were favoured as a food flavouring by the Native Americans. This provided those frontiersmen with a good sense of smell with a way of finding their encampments.

LEEK

Botanical Names: *Allium ampeloprasum var bulbiferum* (wild leek); *Allium porrum* (cultivated leek)
Family: Liliaceae
Gender: Masculine
Planet: Mars
Element: Fire

Allium ampeloprasum var bulbiferum is the wild leek, which is to be found in small populations in the south-west of England (Cornwall and Somerset), in the Channel Islands and, of course, in Wales. Much of the folklore of the plant refers to the wild leek rather than to *Allium porrum*, the cultivated variety. Like the cultivated form, the wild leek can grow up to 1 m high with similar strap-like leaves, but it does not get as thick in the stem as its cultivated cousin.

Myth and Legend

The leek is, of course, one of the national symbols of Wales. Although it is unclear exactly how this came about, various suggestions have been put forward. King James, in the *Royal Apothegies* informs us that it was selected to commemorate the Black Prince, but gives us no explanation of why this should be. One Welsh legend tells that St David fed on a diet of bread and leeks whilst travelling as a monk in Wales, this slightly odd diet being provided by the Christian Celts shortly before they engaged in battle with the pagan Saxons. The Celts won the day, and thereafter adopted the leek as a national emblem. A variation of this story has it that the battle took place in a field of leeks and, because the enemy was to be given no quarter, the Welsh wore the leeks on their clothing so that they could easily identify each other amid the chaos of battle. Some sources have it that St David led the Welsh into battle, whilst others claim that the leader was Cadwalldr. Either way, the whole story appears to owe more to fiction than to fact.

An equally unlikely tale looks back even further, to pre-Christian Wales. Aeddon was the ancient Celtic god of thunder. His followers, who must have had a lively imagination, thought that the odour of the leek was similar to the smell of ozone in thunderstorms. They thus adopted thc leek as a symbol of Aeddon and, when converted to Christianity, this symbol was transferred to St David.

Magic and Lore

Like the onion, the leek was supposed to impart great strength. The ancient Welsh made use of it in this way, rubbing leeks all over their bodies before going into battle in order to make themselves more powerful and to prevent them being injured. Indeed, simply carrying leeks is said to be sufficient to ensure both these objectives, and even today the fans of the Welsh rugby team will tend to brandish leeks rather than daffodils (another national Welsh symbol) at international matches.

On Anglesey there was a custom whereby, on St David's Day, boys and girls would wear leeks. The boys would wear them throughout the morning, up until midday, and the girls would wear them all afternoon. The penalty for not wearing a leek, or wearing one at the wrong time of the day, was to be pinched by the other children. There are very obvious similarities between this custom and the English Oak Apple Day traditions.

The leek is a powerful aid in warding off evil. All you need to do is to bite into one; this will also break any spells cast against you. However, be careful whom you choose to share your leek with, as one tradition tells that where two people eat leeks together they will fall in love. Another Welsh tradition says that if a young woman walks backwards into her garden on Halloween, and places a knife amongst the leeks, she will see a vision of her future husband.

Medicinal

Eating leeks is supposed to be one way in which to ensure continued good health. According to the rhyme:

> Eat leeks in Lide [March],
> And ramsims in May,
> And all the year after
> Physicians may play.

GARLIC

Botanical Name: *Allium sativum*
Family: Liliaceae
Common Names: Ail; ajo; Hecate's supper; poor man's treacle; stinkweed; *toum* (Arabic)
Gender: Masculine
Planet: Mars
Element: Fire
Deity: Hecate
Meaning: 'Go away you evil one'

Sith Garlicke then hath power to save from death,
Beare with it though it make unsavoury breath;
And scorne not Garlicke, like to some that think
It onely makes men winke, and drinke, and stinke.

– Sir John Harrington, *The English Doctor*, 1609

Myth and Legend

Garlic is a close relative of the onion. A Muslim legend tells that when the Devil stepped out of the Garden of Eden, following man's fall from grace, garlic sprang up from where he placed his left foot and onions from where he put his right. An Indian legend claims that Rahu, King of the Asuras, stole the elixir of life from the god Vishnu and drank it. Vishnu took his revenge by cutting off Rahu's head and it was from drops of his blood that garlic sprang.

Garlic was eaten at the festivals of the witch goddess Hecate and was left at crossroads as a sacrifice to her name. Curiously, other writers recommended the use of the herb to counter moon magic, such as that of Hecate. Garlic has been closely associated with the moon and was thought to grow strongest as the moon waned. In myth, Ulysses ate garlic in order to protect himself against the magic of the sorceress goddess Circe.

Magic and Lore

Several religions have, at various times, regarded garlic as 'unclean', including Islam, Hinduism, Zen Buddhism and a number of Christian sects. Some Hindu castes are, in fact, banned from consuming garlic at all, as if they eat it within three months of their death it will cause them to fare badly in the afterlife. Even in ancient Greece those wishing to worship in the temple of Cybele were banned from eating garlic, as the smell on their breath was offensive to other worshippers.

Garlic, like onions and leeks, was thought to give great strength to those who ate it, and to protect against injuries. The Roman writer Pliny claims that the ancient Egyptians treated garlic and onions as deities and would invoke them when making oaths. Other writers say that the builders of the great temples and pyramids of ancient Egypt ate little else but onions and garlic. Bulbs were even put into the tombs of their dead, including that of Tutankhamun, in order to sustain the deceased on their journey through the afterlife. During the reign of the Tutankhamun a healthy male slave could be purchased for 15 lb of garlic.

Garlic was also well known in ancient Rome, and the Roman Empire was won, according to some, by the Roman soldiers, having eaten garlic, driving all enemies before them. There are reports that soldiers in the Middle Ages would wear garlic cloves as a protective amulet. It is also alleged that those who ate it gained greater stamina and, consequently, it was eaten by long-distance runners. When it was eaten by travellers setting out on a long journey at night it would ward off all evil, no doubt due to the smell of their breath. This alleged attribute has been twisted slightly in some writings. Rather than increasing the runner's stamina, they claim that chewing a clove of garlic will prevent other runners from passing, presumably because of the smell.

The power of this plant to give added strength was not limited to mankind; it could also be fed to fighting cocks and horses, and in Hungary garlic cloves would be put into the bridles of racehorses to cause the competition to drop back at the offensive odour. This same offensive odour is supposed to be sufficient to drive moles out of the garden.

All those who are acquainted with the stories of Dracula will be well aware that garlic can be used to keep vampires at bay. Although the vampire legends are recent the power of garlic to ward off evil forces has been recognized since ancient times, and it was used against all evil spirits, monsters, and demons. Midwives in ancient Greece hung garlic in the birthing room to keep evil spirits away. The Sanskrit name for the herb means 'slayer of monsters'. Where it was hung over doorways it repelled 'envious eyes', and was especially powerful as a protection in new homes. Farmers might even add a little garlic to their animals' feeds to protect them from evil influences. In more general terms hanging bunches of garlic over doorways, or placing them on the mantelpiece, was said to bring good luck.

There was a tradition, in some parts of France, of collecting and roasting cloves of garlic on Midsummer's Eve. These would then be given to rural families to keep as a protective amulet. When it was placed beneath a child's pillow at night it would protect them as they slept. If any of the forces of evil did appear it was necessary only to scatter the cloves onto the floor, or bite into one, to drive them from the place. It use to be suggested that brides carry a clove of garlic on their wedding day to ensure good luck and ward off evil. Cooks were also recommended to wipe their pans with a clove before cooking to remove any negative influences that might otherwise contaminate the food. As well as protecting against evil and food poisoning garlic was supposed to protect those sailors who carried it from being shipwrecked. Mountaineers could wear, or carry, a clove of garlic as a protective amulet against foul weather conditions.

This plant has played an important role in ceremonies in the Far East where it was said to be able to bring back lost souls. Another claim of its attributes was that lodestones or magnets rubbed with garlic would lose all their powers. Perhaps the hardest to believe of all the powers ascribed to garlic is that it can be used as an aphrodisiac.

To dream of eating garlic can be an omen of domestic strife, although some sources claim that a man dreaming of eating garlic is an indication that he will discover hidden secrets. Dreaming of having some in the house portends good luck and keeping a clove of garlic beside your bed is supposed to be an aid to dreaming.

Medicinal

Garlic juice is an antiseptic and antibacterial agent. The herb has been in use in medicine for over 3,000 years. It was used as such by the Egyptians, the Romans, the Babylonians, and the Greeks. Hippocrates, the ancient Greek physician who lived around 400 BC, recommended its use for treatment of leprosy, cancer, wounds and infections as well as for digestive disorders. Pliny noted its use in some sixty-one remedies including treatments for the common cold, epilepsy and tapeworms. In general terms it was said to help in maintaining a youthful complexion, and more specifically it was used as a treatment for ailments as diverse as the plague and baldness (being rubbed into the scalp). A concoction, known as 'vinegar of the four thieves' was prepared from several herbs, including garlic, and used by those who robbed plague victims in the belief that it would prevent them from becoming infected. Like the onion, it is said to absorb diseases. The infected area is rubbed with a clove that must then be thrown into running water.

Its use continues into the modern era. In the Second World War, frontline Red Army troops used it in the treatment of wounds. Their reliance on it was such that it became referred to as 'Russian penicillin'.

To treat measles it was said that one should tear a length of homespun linen into nine strips and spread each strip with powdered garlic taken from nine cloves. Wrap these strips around the patient and let it remain there for nine days. At the end of that time remove the strips and bury them in the garden. A full recovery will soon follow. In the treatment of smallpox cloves of garlic were strapped to the feet of the sufferer, possibly to draw out the infection.

Garlic could also be used to prevent hepatitis. Thirteen cloves were strung, and worn as a necklace for thirteen days. On the last day the sufferer had to go to a crossroads at night, throw the garlic necklace behind them and run home without looking back. This may relate to the use of the herb in the rites of the witch goddess Hecate.

Cloves of garlic which had been gathered on Good Friday were usually thought to be the most effective as cure-alls. In the medieval period the herb was seen as a cure for leprosy. The lepers who used it were called 'pilgarlics' as they had to peel their own garlic cloves. The name 'poor man's treacle' probably comes from its wide use as a medicine amongst the common people, 'treacle' being derived from the Latin *theriacus* meaning an antidote.

However, not everyone thought that garlic was a panacea. Horace, it seems, was of the opinion that garlic was more toxic than hemlock.

ALDER

Botanical Name: Alnus glutinosa
Family: Betulaceae
Common Names: Aar; aller: aller-tree; alls-bush; arl; arn; aul; dog-tree; ellar; *fearn*; hasle-bush; Irish mahogany; oller; owler; owlorn; wallow; wullow; whistle wood
Planets: Venus; Mars
Element: Fire
Deities: Bran; Cronus; Ellerkonig; Gwern; Helice; Phoroneus; Proteus; Saturnus

Myth and Legend

The alder is a tree of birth and youthfulness. Like the willow, it is usually found on wet soils or close to waterways. The reason why these two trees should have become so closely associated with water is given in the story of a great feast in honour of the fertility gods. Whilst all other forms of living things, from right across the natural world, celebrated together only the willow and the alder stood apart. They were seen to be gazing longingly into the floodwaters that had brought fertility to the earth. Their lack of acknowledgement annoyed the gods who decided to punish them by leaving them where they stood, that is, staring at the waters. The alder and willow are sometimes described, respectively, as the king and queen of the waterways.

The alder is the third tree of the Celtic lunar year. It was known as *fearn* and the corresponding moon was said to be a fire moon. In the epic poem *The Battle of the Trees* alder is described as the battle-witch of the woods, hottest in the fight and always found in the front lines of the battle. It could be said to be cousin to the hazel and the birch, and in the *Odyssey* it is the first of the three trees of the resurrection (the others being the white poplar and the cypress). These formed a protective wood around the cave of Calypso, daughter of the Titan Atlas. The Greek name for the alder is *clethera*, derived from *cleio*, 'I cling' or 'I confine'. In many legends the island orchard paradises, such as Avalon, are described as being surrounded by alder woods.

It is said that alder is the embodiment of the spirit and power of fire over water. It is associated with the Greek god Phoroneus, the creator of fire. Prometheus, the Titan, stole fire from Phoroneus and gave it to mankind. Its power over water can be viewed in a different ways; for example the timber resists rotting in water, and the piles of buildings, such as Westminster Cathedral and the Rialto Bridge in Venice are made from its wood. It can also be seen in the use of alder wood as a protective amulet against the destructive forces of winter.

The tree was sacred to the ancient British god Bran. Most people would be unaware that the ravens, kept at the Tower of London are sacred to Bran. The head of Bran is said to be buried at the White Mount, site of the Tower of London, facing France. So long as the birds of Bran never leave the site where his head is buried,

England can never fall to invasion. Faeries associated with the alder are said to be particularly protective and when they are required to leave the concealing foliage of the alder tree they adopt the form of ravens.

Magic and Lore

The tree is said to be especially sacred as it bleeds. The timber, when first cut, is white but quickly discolours, reddening. In Ireland the alder is not considered to be a lucky tree, possibly because of the 'bleeding'. In some parts of Ireland it was highly revered to the point that those who cut it down were in danger of having their houses burnt to the ground. It is, perhaps, because of these attributes that witches used the young shoots in their preparations. It is even thought to be unlucky to pass an alder when going on a journey.

Unusually, three different colours of dye can be derived from the alder. Red is obtained from the bark, green from its flowers and brown from the twigs. Traditionally these three dyes represent three of the four elements: fire, water and earth. Some writers claim that a green dye was used to stain the clothing of those who lived or worked in the forest. The colour was called Lincoln green, and the most noted of those who wore it must be the outlaws of Sherwood Forest, including Robin Hood. His character is seen by some as one of the personifications of that fertility spirit known as 'the green man' and the green dye derived from alder was used to colour costumes used in the rites and festivities centred around the green man. (Elsewhere, it should be noted, Lincoln green colouring was derived by dying the cloth blue with woad (*Isatis tinctoria*) then over-dying it yellow with weld (*Reseda luteola*)).

Alder is a faerie tree associated with trickster spirits called the *fear dearg* ('red men'), who were said to help human beings to escape from other worlds. This may stem from the use of the red dye, which was traditionally used to colour the faces of Sacred Kings in ritual. A guardian faerie associated with the alder is called Clethrad, an obvious link to the Greek name for the plant. (*Clethra* is a genus of some thirty trees and shrubs sometimes called white alders.)

The common name is derived from the old English *ealdor* meaning 'chief'; its use in this context is still found in the office of alderman.

In Herefordshire, the alder gave a guide to the start of the trout-fishing season:

When the bud of aul is as big as a trout's eye,
That fish is in season in the River Wye.

Although alder wood makes a poor fuel, its place as a tree of fire is secure, as it does yield the finest charcoal.

Medicinal

Nicholas Culpeper, writing in his herbal, says that when alder leaves are placed under the feet of those 'galled with travelling' they will prove refreshing. He adds that when alder leaves are gathered with the morning dew still on them and placed in a room, they will attract all the fleas present. The whole lot can then be thrown away. The power of alder wood over water can be used in treating rheumatism pains. All that is necessary is that small pieces of the timber be carried in the pocket of the sufferer.

ALOE

Botanical Names: *Aloe* spp; *Aloe vera*
Family: Liliaceae
Common Names: Aloe resin; burn plant; first aid plant; medicine plant
Gender: Feminine
Planet: Moon
Element: Water
Meanings: Acute sorrow; affliction; bitterness

Traditionally aloe, mixed with turpentine, tallow and white lead, would be smeared onto the hulls of wooden ships in order to protect the timber from woodworm and barnacles.

Magic and Lore

In Egypt the plant had a reputation as a protective amulet as well as being a religious symbol. Pilgrims making their journey to Mecca would mark the event by placing an aloe leaf above the doors of their homes. This also, of course, acted as a protection to the house, warding off evil spirits. In Mecca itself, graves have been planted with aloes. The Arabic name is *saber*, meaning 'patience'. This refers to the waiting period between death and resurrection.

Keeping aloe in the house is supposed to ward off evil influences and protect against household accidents. In regions of Mexico wires strung with aloes, pictures of saints, pine nuts and rock salt are hung up in the house to ensure good luck and protect the members of the household.

According to one superstition it is very unlucky to come across an aloe in flower by chance as this is a sure sign that misfortune will soon follow.

Medicinal

The aloe gained the name 'burn plant' from its use in treating burns and scalds in the home. In some parts of the world it is not an uncommon sight on kitchen windowsills, kept there for just this purpose. It has been used for all types of burns; in the 1930s and 1940s it was used to treat radiation burns.

Its use as a medicinal plant has been known for a long time, and it has been claimed that Aristotle had Alexander the Great conquer the Island of Socrato in order to secure sufficient supplies of *Aloe vera* to treat the wounds of all his soldiers.

In the Indian system of Ayurvedic medicine the leaf pulp of *Aloe vera* is used to shrink haemorrhoids. The Greeks seem to have used it for almost every part of the body. Both Greeks and Romans recommended the gel as a treatment for wounds and Pliny suggests that the crushed leaves be rubbed onto ulcerated male genitalia as a cure. The laxative qualities of the plant have led to use as a treatment for intestinal worms.

In China aloe was noted as a elixir for eternal youth, and in ancient Egypt Cleopatra used it to maintain her youthful, unwrinkled complexion.

LEMON VERBENA

Botanical Name: *Aloysia triphylla* (Synonym: *Lippia citriodora*)
Family: Verbenaceae
Common Names: Cedron; herb Louisa; *yerba Louisa*.
Gender: Masculine
Planet: Mercury
Element: Air

The botanical name of the lemon verbena commemorates Maria Louisa, Princess of Parma. This is further reflected in the common names herb Louisa and *yerba Louisa*.

The herb is used in magical ceremonies to increase the potency of any mixtures to which it is added. It may also be used in purification rituals, or even added to bath water to cleanse the bather of evil influences.

When worn, lemon verbena is supposed to make the wearer more attractive to members of the opposite sex. It can, therefore, be put into mixtures used to inspire love in those close by. If it is worn about the neck when retiring to bed it ensures a peaceful night's sleep and prevents the sleeper from dreaming. Alternatively, drinking a small quantity of the juice of the plant will have much the same effect. On the other hand, a little lemon verbena on your pillow at night willl ensure that you enjoy sweet dreams. There is another superstition that suggests that anyone who takes cuttings of lemon verbena will never die unmarried.

GALANGAL

Botanical Name: *Alpinia officinarum*; *Alpinia galanga*
Family: Zingiberaceae
Common Names: Chewing John; China root; colic root; East Indian catarrh root; galingal; galingale; gargaut; ginger lily; India root; low John the Conqueror; Siamese ginger.
Gender: Masculine
Planet: Mars
Element: Fire
Deities: Vulcan

Galangal derives its name from the Arabic word *khalanjan*, which in turn is probably a corruption of a Chinese word, meaning 'mild ginger'. Wearing or carrying the plant will give the bearer protection against ill health and any evil forces. It will also attract good luck, or put into a leather purse it will draw money to it. The plant can be powdered and sprinkled about the house, or carried, in order to promote lust-fulness. Where the plant is burnt it breaks spells and lifts curses.

MARSHMALLOW

Botanical Name: *Althaea officinalis*
Family: Malvaceae
Common Names: Cheeses; guimauve; hock herb; mallard; marsh mallice; mauls; meshmellice; mortification root; schloss tea; sweet weed; white mallow; wymote
Gender: Feminine
Element: Water
Deity: Althea
Meanings: Beneficence; humanity; maternal tenderness

Marshmallow has been cultivated for centuries. The Romans roasted the roots and ate them as a vegetable. The Greeks extolled its medicinal virtues and marshmallow seeds are still used in Chinese medicines.

Magic and Lore

Marshmallow has been used for a long time as a protective herb, and was added to incense burnt to stimulate psychic powers. Where marshmallow is placed on the altar in magical rites, it will also attract good spirits and dispel evil ones.

Medicinal

The botanical name of the marshmallow, *Althaea*, coming from the Greek *Al-thaine* meaning 'to heal' or 'to cure', immediately tells us that the plant has long been used as an herbal medicine. Theophrastus suggests that the herb could be added to sweet wine to make an effective cough medicine. Marshmallow sweets were originally made from candied mallow roots and were given to soothe coughs. Pliny says of it, 'Whoever shall take a spoonful of the mallow shall that day be free from all diseases that may come to him.'

FLY AGARIC

Botanical Name: *Amanita muscaria*
Family: Agaricaceae
Common Names: Death angel; death cap; magic mushroom; redcap mushroom; sacred mushroom
Gender: Masculine
Planet: Mercury
Element: Air
Deities: Bacchus; Dionysus; Wotan

Myth and Legend

Some legends say that fly agaric, like other toadstools, grows from the 'seeds' scattered on 15 June, by St Veit as he rides his blind horse through our woodlands. Scandinavian myth tells a different tale. Wotan was being pursued by devils and the toadstool grew from the red foam that fell from the mouth of Slephir, his six-limbed horse, as he made his escape. The Vikings are said to have eaten fly agaric in order to induce their fighting frenzy, known as 'berserk'.

Some sources have suggested that it is the *soma* described in the *Rig-Verda*, that is the 'god plant' of the Hindus.

Magic and Lore

Despite some of its common names this is not the toadstool usually called death cap, nor is it what is normally referred to as the magic mushroom. It is, however, extremely poisonous. It is thought to have been central to several of the mystical religions of classical times. Considering its high toxicity it is surprising that it was used in fertility rites, being placed on the altar or in the bedroom.

It is possibly the most instantly recognizable of all British fungi and it is the one most closely associated with the world of pixies and faeries.

In Lapland there have been those who sought to use fly agaric as a hallucinogenic drug. In view of its toxic nature, it was first fed to reindeer, and the herders would then drink the reindeer urine, which contains much of the hallucinogenic agents of the mushroom, but with less of the toxicity. Therefore, if you see a figure dressed in the red and white colours of the toadstool coming down your chimney at Christmas, perhaps you may have consumed one glass of reindeer urine too many!

AMARANTH

Botanical Names: *Amaranthus caudatus*; *Amaranthus cruentus*; *Amaranthus hypochondriacus*
Family: Amaranthaceae
Common Names: Cock's comb; floramor; flower gentle; flower of immortality; flower-velure; huauhli; love-lies-bleeding; prince's feather; tassel flower; velvet flower
Gender: Feminine
Planet: Saturn
Element: Fire
Deity: Artemis
Meanings: Desertion; hopeless and heartless; immortality; unfading love

The name derives from the Greek *amarantos*, which means 'unfading', referring to the flowers that retain their colour even when cut.

> Immortal amarant – a flower which once
> In Paradise, fast by the Tree of Life,
> Began to bloom; but soon, for man's offence
> To heaven removed, where first it grew, there grows
> And flowers aloft, shading the fount of life …
> With these, that never fade, Spirits elect
> To bind their resplendent locks in wreath'd with beams.
>
> – Milton, *Paradise Lost*

In Aesop's Fables the rose is compared to the amaranth illustrating the difference in fleeting and everlasting beauty.

Myth and Legend

This plant has long been a symbol of immortality. It was made into wreathes to crown statues of the gods, and was used in pagan funeral rites, being placed as decoration on tombs. Indeed, at the funeral rite of Achilles, we are told, the Thessalians wore crowns of amaranth in honour of the dead hero. This may have caused some difficulties as other writers claim that the wearing of an amaranth crown renders the wearer invisible. Clement of Alexandria notes the plant's symbolic significance with the words '*Amarantus flos, symbolum est immortalitatis*.' The goddess Artemisa is linked to the plant, as one of her alternative names is Amarynthia, from the festival held at Amarynthus each year in her honour.

In view of the widespread use of amaranthus in the religious rituals of the Aztecs, its cultivation was for a time outlawed in Mexico by the Spanish governing authorities. The alternative common name for this plant, 'love-lies-bleeding', is rather appropriate as the Aztec religion involved human sacrifices.

The cock's comb amaranth (*Celosia argentea)* is a related species. Korean legend tells the story of Kim and Choi, neighbours in a small village. They played draughts and when Kim lost Choi challenged him to put up his prize golden cockerel in a cockfight. Kim was obliged to accept the challenge. Choi armed his own bird with sharp spurs concealed amongst the feathers on the bird's legs. The fight was somewhat one-sided and Kim's bird was soon killed. Kim and his wife solemnly buried their cockerel in the garden. Next morning, while sweeping his yard, Kim found the cock's comb flower, *Celosia*, growing where the bird had been buried.

Magic and Lore

The amaranth can be used to produce a very practical form of defence. It must be pulled up by the roots, on a Friday when the moon is full, and an offering left in its place. The whole of the plant should then be folded into a white cloth and worn against the chest as a bullet-proof vest.

In 1653 Queen Christina of Sweden renounced the throne, and sought to attain immortality by dedicating herself to a life of letters. To mark her sacrifice she instituted a new order of chivalry, the Knights of the Amaranth. This was not immortal, however, as it ceased to exist after her death.

Amaranth is the birthday flower for 31 March.

Medicinal

As well as conferring invisibility on the wearer a crown of amaranth flowers is supposed to aid the healing, and the dried flowers have been used to 'call forth the dead', or carried to mend a broken heart. Amaranth was widely used by the Chinese for its healing proporties, treating such ailments as infections, rashes, and migraine.

SCARLET PIMPERNEL

Botanical Name: *Anagalis arvensis*
Family: Primulaceae
Common Names: Adder's eyes; bird's eye; bird's tongue; change-of-the-weather; cry baby; cry baby crab; drops-of-blood; eyebright; grandfather's weatherglass; John-go-to-bed-at-noon; ladybird; laughter bringer; little Jane; little peeper; numpinole; old man; old man's friend; old man's eyeglass; old man's weatherglass; owl's eye; pheasant's eye; ploughman's weatherglass; poor man's weatherglass; red bird's eye; red chickweed; red weed; shepherd's calendar; shepherd's clock; shepherd's delight; shepherd's dial; shepherd's glass; shepherd's joy; shepherd's sundial; shepherd's warning; shepherd's watch; shepherd's weatherglass; snap jack; sunflower; tom pimpernel; twelve o'clock; weather flower; weather glass; weather teller; wink-a-peep; wink-and-peep
Planet: Sun
Element: Air
Meanings: Assignations; change; childhood; faithfulness; meeting; opportunity of a rendezvous

The botanical name, *Anagalis*, is derived from the Greek, *anagela*, meaning 'to laugh', because the plant was supposed to lift depression and, according to Pliny, bring mirth (especially to those it has cured).

Magic and Lore

The scarlet pimpernel, or poor man's weatherglass, is a combination clock and barometer. The flowers open with great regularity at about 8 o'clock each day, and close again around 3 o'clock in the afternoon. The flowers are also very sensitive to humidity and temperature, and so can be read to forecast imminent changes in the weather; when the flowers are fully opened it is a sign that fine weather can be expected.

In Ireland the scarlet pimpernel is a blessed herb and is supposed to be a witch repellent. If held it gives the bearer the gifts of second sight and hearing. They will also be able to understand the speech of birds and animals (in view of this it comes as no surprise that it dispels melancholy and brings joy). When gathering pimpernel the following should be repeated:

> Herbe Pimpernell, I have found thee,
> Growing upon Jesus Christ's ground:
> The same gift the Lord Jesus gave unto thee,
> When he shed his blood upon the tree,
> Arise up, Pimpernell, and goe with me,
> And God bless me,
> And all that shall wear thee. Amen.

If one says this twice a day, for fifteen days in succession, whilst observing a fast in the morning (but eating in the evening) one can expect good luck to follow.

It has also been claimed that if the flower is dropped into running water it will float upstream, against the current.

Scarlet pimpernel is the birthday flower for 19 August.

Medicinal

In medical treatments the scarlet pimpernel was once considered as something of a panacea.

> No heart can think, no tongue can tell
> The virtues of the Pimpernel.

Its cleansing properties led to it being used to draw out splinters and thorns caught in the flesh.

PINEAPPLE

Botanical Name: *Ananas comosus*
Family: Bromeliaceae
Gender: Masculine
Planet: Sun
Element: Fire
Meanings: Welcome; 'You are perfect'

Myth and Legend

Ananas, the botanical name, is derived from the native name for the plant on the Pacific islands. In 1493, Columbus came across the pineapple on the island of Guadeloupe: the natives believed them to have been introduced to the island from the Amazon. This may be true, as pineapple-shaped jars have been found in pre-Incan burial sites in Brazil. The natives who cultivated these fruits called them *ananas* and believed that they had been brought from the Amazon many generations earlier by the warlike Caribs.

It is often said that members of the British East India Company brought the pineapple to Britain in the seventeenth century. They would have had to have been kept in hothouses and so could only be afforded by the wealthy. They therefore became symbols of opulence and high living.

Magic and Lore

Wild pineapple features in a somewhat macabre legend from Peru. There it is said that the dead rise up from their graves each night to eat *guabana* (or *guarabana*), the Peruvian name for the wild pineapple.

In magical rites pineapple is used to attract prosperity and wealth. The dried flesh of the fruit can be added to sachets for this purpose, or even dropped into bath water to bring the bather good luck. The juice, it is claimed, has a totally different influence. It is suppose to suppress the 'lustful desires' of anyone who drinks it.

The pineapple gains its common name from its resemblance to a pine cone. Many types of fruit have, over the years, been referred to as 'apples' even though they are clearly unrelated to that fruit. It came to symbolize hospitality, prossibly because of its exotic rarity, and so stone pineapples frequently adorn the gateposts of houses. In New England, USA, there are accounts of sea captains who, returning from the Caribbean and the Pacific, placed a pineapple outside their homes as a symbol of their safe return.

Medicinal

South American Indians used pineapple poultices to reduce inflammation in wounds and other skin injuries. They also drank the juice to aid digestion and to cure stomach aches. In 1891 an enzyme that broke down proteins (bromelain) was isolated from the flesh of the pineapple, accounting for many of the pineapple's healing properties. It has been found that bromelain can also break down blood clots, which consist mainly of protein. Research continues. This enzyme may well play a major part in heart attack treatment in the near future, as well as in the treatment of burned tissue, abscesses and ulcers.

ROSE OF JERICHO

Botanical Name: *Anastatica hierochuntica*
Family: Cruciferae
Common Names: Resurrection flower; rose of the Madonna

This strange plant is a native of the deserts of Egypt, Arabia and Syria. In drought, when its flowers and leaves have withered and fallen off, the branches dry up, bending inwards to form a ball. The roots wither and the plant, torn from the ground by the wind, is blown across the sands. If it lands on an area of dampness or is moistened by rain, the ball uncurls and the plant's seeds, which were hidden away inside, are deposited to germinate.

This plant was thought to be propitious to nativity, and hence the name 'Rose of the Madonna'. One legend claims that it first flowered when Christ was born, closed when he was crucified, and reopened again on Easter Sunday. At times of childbirth it would be collected and put into a glass of water at the side of the bed. Once the plant had fully opened it was taken as a sign that the birth was imminent.

WOOD ANEMONE

Botanical Name: *Anemone nemerosa*
Family: Ranunculaceae
Common Names: *Bainne bo bliatain*; bowbells; bread and cheese and cider; Candlemas caps; chimney smocks; crowfoot; cuckoo; cuckoo flower; cuckoo spit; darn grass; drops of snow; Easter flower; enemy; emony; evening twilight; faeries' windflower; flower of death; granny's nightcap; granny-thread-the-needle; jack-o'-lantern; lady's milkcap; lady's nightcap; lady's petticoat; lady's purse; lady's shimmy; milkmaids; Moll of the woods; moon flower; Nancy; nedcullion; nemony; shame faced maiden; silver bells; shoes and slippers; smell fox; smell smock; snakes and adders; snakes' eyes; snake flower; soldiers; soldier's buttons; star of Bethlehem; thunderbolt; white soldiers; wild jessamine; wild plant; windflower; woolly heads
Planet: Mars
Element: Fire
Deities: Adonis; Aphrodite; Flora; Venus
Meaning: Forlornness

Anemones used to be called windflowers, perhaps because they grew in areas of wind; the Greek word *anemos* means 'wind'. For the gardener the appearance of the little white flower heralds the start of spring. It has from classical times been a favourite for making up chaplets and garlands.

Myth and Legend

In mythology, Anemone was a nymph beloved by Zephyr, the god of the west wind. This aroused the jealousy of the goddess Flora, who banished Anemone from court and transformed her into the early spring flower that bears her name. Even Zephyr abandoned her to the rough north wind, Boreas. The north wind failed to win her love but disturbed her to such an extent that she flowers too early in the spring, and so fades more quickly.

Greek mythology also tells us of the story of Adonis, who was much loved by Aphrodite. Fearing that he would be injured whilst out hunting, she hid him in the Underworld. Her caution was to no avail as, while hunting, he came across a particularly large wild boar, which turned on him and gored him to death. From his blood sprang up the Adonis flower, and anemones grew from the tears of Aphrodite as she wandered the woods weeping after his death.

> Where streams his blood there blushing springs a rose,
> And where a tear has dropped, a windflower blows.
>
> – Ester Singleton, *The Shakespeare Garden*

An alternative version of the story tells that Aphrodite found Adonis just as he died. She wept over his lifeless body. From where his blood soaked into the ground, Zeus caused a flower to grow, the anemone.

Magic and Lore

Pliny describes anemones as daughters of the wind, which would only open their flowers when the wind was blowing, an allusion which has led to it being named 'windflower'. In Staffordshire it was often known as 'thunderbolt', as picking it was supposed to cause thunderstorms. It is said that when the anemone closes its petals rain is on the way. Traditionally it is thought to be a plant of luck and a protection against evil.

To dream of anemones suggests that your lover is unfaithful and should be spurned in favour of another.

Medicinal

In folk medicine wood anemones are used to treat headaches. Poultices containing the plant leaves are pressed to the head as a cure.

DILL

Botanical Name: *Anethum graveolens*
Family: Umbelliferae
Common Names: Anethum; aneton; common dill; dill weed; dilly; *fructus anethi*; garden dill.
Gender: Masculine
Planet: Mercury
Element: Fire

In ancient times the Greeks considered dill to be a sign of wealth. They would flaunt their wealth by burning dill-scented oils. In the Middle Ages dill found use in love potions and as a protective herb, countering the effects of witchcraft.

> Therewith he vervain and her dill,
> That hindereth witches of their will.
>
> – Drayton, *Nymphidia*

Sprigs of the herb would be hung above doorways to prevent any evil from entering, or placed into the cots of babies to protect them whilst they slept. Small quantities of the herb might also be carried in sachets to protect the bearer from harm. Presumably it is due to its use as a medicinal and culinary herb that it is thought of as being a plant of good luck. Brides would wear or carry a sprig of dill on their wedding day or put a small sprig into their shoes, with a little pinch of salt for luck.

Dill is mentioned in the Bible (Matthew 13:23), although in some versions the translator has incorrectly rendered the original Greek *Anethon* as 'anise'. However, dill is now commonly accepted as the *Anethon* referred to by Dioscorides.

Burning dill is supposed to clear thunderstorms.

Medicinal

In herbal medicine dill is valued as a calmative and, according to *The Popular Names of English Plants* by Prior, it is from the Old Norse word *dilla*, meaning to lull, that the name is derived. At one time nursing mothers would rub dill onto their breasts to lull their babies to sleep after suckling them.

Dill is another plant said to be an aphrodisiac. For this purpose it can be added to foods and eaten. However, the very smell of the herb is supposed to cause lust, and so it can be added to bath water in order to make the bather more attractive to the opposite sex. The smell of dill, especially when it is boiled in wine, is also claimed to be a sure cure for hiccups and a treatment for rickets, which might be some consolation if it fails to live up to expectations as an aphrodisiac. The emperor Charlemagne, aware of its properties, ordered a crystal vial of it to be placed on banqueting tables to be used to cure hiccups if guests ate or drank too much. Generally dill is best known as a calmative. Dill tea was drunk as an aid to sleep.

In addition to all of this it has been used in spells to attract money, presumably because of the abundance of the seed that the plant produces. The early settlers in America called the seeds of dill 'meeting house seeds' because they chewed them through the long sermons of their preachers in order to relieve the boredom. Chewing dill has the additional benefit of sweetening the breath.

ANGELICA

Botanical Name: *Angelica archangelica*
Family: Umbelliferae
Common Names: Angel plant; archangel; masterwort; root of the Holy Ghost; wild parsnip
Gender: Masculine
Planet: Sun (in Leo)
Element: Fire
Deities: Sophia; Venus
Meanings: Inspiration; 'You are my perfect inspiration'; 'Your love is my guiding star'

The plant is said to flower on the feast day of the Apparition of St Michael (8 May) and was supposed to have been revealed to mankind by the archangel as a cure for the plague. It was, of course, St Michael who drove Lucifer out of heaven and consequently all parts of this plant are seen as protecting against witchcraft and evil. It may be grown in the garden to protect the household or added to the incense burned during exorcisms. A quick method of driving evil forces from a house is to sprinkle a little angelica at the four corners of the perimeter. Where it is added to bath water it will remove all spells or curses cast against the bather.

It was revered long before it became associated with St Michael the Archangel. It was seen as a panacea, curing every conceivable ailment, and especially useful in the treatment of contagious diseases. The seeds were chewed during the worst days of the plague in order to avoid catching the disease. It was added to incense and mixtures used to aid healing. In parts of Pomerania and East Prussia it was customary for the peasants to collect angelica, march into the towns and offer it for sale. The collection and marching would be accompanied with chanting. The words used in the chants were learned in childhood, and so ancient in origin as to be unintelligible, even to those chanting them. This probably reflects an association with a pagan festival that was lost following the spread of Christianity.

The plant is also said to be useful in the treatment of alcoholism as it is said to cause distaste for alcohol in those who drink too much.

The Norwegians and Sami used the roots in their bread making. The Sami also chew, and smoke, the roots like a tobacco in the belief that it prolongs life. Native Americans carry the root as a lucky talisman, particularly when gambling.

Medicinal

Culpeper advises that:

> It is an herb of the Sun in Leo; let it be gathered when he is there, the Moon applying to his good aspect; Let it be gathered either in his hour, or in the hour of Jupiter; let so, be angular; observe the like in gathering the herbs of other planets, and you may happen to do wonders.

Gerard claims that it 'cureth the bitings of mad dogs and all venomous beasts'.

COW PARSLEY

Botanical Name: *Anthriscus sylvestris*
Family: Umbelliferae
Common Names: Bad man's baccy; bad man's oatmeal; bird's nest; bishop's lace; break-your-mother's-heart; dead man's flesh; de'il; de'il's meal; de'il's oatmeal; devil's parsley; devil's porridge; dog's flourish; kill-your-mother-quick; lady's lace; mother-die; my lady's lace; Queen Anne's lace; shit parsley; stepmother's blessing; wild carrot; wild chervil
Deity: Mary
Meanings: Haven; sanctuary; self-reliance

Myth and Legend

The name 'Queen Anne's lace' is supposed to be derived from the wife of James I. Legend has it that friends challenged Queen Anne to create lace as beautiful as a flower. In her attempt to do this she pricked her finger and the purple-red in the centre of the Queen Anne's Lace represents a droplet of her blood.

Magic and Lore

Cow parsley, like many other white-flowered plants, has usually been considered to be a plant of ill fortune if taken into the house. In most cases no specific reason is given for this, but in the case of cow parsley it is said that snakes will follow it.

It shares the name 'mother-die' with several other wild flowers, the warning being that if it is picked and used to decorate the home it will result in the death of your mother; presumably in this case from snakebite.

Cow parsley has been associated with the Virgin Mary, probably from the whiteness of its flowers. White-flowered wild plants were, for the most part, sacred to the White Goddess in ancient Britain and this reverence would have been transferred to the Madonna with the acceptance of Christianity.

SNAPDRAGON

Botanical Name: *Antirrhinum majus*
Family: Scrofulariaceae
Common Names: Bulldogs; calf's snout; dog's mouth; lion's snap; rabbit's mouth; toad's mouth; weasel's snout
Gender: Masculine
Planet: Mars (in Gemini)
Element: Fire
Meanings: 'I cannot care for you'; indiscretion; 'no!'; presumption; refusal

Myth and Legend

A Russian tale tells of a poor woodcutter whose home was set amid a field of snapdragons. One day, as he went home for his lunch, he met a small man who asked him if he could spare a little food. The stranger said that he had travelled far and was very hungry. The woodcutter offered to share his meagre lunch of black bread and apologized that he had no butter to spread on it. The stranger said that he had a solution and went out to gather some snapdragon seeds. He squeezed the oil from the seeds onto the bread, which he then ate. Having thanked the woodcutter for his hospitality the stranger went on his way. The woodcutter learned quickly, and thereafter took to extracting the oil from the snapdragons, which he then sold in the local town. This greatly improved his finances.

Magic and Lore

The snapdragon is a close relative of the toadflax. Like the latter it has been valued since ancient times as a protective herb, warding off witchcraft and evil forces, and providing a shelter for elves. Gerard says of the plant: 'They report (saith Dioscorides) that the herb being hanged about one preserveth a man from being bewitched, and that maketh a man gracious in the sight of people.'

Some writers claim that it must be the seed that is worn for this effect. Elsewhere it is claimed that wearing any part of the plant increases your powers of perception, preventing others from deceiving you, whereas concealing a snapdragon will make you appear fascinating and cordial.

If, when you are outside, you feel that you are in the presence of evil forces it is only necessary to step onto a snapdragon, or pick the flower and hold it against you, and you will feel the evil pass. In German legend a housewife was kidnapped by a malevolent elf. Whilst leading her away from the house the elf ordered her to lift the hem of her skirt so that she should avoid treading on origanums and snapdragons. The woman immediately stepped on them, crushing them, and breaking the power of the elf over her so that she could escape.

When a spell or curse has been made against you snapdragon can be used to send it back onto the person that cast it. To do this a small vase of snapdragon flowers should be placed onto an altar, with a mirror behind them. Indeed putting snapdragons onto the altar will help in any magical rites involving protective magic. Planting snapdragons on your roof might make you something of a talking point amongst your neighbours but it is sure to protect your house from lightning and fire, and to bring you good luck.

Snapdragon is the birthday flower of 28 June.

CELERY

Botanical Name: *Apium graveolens*
Family: Umbelliferae
Common Names: Smallage; wild celery
Gender: Masculine
Planet: Mercury
Element: Fire

Celery is well known as an aid to those wishing to lose weight. It is one of the very few foods that require more calories to eat than it actually contains. Some people may have used incense formed from of celery seed and orrisroot, to increase their psychic powers. Few people will, however, have considered chewing the seeds, which is supposed to aid concentration, and prevent dizziness. Witches, it is said, will chew celery seeds before mounting their broomsticks and flying to their Sabbats, to ensure that they do not become dizzy and fall off.

Another alleged attribute of celery, of which dieters are most likely unaware (and witches very aware), is that it is an aphrodisiac. The seed and stalks of celery can be eaten for this purpose. This may be seen to be somewhat at odds with the claim that celery seeds are also a sedative, and were added to 'spell pillows' to induce sleep.

BITTER ROOT

Botanical Name: *Apocynum androsaemifolium*
Family: Apocynaceae
Common Names: Dog's bane; rheumatism weed

Bitter root has been used in herbal remedies as a treatment for gallstone, and to 'correct bile flow'. The reason for its inclusion here is that the flowers of the bitter root are said to induce love. It is because of this that they have been included in mixtures used for that purpose.

WOOD ALOE

Botanical Name: *Aquilaria agallocha*
Family: Thymelaeaceae
Common Names: Lignum aloe
Gender: Feminine
Planet: Venus
Element: Water

The wood aloe is a tropical tree and is most probably the aloe referred to in biblical texts. Until the end of the nineteenth century is was greatly valued for its resinous wood, which was used in incense but is now largely unobtainable. In Egypt the incense made from wood aloe was burned to bring good fortune. Modern users value it for its high spiritual vibrations and, when available, add small quantities to their mixtures in order to intensify their strength.

It may have been because of its potency that wood aloe was used by magicians, during the Renaissance period, in evocation rituals. It is now more often carried or worn to attract love.

COLUMBINE

Botanical Name: *Aquilegia vulgaris*
Family: Ranunculaceae
Common Names: Ackely; baby's shoes; bachelor's buttons; Cains and Abels; bonnets; boots and shoes; culverwort; Dolly's bonnet; Dolly's shoes; doves-at-the-fountain; doves-in-the-ark; doves-round-a-dish; folly's flower; fool's cap; grandmother's bonnets; granny's bonnets; granny's hoods; granny-jump-out-of-bed; granny's nightcap; hen and chickens; lady's petticoat; lady's shoes; lady's slippers; lion's herb; nightcaps; Noah's ark; old lady's bonnets; old maid's baskets; old maid's bonnets; old woman's bonnets; old woman's nightcap; rags and tatters; shoe and stockings; skullcaps; snapdragon; soldier's buttons; thimbles; two-faces-under-a-hat; widow weeds
Gender: Feminine
Planet: Venus
Element: Water
Meanings: Anxiety and trembling (red); constancy; desertion; folly (wild columbine); resolution; resolve to win (purple)

Although it is not at first obvious, both the common and the botanical names of the columbine carry references to birds. *Aquilegia* is derived from the Latin *aquila*, meaning 'eagle', as the flowers are supposed to resemble the great talons of the bird. Columbine is also from the Latin, *Columba*, 'pigeon' or 'dove', the flowers being said to resemble a group of these birds in flight. (*Culfre* was the Saxon for pigeon, hence 'culverwort' – pigeon plant). It is, perhaps, ironic that the botanical name is an eagle, the symbol of war, whilst the common name is a dove, the symbol of peace. One other suggestion for the derivation of the botanical name is that it is from *aqua* meaning water, as the pieces of the flower resemble vessels for carrying water. The ancient name of *herba leonis* (*lion's herb*) comes from a medieval belief that it was a plant favoured as a food by lions.

Myth and Legend

A legend from the Iroquois tribe of Native Americans claims that the five-petalled native columbine (*Aquilegia canadensis*) came about from five tribal chiefs who were transformed into the flower by the Great Spirit. These chiefs had neglected their lands and people and gone off to search for a sky maiden, with whom they had each fallen in love during a dream. The flower shows these five chiefs huddled in a circle, wearing yellow moccasins and doe skin shirts dyed red and yellow.

This theme of forlornly chasing love is also found in British symbolism. Columbine was, in former times, used to symbolize deserted lovers. This role is now largely given over to the willow.

> The Columbine by lonely wand'rer taken
> Is there ascribed to such as are forsaken;
>
> – Browne, *Britannia's Pastorals*, 1613–1616

Magic and Lore

In religious art the dove-shaped petals were used as a symbol of the seven Gifts of the Spirit, i.e. wise council, knowledge, piety, fear, strength, understanding and wisdom. As wild columbine has five petals the early artists would paint seven flowers on the flowering spike.

Carrying the flowers of the columbine, or rubbing a little of the sap onto the hands, was supposed to induce courage and fearlessness (a piece of sympathetic magic, as in the courage of a lion). Rubbing the pulverized seed into the hands or elsewhere on the body was said to attract love, the scent being a 'love perfume'.

Columbine has been a badge of the Derby family and, like the red rose, of the House of Lancaster. It is the birthday flower of 1 April. The flower was also taken as an emblem of cuckoldry.

MONKEY PUZZLE TREE

Botanical Name: *Araucaria araucana*
Family: Araucariaceae

The monkey-puzzle tree was introduced into Britain from Chile during the eighteenth century. Over the years, for whatever reason, it has gained a reputation as a plant of misfortune and is associated with the Devil. One superstition recommends that you should stay silent whenever you stand beneath the tree as to break the silence will lead to three years bad luck. Although in Cambridgeshire the tree was thought to be unlucky, in many other parts of The Fens, it was recommended that it should be planted near to graves, because it was believed that where it grew in a graveyard it would act as an obstacle to the Devil if he should try to watch the funerals taking place.

NORFOLK ISLAND PINE

Botanical Name: *Araucaria heterophylla*
Family: Araucariaceae
Gender: Masculine

The Norfolk Island pine has become quite a popular houseplant in the UK but few of those who cultivate it will be aware of the benefits that this plant is bringing to their homes. Traditionally, where it is grown near a house (which presumably includes inside it) it will ward off all evil and protect the occupants from ever suffering hunger.

STRAWBERRY TREE

Botanical Name: *Arbutus unedo*
Family: Ericaceae
Gender: Masculine
Deity: Cardea

The strawberry tree, like the Norfolk Island pine, has a reputation for being able to dispel evil forces. This has led to it being used in exorcism rites since the times of the ancient Greeks. The Romans also used it as a protection against evil, particularly applying it to protect young children.

BURDOCK

Botanical Names: *Arctium lappa*; *Arctium minus*
Family: Compositae
Common Names: Bachelor's buttons; bardana; bardane; bardog; bazzies; beggar's buttons; Billy buttons; Bobby buttons; buddy weed; burdocken; burrseed; butter dock; cleavers; clitch button; clite; clot-burr; cloud-burr; clote; clog-weed; clots; clouts; cluts; cockle bells; cocklebur; cockle buttons; cockle dock; cockles; cradan; credan; cuckoldy-burr-busses; cuckold; cuckold butter; cuckold dock; cuckow; donkeys; eddick; flapper bags; fox's cote; gipsy comb; gipsy-rhubarb; great bur; great burdock; happy major; hardock; har-dokes; hareburr; hayriff; hor-docks; hurrburr; kisses; kiss-me-quick; lappa; lopper major; love leaves; old man's buttons; personata; philanthropium; pig's rhubarb; snake's rhubarb; soldier's buttons; sticky bobs; sticky buttons; sticky jacks; sweethearts; thorny burr; touch-me-not; turkey burrseed; turkey rhubarb; tuzzy-muzzy; wild rhubarb
Gender: Feminine
Planet: Venus
Element: Water
Meanings: Importunity; persistence; 'Touch me not'; 'Your suit is rejected'

The botanical name is derived from the Greek, *arktos*, 'a bear', a reference to the rough coated seeds and *lappa*, 'to seize', no doubt from the way in which it hooks into hair and clothing. The old English name for the plant was *herrif*, from the Anglo Saxon *hoeg*, a hedge, and *reafe*, a robber or *reafian* to seize. The English name burdock comes from a contraction of the French *bourre* or the Latin *burra*, meaning a lock of wool, and *dock*, a reference to the large leaves.

Magic and Lore

Burdock has long been recognized as a plant with protective and restorative powers. The root of the plant must be collected during the waning moon. It is then dried, cut into small pieces, and strung as a necklace to be worn to ward off evil.

If you consider that a necklace of burdock root is not the sort of fashion statement you wish to make then simply scatter the pieces of dried root about the house to dispel negativity and evil forces, or add them to incense and burn them for the same effect.

In Cornwall it is said that the pixies like to amuse themselves by riding young horses around the paddocks at night and tangling the burrs from burdock in their manes and tails. In many parts of England children would throw the burrs onto the backs of their friends. If the burr stuck it showed that the person had a sweetheart. If it fell off it indicated that their love was not reciprocated. Burrs were also a feature of the costume of the 'burry-man', who in South Queensferry wanders through the town each year, on the second Friday in August, receiving greetings and gifts from the people.

An entertaining spell, presumably a fertility potion, required the hooked seed heads of burdock to be collected and ground down with a mortar and pestle before being mixed with the 'private parts' of a billy goat and hair that had been snipped from a white puppy on the first day of a new moon, then burned on the seventh. To this mixture a little brandy was added and the resulting concoction left, uncovered in the open, to be impregnated by astral forces. The mix was then cooked until thick before the final ingredient, a few drops of crocodile semen, was added. Unfortunately, the difficulties in acquiring the various ingredients must have severely limited its usefulness.

Dandelion and burdock, the traditional soft drink, was made from the fermented roots of the two plants and so was naturally fizzy. Its modern manufactured counterpart is flavoured and may contain no trace of either plant. There are a number of tales relating to the origins of this drink, one of which says that St Thomas Aquinas, after spending the whole night in prayer, went out into the countryside trusting that the good Lord would provide for his needs. He therefore made a drink from the first plants that he found, and discovered that this drink aided his concentration when he went on to formulate the logic of his arguments in his great work, *Summa Theologica*.

Medicinal

Burdock can be useful eaten as a vegetable, as it contains iron and beneficial oils, but the most interesting claim for its powers as a restorative and cure is in its treatment of gout. It has been used as a laxative, but it also contains polyacetylenes, which are known to be antibacterial and antifungal, enhancing the overall performance of organs such as the liver and kidneys. Thus it clears waste from the body, improving general health and correcting digestive disorders. Some of the older herbals claim that simply placing burdock leaves onto the soles of the feet will cure gout. Other writers recommend wearing a small bag of herb about the neck as a remedy for rheumatism.

It was once thought that eating the burrs of the plant would help things to stick in one's mind.

UVA URSA

Botanical Name: *Arctostaphylos uva-ursi* (Synonym: *Arbutus uva-ursi*)
Family: Ericaceae
Common Names: Arberry; bearberry; bear's grape; blanchnog; brawlins; burren myrtle; craneberry wire; creashak; creashat; dog berry; gnashicks; hog cranberry; kinnikinnick; mealberry; moanagus; mountain box; mountain cranberry; nashag; rapper-dandy; red bearberry; sagackhomi; sandberry.

This plant has been used by the natives of North America in their religious ceremonies. In more recent times modern occultists have used it to increase their psychic powers.

Uva ursa is found in Britain, along with *Arctostaphylos alpina*, the black bearberry. This latter plant, found only in the barren mountain regions of northern Scotland, is the badge of the clan Ross.

BETEL NUT

Botanical Name: *Areca catechu*
Family: Palmae
Common Names: Areca nut; betel palm; catechu; pinang; pinang siri; pinglang; supari

Betel nut is a tree cultivated in some of the warmer regions of Asia. The nuts are grated, or cut into strips, mixed with various spices and then rolled into betel pepper leaves. The resulting 'cigar' is wiped with lime juice and then chewed. The effect is actually beneficial as it kills internal parasites, but it stains the teeth and gums dark red or black. Consequently, on some of the Pacific islands, black teeth are a sign of high social standing. This practice has been followed for over 2,000 years and is referred to in Sanskrit and Chinese texts.

Men, in some Tantric sexual rituals, chew betel nuts. The Maro dance of West Seram in Indonesia was danced by the men of the 'nine families of man', in a nine-fold spiral for nine days and nights. Women at the centre of the spiral would hand out betel nuts to the men.

In Sumatra it was customary to place small coins and betel nuts on young trees when they were planted to replace larger trees that had to be felled. This was to compensate the tree spirit for its loss and the young tree afforded the spirit a new home.

SNAKEROOT

Botanical Name: *Aristolochia serpentaria*
Family: Aristolochiaceae
Common Names: Pelican flower; radix viperina; serpentary radix; serpentary rhizome; snagree; snagrel; snakeweed; virginian snakeroot

'Snakeroot' and 'serpentary' are overused common names and have been applied to several different plants. Some care must, therefore, be taken in identifying the appropriate plant in each case. The plant has been used to stupefy snakes. Egyptian jugglers would make use of it to calm, or sedate, their snakes before they handled them.

The root is also supposed to be effective as a good luck charm. It may be carried as an amulet to bring good fortune and break curses, or to lead the bearer to find money. Under the Doctrine of Signatures the plant found use in aiding childbirth, and was said to resemble the position of the foetus prior to birth.

HORSERADISH

Botanical Name: *Armoracia rusticana*
Family: Cruciferae
Common Names: Great raifort; mountain radish; red cole
Gender: Masculine
Planet: Mars
Element: Fire

Horseradish is not a British native, but it is not certain when it was introduced. It is one of the bitter herbs eaten by Jews at Passover to remind them of the hard times suffered by the Jewish peoples under Pharaoh. Others herbs used include nettle, lettuce, coriander and horehound. It is probable that it was first used by the Germanic and Slavic people in their pickles and spices.

Magic and Lore

The dried, powdered, root can be used to dispel evil forces. To this end it is sprinkled into the corners of the house and at all doors and windows, which forces out any evil present and lifts curses or spells cast against you. It may also be added to the incense that is used during exorcisms.

If a couple is expecting a child, and want to know whether it will be a boy or a girl they should each cut a small piece of horseradish root and place it beneath their pillow before going to bed. These pieces should be compared daily, and if the one under the man's pillow turns black before that of his partner it indicates that the child will be a boy. If the woman's turns black first, then it will be a girl.

Medicinal

Horseradish was valued for its medicinal uses long before it became used as a condiment. The Greeks had a medicinal plant they called *rahpnos agrios*, wild radish, which is thought to have been horseradish. By the Middle Ages in Britain the fresh root was being used for poultices in the treatment of sciatica, rheumatism and gout. In the *Art of Simpling* (1657) by William Coles a poultice of horseradish is even recommended for the treatment of worms in children.

SAGEBRUSH

Botanical Name: *Artemisia* spp.
Family: Compositae
Gender: Feminine
Planet: Venus
Element: Earth
Deities: Artemis; Diana

Artemisia is a large genus of plants containing some 180 different species. The whole genus derives its name from Artemis the Greek goddess of chastity.

It has long been used in the rites and ceremonies of American Indians. It is another herb associated with warding off evil spirits, and bathing in water to which sagebrush is added will cleanse all past evil actions and negativity from you. Burning the herb will likewise drive evil away from the site.

SOUTHERNWOOD

Botanical Name: *Artemisia abrotanum*
Family: Compositae
Common Names: Appleringie; boy's love; *garde robe*; lad's love; maid's ruin; old man; stabwort
Planet: Mercury
Element: Air
Deities: Artemis; Diana
Meanings: Bantering; jest

Magic and Lore

Once upon a time it was customary for ladies to carry a small bunch of southernwood and balm into church with them, so that their sweet scents might prevent drowsiness (which does not say much for the quality of the sermons preached!). Southernwood would also be planted near to doorways so that the skirts of ladies entering and leaving the building might catch it, causing the scent to fill the air.

The lust-inducing qualities of the plant have been acknowledged since ancient times, though they are sometimes attributed to different usage of the plant. It might be carried, or placed in a bedroom, to attract love. In ancient Greece and Rome it was placed beneath mattresses in order to induce lust in those lying on it. 'Plinie writeth that if it be layde under the bedde, pillow or bolster, it provoketh carnall copulation, and resisteth all enchantments, which may let or hinder such businesse and the inticements to the same,' wrote Lyte in 1578. Furthermore there was a belief that it prevented impotence and could be used by men to boost their virility.

In *The Popular Names of British Plants* by R.C.A. Prior it was claimed that young men used to rub their chins with the sap of southernwood, or an ointment made from the ashes of the leaves, to induce beard growth. This made them appear manlier and so more attractive to women. Surprisingly, another suggestion for the origin of the name 'lad's love' or 'boy's love' comes from its use as a moth repellent. Young men would give small muslin bags containing dried southernwood and other herbs to their sweethearts as love tokens. These were placed amongst clothes to scent them and keep moths away. It is from its use as a moth repellent that southernwood earns its French name of *garde robe*. Elsewhere, it is said that young men would wear lad's love in their buttonholes in order to attract a partner, as it showed that they were thinking of matrimony. The man would then give it to the girl of his choice and, if she accepted it, it indicated that she was prepared to marry him.

If a girl put a sprig of southernwood down the back of her blouse or dress before leaving home in the morning she would ultimately marry the first eligible man that she met. Placing it beneath her pillow before going to bed would lead her to dream of her future spouse. Other writers claim that the plant was called 'lad's love' because it was included in the posy, given by lads to their sweethearts in the hope of seducing them.

Amongst Slavic nations it was believed that southernwood could be burned as an incense to ward off thunderstorms.

In Lithuania it was used in a form of divination. Three short sections of the plant, about the length of a finger, were thrown into the air. In a similar way to runes or I Ching, the way they landed would predict a good, or bad omen.

Medicinal

The plant was once considered to be a cure for all sorts of contagious diseases and was generally thought to ward off infections. In the early part of the nineteenth century, at the Court of Assizes, a pot of herbs, including southernwood and rue, would be placed upon the judge's bench to ward off typhus, commonly called gaol fever.

Its antiseptic properties were considered invaluable in the treatment of diseases caught during 'careless lovemaking'. Perhaps this is another reason for it being called 'lad's love'.

Barley meal, boiled with southernwood, might be prescribed to remove pimples, and southernwood has also been recommended as a treatment for snakebites. Culpeper reports Dioscorides' claim that wine made with the seeds of southernwood will make an antidote to snakebite. Elsewhere it is said that the burning of the herb in a room drives out all snakes, frogs and toads that might have entered.

WORMWOOD

Botanical Name: *Artemisia absinthium*
Family: Compositae
Common Names: Absinth; absenthe; crown for a king; green ginger; mugwort; old woman; vermouth; wermout; wermud; wormit; wormod
Gender: Masculine
Planet: Mars
Element: Fire
Deities: Artemis; Diana; Iris
Meanings: Absence; 'Even the best of friends must say farewell'; heartache; sorrowful parting

Myth and Legend

Wormwood is an attractive and heavily scented garden plant. Tradition has it that it sprang up from the tracks left by the serpent when it was driven out of the Garden of Eden. This may account for it being used in the Bible as a symbol of calamity and intense sorrow. It is second only to rue in bitterness, and rue of course shares this symbolic usage.

Magic and Lore

Wormwood has, for centuries, been used as an ingredient in herbal wines and liquors, giving its name to both absinthe and vermouth. Vermouth, however, is more correctly the Roman wormwood, *Artemisia pontica*, which is the least powerful of the artemisia group. The word 'vermouth' (or *wermuth*) means 'preserver of the mind' and refers to the plant's use as a mental restorative. Absinthe, which is made from wormwood, has been outlawed in several countries because of its addictive nature. Over-indulgence can lead to the symptoms of absinthe overdose, vertigo, cramps, and delirium. The Romans piled wormwood on altars at the Latinae Festival and gave a beverage made from it to the victors of their games and chariot races in order that they should enjoy good health and long life. Pliny says that this beverage is the most ancient of herbal drinks.

In magic wormwood is used to heighten psychic powers. It may be worn, or burned as incense. According to old grimoires wormwood incense, possibly with some sandalwood added, could be burnt in graveyards during necromantic rites. This would summon the spirits of the dead, who could then be compelled to answer questions put to them.

A much more modern use of the plant is as a lucky amulet, hung from the rear-view mirror of a car, to protect the driver from road accidents.

A totally different application is in love magic. It could be included in love potions, or hidden beneath a bed in order to 'draw love to it'. Wormwood is also an ingredient in the St Luke's Day charm. The charm, as set out in *Popular Rhymes and Nursery Tales*, requires that you 'take Marigold flowers, a sprig of Marjoram, Thyme, and a little Wormwood; dry them before a fire, rub them to a powder; then sift through a fine piece of lawn, and simmer it over a slow fire, adding a small quantity of virgin honey, and vinegar.' You must then anoint yourself with the mixture before going to bed on St Luke's Day (18 October), saying the following spell three times:

> St Luke, St Luke be kind to me,
> In dreams let me my true love see.

This will result in you dreaming of your future partner in life. Dreaming of wormwood, or just seeing it in a dream, is usually considered to be an omen of good luck.

Medicinal

Like southernwood, its close relative, wormwood has long been regarded as useful as an insect repellent.

> While Wormwood hath seed get a handful or twaine
> To save against March, to make flea to refraine:
> Where chamber is sweeped and Wormwood is strowne,
> What saver is better (if physick be true)
> For places infected than Wormwood and Rue ?
> It is a comfort for hart and the braine,
> And therefore to have it it is not in vaine.
>
> – Tusser, *July's Husbandry*, 1577

It is also reported to have the power to counter many types of poisoning, including from hemlock, toadstools, or the 'bite of sea serpents'.

TARRAGON

Botanical Name: *Artemisia dracunculus*
Family: Compositae
Common Names: Dragon mugwort; fuzzy weed; little dragon; *herbe au dragon*
Planet: Mars
Meaning: Share

The name 'tarragon' is a corruption of the French word *esdragon*. This, in turn, is derived from the Latin *dracunculus*, meaning 'little dragon'. Like the other 'dragon' plants, and other species of artemisia, it is alleged to have the power to heal bites caused by dragons. Like wormwood, to which it is closely related, is used as a treatment for stings and the bites of venomous creatures or rabid dogs.

Another attribute shared with wormwood and southernwood is its power to attract love and affection. Native Americans would rub it into their clothing, and onto their bodies as a perfume. Tarragon, it seems, not only attracts the opposite sex but also wild beasts. It was carried or worn when hunting, in order to ensure a successful outcome.

MUGWORT

Botanical Name: *Artemisia vulgaris*
Family: Compositae
Common Names: Apple pie; Artemis herb; bowlocks; bulwand; *Cingulum Sancti Johannis*; docko; dog's ears; fat hen; felon herb; gall-wood; green finger; grey bulwand; herb of St John the Baptist; maiden's wort; migwort; mogvord; mogworte; moogard; motherwort; mugger; muggert; muggert kail; muggons; muggurth; mugweed; mugwood; naughty man; old man; old uncle Harry; old uncle Henry; sailor's tobacco; St John's plant; smotherwood; wormwood
Gender: Feminine
Planet: Venus
Element: Artemis; Diana
Deity: Earth
Meanings: Happiness; tranquillity

Magic and Lore

Mugwort is another close relative of wormwood and southernwood and, like them, it has been used as a moth repellent. Some writers have gone so far as to suggest that this usage led to the common name of 'mugwort', saying that it is derived from *moughte*, as in maggot or moth. Another suggestion is that it comes from an Old English name *muggiawort*, meaning 'midge plant', as it was used as a mosquito repellent (from this it is suggested that it could be added to cut flowers in the house in order to keep flies away). These suggestions are not universally accepted however, as other writers tell us that the name comes from its use in flavouring beer before the use of hops.

Some of the common names applied to the plant have come from its association with St John the Baptist. In the Middle Ages it was called *Cingulum Sancti Johannis*, and it was believed that the saint wore a girdle of the herb during his time in the wilderness. In Holland and Germany mugwort is known as St John's plant and traditionally, when it is collected on St John's Eve it will give protection against all illnesses, evil and misfortune.

There was a belief that if you dug under mugwort on Midsummer's Eve you would find a 'coal'. Carrying this coal protected the bearer from plague, ague, carbuncles, burning and lightning strike. It also protected them from witchcraft. Wearing a crown formed from the plant on St John's Eve is a protection against possession by evil spirits. Indeed, in both Eastern and Western traditions, mugwort is seen as a protective herb hung up in the home to prevent evil from entering.

Eldest of worts
Thou hast might for three
And against thirty
For venom availest
For flying vile things
Mighty against loathed ones
That throughout the land rove.

– *Lacnunga*, 10th century

In Japan it is believed that bunches of the herb can be used to exorcise the spirits that are responsible for causing illnesses. Western traditions merely require that the plant be carried to cure diseases and protect against evil spirits. Amongst some of the Native American peoples, mugwort was likewise used in 'witchcraft medicine'. Rubbing it onto the body would keep ghosts away, and wearing a necklace of the herb prevented the wearer from dreaming of the dead.

Michael Drayton in *The Muses Elisium* says of mugwort:

This is my moly of much fame,
In magic often used;
Mugwort and nightshade for the same,
But not by me abused.

As this suggests, it is used in magical rites, not just as a protective herb. Its main use is by those interested in fortune telling, particularly scrying (crystal gazing). An infusion can be used to clean magic mirrors and crystal balls before they are used. Burning incense made from mugwort and sandalwood helps the fortune teller to concentrate. A few leaves of the plant should also be placed at the base of, or beneath, the crystal ball itself. Where a bunch of the herb is put at the side of the bed at night, it will aid astral projection. Sleeping on a pillow stuffed with the herb induces prophetic dreams and protects against nightmares.

In Britain, mugwort has a particular association with the Isle of Man. It is thought that it might have been the King's emblem and it is said that soldiers attending the Tynwald (parliament) would wear a small sprig of the herb to indicate their allegiance to their king.

Medicinal

The herb is dedicated to the goddess Artemis, sister of Apollo and the Hellenic goddess of forests, hills and the hunt, but also of childbirth, virginity and fertility. It was therefore used to treat problems of the female reproductive system.

To treat illness it was said to be necessary merely to have the patient sleep on a pillow containing the herb. If, however, they were unable to get to sleep the prognosis was far from favourable.

Carrying the herb is said to protect against sunstroke, the attacks of wild beasts, madness, evil spirits and backache. If that were not enough, carrying it will also increase fertility and lustfulness.

Generally, mugwort it said to be a useful aid to weary travellers. To prevent fatigue during long walks collect a little mugwort, before sunrise, saying; '*Tollam te artemisia, ne lassus sim in via.*' It must then be placed inside your shoes before setting out. The Romans may have placed mugwort in their sandals to prevent aching feet.

> And if a Footman take Mugwort and put it into his
> Shoes in the Morning, he may goe forty miles
> Before Noon and not be weary.
>
> – William Coles, *The Art of Simpling*, 1656

In China it is called *Moxa*. Moxabustion is the burning of the herb and the application of the heat generated to the acupuncture points.

In Scotland and northern England there is a saying:

> If nettles were used in March, and Muggins in May
> Many a bra' lass wudna' turn to l'clay.

There is no explanation as to how the nettles and muggins (mugwort) might be used. It is at least fairly certain that it has nothing to do with treating opium addiction, something else for which mugwort was once recommended.

LORDS AND LADIES

Botanical Name: *Arum maculatum*
Family: Araceae
Common Names: Aaron's rod; adder's food; adder root; adder's tongue; angels and devils; aron; babe in the cradle; bloody fingers; bloody man's fingers; bobbin and joan; bobbin joan; bobbins; bullocks; bulls and cows; bulls: calve's foot; cocky barn; cows and calf's; cows and kies; cow's parley; cuckoo cock; cuckoo-flower; cuckoo pint; cuckoo point; dead man's finger; Devil's ladies and gentlemen; Devil's men; Devil's men and women; dog bobbins; dog cocks; dog's dick; dog's dibble; dog's spear; dog's tassel; faeries; faerie candles; faerie lamp; fly catcher; friar's cowl; frog's meat; gentlemen and ladies; gentlemen's and ladies' fingers; gentlemen's fingers; gethsemane; great dragon; hobble-gobbles; jack-in-the-box; jack-in-the-green; jack-in-the-pulpit; kings and queens; kitty-come-down-the-lane-jump-up-and-kiss-me; knights and ladies; ladies' lords; lady's finger; lady's key; lady's slipper; lady's smock; lamb-in-a-pulpit; lamb's lakens; lily; lily grass; long purples; lord's and ladies' fingers; man-in-the-pulpit; men and women; moll-of-the-woods; nightingale; parson and clerk; parson-in-the-pulpit; parson's billycock; preacher-in-the-pulpit; priesties; priest-in-the-pulpit; priest's pilly; priest's pintle; poison fingers; poison root; pokers; quaker; ramp; ram's horn; ramson; red-hot poker; school master; shiners; silly lovers; small dragon; snaker food; snake's food; snake's meat; snake's victuals; soldiers; soldiers and angels; soldiers and sailors; stallions;

stallions and mares; standing gusses; starchwort; sucky calves; toad's meat; wake robin; white and red; wild lily

Meanings: Ardour; 'My heart is aflame with passion'; zeal

Myth and Legend

In the fenland region of East Anglia this plant has become associated with St Withburga. It was believed that when nuns came to Thetford from Normandy to build a convent they brought with them the body of the saint and the lords and ladies plant. The party carrying the remains of the saint had reached East Dereham when monks from Ely stole the body and carried it back to their monastery. On their journey they rested by the Little Ouse at Brandon and it was here that the nuns from Thetford caught up with them. The nuns covered the saint's body with the white flowers of lords and ladies and, during the journey by barge up the river, several of the flowers fell into the waters. Where they washed up on the banks they took root, and within an hour the whole of the river bank, as far as Ely, was covered with these white flowers. At night the flowers were said to glow with their own radiance.

The leaves of the plant are marked with dark lines and, like the early purple orchid, it is said that this plant grew at the base of the cross, the staining on the leaves being caused by drops of Christ's blood when he was crucified, hence the name 'Gethsemane'.

> Those deep unwrought marks,
> The villager will tell you,
> Are the flower's portion for atoning
> On Calvary shed beneath the cross it grew.
>
> – Edward Berboe, *The Browning Cyclopaedia*

German folklore tells that wheresoever *Aronswurzel* ('Aaron's rod', arum) grew and flourished the wood sprites would rejoice. They give it the name Aaron's rod from the tale that, on leaving their Egyptian exile, one of the items carried by the Israelites was a long pole or staff. This staff was used to support the heavy bundles of grapes they picked for food during their journey. It was called Aaron's rod in allusion to the brother of Moses as it was said that, at one point where they rested, Aaron pushed his staff into the earth and up grew arums to indicate the soil's fertility. Modern farmers will say that if the arums have big flowers with large spadices the harvest will be especially good.

Magic and Lore

Writing in his herbal of 1597 John Gerard reports a tale, which he attributes to the philosopher Aristotle and others. Bears, having hibernated through the winter months, emerge in the spring from their dens. They are half starved but unable to eat because their stomachs have shrunk. They seek out this arum because, when they eat it, the gaseous nature of the plant causes their stomachs to expand, enabling them to eat normally. Another odd belief was that adders would eat the berries of this plant and absorb the poison from them, so as to increase the venom in their fangs.

Gerard also reports that a starch made from the roots of the plant was used to stiffen ruffs. However, it proved so caustic that it caused the hands of the laundresses to roughen, chap, blister and smart. Traditionally, if a man placed a small piece of the spadix of lords and ladies in his shoe, when on his way to a dance, while saying the following, he could be sure of the pick of prettiest girls to partner him.

> I place you in my shoe,
> Let all the young girls be drawn to you.

In some works, and occasionally in early art, the form of the flower was used as a symbol of copulation, presumably because of the shape of the flowers. It may also be assumed that its reputation as an aphrodisiac results from the same cause. Lords and ladies was believed to be so powerful that it was necessary for a woman merely to touch the plant for her to become pregnant.

It is the birthday flower of 19 May.

Medicinal

In Cambridgeshire it was traditionally believed that the plant should never be picked and taken into the house, as it would cause all who came into contact with it to suffer from tuberculosis.

MILKWEED

Botanical Names: *Asclepias acida*; *Asclepias curassarica*
Family: Asclepiadaceae
Common Names: Ambrosial tree; amrita; blood flower; indian root; matal; swallow wort
Deity: Soma

The botanical name of *Asclepias* comes from the name of the Greek god of medicine Askelpios (in Latin, Aesculapius). Askelpios is normally depicted flanked on either side by Hygiena, goddess of wise living, and Panakeia, goddess of cure-alls.

In Hindu mythology the plant personifies Soma, the Hindu equivalent of the god Bacchus, born of Brahma and one of the most important of the Verdic gods. The plant sap, prepared in a sacred ceremony, could be drunk by both men and the gods, as the true home of the plant is in heaven. It was Soma who first nourished mankind and sustained life according to the myth. It was under the influence of this 'soma juice' that Indra created the universe and set the worlds in their respective positions. In post-Vedic writings, Soma is the name given to the moon, which is still drunk by the gods. The 'drinking' of the moon results in its waning, only to be refilled by the sun. In *Rig Veda*, and *Zend Avesta*, Soma is the King of the Planets, and a medicine which removes death, giving health and a long life.

ASPHODEL

Botanical Names: *Asphodelus* spp; *Asphodeline* spp.
Family: Asphodelaceae
Planet: Saturn
Element: Earth
Deity: Persephone
Meanings: Death; ' My regrets follow you to the grave'; 'Our love shall endure after death'

White asphodel, *Asphodelus alba*, is native to southern Europe and has been a symbol of death for centuries. It was planted near to graves in the belief that the roots would nourish the ghosts of the departed. In Greek mythology there were said to be great fields of asphodels growing beyond the river Acheron, one of the five great rivers of the Underworld, and it was in these fields that the shades of the departed wandered.

It may be the yellow asphodel, *Asphodeline lutea*, which has been confused with the daffodil in the stories of Persephone. Persephone was abducted by Hades, whilst out picking asphodels, and taken off to become his queen of the Underworld.

ASTER

Botanical Name: *Aster* spp.
Family: Compositae
Common Name: Starwort
Planet: Venus
Element: Water
Deity: Astraea
Meanings: 'I partake of your sentiments' (double); 'I will think of it'(single)

Myth and Legend

The name 'aster' comes from the Greek, meaning 'a star'. In mythology, Astraea, the goddess of innocence, dwelt on the Earth during the golden age of creation, before evil and hardship had entered the world. Once evil had come she could no longer remain amongst mankind and so she was metamorphosized into the constellation of Virgo. Zeus was angered by the evil ways of mankind and so caused a great flood that encompassed the entire globe, except for the peak of Mount Parnassus. It was there that the only two people who were spared, Deucalian and his wife Pyrrha, stayed until the waters receded. After the water level had dropped sufficiently Deucalian and Pyrrha walked amid the mud guided by starlight, a gift of Astraea. It was Astraea's tears that fell to earth as stardust and were there transformed into the flowers, asters and starwort. An alternative story of their origin is that the fields bloomed with asters when Virgo scattered stardust on the earth.

In the myth the Minotaur was the offspring of Pasiphae and a bull sent to her by the sea god Poseidon. It had the head of a bull but the body of a man. King Minos, Pasiphae's husband, imprisoned it in a labyrinth on Crete. The monster fed on human flesh and Minos demanded that Aegeus, King of Athens, send seven youths and seven maidens annually from Athens to feed it. One year Theseus, Aegeus' son, volunteered to be one of the youths, believing he could kill the Minotaur. Before setting sail for Crete he told his father that when he returned he would hoist white sails on the ship to indicate that he had been successful, instead of the black ones that were raised when the ship left. When Theseus arrived in Crete he fell in love with the king's daughter, Ariadne and with her help, he entered the labyrinth and killed the Minotaur. On returning to Athens, however, Theseus forgot to hoist the white sails. On seeing the black sails Aegeus, believing his son was dead, killed himself. Where his blood flowed onto the ground purple asters sprang up, the result of a spell put on him by the sorceress Medea, who had once been his wife.

Among the Cherokee of the southern USA there is another legend regarding the origins of this plant. Two tribes were at war over a particularly good hunting ground. The fighting was intense and ranged up a hill, down the valley, over a creek and into a village. All those living in the village were killed, with the exception of two young sisters who hid themselves in a wood. These two girls then sought out a wise woman who lived in the next valley. That night, while the girls slept, the woman looked into their future and saw that they would be tracked down and killed by their enemies. She therefore went out and collected herbs, from which she brewed a magic potion. When she sprinkled the potion onto the girls they were transformed into leaves. In the morning there were two new plants growing in the wood. One sister, who had been wearing a doeskin dress which had been dyed blue, had been transformed into a lavender blue aster. The fringe of her dress had become the ray petals of the flower. The other sister, whose dress had been yellow, was transformed into the beautiful goldenrod (*Solidago*).

The Michaelmas daisy, which usually now refers to *Aster novi-belgii* and its cultivars, was a common name once given to *Aster amellus* (Italian starwort) although the English applied the name Michaelmas daisy to *Aster tripolium*. The plant was supposed to flower at the time of Michaelmas, the feast day of St Michael and All Angels (29 September).

> The Michaelmas Daisy, among dead weeds,
> Blooms for Saint Michael's valorous deeds.
>
> – *The Wander Garden*

Magic and Lore

In ancient times it was believed that the perfume from burning Aster leaves would drive away evil serpents.

See also China aster (*Callistephus*).

MASTERWORT

Botanical Name: *Astrantia* spp.
Family: Umbelliferae
Gender: Masculine
Planet: Mars
Element: Fire

Masterwort was once cultivated in Britain as a potherb. Wearing the plant as an amulet promises the wearer many benefits. As well as calming the emotions and warding off evil forces, it grants the wearer increased physical strength.

This is one of several plants that have been used during magical rites to conjure up spirits. In such ceremonies it should be liberally sprinkled about the site.

DEADLY NIGHTSHADE

Botanical Name: *Atropa belladonna*
Family: Solanaceae
Common Names: Banewort; *belladonna*; black cherry; daft berries; death's herb; Devil's berries; Devil's cherries; Devil's herb; Devil's rhubarb; divale; doleful bells;

dog berries; dwale; dwaleberry; dwayberry; fairlady; great morel; Jacob's ladder; Jacob's stee; naughty man's cherries; satan's cherries; sorcerer's berry; witch's berry
Gender: Feminine
Planet: Saturn
Element: Water
Deities: Atropos; Bellona; Circe; Hecate
Meanings: Deception; imagination; loneliness; silence; truth

Myth and Legend

This extremely poisonous plant is named after Atropos, the eldest of the three Fates of Greek mythology. The Fates governed the major events in human existence, and held the shears to sever the thread of life, without respect for age, gender or any other factor. Considering the toxic nature of the plant this is very appropriate. In ancient Rome, we are told, the priests of Bellona (goddess of War) drank an infusion of the plant before worshipping her or imploring her aid. It has been suggested that the botanical name was derived from this usage.

Belladonna gives its name to beautiful women of course. This comes from the practice of dripping the sap of the plant into ladies' eyes, causing the pupils to dilate and the eyes to have more lustre. An alternative legend relates that it was used by the Italian, *Leucota* to poison beautiful women.

One story tells us that Truth, a being of divine origin, lived in the bottom of a well. There she dispensed her blessings upon those who sought her out, but mixed them with a little bitterness. As truth and nightshade preferred the shade rather than full sunlight, the plant became the emblem of the truth. It is a symbol of deception either because of its attractive appearance and deadly nature, or because of its use by women wishing to appear more attractive.

The poisonous nature of the plant has been used to great effect. It is reported that this is the poison used to kill the troops of Marcus Antonius during the Parthian Wars. Closer to home Buchanan, in the *History of Scotland* (1582), tells us of an event which happened that during the reign of Duncan I. Macbeth had his troops poison an invading army of Danes with a liquor mixed with 'dwale', during a period of truce. As a result the invaders slept soundly and were easily overpowered and slain.

Magic and Lore

Deadly nightshade is the Devil's plant, which he tends at his leisure. The only time that he is diverted from this task is during the preparations for the witches' sabbats on Walpurgis Night (30 April, May Eve). Witches might then use the belladonna, together with hemlock and wolfsbane, to make their flying ointment. Nightshade causes delirium and wolfsbane induces irregular heart flutters. This, according to experts, could give a sensation of flying. It has been suggested that the plant might be used on its own to aid astral projection, but its very poisonous nature must cause concern amongst those who might use it.

In magic it found use in those spell cast to induce madness and cause death. A collar formed from nightshade could be used to protect cattle from enchantment and hag-ridden horses could be saved by placing a wreath of nightshade and holly about their necks. Even humans could benefit from the use of this plant, as wearing a wreath of nightshade on the head would counter the effect of any spells cast against the wearer. Belladonna is also said to be a favourite plant of clairvoyants as it can be used to gain the gift of second sight.

Nightshade is the birthday flower for 15 July.

OATS

Botanical Name: *Avena sativa*
Family: Gramineae
Common Names: Groats; oatmeal
Gender: Feminine
Planet: Venus
Element: Earth
Deity: Brigid
Meanings: The witching soul of music; 'Your voice is music to my ears'

Oats are used in magical spells to bring money and prosperity. On a somewhat different note it is claimed that oats are a 'naturalizer' to the sexual system, and eating them promotes youthful vigour.

Across Britain there are several different guides to the appropriate dates for sowing oats. In Cardiganshire it was traditional to sow them at the time of the Caron Fair, or Ffair Garon, which was 15 March. In the colder parts of North Wales a later date was favoured: '*Tridiau y deryn du, a dau lygad Ebrill (The three days of the blackbird, and the two eyes of April)*.' That is the last three days of March and the first two days of April, by the old calendar (around 19 March in the modern calendar). In Worcestershire it was said:

Who in January sows oats,
gets gold and groats.

On the other side of Britain, in Suffolk, the rhyme instructed:

Upon St David's Day
Put oats and barley in clay.

Another of these old rhymes seems, at first view, to make little sense. It says:

Cuckoo oats and woodcock hay,
Make the farmer run away.

It refers to a 'backwards' spring in which the cuckoo could be heard during the time when oats were being sown, and the hay was being cut when woodcocks could be seen. This would indicate a bad year and severe losses for the farmer.

BAMBOO

Botanical Names: *Bambusa ventricosa*; *Bambusa vulgaris*
Family: Gramineae
Common Names: Common bamboo; feathery bamboo; *ghrab*; *hizaran*; ohe
Gender: Masculine
Deities: Hina; Thoth

The Chinese regard bamboo as a symbol of long life because shoots are always sturdy and green.

Myth and Legend

A creation myth from the Philippines credits the birth of humanity to the bamboo stem: The first man and woman came from a bamboo stem and began the world's progeny. Likewise in the creation story of the Andaman Islanders of the Indian Ocean, the first man is born inside a large bamboo stem, and a Malaysian legend tells of a man finding a beautiful woman, the love of his life, inside a bamboo stem.

Magic and Lore

It is no surprise that bamboo is much more important in Eastern folklore and magic than it is in the West. In China, bamboo has long been used to make I Ching sticks, which are thrown by the priest, who foretells the future by the way in which they fall. Bamboo is also used for making amulets to keep away evil, and for the flutes used to summon up good spirits. Some cynics might note that it is a popular material for making flutes, regardless of the use to which they might be put.

In Borneo, the Dyaks used bamboo on the altars to their gods at harvest. The leafy tops were place on each of the corners of the altar. The Ainu people in Japan made long, sacred wands called *inabos* from pieces of bamboo. Spiral designs were cut into the stem and, once again, the leaves were left attached.

The use of bamboo as a protective plant has become widespread. Simply growing the plant in the garden is considered to bring good fortune to the owner; this has the added advantage that it will break spells or curses cast against you. Placing a short length of bamboo over doorways conveys good luck to the household, as long as the wood retains its colour. To prepare a protective amulet for the house, carve a suitable symbol, such as a pentagram, into the bark of a bamboo shoot then plant it in the garden. Similarly carving, or marking, a wish into the bark of a bamboo twig then burying it in a secluded part of the garden will ensure that the wish will come true. The traditional belief amongst Philippine Islanders was that bamboo crosses in their fields would ensure good crops.

In some countries species of bamboo have been used by witchdoctors to induce the states of ecstasy needed to enable them to predict the future.

Nandina domestica, a member of the barberry family, Berberidaceae, is commonly called sacred bamboo or heavenly bamboo, even though it is a completely unrelated plant. Planting this plant in the garden is said to keep witches out of the garden. In Japan people will grow heavenly bamboo close to the house. There is a superstition that if you tell your nightmares and bad dreams to the plant it will ensure that they will never come true.

DAISY

Botanical Name: *Bellis perennis*
Family: Compositae
Common Names: Baby's pet; bairnwort; baiyan flower; banewort; banwood; bennergowan; bessy bairnwort; bessy banwood; billy button; bone flower; bruisewort; cat's posy; curl doddy; day's eyes; dog daisy; ewe-gan; ewe gollan; ewe gowan; eyes; field daisy; flower-of-the-spring; golland; goose flower; gowan; gracy daisy; harbinger of the spring; hen and chickens; innocent; kokkeluri; little-open-star; little star; *llygad y dydd* (eye of the day); lucken gowan; *marienblümchen*; Mary gowlan; Maudin daisy, may gowan; midsummer daisy; miss modesty; nails; open eye; shepherd's daisy; silver pennies; twelve disciples; white frills
Gender: Feminine
Planet: Venus (in Cancer)
Element: Water
Deities: Artemis; Freya; Thor
Meanings: Abundance; adoration; innocence; 'I will never tell'; faithfulness; gentleness; loyal love; purity; romance; unconsciousness; virginity; wealth (wild); temporization (field); participation; 'I partake of your sentiments' (doubles); beauty (part-coloured); 'I will', 'You have as many virtues as this flower has petals' (white); 'I will think of it' (wreath of wild daisies)

Myth and Legend

Some writers claim that the name *Bellis* is derived from *bellus*, meaning pretty or charming, whilst others say that it is taken from mythology and was the name of a dryad, Belidis. The plant was dedicated to the goddesses Artemis and Aphrodite, but with the spread of Christianity it became dedicated to St Mary Magdalene, hence the name 'Maudlin daisy'.

Admetus loved Alcestis, the daughter of King Pelias, and sought to marry her. Pelias was not easily convinced that he should allow his daughter to be married, and swore that he would only consent to her marrying a man who came to collect her in a chariot drawn by a team of lions and boars. Admetus achieved this task and, true to his word, Pelias consented to the wedding.

Unfortunately, some time after their wedding, Admetus became seriously ill and Alcestis feared that he would die. Even the god Apollo was unable to cure him, but did appeal to the Fates to spare his life. Although prepared to let Admetus live the Fates required that someone would have to die in his place. Alcestis gave up her life in order to save that of her husband. To mark her goodness she was 'restored' to the world in the form of the daisy.

Bellis was a dryad, the great granddaughter of Danaeus, King of Argos (Danaeus had fifty daughters called the Danaeids). Bellis was engaged to Epigeus, the deity of rural areas, but because of her beauty she was noticed by Vertumnus, the guardian god of orchards. She found it difficult to avoid his advances and so, eventually, appealed to the gods for aid. They transformed her into the little flower that bears her name.

In Christian legend, daisies represent the tears of Mary. A tale tells that, while picking daisy flowers to give to the infant Jesus, Mary pricked her finger and a little of her blood was left on the flowers. This accounts for the red tinge to the otherwise pure white petals, and for the name of *marienblümchen*.

According to Celtic legend, daisies sprang from the spirits of children who had died at birth. God, to gladden the hearts of their parents, sprinkled these flowers across the earth. This legend explains why daisies have the meaning of innocence. Daisy chains must always have their ends joined up when completed as they represent the sun and the circle of life.

Magic and Lore

Daisy is 'the day's eye', because it opens its flowers with the appearance of the sun, and closes them again when the sun goes down. It has been considered to be a protective plant because it is a symbol of the sun.

> That well by reason men it call maie
> The Daisie, or else the Eye of the Daie.
>
> – Chaucer

In other countries some of the names associated with this plant connect it with geese, from a belief that it grew from goose droppings.

Throughout the Middle Ages the daisy was considered to be a symbol of humility. It was dedicated to St Margaret of Cotona and knights would wear the flowers as a sign of their fidelity. In Shakespeare's *Hamlet* Ophelia gives the queen a daisy to signify, 'that her light and fickle love ought not to expect constancy in her husband.'

During the Age of Chivalry if, at the knightly tournaments, a lady allowed her knight to emblazon a double daisy on his shield it indicated that his love for her was returned.

A traditional love charm from Norfolk tells that when a girl walks in a meadow, in which daisies grow, reaches down with her eyes closed, and grasps a handful of daisies, she must count them. The number of daisies she holds is the number of years that will pass until she marries. She should be careful, however, as it is also said that the person who picks the first daisy of the spring will have a 'spirit of coquetry' beyond control. Sleeping with the root of a daisy beneath the pillow will ensure the return of an absent lover, and wearing a daisy flower will always attract love to you.

Perhaps the best known folk use of daisies is in love divination. To discover a lover's fidelity simply pick a daisy and pull off the petals, one at a time, saying:

She loves me,
She loves me not.

Or alternatively, you can say:

Does he love me?
Much,
A little,
Devotedly,
Not at all.

To discover when you will marry, pull off the petals with the words, 'this year, next year, sometime, never.' In all cases, it is the final petal pulled off the flower that gives the answer.

Another odd belief associated with this little flower is that if it was eaten it prevented growth. It was said that if puppies were given milk in which daisies had been boiled they would grow no bigger. One story tells of a child fed on daisies by his foster mother, Milkah, to prevent him growing taller than the plant.

She robb'd dwarf-elders of their fragrant fruit,
And fed him early with the daisy's root,
Whence through his veins the powerful juices ran,
And form'd in beauteous minature the man.

– Thomas Ticknell, *Kensington Gardens*

Later writers have gone so far as to suggest that if a child so much as digs up the daisy plant it will be sufficient to stunt their growth. The significance of all this may lie in the belief that the faeries would steal human children, leaving their own offspring in their place. Then, using daisies, they could prevent the human child from growing to full size and becoming too big.

It is a protective herb and placing daisy chains on the heads, or about the necks, of infants was believed to be sufficient to prevent their being carried away by faeries. Ironically, the daisy is supposed to be a plant to be used when invoking faeries or elves, and is used in faerie magic. It is said the Scottish name of 'bairnwort' is a testimony to the joy children have in making daisy chains.

There is a traditional saying in some parts of England that 'it is not spring until you can put your foot on twelve daisies.' The number of daisies referred to varies from region to region, with figures between six and twelve being quoted. Treading on the first daisy of spring is often said to be lucky, but elsewhere it is said that you ought to tread lightly on the first daisies of spring, or else daisies will grow over your grave or that of a loved one before the year is out.

Daisies can apparently be used by anglers to ensure that they will catch plenty of fish.

To dream of daisies is an omen of good luck if the dreams are in the spring and summer. A dream of daisies in the autumn is considered to be greatly unlucky.

The daisy is the birthday flower of 17 April, but was used in decorations at other points in the year. Commonly it was used on St Barnabas's Day, 11 June and on 6 May, the feast day of St John the Evangelist, as well as at Midsummer, or the feast day of St John.

Medicinal

One superstition claims that you should carry the root of the daisy as an amulet to protect you from accident and illness. In the Middle Ages daisies were used to treat battle wounds. Bandages containing crushed daisies were thought to give pain relief and aid healing.

Gerard, in his herbal, refers to the daisy as 'bruisewort' as, in common with the ox-eyed daisy, it had a reputation as a cure for pains and wounds. At one time, the daisy was valued as an aid in the treatment of broken bones. It is from this usage that names such as 'banewort' arise. Generally, it was reputed to be valuable for the treatment of 'female illnesses'.

In a treatment that must owe its origin to the Doctrine of Signatures it is claimed that the 'day's eye' can be used as a treatment for bloodshot eyes.

BARBERRY

Botanical Name: *Berberis vulgaris*
Family: Berberidaceae
Common Names: Berber; guild: guild tree; holy thorn; jaunder's berry; jaundice berry; jaundice tree; maiden's barber; pepperidge; pipperidge bush; piprage; ritts; sowberry; woodsore; woodsour
Planet: Mars
Meanings: Hot temper; sharpness; sourness of temper; tartness

Berberis derives its name from Arabic, signifying a shell. This is thought to be an allusion to the gloss of the leaves resembling the inside of an oyster shell.

Myth and Legend

This well-known and greatly valued garden plant has vicious thorns. It is commonly seen used in barriers and hedges. These thorns have led to it being considered as the plant used to form the 'crown of thorns' at the crucifixion of Jesus, particularly in Italy, where it is called the 'holy thorn'.

Medicinal

The names 'jaundice tree' and 'jaundice berry' relate to the use of the plant in treating this ailment. The plant has a very yellow under-bark and, being yellow, was recommended in the Doctrine of Signatures as a remedy for a disease causing yellowing of the complexion. In the Far East berberis and other plants containing berberine have been used in the treatment of dysentery and diarrhoea.

BRAZIL NUT

Botanical Name: *Bertholletia excelsa*
Family: Lecythidaceae
Gender: Masculine
Planet: Mercury
Element: Air

In the UK Brazil nuts tend to be associated with the mixed nuts bought in at the start of the Christmas holidays. They can, however, be used as lucky amulets, especially in 'affairs of the heart'.

BEETROOT

Botanical Name: *Beta vulgaris*
Family: Chenopodiaceae / Cruciferae
Common Name: Garden beet
Gender: Feminine
Planet: Saturn
Element: Earth

Magic and Lore

The humble garden beetroot has been in cultivation for several hundred years. Indeed, unlikely though it may seem, it was among the offerings made at the temple of Apollo at Delphi.

It is also rather hard to consider the beetroot as an aphrodisiac, but it has been used for just that purpose, the juices being used in love potions and other forms of love magic. There is a tradition in folklore that where a couple eat from the same beetroot they will fall in love.

Medicinal

Writing in his herbal Culpeper asserts that the beetroot is a powerful cure and is effective 'against all venomous creatures ... and with a little alum put to it is good for St Anthony's Fire'. Elsewhere it is said that if you eat a little red beetroot each day you will never have to worry about cancer.

BIRCH

Botanical Name: *Betula pendula*
Family: Betulaceae
Common Names: Beithe; bereza; berke; beth; bouleau; lady of the woods; white birch
Gender: Feminine
Planet: Venus
Element: Water
Deities: Berkana; Thor
Meanings: Gracefulness; meekness

The name is possibly derived from *bhurg*, a Sanskrit word relating to the continuous phases of life. It has been associated with festivals at May Day, Whit, Midsummer, Samhain, Lammas (Harvest or the Festival of First Fruits) and Yule.

Myth and Legend

The tree is sacred to Thor, the Norse god of thunder, and is supposed to protect any building near to where it is growing from being struck by lightning. Scandinavian myth also dedicates the birch to Freya and Friggaa. Friggaa is the Norse goddess of married love, and of the sky and clouds. She had seven handmaids who assisted her in caring for humanity and, it was said, it was her seven mortal sons who founded the seven kingdoms of Saxon England. It is, perhaps, because the birch was sacred to Friggaa that it has been described by some as a tree of the sun and sky, growing at the very gates of Paradise. Norse myth tells that it is about a great birch tree that the final battle for the existence of the world will take place.

In the Arthurian legends the Taliesin is represented by the birch. 'Taliesin' was the title given to the chief druidic bard and passed from person to person through history. Although there are many different stories relating to his birth and background it is usually accepted that Merlin, son of Cer and Cerridwen, was the Taliesin.

In the story of Diarmid and Grainne, in Irish legend, the birch plays an important part as the shelter for the lovers, Diarmid, the King of All Ireland, and Grainne, a goddess. They were pursued by Finn MacColl, who chased them for a year and a day. During their short stops at Cork, Kerry, Limerick and Tipperary they made their homes in shelters made from birch twigs. Unfortunately, Finn MacColl eventually caught up with them. In the form of a great bear, he killed Diarmid and claimed Grainne as his own.

Magic and Lore

The graceful birch, the 'white lady of the woods', is a symbol of spring, being one of the first trees to come into leaf each year. It was seen as the embodiment of graceful beauty and was closely associated with the spring festival celebrations on May Day. In some parts of Britain a birch would be used as a living maypole, at the centre of all the activities on that day. Beltane, the eve of May Day, is supposed to be one of the most important dates in the witches' calendar. The great bonfires, usually made up predominantly of oak wood, were lit with branches of birch. Birch woods were the favoured place for the celebrations and love making which followed.

The country people might put a cross of birch twigs over their doorways, at this time, to encourage good luck and ward off witches. In Hertfordshire, a young birch tree would be cut down on May Day and taken into the stable yard, where it was dressed in red and white rags before being propped up against the stable doors. This was done to form a protective amulet to prevent the horses being ridden by hags in the night or having knots tied in their tails and manes. A similar tradition exists in parts of Russia, where a young birch was cut down on the Thursday before Whitsunday. The tree was then dressed either in rags or in woman's clothing and was a 'guest' at the Whit festivities over the next three days. On the Sunday it would be thrown into a nearby stream. The dressing of the tree was its personification and when it was thrown into the waters it becomes a sacrifice, possibly in substitute for a human sacrifice. Perhaps this ritual shares its origin with the Russian belief that tying a red ribbon around the trunk of a birch tree will rid the person who ties it there of the 'evil eye'. In Britain, the birch was used to decorate churches and chapels on Whitsunday. The young growth of the tree was emblematic of the renewal of life and the movement of the air in the leaves symbolized the rushing winds of the Holy Spirit when it came upon the apostles at Pentecost.

In Ireland it was believed that the birch was disliked by the 'little people', and so could be used to avert mischief caused by faerie folk. Placing a cross cut from birch wood at the main entrance to the home would prevent any evil from entering, and would protect the occupants from misfortune and enchantments. Carrying a little piece of the wood could prevent the bearer from being kidnapped by faeries. Elsewhere the birch was taken as the 'hallmark' of the land of Faerie. In some parts of Britain there is a legend of a faerie spirit that inhabits the birch trees and is simply called 'the one with the white hand'. If the hand touches you on the head it will leave a bright white mark, and cause madness. If, however, it touches your heart then you will surely die.

In Scotland the faerie guardian of the tree is called Ghillie Dhu. This character lives in the birch thickets and wears clothing made from moss and leaves. Traditionally the bark of birch trees was used in ancient times for writing on, as it is very durable. According to the Hanes Taliesin, from the tales in the thirteenth-century *Red Book of Hergest* (one of the most important medieval Welsh manuscripts), 'On a switch of birch was written the first Ogham inscription in Ireland, namely seven Bs as a warning to Lug, son of Ethliu, to wit, "Thy wife shall be seven times carried away from you to faerieland unless birch be her overseer."'

In Russian myth, birches are the favourite trees of mythical creatures that inhabit the forests. The creatures, called 'forest devils' or 'genii of the forest', view the world from the tops of the trees. They are able to transform their size and shape at will, so that when they are in the forest they are the same height as the trees but on open land they can become as small as a blade of grass. If you want to make them appear you must cut birch branches and place them in a circle with the tips towards the centre. The pendulous or weeping habit of the tree may have made it a natural emblem of sorrow and grieving: 'Weeps the birch of silver bark with long dishevelled hair.'

To some extent, it has also been associated with the spirits of the dead.

In addition to being important as a symbol at Beltane birch also featured at Samhain (31 October), the turning point of the pagan year, better known now as Halloween (All Hallows Eve). This is the point in the year when the veil between this world and the next is supposed to be at its thinnest, and it is a time for looking back at what has passed and planning for the future. If this is a time when witches and spirits are abroad the birch will be essential as it is said to cast off all malignancy.

Birch also features at the end of the calendar year, as it was commonly used for Yule logs. Typically these logs were chosen from either birch or ash wood. Burning of the Yule log was once one of the most firmly adhered to of all the Christmas customs. It is thought to have derived from the druids who burned great fires of evergreen logs at the winter solstice in order to draw back the sun. Traditionally the Yule log should be taken from one's own land or cut from a neighbour's wood, as buying one was considered to be unlucky. The log had to be a substantial piece of wood as, once lit, it had to burn for at least twelve hours. It was brought into the home on Christmas Eve and laid on the hearth. It should be lit from a piece saved from the remains of the previous year's log, which might have been kept under the bed as a protective amulet for the year to ward off fires and lightning strikes. Both the remains of the old log and the new had to be handled with clean hands as a mark of respect. If the log failed to light first time it was a sign that the family would suffer misfortune within the coming year. Once burning, it could be tended throughout the Christmas Eve supper provided that food remained on the table or that someone was still eating.

Birch also features at the festival of Rogation, a once important celebration in the Christian calendar. Rogation week is the week in which Ascension Day falls and it was at this time that the practice of 'beating the bounds' was carried out. The birch rod would have been used to beat the parish boundaries, much as it was used to beat out the old year. The association with casting out the year goes further as it has been said that the New Year cannot start until the old year has been birched out.

Birch was the traditional wood used for making the heads of besoms, or witches' broomsticks. Although associated with witches, besom brooms were also believed to avert evil. One tradition that used them was jumping the broomstick. A besom was held across the doorway into a house and a couple who wanted to be married would, one at a time, jump over it. Having done so they could be considered married for the period of a year and a day, after which time they could 'remarry' or separate. There may be an echo of this practice in the Welsh tradition of couples exchanging the gift of a wreath of birch wood. A young man would weave the wreath from birch twigs then give it to the woman of his choice as a sign of his love. If she returned his affections she would present him with a similar birch wreath.

Children's cradles were also once commonly made from birch wood, as it would protect the child from harm. If the child was ill, dried birch leaves might be put into the cradle as a charm to give him or her strength and help to shake off the illness. Birch leaves were also added to the warmed beds of those suffering from rheumatic pains or arthritis as a cure. Alternatively birch sap, sometimes called 'birch blood' or 'birch water', was given as a medicine.

The birch is associated in some parts of the world with corporal punishment. This use of birch as a whip probably originates from its use in exorcism. It was believed that a person could be freed from demonic possession by striking them with a birch whip, the protective powers of the birch driving the demon out. After all, why would anyone commit a crime unless an evil spirit possessed them? So the punishment by birching was done for the offender's own good.

Birch is a 'pioneer species'. That is to say it is one of the first tree species to shoot in areas of open ground. Once the birch woodland begins to mature, oaks will often being to appear amongst the birch trees. This has led to them being seen as 'husband and wife'.

BORAGE

Botanical Name: *Borago officinale*
Family: Boraginaceae
Common Names: Bee bread; bugloss; burrage; corago; euphrosynon; herb of gladness; nepenthe; tailwort
Gender: Masculine
Planet: Jupiter (in Leo)
Element: Air
Meanings: Bluntness; brusqueness; courage; energy; roughness of manners; talent

Some writers claim that the name is derived from *barrach,* a Celtic word meaning 'man of courage'. Alternatively, it could be a corruption of the Latin *cor* ('heart' and *Ago* 'I rouse'), as it was used in a stimulant drink.

Myth and Legend

Polydamna, wife of Thonis, sent nepenthe (borage) to Helen of Troy to ease her heartache, with the claim that it was 'of such rare virtue that when taken steeped in wine, if wife and children, father and mother, brother and sister, and all thy dearest friends should die before thy face, thou could'st not grieve or shed a tear for them'.

Magic and Lore

In his herbal of 1597 John Gerard writes of borage:

> Pliny calls it Euphrosinum, because it maketh a man merry and joyfull: which thing also the old verse concerning borage doth testifie:
>
> Ego Borago Gaudia semper ago.
> I, Borage bring alwaies courage.
>
> Those of our time do use the flowers in sallads to exhilarate and maketh the mind glad.

Dioscorides supports these sentiments, saying that borage is the plant referred to as 'Nepenthe' by Homer which, when drunk, brought forgetfulness. It is perhaps this forgetfulness that dispelled melancholy and brought joy, as people forgot their troubles. Then again, perhaps as borage is also supposed to be a sexual stimulant it might be that it gives us other things to think about.

Borage is one of the cardinal plants of the ancients, and a tea made from the herb is said to promote the psychic powers. It also has protective qualities and, to benefit from these, the flowers of the plant should be worn whenever out walking.

FRANKINCENSE

Botanical Name: *Boswellia thurifera*
Family: Burseraceae
Common Names: Incense; olibans; olianum; olianus
Gender: Masculine
Planet: Sun
Element: Fire
Deities: Baal; Bel; Ra
Meaning: The incense of a faithful heart

The only thing that most people know about frankincense is that it was one of the gifts, brought by the magi, to Christ at the Nativity, but it has been used in religious ceremonies since ancient times. The Egyptians burned it at their sunrise ceremonies in honour of the god Ra, and Herodotus informs us that a thousand talents' worth of frankincense was offered annually, at the festival of the god Bel, on the great altar in the temple at Babylon. Herodotus also relates that the Arabs brought a thousand talents worth of frankincense to King Darius each year as a tribute.

This incense is still used today in Catholic churches. Generally, it is considered that its burning will dispel negativity and evil. It is also said to aid meditation and spiritual growth, and bring good fortune. It has been used in the ceremonies of consecration, purification, and exorcism. The 'perfect' incense, it has been claimed, used to be made from four ingredients, symbolizing the four elements. God revealed the recipe for this mixture to Moses but, unfortunately, over the passage of time it has been lost.

MOONWORT

Botanical Name: *Botrychium lunaria*
Family: Ophioglossaceae
Common Names: Fern grape; un-shoe horse
Gender: Feminine
Planet: Moon
Element: Water
Meaning: Forgetfulness

Moonwort is a fern and, like many ferns, was considered by the ancients to have magical attributes. It was used by magicians, necromancers and witches in magical rites. In order to be of value for magical purposes it must be collected by the light of the moon. John Parkinson, writing in the early seventeenth century, tells us that the alchemists used the plant in their attempts to create silver from mercury (quicksilver). This last supposed attribute has survived in modern folk magic, as the herb is one of those used in spells to bring wealth. It is also used in spells to attract love and affection.

Culpeper reveals the reason for moonwort being called 'un-shoe horse'. He says, 'moonwort (they absurdly say) will open locks and unshoe such horses as tread upon it; but some country people call it unshoe the horse.' In addition to opening locks when placed into the keyhole, it was said that moonwort could break chains if it merely touched them. Presumably the reason that the horseshoes fell off is because the plant caused the nails to drop out. A folk tale claims that when woodpeckers find a nail obstructing their nests they will seek out moonwort. If, at Midsummer, they rub their beaks on the leaves it will enable them to pull out the offending nail with ease.

RAPE

Botanical Name: *Brassica napus*
Family: Cruciferae
Common Names: Cole; colewort; colza

It cannot be a great surprise that rape was once regarded to be of great medicinal value when one considers that other brassicas have been esteemed for their healing properties. Whereas mustard was used to 'revive the spirits', rape was valued as a treatment for 'those with dim eyes', at least according to Gerard. It was also recommended that it be eaten before meat when dining, to prevent drunkenness.

That is not the limit of the medicinal properties of this plant; it might also be added to wine and the resulting liquor drunk as a cure for snakebites, and the seeds might be eaten to remove freckles.

MUSTARD

Botanical Names: *Brassica nigra* (black mustard); *Sinapis alba* (white mustard)
Family: Cruciferae
Gender: Masculine
Planet: Mars
Element: Fire
Deity: Aesculapius

Today we tend to take mustard somewhat for granted but in bygone days it was much more greatly valued. John Evelyn, writing in 1699, says of this herb, 'Mustard, especially in young seedling plants, is of incomparable effect to quicken and revive the spirits, strengthen the memory, expelling heaviness.' Perhaps it was this expulsion of 'heaviness' that led to the use of mustard seeds by Hindu mystics to induce levitation.

Given the heat associated with mustard it is perhaps no surprise that it is said to ward off colds. To be effective it need not be eaten, but the seeds should be carried, wrapped in a red cloth. This is also supposed to increase the mental faculties. In Italy, peasant families have used mustard seeds as a protection against evil spirits. Scattering the seeds on the doorstep, or burying them beneath it, is sufficient to keep all supernatural entities at bay. On a more mundane level, chewing mustard was said to cure toothache and it could be rubbed onto the skin to draw out splinters.

CABBAGE

Botanical Name: *Brassica oleracea capitata*
Family: Cruciferae
Common Name: Kell
Gender: Feminine
Planet: Moon
Element: Water
Meanings: Gain; profit

Mankind has cultivated the cabbage for many centuries. According to the Roman writer Pliny, it was looked upon with a degree of reverence by the Greeks, Romans and ancient Egyptians, being used both as a medicine and as a food. Perhaps, due to the reverence with which it was once considered, cabbages are usually regarded as symbols of good fortune. It seems absolutely appropriate that, in the language of flowers the cabbage symbolizes profit and that cabbage is a slang word for money.

Magic and Lore

The theme of profit and good luck is found repeated in many of the superstitions and stories associated with the cabbage. One example is the tradition that a couple can ensure good fortune in their married life if the first thing they do, following their wedding, is to plant a cabbage in their garden. Another example is that if two shoots appear from the same cabbage root it will bring good luck to the person who grew it. At Halloween a cabbage should be placed on the doorstep of every house that you want to enjoy abundance (in the Shetland Isles, cabbages were tossed though the doorways on that night). A cabbage head fastened to the ceiling and stuck with pins at Christmas will bring good luck for the following year but bad luck is supposed to follow if you have a cabbage in the house at New Year.

If you should dream that you are eating cabbages it signifies sorrows, sometimes indicating a loss of money or sickness amongst loved ones, with little hope of their recovery. Dreaming of growing cabbages is usually taken as a sign of good fortune and a dream of cutting cabbages indicates that a partner is jealous.

A girl might hang the stems, or runts, of cabbages over the main doorway to the house. Each runt would be given a number, and the name of one of the girl's suitors. Hence, the name 'John' might be given to the third runt. If John were the third person to enter through that door it was taken as a very good omen for their future.

To discover the nature of the man she will marry a young woman should walk blindfolded into a field of cabbages at Halloween and catching hold of one pull it up out of the ground. As she does so she should say:

Hally on a cabbage stock, and hally on a bean,
Hally on a cabbage stock, tomorrow's Hallowe'en.

If the stalk of the cabbage is long, then her future husband will be tall, but if it is stumpy then he will be short. If the stalk is crooked she will marry a man who is twisted in mind and body. A hard stalk signifies that her husband will be strong-willed, determined, and self-confident. A soft stalk shows that he will lack confidence and be weak willed. If soil clings to the roots she and her husband will enjoy great wealth, but if the roots are clean then she should expect poverty. The taste of the cabbage heart is also said to give an indication of the nature of her future spouse, its sweet or bitter flavour showing whether he will have a kind disposition or not. Another reference to cabbage hearts may sometimes be found copied into valentine cards:

My love is like a cabbage
Often cut in two,
The leaves I give to others
The heart I give to you.

Superstitions abound regarding the cultivation of cabbages. Traditionally gardeners were advised to plant cabbages on the day of a full moon (some say a new moon), or on the day following. Planting at the time of the twins (Gemini) would ensure that two heads could be harvested from every plant. If planting was done on a Friday, in the new moon, then the cabbages would never be troubled by frosts. Hoeing through the plants during the dog days (3 July to 11 August) would lead to the cabbages being eaten by insects. To prevent this they might be interplanted with sage and thyme in order to keep the pests away. Bolting, the premature running to seed of the plants, was said to predict a death within the household. One last thought on the subject of cabbage cultivation, as a message of hope to gardeners: if you kill the first cabbage white butterfly you see in the spring it acts as a warning to the rest and will give your garden some protection from this pest for the whole season.

Medicinal

Cabbages are used in a range of medicinal treatments. A cabbage leaf, a kell-bled, heated up and placed on a septic wound will draw out all foreign bodies. Binding the leaves about swollen legs will reduce the swelling and, if fresh, will help to treat varicose veins. Heated cabbage leaves applied to the soles of the feet will ease fevers and the leaves should be chewed to cure fevers, colds and the flu. Even the water in which cabbage has been cooked has its value; it can be saved and drunk as a remedy for hangovers.

TURNIP

Botanical Name: *Brassica rapa*
Family: Cruciferae
Gender: Feminine
Planet: Moon
Element: Earth
Meaning: Charity

Myth and Legend

Jack, a blacksmith, was enjoying a drink at his local pub one evening when he met with the Devil. Having run out of funds Jack offered to sell his soul in exchange for a drink. The Devil turned himself into a sixpence so that Jack could buy another drink, but Jack pocketed it instead. The Devil was trapped there, as Jack was also carrying a cross in his pocket. Before he released him Jack made the Devil promise not to take his soul for ten years. Exactly ten years later the Devil came to collect Jack's soul and met him on a country road. Jack asked the Devil to help get an apple off a nearby tree, saying that afterwards he would peacefully accompany him into Hell. As soon as the Devil jumped onto Jack's shoulders, in order to reach the apple, Jack drew a knife from his pocket and cut a cross into the tree bark, trapping the Devil once again. This time he extracted a promise that the Devil would never take his soul.

When Jack died his soul could not enter Heaven because of his sinful life, but nor could he go into Hell. When he asked the Devil what he was supposed to do the Devil told him to return whence he came. Jack asked for light so that he could find his way but the Devil threw a hot coal at him. Jack put the burning coal inside a turnip he had been eating so that it would not be extinguished by the wind. Ever after, as jack-o'-lantern, he was doomed to wander through the darkness until Judgement Day.

Magic and Lore

Today it is usually the pumpkin that we associate with Halloween lanterns but at one time turnips were frequently used for this purpose. They were hollowed out, and a candle lit inside. Then they were placed on windowsills to ward off evil spirits. Those who had to leave the relative safety of their homes, during the hours of darkness, on the night when the forces of evil were supposed to be at their strongest, would also carry these turnip lanterns. It may well be because of this usage that turnips are still said to be able to dispel negative forces from the home.

This power to dispel negativity is not limited to 'the forces of evil' according to the 'old wives'. If you should find that you are troubled by unwelcome advances from a person for whom you do not care simply serve them a dish of turnips. This will be sufficient to ensure that they never trouble you again.

There are several pieces of guidance for those who wish to grow turnips. It was recommended that sowing should be carried out before the feast of St Columcille (15 June) and the plant ought to be thinned, or singled, by the date of Crewkerne Fair (4 and 5 September).

In the somewhat unlikely event that you should dream of turnips it is a warning that some task on which you have embarked will prove to be fruitless. A dream of eating turnips implies that you should expect family problems.

QUAKING GRASS

Botanical Names: *Briza maxima*; *Briza media; Briza minor*
Family: Gramineae
Common Names: Bloody thumbs; capuchin's lice; cow quakes; cow quakers; didder grass; diddery grass; dithering grass; dithery dother; dodder grass; doddering dickies; doddering dillies; doddle grass; dother grass; dothering dick; dothering dickies; dothering jockies; dothering Nancy; dothery; earth quakes; faerie grass; hare's bread; hat shakers; hen fleas; horses and chariots; hungry grass: *Joppchen stohn still* (little Jacob stay still); lady's hair; lady's hands; lady's shakes; maiden's hair; Mary's tears; nodding Isabel; quake grass; quaker grass; quakers; rattle baskets; rattle grass; shackle baskets; shackle boxes; shackle grass; shaking grass: shaky grass; shickle-shacklers; shiver grass; shivering Jimmy; shiver-shakes; shivery shakeries; shivery shakes; siller tassles; silver ginglers; silver shackle; silver shakers; silver shekels; totter grass; tottering grass; totty grass; trembling grass; trembling jockies; trimmling-jockies; waggering grass; wagtails; wag-wafers; wagwams; wag-wantons; wag-wants; wiggle-waggles; wiggle-waggle-wantons;

wiggle-wants; wiggle-woggles; wigwag-wantons; wig-wagons; wing-wangs; woman's tongue

Some of these names came about as a result of a tradition in Wiltshire; parents would tell their children that should the flower spikes of this grass stop trembling they would change into silver sixpences or shillings. In France, the plant is called St John's herb, and is collected on St John's Eve, for purification in the smoke of a bonfire.

An Irish tradition tells of the *fairgurtha*, or *fairgorta*, the hungry grass. Some people think that this hungry grass is the same as quaking grass. When trodden on the grass was supposed to cause a violent hunger or an abnormal craving for food. This could be prevented by the traveller carrying a small amount of food with them, or by the eating of just one grain of oats.

The plant was once believed to grow where a young girl had drowned herself, usually as a result of some unsuccessful love affair. The grass sprang up where the body had been pulled from the waters and laid on the bank. Cows, it is said, would never stand on or eat quaking grass.

In some parts of Britain is was said to be unlucky to take the plant into the house, although, elsewhere it was thought that placing a dried bunch on the mantle would prevent mice from entering the house.

> A Trimmling Jock i' t' house
> And you weeant hav' a mouse.

Quaking grass was used to treat ague and fevers, probably as an application of the Doctrine of Signatures.

BRYONY

Botanical Names: *Bryonia alba*; *Bryonia dioica*
Family: Cucurbitaceae
Common Names: Ache; Canterbury jacks; cow's lick; dead creepers; death warrant; dog's cherries; elphamy; English mandrake; gout root; hedge grape; hop; jack-in-the-hedge; lady's seal; mad root; mandrake; murren; old vine; our lady's seal; poisoning berries; rowberry; snake berry; snake grape; tamus; tetter berry; tetterbury; vine; white bryony; wild cucumber; wild hop; wild nep; wild vine; wood vine
Gender: Masculine
Planet: Mars
Element: Fire
Meaning: Prosperity

Bryony has long been used as a substitute for mandrake, which is rarer and consequently more expensive. Like mandrake, the roots of the bryony plant are supposed

to resemble a human form and the use of the plant as an aphrodisiac probably stems from this. Indeed, so powerful are the aphrodisiacal properties of this plant that it is said to be only necessary to use a piece of root the size of a 5p piece (or old sixpence). This is just as well considering its highly toxic nature. Men wishing to use bryony in this way must, obviously, choose a piece from a root that resembles a woman. Likewise, a woman must seek out a root resembling a man. In Italy these root figures were called *Mammettes* and *Pappettes*, and were prized by women wishing to have children; they would be purchased as a magical way of increasing fertility. Another way to use its powers was to add it to food;

> If to childe-bede thou wouldst goe,
> Dust thy food with brionie.

In the fourteenth century the plant was known as 'wild nep' and was used as a treatment for leprosy. The French called it *navet du diable*, or 'Devil's turnip', because of the danger of poisoning. Despite the risk from its high toxicity, it was once used to flavour a drink, known as 'hop bitters'.

Bryony is reputed to have other attributes. By hanging the plant up in the house, or garden, you can protect the area from the effects of extreme weather conditions. The writer Bartholomew claims that the Roman Emperor Caesar Augustus would wear a wreath of bryony during thunderstorms to protect himself from lightning. As there are no records of the emperor having been struck by lightning we can presume that it worked.

Bryony is said to attract wealth. When money is placed near to the roots of a bryony plant more will, inevitably, be drawn.

(Black bryony is *Tamus communis*).

COPAL

Botanical Names: *Bursera odorata*; *Bursera microphylla*
Family: Burseaceae
Gender: Masculine
Planet: Sun
Element: Fire

Copal has quite a specialized usage in magical rites. The poppets, human figures formed from clay or the roots of some plants (like bryony and mandrake), might be given a heart of copal. Small pieces of the plant were inserted into the figure to represent the heart and so make the figure more effective.

This is the main, but not the only, use of copal. It has also been added as an ingredient of the incense burned during purification rites, or in order to attract love.

BOX

Botanical Name: *Buxus sempervirens*
Family: Buxaceae
Common Names: Bush tree; dudgeon
Deities: Pluto
Meanings: Firmness; stoicism

Box is one of the shrubs that have traditionally been used by gardeners for hedging and topiary. Francis Bacon, however, had a warning for those who grew it in their gardens, saying that it produced a honey 'of such a nature that men were driven mad by it'. It may be this ability to induce madness that led to box being recommended as a treatment for the bites of mad dogs. This would be much as we might use an inoculation in modern medicine.

In Lancashire it was traditional to place a small dish of box sprigs in the room where a dead person was laid out. When the mourners came to pay their respects they would take one of these pieces and later drop it onto the coffin when it was lowered into the grave. It would seem that box was used where rosemary, which would more commonly be used in this way, was not available. The belief that box is a plant of ill fortune probably stems from this usage. Taking box into a house was unlucky because:

> Take box into a house,
> Carry a box out.

Box has been used as a substitute for palm fronds at Easter and at the Feast of the Tabernacles. It is not surprising, given the name, that boxwood has been used for making boxes for containing particularly precious items.

If you should recognize a box plant in your dream it is a very fortunate sign, denoting a long life, prosperity, and a happy marriage.

MOLUCCA BEANS

Botanical Names: *Caesalpinia* spp.; *Entada* spp.
Family: Leguminosae
Common Names: Faerie's eggs; lucky beans; Mary's beans; sea beans; Virgin Mary's nuts

Molucca beans are the seeds of several species of tropical plants that are carried by the Gulf Stream, all the way from the Caribbean and northern South America, and deposited on the shores of Western Europe, although their name would suggest that they originate from Indonesia. Some would, inevitably, be washed up on the coast of the British Isles. For the scientists the origin of these beans is very straightforward, but for the local inhabitants there was a degree of mystery.

In Cornwall these seeds were used to ease the pains of childbirth, and elsewhere they were believed to prevent illnesses and avert misfortune. Where the seeds were worn it was thought that they would attract good luck, hence they were mounted as brooches, lockets and other pieces of jewellery.

The seeds, as well as those of Merremia, a member of the bindweed family, have a characteristic cross-shaped marking on them. They were highly prized in the Scottish Highlands for aiding childbirth, being clenched in the hand during labour. The seeds would afterwards be strung onto a thread and hung about the necks of infants to reduce the pains of teething.

MARIGOLD

Botanical Name: *Calendula officinalis*
Family: Compositae
Common Names: Bride of the sun; drunkards; *flore d'ogni mese*; goldes; gold ruddles; holigolde; husband's-dial; husband-man's-dial; jackanapes-on-horseback; marybud; marygolds; Mary gowles; measles flower; *oculus christi*; pot marigold; ruddes; ruddles; Scotch marigold; *solis sponsa*; *spousa solis*; summer's bride; summer pride
Gender: Masculine
Planet: Sun (in Leo)
Element: Fire
Deity: Artemis
Meanings: Anxiety; chagrin; constancy in love; cruelty; despair; foreboding; grief; honesty; jealousy; love; misery; pain; sacred affection; uneasiness

Marigolds are dedicated to the sun and are symbolic of love, hot passion and sexuality. In folklore they are linked to the sparks given off by lightning during a thunderstorm.

Myth and Legend

Classical mythology tells us that marigolds were originally a group of nymphs who were transformed into the flowers by the goddess Artemis. A more recent legend tells the story of a young girl by the name of Mary-Gold who would spend her whole day just sitting watching the sun. One day she simply disappeared and where she had been sitting people found marigolds growing. Her friends said that she had been turned into the flower, as she loved the sun so much.

In a Christian legend, during their escape into Egypt the Holy Family was stopped by thieves, who stole Mary's purse. When they opened it, marigolds fell out. Early Christians called marigolds 'Mary's gold', and placed the flowers around statues of the Madonna, offering the blossoms in place of coins. The blooms are representative of the rays of glory that are often depicted around the head of the Madonna and so the flowers have been used at the festivals of the Virgin Mary.

In *The Magic Art* Sir James Fraser refers to the marigold as a symbol of mourning, but also relates another aspect of the folklore of this plant. The story he tells is that of a slave girl who ventured out on St Luke's Day (18 October) to collect soil from the footprints of the man she had fallen in love with. She used this soil to fill plant pots, and into them planted marigolds. The plants grew and flourished, and as they grew so did his love for her.

Magic and Lore

The botanical name is taken from the Latin *calendae*, the calends being the first days of each month, and is sometimes translated as 'little calendar' or 'little clock'. The marigold takes its name from these because it is said that it can be in flower at every calend, or at least throughout the months from spring to autumn.

> For being once sowen, they
> Afterwards grow of themselves; and beare
> Flowers in the Calends of every moneth of
> The yeare, as well in sommer as in winter,
> For which cause the Italian cal them
> The floure of al the moneths.
>
> – Anthony Askham, *A Lyle Herball*, 1550

Not only is the long flowering period of the marigold noted, but also the regularity with which the flowers open and close. Linnaeus, the 'father of botanical nomenclature', observed that the flowers opened at 9 o'clock in the morning and closed again at 3 o'clock in the afternoon. Shakespeare's *The Winter's Tale* suggests that the opening and closing of the flowers is tied to the appearance of the sun:

The Marigold that goes to bed wi' the' sun,
And with him rises weeping.

For the most part the marigold is thought of as a lucky flower. Like many other yellow flowers it is said to reflect the light of the sun. It is dedicated to Lady Day (25 March), as it typifies piety and devotion. It is the birthday flower of 15 January.

In Mexico and Germany this flower of the sun is associated with death. In ancient Greece marigolds would be included amongst the flowers given as funeral offerings. At weddings marigolds could be added into the bride's bouquet as a love charm, one of the other attributes of which marigold is symbolic being 'constancy in love'.

There are a number of uses of marigolds in love charms, like the one described under wormwood.

A similar superstition was that if the maiden put marigold flowers beneath her pillow at Halloween, she was assured of dreaming of the 'man of her dreams'. William Blake, in *Daughters of Albion*, refers to another tradition where a maid would place a marigold flower between her breasts in the hope that the flower, and her sweetheart's love, would never fade.

The Ooothoon pluck'd the flower saying
Sweet flower, and put thee here to glow between my breasts
And thus I turn my face to where my whole soul seeks.

If you picture marigolds in your dream it is a very favourable omen as it means prosperity, success and a happy marriage.

You cannot, of course, simply go out and pluck a marigold from your garden if you intend to use it for magical purposes. For the flowers to strengthen the heart and drive out melancholia, you must pick them at noon. This is when the sun is at its hottest and the plant most potent. If you have been robbed, and wish to discover the identity of the thief, you should collect marigolds when the moon is in Venus. Jupiter must not be in the ascendant, or else all the power of the plant will be lost. At the time of collecting the plant you must say three paternosters and three aves. Gathered in this way it will cause you to have a vision in which you will see the thief, but only if you are a truly virtuous person.

In Wales it was said that where the flowers of the marigold failed to open before 7 o'clock in the morning there would be thunderstorms. In Wiltshire and Devon the superstition dictated that if you were to pick the flowers it would cause thunderstorms.

It is not all good news of course. A West Country superstition tells that staring at a marigold flower for too long will cause you to become a heavy drinker, and hence the common name of 'drunkards'. A similar superstition claims that if you pick a bunch of marigolds before dawn you risk becoming an alcoholic.

There are many other superstitions associated with this plant. For example, if marigold is added to bath water it will help the bather to win the respect and admiration of all those they meet. Wearing a marigold flower in court will ensure that 'justice smiles on you'. Perhaps the oddest superstition is that if a young woman touches the petals of a marigold with her bare foot she will be able to understand the language of the birds.

Scattering marigolds beneath the bed before retiring to sleep will give protection from evil through the night, and will induce prophetic dreams. Marigolds can be further used to protect the household by hanging garlands of the flowers at the doorways into the house. This will prevent any evil from entering.

Medicinal

The marigold has an ancient history as an herbal remedy. It was known to the Greeks, the Romans and the Egyptians as a remedy for stomach disorders and skin complaints. During the Elizabethan period it was believed that the flowers had restorative properties. They could be dried and added to soups, broths or salads, either as a restorative or in order to relieve depression. They have also been recommended as a painkiller. Rubbing the flower or the juice onto the skin is said to remove pain and it can, for example, be used for treating insect stings and sunburn. It was even used in the American Civil War as part of a treatment for wounds. An added bonus from this treatment was that it is an aid to improving the complexion. William Turner, in the *New Herbal*, informs us that the flowers were at one time used to make a hair dye: 'Some use to make their heyre yellow with the floure of this herbs, not beyne content with the natural colour which God hath gyven them.'

According to Odo Magdunensis, in the medieval Latin poem '*Macer Floridus De Viribus Herbarum*', just gazing upon the flowers is sufficient to draw out wicked humours from the head, and his suggestion that this will also improve eyesight passed into popular superstition.

CHINA ASTER

Botanical Name: *Callistephus chinensis*
Family: Compositae
Common Names: Aster; Michaelmas daisy; starwort
Gender: Feminine
Planet: Venus
Element: Water
Deities: Astraea; Venus
Meanings: Afterthought; variety

This is one of many plants that have been considered sacred to the gods. This 'star flower' was used at several religious festivals where it would often be foremost amongst the flowers selected to be placed, or strewn, on to the altars in the temples.

Myth and Legend

The myths relating to the origin of this plant are shared with those of the aster: when men learned to make tools and weapons of iron, the great god Jupiter was angered to see how much destruction was being wrought, and decided to destroy mankind with a mighty flood. The gods, who then resided amongst men, fled from the earth. The last to leave was Astraea, the goddess who symbolized innocence and justice, who was so saddened by events that she asked to be transformed into a star and now looks down from the heavens as the constellation of Virgo. Two mortals who had been faithful to the gods survived the deluge by ascending Mount Parnassus, where they were saved by Jupiter. When the floodwaters receded all that was left for these two people to inherit, was a world covered in mud and slime. Astraea, viewing their plight, wept for them. Her tears fell as stardust which, when they reached the earth, blossomed into flowers, the asters.

Magic and Lore

One of the powers attributed to this flower is the ability to attract love. They are consequently used in all types of love magic. Carrying the flowers, or growing them in the garden, will help all those who seek the affection of others. Another use to which they have been put is in warding off serpents. In medieval Europe the leaves were burned to drive snakes away. The crushed roots were also used as a medicine for bees that were in poor health.

The name 'Michaelmas daisy' has been applied to this plant because it is supposed to flower on Michaelmas Day, in September.

HEATHER

Botanical Name: *Calluna vulgaris*
Family: Ericaceae
Common Names: Bazzom; bissom; blackling; broom; dog heather; hadder; hedder; griglans; griglum; he-heather; ling; mountain mist; red heather; Scots heather
Gender: feminine
Planet: Venus
Element: Water
Deities: Aphrodite; Cybele; Eryx; Isis
Meanings: 'We shall be lucky in our life together' (white); admiration; good luck; protection; solitude; wishes will come true

Myth and Legend

Legend has it that the Picts in Scotland made a wonderful brew, a much-coveted heather mead. The recipe was a closely guarded secret known only to a select few and, when other tribes conquered the Picts, it was all but lost. It is said that eventually there remained only two men, a father and son, who knew the secret of the heather mead and neither would divulge the recipe to their conquerors. In an attempt to frighten the father into revealing what he knew the son was put to death. Still the father refused, saying that he was glad his son had been killed first as the youngster, being more gullible, might have been coerced into speaking. For his part the father never did reveal the secret of the recipe and it died with him so that the conquerors never enjoyed the beverage of the Picts.

The Ossian cycle of poems ('translated' by James MacPherson) tells that long ago in Scotland, Ossian had a daughter named Malvina, who was beautiful and sweet natured. She won the heart of a handsome warrior called Oscar and they became betrothed. However, Oscar left her and went off in search of fame and fortune. Malvina pined for her lost love and sought solace by telling her father how much she loved the brave warrior. One glorious autumn day, Ossian and his daughter were sitting on a Highland hillside when a messenger came to them bringing the sad news that Oscar had been killed in a great battle. The messenger gave Malvina a spray of purple heather, a last gift from Oscar, and said that he had died calling her name and pledging his love. In her grief, Malvina ran weeping across the hillside. And where her tears fell, the purple heather turned pure white. On seeing this, she said, 'May white heather forever bring good luck to all who find it.'

In Scotland the virtues of the white heather as a lucky amulet are less clearly defined than elsewhere. Some claim that it is lucky as it is free from the blood of the Picts, whereas the purple heather is stained with their blood. Among some of the Highland clans, however, there is a tradition that white heather is far from lucky. This stems from the time of the Jacobite rebellion of 1745, when Bonnie Prince Charlie landed in Scotland in hope of claiming the throne. Tradition has it that he was given a sprig of white heather when he stepped ashore, in order to bring him luck. The rebellion failed and Bonnie Prince Charlie was lucky to escape alive.

Magic and Lore

Heather is commonly thought of as being a plant of good luck. This superstition seems to have reached England as part of the Victorian enthusiasm for Scottish traditions, and is now known everywhere. Anyone who was unaware of this has never been accosted by a gypsy selling 'lucky white heather' and clothes pegs. (Unfortunately the 'heather' sold by gypsies is now, all too often, florists' statice, *Limonium* spp.). The tradition is, of course, that it is white heather that will bring the bearer good luck and not any other colour. It also used to be said that for the white heather to prove effective as a lucky amulet it must be come upon by chance.

Bringing luck may be the best-known attribute associated with this plant, but it is not the only one. Heather is also said to have the power to protect people against physical harm, especially against violent crimes such as rape. In magic the plant was used in conjuration rites when trying to raise the spirits of the dead. It is also, according to some authors, one of the materials used in the making of a witch's broom. One other old superstition relating to heather is that if it is burned out of doors, with ferns, it will cause rain to fall.

In faerie lore heather stalks are said to provide food for faerie folk and a field of heather might contain the portal through into the faerie realm.

MARSH MARIGOLD

Botanical Name: *Caltha palustris*
Family: Ranunculaceae
Common Names: Bachelor's buttons; bee's rest; big kingcup; billy buttons; bludda; blugda; blugga; Bobby's buttons; bog daisy; boots; bull buttercup; bullcup; bull cloga; bull flower; bull-rushes; bull's eyes; butter bleb; buttercup; butter flower; carlicups; chirms; claut; cow cranes; cow lily; cow slops; crazy; crazy bet; crazy betsy; crazy betty; crazy ladies; crow cranes; crow flowers; cup and saucers; dale cup; down scwobs; drunkard; fiddle; fire o' gold; gilcup; gypsy's money; guilty cup; golden cup; golden knob; golden kingcup; goldilocks; golland; gowan; grandfather's buttons; halcup; horse-blob; horse buttercup; horse-hooves; Johnny cranes; Johnny Georges; king's cob; kingcup; leopard's foot; livers; *lus y voaldyn*; mare's blobs; marigold; marsh lilies; Mary's gold; mary buds; maybout; may blobs; may blubs; may bubbles; may buds; may flower; meadow bout; meadow bright; meadow routs; moll blob; Molly blob; monkey bells; old man's buttons; policeman's buttons; publicans; soldier's buttons; solsquia; *Sponsa solis*; *Verrucaria*; water babies; water blabs; water blebs; water blobs; water blubbers; water buttercup; water dragon; water geordies; water goggles; water golland; water gowan; water gowland; water lily; yellow blobs; yellow boots; yellow crazies; yellow gowlan

Magic and Lore

This plant, like the English marigold, is dedicated to the Virgin Mary and in the Middle Ages was commonly used in church decorations on festival days. It became much more widely used after the calendar was changed in 1752, when it was used as a substitute for other May flowers which were then not yet in flower. In the May Day festival it would be incorporated into garlands hung over cottage doorways, or the flowers might be strewn onto the doorsteps. Incidentally, where bunches of the flowers are used in decorations they should always be hung with the stalk upwards. In Ireland and the Isle of Man the marsh marigold was used as a protective herb at Beltane, May Eve, against the malice of witches and faeries. The Celtic name, *lus y voaldyn*, means 'herb of Beltane'. In some regions, however, it was believed that it was unlucky to bring the plant into the house before 1 May. The plant has also been said to protect against being struck by lightning during the May thunderstorms.

There is a grouping of plants that is called 'publicans and sinners', this is *Caltha palustris* planted side by side with *Ranunculus*.

Medicinal

The common name 'kingcup' is mirrored in the botanical name of *Caltha*, which is derived from *Calathos*, meaning a cup or goblet. The flower gets the name '*Sponsa solis*' because it is said to open and close with the sun and *Verrucaria* for it being used to treat warts. This is not the limit of the plant's medicinal properties. One writer reports: 'It would appear that the medicinal properties may be evolved in the gaseous exhalations of the plant and flowers, for a large quantity of the flowers of the Meadow rout being put into the bedroom of a girl who has been subject to fits, the fits ceased.'

CAMELLIA

Botanical Name: *Camellia japonica*
Family: Theaceae
Common Name: Rose of Japan
Gender: Feminine
Planet: Moon
Element: Water
Meanings: Perfect loveliness; purity; unpretending excellence (white); 'beauty is your only attraction'; 'how radiant is your beauty'; steadfastness; unpretending excellence (red flowered)

Camellia is named after Georg Josef Kamel, known as Camellus, a Jesuit missionary and botanist, not the nineteenth-century courtesan Marie Duplessi, called Camille in the novel *The Lady of the Camellias* by Alexandre Dumas, Jr. Camille wore a fresh camellia each day. On most days they would be white, but on five days of the month they would be red to indicate to her clients the period of her menses.

It seems utterly appropriate that the beauty and elegance of the camellia is recognized by it being the symbol of loveliness in the Language of the Flowers. It is then, no surprise, that the plant is used in magic spells to bring wealth and luxury. Where a ritual is being performed in order to gain riches a small vase containing a fresh camellia flower should be placed on the altar.

The wild, single flowered, forms tend to drop their heads suddenly and are therefore said to suggest loss of life. As the fragility of the flower symbolizes death, it ought never to be given as a gift. In Eastern countries it might be used at a funeral but never at a wedding.

TEA

Botanical Name: *Camellia sinensis*
Family: Theaceae
Common Name: Tea tree
Gender: Masculine
Planet: Sun
Element: Fire

Myth and Legend

In legend, Dharma, an Indian prince and mystic who was the personification of truth and justice, travelled throughout China teaching philosophy. He became tired by his long journeys through unfamiliar lands and his constant fasting, prayers and devotions. One night, to atone for his weaknesses and inadequacies he cut off his eyebrows and threw them on the ground. During the night, as he slept, the eyebrows took root and from them grew the first tea plants. Dharma discovered that if he drank an infusion made from the leaves of the plant it stimulated his senses, promoting alertness and enabling him to continue his teaching.

Magic and Lore

Those people who know and love the beautiful camellia may never have associated it with their morning 'cuppa', but the flowering camellia and the tea plant are closely related. Like the other camellias, tea can be used to attract wealth. Leaves from the plant should be added to mixtures used to bring riches, and burning the leaves of the tea plant will also ensure that you enjoy wealth in the future.

For most people tea is merely a beverage, and from a purely practical viewpoint, amongst the practitioners of magic it does make an excellent base to which aphrodisiacal infusions can be added. In its own right, however, it can be used in charms that aim to give the bearer greater courage and strength. Of course, if all else fails, you can always read the tea leaves.

CANTERBURY BELLS

Botanical Name: *Campanula medium*
Family: Campanulaceae

In one legend a priest turned three particularly wicked young men into swans. Their punishment was that they must fly for 1,001 years. One of their flights took them over the top of Canterbury Cathedral. As they passed over it the bells were being rung, which broke the spell and they fell to earth. They were found wandering, lost and bewildered, by St Augustine. He led them into the cathedral, and as they entered little bell-shaped flowers sprang up from the ground wherever they stepped. These flowers were dedicated first to St Augustine and then later to St Thomas à Becket.

Another traditional tale says that the flowers gained their name from the pilgrims making their way to Canterbury, who carried poles on which were hung horse-bells, cowbells, or small handbells. The ringing of these bells signified that the folk were harmless pilgrims. The flowers gained their name from their similarity to the bells carried by the pilgrims.

RAMPION

Botanical Name: *Campanula rapunculus*
Family: Campanulaceae

Rampions are close relatives of Canterbury bells and other *Campanula* plants. The leaves are the part of the plant that are most commonly used, but they also produce a large, turnip-like, root tuber. There is an Italian superstition that rampions will cause children to quarrel. An old Calabrian story tells of a maid, who uproots a rampion growing in a field and finds beneath its roots a staircase leading down to a great palace under the earth.

The most famous story among the faerie tales involving rampions, is that of Rapunzel, whose very name was taken from that of the plant. In one version of the tale Rapunzel stole rampions from the garden of a witch and was imprisoned in a high tower for her crime. In the better-known version a couple who had long wanted a child lived next to the walled garden belonging to an enchanteress. The wife became pregnant and, as a result, craved for a *rapunzel* (rampion) plant she had noticed growing in the walled garden. On two nights her husband went out into the garden to steal some for her, but on the third night, as he scaled the wall to return home, he was caught by the enchantress. He begged her for mercy, and the old woman, called Dame Gothel agreed to be lenient, on the condition that the child was surrendered to her at birth. The man had no choice but to agree. Therefore, when the girl was born, she was given to the enchantress to raise, and named Rapunzel. When she reached the age of twelve, the enchantress shut her away in a tower in the middle of the woods, which had neither door to enter nor stairs to ascend, and had only one room with one window. Whenever the enchantress visited Rapunzel, she would stand beneath the window and called out: 'Rapunzel, Rapunzel, let down your hair, so that I may climb the golden stair.' Upon hearing these words, Rapunzel would wrap her long, blond hair, that had never been cut, around a hook positioned beside the window, and drop it down to the enchantress. The enchantress would then climb up the hair to reach Rapunzel's room.

HAREBELL

Botanical Name: *Campanula rotundifolia*
Family: Campanulaceae
Common Name: Bellflower; blaver; blawort; bluebell; bluebells of Scotland; blue blavers; blue bottle; cuckoos; dead man's bells; Devil's bell; ding-dong; faerie

bells; faerie cap; faerie cup; faerie ringers; faerie thimbles; gowk's thimles; gowk's thumles; granny's tears; harebells; lady's thimbles; milk-ort; old man's bell; school bell; scottish bluebells; sheep bells; thimbles; witch bells; witches' thimbles

This is a plant of faeries, goblins and the Devil. It is the faerie plant of the south-west of England; clumps of harebells were believed to offer shelter to the faeries. They could help mortals to see faeries or look into their realm, but they were often considered to be unlucky, as they might attract malign spirits, including the Devil himself. Where it grew as a garden weed it was often left unpulled for fear of offending the faeries or the Devil.

It is said that the name 'old man's bells' refers to the Devil, and as his plants they should never be picked. 'Old man' was a way of referring to the Devil without the danger of invoking him by calling his name. In some parts of Scotland, the 'auld man' is the resident ghost of a cemetery that comes out on stormy nights. If anyone should hear the ringing of the 'auld man's bell' above the noise of the storm, their death will follow within a fortnight.

The juice from the harebell could be used by witches as part of their flying ointment. It was also supposedly used by witches to transform themselves into hares.

CANNA LILY

Botanical Name: *Canna indica*
Family: Cannaceae
Common Name: Indian shot

A Burmese legend claims that Dawadat was jealous that he did not have a loyal following like that of his cousin, the Buddha, and so decided to destroy him. One day, when the Buddha was meditating near the bottom of a steep hill Dawadat rolled a huge rock down the hill hoping to crush him. Miraculously the rock broke into harmless little pieces. One of these pieces cut the Buddha's toe causing it to bleed. Where the blood fell on the ground it was turned into the brilliant, and very beautiful, bright red flower we call Canna. In Burma the plant is called *Bohdda tharanat*.

HEMP

Botanical Name: *Cannabis sativa*
Family: Cannabidaceae
Common Names: Cannabis; chanvre; gallowgrass; ganeb; ganja; grass; hanf; kif; marijuana; neckwede; tekrouri; weed
Gender: Feminine
Planet: Saturn
Element: Water
Meaning: Fate

Hemp was once commonly available, and commonly used in magical rites. It was often added to incense, particularly those used to aid scrying, in order to increase psychic awareness. One example of this is the incense made from hemp and mugwort, which was burned in front of magic mirrors to help the seer perceive prophetic visions.

These are not the only types of vision that hemp could be used to create. The 'hemp seed spell' was another of those methods by which a young woman might gain a vision of whom she would eventually marry. One version of this spell instructs the young woman to scatter hemp seeds in a field at Halloween. As she sows the seed she must chant:

Hemp seed I sow thee,
Come after and show me.

If she then turns around quickly she will see a vision of her future husband. Another version of this same spell tells that the girl must take two handfuls of hemp seed into a church at midnight, as Midsummer begins. She must walk around the church nine times, sprinkling the seeds as she walks, and repeat the lines;

Hemp seed I sow, Hemp seed I grow,
Who will come after and me and mow?

This will cause her to see the vision of her future love.

There are several variations to this spell around Britain. In Oxfordshire, for example, on Midsummer Day a girl would enter the churchyard carrying a rake on her left shoulder. As she walked, she would throw the seed over her shoulder saying:

I sow hempseed, hempseed I sow,
He that is to be my husband
Come after me and mow.

She would then see the image of her future husband walking behind her, carrying a scythe, in the act of cutting the hemp. Another variation suggests that the ritual should be carried out on Christmas Eve when the girl should toss the seed over her left shoulder saying the same rhyme as above but with these lines added:

Not in his best or Sunday array
But in the clothes he wears every day.

In Derbyshire the hempseed spell is conducted on St Valentine's Day. In this case the poor young woman is required to enter a churchyard at night and run around the church twelve times, scattering the seeds, just as the clock strikes midnight. She will then see an apparition of her future spouse. If no such vision appears she will die unmarried. Occasionally, it is said, a spectral figure might be seen following the woman. This is Death and shows that she will die young and unmarried. The Devonshire version of the spell is slightly different again. It follows much the same form as in Derbyshire and, likewise, takes place on St Valentine's Day. Where it varies is in not requiring the young woman to sprinkle the hemp seeds as she makes her way around the church. Instead, she must say:

Hemp seed I sow, Hemp seed I grow,
He that will my true love be,
Come and rake this hemp seed after me.

Then, at thirty minutes past midnight, the girl must walk from the church porch and make her way home. She scatters the seeds as she goes and the image of her husband to be should be seen behind her raking them up.

Other dates when similar ceremonies have been conducted include St Mark's Eve (24 April) and St Martin's Eve (10 November). In all these cases, however, it must be said that today the images that appear may prove to have more substance than the young woman would really like, and he may well be wearing a police uniform.

The young woman must on no account walk in a field of hemp lest she should become barren.

CAPERS

Botanical Name: *Capparis spinosa*
Family: Capparidaceae
Common Names: Common Caper
Gender: Feminine
Planet: Venus
Element: Water

Capers have been used in magical preparations for their aphrodisiacal properties. They were added to love and lust potions to promote sexual attraction and virility. This has also led to them being eaten by men, as a cure for impotence. This usage has, I assume, no connection with the suggestion by some authorities that the hyssop that is mentioned in the Bible was actually the caper plant.

SHEPHERD'S PURSE

Botanical Name: *Capsella bursa-pastoris*
Family: Cruciferae
Common Names: Bad man's oatmeal; blind weed; case weed; casewort; *clappedepouch* (Irish); crow pecks; fat hen; gentleman's purse; guns; hen and chickens; lady's purses; mother's heart; naughty man's plaything; pepper and salt; pick-pocket; pick-pocket-to-London; pick purse; pick-your-mother's-heart-out; poor man's pharmacettie; rattle pouches; riffle-the-ladies'-purses; St James's wort;

sanguinary; shepherd's bag; shepherd's scrip; shepherd's sprout; snake flower; stony-in-the-wall; tacker weed; witches' pouches; woman's bonnet
Planet: Saturn
Meaning: 'I offer you all'

The heart-shaped seed heads of this common garden weed resemble old-fashioned scrip purses and it is from these that the plant derives many of its common names. In Europe the names are very similar. The French call it *bourse de pasteur*, and in Germany it is known as *hirtentasche*. The Irish name of *clappedepouch* alludes to the bags used by lepers as they begged. They would stand at crossroads ringing their bell, or clapper, and receive alms in a pouch held at the end of a long pole.

Magic and Lore

Picking the little heart-shaped purses can give some indication of what the future holds for you. When you open the 'purse', if the seed inside is yellow it shows that you will be rich. If the seed is still green it indicates that you will be poor. Unfortunately, picking the purses is not without some risk. Superstitions from all across the country warn that in pulling the hearts off the plant you are plucking your mother's heart out, which will cause her to die.

There is a children's game, once commonly played in Britain and Germany, which was based on the accusation of theft. One child invites another to pick off a seed head. As the coins (seeds) pour out there is the chant of,

> Pick pocket to London
> You'll never go to London.

or

> Pick Pocket, Penny nail,
> Put the rogue in the jail.

Medicinal

This plant has been recommended as a treatment for both internal and external bleeding by various herbalists. It is from this that it derives the name 'sanguinary'.

PEPPERS

Botanical Names: *Capsicum annum var. annum*; *Capsicum frutescens*
Family: Solanaceae
Common Names: Capsicum; Cayenne pepper
Gender: Masculine
Planet: Mars
Element: Fire

Peppers are used in love magic to prevent a lover from straying. If you should suspect your partner of infidelities then you must buy two large chilli peppers and tie them together in the form of a cross using a red or pink ribbon. For the charm to take effect they must then be hidden beneath your pillow. When you have done this you can have every confidence that your partner will not prove unfaithful to you. If you wish to spice up your own love life then peppers can be added to love potions that will inflame your lover's passions.

Peppers may also be used in protective magic. Scattering pieces of red pepper about the home will break any curses, or malicious spells, cast against you.

LADIES' SMOCKS

Botanical Name: *Cardamine pratensis*
Family: Cruciferae
Common Names: Apple pie; bird's eye; bog spink; bonny bird's eye; bread and milk; carsons; cuckoo; cuckoo bread; cuckoo flower; cuckoo pint; cuckoo pintle; cuckoo's shoes and stockings; cuckoo spice; cuckoo spit; garden cress; headache; lady's cloak; lady's flock; lady's glove; lady's milk sile; lady's pride; lady's sile; lamb's lakens; laycocks; lonesome lady; lucy locket; may blob; may flower; meadow cress; meadow pink; meadow kerses; milkies; milk girls; milking maids; milky maids; my lady's smock; naked ladies; nightingale flowers; paigle; pick folly; pigeon's eyes; pig's eyes; pink; smell smock; smick-smock; water cuckoo; water lily
Meaning: Paternal error

Magic and Lore

The flowers of this plant, when seen at a distance in the meadows, were said to resemble linen left out to whiten on the grass, 'when maidens bleach their summer smocks'. The name of 'cuckoo flower' comes from the traditional belief that the plant will be in flower throughout the period when the cuckoo can be heard.

This is traditionally a plant associated with faeries. It may be because of this that it is not always considered to be a terribly lucky plant. For ladies' smocks to be included in the garlands carried on May Day would be disastrous, and the whole garland would have to be split apart and remade. In common with other white-flowered plants, ladies' smocks should never be taken into the house, as it will bring bad luck for the whole household.

Medicinal

Galen claims that ladies' smock has many of the same virtues as watercress, in that it can be used as a simple stimulant and diuretic. Elsewhere it is recommended for treating nervous hysteria and St Vitus Dance.

THISTLE

Botanical Name: *Carduus* spp.
Family: Compositae
Common Names: Lady's thistle; thrissles
Gender: Masculine
Planet: Mars
Element: Fire
Deities: Minerva; Thor
Meanings: Austerity; independence; retaliation

Myth and Legend

The thistle is a plant of sin and disgrace, at least according to the biblical story of creation. When man was excluded from the Garden of Eden the ground was cursed by God, to bring forth thistles and thorns. From this beginning it may seem a little strange that it is usually considered to be a plant of good fortune and protection from evil.

The thistle is, of course, the national symbol of Scotland. The cotton thistle, *Onopordum,* is now often associated with this usage but it is quite probable that one of the common species of thistle was originally used, possibly a species of *Carduus,* or even the common spear thistle *Cirsium vulgare.*

Magic and Lore

English wizards would at one time select stems from the tallest thistles to use as walking sticks and magic wands. This may be because carrying a thistle, or part of a thistle, was said to bestow strength and energy. Carrying a thistle flower in your pocket will also give protection against evil and help to dispel melancholy. It is also said to aid masculine virility. Planting thistles in the garden will ward off thieves and burglars. A pot of thistles put onto the doorstep will prevent evil from entering the house, and strewing thistles about the home will drive away any evil already present. Wearing a shirt made from the spun fibres of the plant will break all spells cast against you, and where poppets are to be used to break spells these should also be stuffed with thistles. The thistle also has the power to dispel melancholy; placing a bowl filled with the plants into a room will raise the morale of all those present. The use of the plant in a healing spell may result from the idea of driving out the evil spirits that have caused the illness.

The thistle can be used in the conjuration of spirits. For this purpose the thistles are placed into boiling water, which should then be removed from the heat. You must then sit, or lie, down beside the bowl and wait for the spirits to appear in the steam as it rises. These spirits may then answer any question that you put to them.

If you should see yourself surrounded by thistles in your dreams it is a sign of good fortune as it means that a pleasant event will soon follow. It may also, however, indicate that someone in whom you have placed your trust will prove to be disloyal to you.

There is a traditional warning to gardeners regarding thistles. One country rhyme says:

Cut your thistles before St John [24 June],
Or you'll have two instead of one.

An alternative version says:

Cut dashels [thistles] in June – it's a month too soon.
Cut in July – they're sure to die.

In Devon it was put yet another way:

Speed them in May,
They are up the next day.
Speed them in June,
They will come again soon.
Speed them in July,
They will soon die.

See also Scottish thistle (*Onopordon acanthium*).

PAPAYA

Botanical Name: *Carica papaya*
Family: Caricaceae
Common Names: *Mamaeire*; Melon tree; papaw; paw-paw
Gender: Feminine
Planet: Moon
Element: Water

The tropical papaya fruit is considered by some to be an aphrodisiac that will intensify the feelings of love between those who share one. This is only one of the powers with which the plant is supposedly endowed. In parts of India pregnant women were forbidden to eat the papaya fruits as the black seeds it contains were believed to cause miscarriages. The plant also has protective powers. Hanging a small bundle of papaya twigs above doorways will keep evil forces out of the home. Finally, it can be used to ensure that your wishes come true; wrap a piece of rag around the branch of a papaya tree whilst visualizing your wish.

CARLINE THISTLE

Botanical Name: *Carlina vulgaris*
Family: Compositae
Element: Fire

The original name of this plant was *Carolina*, as it derives its name from the Emperor Charlemagne. Legend tells that a plague was spreading through his army, killing thousands of his troops. Charlemagne prayed fervently, and while he slept an angel appeared to him, who fired an arrow from a crossbow, instructing the emperor to note carefully the plant on which the arrow landed, as it could be used to cure the illness of his troops. Needless to say the plant was the carline thistle.

In some areas of Europe, the flowers are nailed over doorways to be used as a hygrometer, because the flowers will only open in dry weather.

CARAWAY

Botanical Name: *Carum carvi*
Family: Umbelliferae
Gender: Masculine
Planet: Mercury
Element: Air

Our folklore is full of stories of ghosts and ghouls, boggarts and abbey lubbers. One such monster of the night is Lilith, a vampire, who haunts the wilderness areas, particularly during stormy weather, and attacks those foolish enough to venture out. It has a particular fondness for children and one way to protect them, and yourself, from its attentions is to carry caraway seeds. This will also ward off all other evil forces.

Another characteristic of caraway is its power of 'retention'. It is claimed that those things that contain caraway cannot be stolen; any burglars foolish enough to try would find themselves unable to leave the scene of the crime with their loot, as they would be held until the authorities arrived. This attribute can be applied elsewhere. A wife may feed her husband caraway, or put a little into his pockets, to prevent him from leaving her for another woman. This may have additional benefits, as eating caraway is said to prevent hair loss and loss of memory. Doves and pigeons can be fed with caraway seeds to prevent them from flying away, perhaps because they remember their way home. Dioscorides recommended women to eat caraway, presumably so that they did not lose their good looks but improved their complexion. Chewing caraway seeds, using them in love charms, or wearing them in a sachet are all said to ensure that you will gain the affection of the one you desire. Alternatively you can bake them in breads and biscuits to induce lust.

CASSIA

Botanical Name: *Cassia fistula*
Family: Leguminosae
Common Names: Golden shower tree; Indian laburnum; pudding pipe tree; purging cassia
Planet: Moon

In Chinese legend it is said that a cassia grows at the middle of the moon and because of this the moon is known as Kueilan, the disc of cassia. The story is that a man from Si-Ho, named Kang Wou, found a genie but abused the powers it gave him. In punishment for the crimes he had committed he was sentenced to spend eternity trying to cut down the cassia on the moon.

CEDAR

Botanical Name: *Cedrus* spp.
Family: Pinaceae
Gender: Masculine
Planet: Sun
Element: Fire
Deities: Arinna; Baalat; Osiris
Meanings: Faithfulness; incorruptibility; strength

There has been some confusion regarding cedarwood as several different species of plant have been given 'cedar' as a common name. The red cedar, for example, is a species of juniper. Western red cedar is *Thuja occidentalis*, and white cedar is *Cupressus*.

Myth and Legend

Winabojo, described as a man who wears a cedar tree as an ornament, with its roots twisted all around him, appears in a long and twisting Native American story about a man whose only daughter died. The father felt that he could not live without her and told his friends that he wanted to visit the spirit world in order to find her. In order to discover the way he spoke to the Grand Medicine Men of the Chippewa. They travelled with him to an island in Lake Superior where they located Winabojo who pointed them on their way.

Magic and Lore

This is a sacred tree and, in common with other trees considered to be sacred, anyone damaging it could expect bad luck to befall them. It is a tree of eternity and righteousness, sometimes called the tree of paradise or the tree of good fortune.

Cedarwood, and the oil distilled from it, were highly prized commodities in the ancient world. It was considered to be incorruptible, and would preserve anything enclosed within it. The Egyptians used cedar oil among their embalming and mummifying oils, and the wood for making mummy cases. Expensive books would be kept in cedar boxes, and we are told in the Bible that the Great Temple of Solomon was lined with cedarwood panels, the scent of the wood being sufficient to keep away all worms and spiders.

The fragrance of cedar has a soothing and calming effect. The wood can be added to incense used to scent rooms and purify the atmosphere. The Native Americans would place some on the hot stones in their sweat baths to help in their cleansing and purification. This burning of cedar is also said to prevent colds, increase psychic awareness and attract wealth, although elsewhere it is said that burning cedar wood will bring bad luck. Concealing a small piece of the timber in your purse will also ensure that money finds its way to you. Hanging a small piece of cedarwood up in a house will give protection against lightning. If a trident is carved from cedarwood and placed prongs down in the ground near the house, it will ward off all evil forces.

Planting a cedar in the garden is supposed to ensure good luck for the person who plants it, although some sources claim that if you put a cedar tree into your garden, 'you will never sit in its shade'. If the tree is to flourish, it must be given the same orientation as before it was moved. It might be worth getting someone else to do the work for you however, as one superstition says that the transplanter will die when the lower limbs of the tree reach a sufficient length to make his or her coffin.

CORNFLOWER

Botanical Name: *Centaurea cyanus*
Family: Compositae
Common Names: Bachelor's buttons; blaver; blawort; blue blaw; blue blow; blue bobs; blue bonnets; blue bothem; bluebottle; bluebow; blue buttons; bluecap; blue jack; blue poppy; bluet; bobby's buttons; bottle of sorts; broom and brushes; brushes; corn blinks; corn bottle; cuckoo hood; hurtsickle; knob weed; knotweed; ladder love; logger-heads; miller's delight; pincushions; ragged sailors; thumble; witch bells; witch's thimbles
Planet: Sun (Saturn, according to Culpeper)
Meanings: 'A dweller in heavenly places'; 'Be not over impetuous, my heart cannot be stormed'; delicacy; sensitivity

Myth and Legend

The generic name *Centaurea* is a reference to the centaur, Chiron, who taught the skills of herbal medicine to Achilles. Hercules wounded Chiron with an arrow dipped in the poisoned blood of the Hydra and Chiron used the cornflower to heal the wound. This same tale is told elsewhere with the healing properties attributed to a totally unrelated species, *Centaurium,* a member of the African violet family (Gesneriaceae).

Cyanus was a youth devoted to the goddess Flora, and the cornflower was a particular favourite of his; indeed he spent much of his time picking them and making wreaths. One day he was found lying dead amongst the cornflowers he had picked. Flora transformed him into a cornflower in acknowledgement of his veneration of her.

In Russia the cornflower is known as *Basilek*, or 'flower of Basil'. In a folk tale Rusalka lures the young Basilek, or Basil, into a cornfield and transforms him into a cornflower.

Magic and Lore

The cornflower is what gardeners and farmers call a gross feeder, that is say that it draws a great deal of nutrition out of the earth. It is also called 'hurtsickle' because the thick stems would blunt reaping hooks during harvest.

> Thou blunt'st the very reaper's sickle and so,
> In life and death becom'st the farmer's foe.

This is another flower sometimes called 'bachelor's buttons' and young men might carry it in their pocket to see whether they would marry their sweetheart. If the flower lived all would be well, but if it died he should find someone else. It may have been given this name because of a resemblance to buttons on sixteenth-century clothing. Wearing a bachelor's button was an indication that a man was unmarried. Elsewhere it is said that girls would wear bachelors' buttons to show that they were unwed. If a girl concealed a cornflower beneath her apron she would be able to have the bachelor of her choice.

If the flowers are picked with the dew still on them, and worn for twenty-four hours, it will ensure success in courtship so long as the flowers remain in good condition.

GREATER KNAPWEED

Botanical Name: *Centaurea scabiosa*
Family: Compositae
Common Names: Boltsede; bottle weed; bull weed; churl's head; cowede; hardhead; hard irons; horse knops; ironhead; logger head; mat felon; matte felon
Planet: Saturn

The flower of the knapweed is a solid, hard mass and it is from this that it derives such names as 'hardhead' and 'ironhead'. The name 'knapweed' also comes from this feature, 'knap' being derived from 'knop' or 'knob'. The name 'mat felon' refers to its use to treat felons, or whitlows, small swellings at the side of the fingernail. In some parts of Britain it was common for the flowers to be given to the bereaved at funerals.

YELLOW STAR THISTLE

Botanical Name: *Centaurea solstitalis*
Common Names: St Barnaby's Thistle
Element: Fire

The yellow star thistle is rare in Britain, having been introduced into the country. It is said to flower on St Barnabas' Day, which is 11 June. This is reflected in the species name of *solstitalis*, which is derived from it being in flower at the summer solstice.

CENTAURY

Botanical Name: *Centaurium erythraea* (Synonym: *Erythraea centaurium*)
Family: Gentianaceae
Common Names: Centaury gentian; centory; Christ's ladder; felwort; *fel terre*; feverwort; filwort; gall of the earth; red centaury
Gender: Masculine
Planet: Sun
Element: Fire
Meanings: Delicacy; felicity

Myth and Legend

The genus was previously called *Chironia* after the centaur Chiron. In Greek myth, Chiron is said to have made use of this herb to treat his own wound, caused by arrows poisoned with the blood of Hydra, accidentally fired by Hercules (see Cornflower).

Magic and Lore

This plant was known to the ancients as *fel terre*, or 'the gall of the earth', because of its bitter nature. In Worcestershire the name 'centaury' was corrupted to become 'the centre of the sun'. Centaury is included in the fifteen magical herbs of the ancients, as given in *Le Petit Albert,* which was translated into English in 1619.

> The eleventh herbe is named of the Chaldees, Isiphon … of Englishmen, Centory … this herbe hath a marvellous virtue, for if it be joined with the blood of a female lapwing, or black plover, and put with oile in a lamp, all that compass it about shall believe themselves to be witches, so that one shall believe of another that his head is in heaven and his feete on the earth; and if the aforesaid thynge be put in the fire when the starres shine it shall appear yt the starres runne one agaynste another and fyghte.

A different source makes the even more bizarre claim that anyone in the room in which the lamp oil mix is being burnt will see themselves upside down with their feet in the air.

Centaury can also be used, in connection with other herbs, to identify witches. A cross is made from centaury, lovage and ground ivy.

One superstition says that if you put centaury under someone's nose it will cause them to run away as fast as they possibly can.

Medicinal

A tea brewed from dried centaury may be given as an appetite stimulant, to aid digestion and ease heartburn. Then again it might be best avoided, as all reports are that it is extremely bitter.

ST JOHN'S BREAD

Botanical Name: *Ceratonia siliqua*
Family: Leguminosae
Common Names: Algaroba; *bharout*; carob; John's bread; locust pods; locust tree; sugar pods

A tale from the Talmud tells that Rabbi Chomi lived in a small village with his wife and son. He was a good and wise man to whom the villages often turned for help and advice. One bright and sunny day, tired of listening to the troubles of others, he decided to go out walking in the countryside. After walking for some time, he met an old man by the side of the road, digging a hole. Rabbi Chomi asked what he was doing and the old man told him that he was planting a carob tree. Rabbi Chomi told him he would never live long enough to enjoy the fruits of the tree, as it would be thirty years before it would reach maturity. The old man agreed, but said that he planted the tree for the benefit of his children and grandchildren. The Rabbi asked him why he chose a carob tree and the old man narrated the story of Rabbi Shimon bar Yohai, who rebelled against the Roman government. When he and his son had to flee they hid near a cave, which had a carob tree growing nearby. The fruits of the tree saved them from starvation. The old man said that he was planting the carob tree in honour of that story.

This plant gains the name 'St John's bread' from the tradition that it was one of the main sources of food for St John the Baptist, during his time wandering in the wilderness. Unfortunately, the fruit has a powerful laxative effect.

The seeds have another claim to fame: it is from them that the carat weight, used by jewellers, is derived.

RED SPUR VALERIAN

Botanical Name: *Centranthus ruber*
Family: Valerianaceae
Common Names: Bouncing Bess; bovis and soldier; delicate bess; drunken sailor; fox's brush; jupiter's beard; kiss-me-quick; mate-bate; pretty Betsy; setewell

Red spur valerian should not be confused with the true valerian (*Valeriana officinalis*). However, because of confusions arising from the use of the same common name it is only to be expected that some of the qualities of the true valerian have been incorrectly attributed to this plant. The red spur valerian has none of the medicinal qualities of its namesake.

One of the powers with which this plant is supposed to be endowed is the ability to 'disturb' sexual desires. In consequence of this, young women carry the flowers in the hope of attracting lovers to them. The disruptions caused by the plant have also led to it gaining the name 'mate-bate', as it can cause arguments. It is said that if a root of the red spur valerian is concealed in the bed of a married couple it will cause them to start quarrelling before the night is over.

Like other species of valerian, this plant is used as a protective talisman. It could be hung above doorways to prevent evil from entering. In Germany, it was seen as a powerful protection against the mischief or malice of imps.

JUDAS TREE

Botanical Name: *Cercis siliquastrum*
Family: Leguminosae
Common Names: Love tree; red bud
Meanings: Betrayal; unbelief

It is likely that the first recorded use of the name Judas tree was in *Gerard's Herbal* of 1597. The name may have come from a legend that the flowers were originally white but blushed pink with shame. It is more likely that the name is a corruption of 'Judaea tree'; in France the plant is still called *Arbre de Judea*.

The Judas tree is popularly thought of as having been so called because it was the tree on which Judas Iscariot ended his life, but a glance through the entries in this book will discover several other contenders for this dubious honour.

MOUNTAIN MAHOGANY

Botanical Name: *Cercocarpus ledifolius*
Family: Rosaceae
Gender: Masculine
Element: Fire

There is a tradition amongst the Native Americans that this plant will give protection against being struck by lightning, especially at high altitude. This is because the plant naturally flourishes at high altitude, where the thunder and lightning 'live'. Climbers are therefore well advised to carry a small piece with them during their time on the mountain.

CACTUS

Botanical Names: *Cereus grandiflorus*; *Cereus* spp.; *Echinocactus* spp.; *Opuntia* spp.
Family: Cactaceae
Planet: Mars
Element: Fire
Meanings: Ardour; endurance; 'I burn'; 'Our love shall endure'; warmth

The Mexican coat of arms depicts an eagle holding a snake and perched on a cactus. This is an image central to the Aztec creation myth. Tenochtitlan (the earlier name of Mexico City) translates as 'place of the sacred cactus'. Pictorial representations, drawings and sculptures of cactus-like plants can be found among the remains of the Aztec civilization. Many of these resemble the 'Mother-in-law's cushion' cactus, *Echinocactus grusonii*, which has great ritual significance as it was on it that human sacrifices were carried out.

Cacti of all types have enjoyed some popularity as houseplants. Apart from being easy to grow and having attractive flowers – when they do eventually flower – they have another benefit. They are protective plants and will help to protect property from burglars and other unwelcome intruders. They also dispel negativity in the area where they are being grown. If they are planted outdoors, their protective qualities can be amplified by placing one at each point of the compass around the house.

The spines of the cactus are used in witchcraft to mark words and symbols onto the roots and candles used during magical rites.

COMMON CHAMOMILE

Botanical Name: *Chamaemelum nobile* (Synonym: *Anthemis noblis*)
Family: Compositae
Common Names: *Athair talanh* (father of the ground); camel; camil; cammany; camomine; camomyle; camovyne; chamaimelon; ground apple; *heermaachen*; low chamomile; *manzanilla*; *maythen*; Roman chamomile; Whig plant
Gender: Masculine
Planet: Sun
Element: Water
Meanings: Energy in adversity; fortitude; 'I admire your courage, do not dispair'; initiative; love in austerity

> The Chamomile shall teach the patience
> That rises best when trodden most upon.

Magic and Lore

The name 'chamomile' comes from the pleasant odour, reminiscent of apples. The Greeks called it 'ground-apple', *kamai,* meaning 'the ground', and *melon*, 'an apple'.

The species name *nobile* refers to the large flowers on a small plant. To the Saxons it was *maythen*, one of the nine herbs given to the world by Wotan, and considered as a sacred herb alongside fennel, watercress, crab apple, chervil and plantain. It became known as 'Roman Chamomile' after it was found, by a sixteenth-century German writer, growing in Rome. The Spanish name for it, *manzanilla*, meaning 'little apple', has subsequently been used for a light sherry flavoured with the herb.

Gamblers might choose to use chamomile, especially as a handwash, as it is supposed to attract money and ensure winning. Adding a little to bath water will attract love, and sprinkling an infusion of the plant about property will remove any curses cast against those who live there.

Chamomile is the birthday flower for 17 December.

Medicinal

Chamomile has been grown for centuries, and is valued for its medicinal properties. The ancient Egyptians revered it, using it for treating fevers, and dedicated it to their gods. It has been considered to be one of the finest medicinal herbs and has been used as an anti-inflammatory and as a sedative in the treatment of nervous disorders. In homeopathic medicines it is used to treat 'inner turmoil', that is to say anxiety, anger convulsions, earaches, teething, coughs and diarrhoea.

Perhaps the best known use of chamomile is as a soothing herb. It may be added to incense burned to aid restful sleep or encourage peaceful meditations. Turner says: 'It will restore a man to hys color shortly yf a man after longe use of the bathe drynke of it after he is come out of the bathe.'

It is not only human health that can benefit from chamomile; it has been recommended as a 'plant doctor'. If planted near to choice plants which are ailing it is said to help them regain healthy vigour.

WILLOWHERB

Botanical Name: *Chamerion angustifolium* (Synonym: *Epilobium angustifolium*)
Family: Onagraceae
Common Names: Apple pie; blood vine; blooming Sally; cat's eyes; eyebright; fireweed; flowering willow; flowering withy; French saugh; french willow; mother-die; persian willow; purple rocket; ranting widow; rose bay willow herb; tame withy; wickup; wicopy; wild snapdragon
Planet: Saturn
Meanings: Celibacy; pretension

It may now be hard to believe, but this common weed was something of a rarity when Gerard published his herbal in 1597. He tells of receiving some plants from Yorkshire, and introducing them in his garden. Modern gardeners now spend considerable amounts of time and effort trying to rid their gardens of this weed. Hand weeding may have its dangers, however, as to pull up willowherb is said to cause thunderstorms or the death of your mother.

It grows well where the soil has been disturbed, possibly because dormant seeds are being brought to the surface where they can germinate, and where there have been fires. It was much in evidence in London after the Blitz, and was one of the first plants to grow on Mount St Helens, in the USA, following its eruption.

GREATER CELANDINE

Botanical Name: *Chelidonium majus*
Family: Papaveraceae
Common Names: Celydoyne; chelidonium; devil's milk; figwurt; garden celandine; Jacob's ladder; Kenning wort; kill-wart; St John's wort; swallow herb; swallow wart; swallow wort; tetterwort; wart flower; wart plant; wartweed; *warzenkraut*; witch's flower; yellow spit
Gender: Masculine
Planet: Sun
Element: Fire
Meanings: Joys to come; reawakening; 'Let this harbinger of spring speak to you of my love'

Myth and Legend

This is the plant that most would consider to be the 'true' celandine, and it is unrelated to the lesser celandine *(Ranunculus fricaria)*. It is said to come into flower as the swallows arrive in the summer, and finish flowering at the time when they start their migration. Pliny says that it was the swallow which first discovered that celandine could be used to help eyesight, hence the name swallow wort. There is a traditional story that the mother swallow uses the plant sap to restore sight to her blind nestlings.

Magic and Lore

Wearing a fresh celandine flower against the skin is alleged to enable the wearer to avoid any traps set for them and escape unwarranted imprisonment. This may be due to another power attributed to it, that of ensuring the wearer wins the favour of judge and jury when appearing in court.

Medicinal

'Kenning wort' is a further reference to the use of the herb in the treatment of eye complaints, the cloudy spots on the eyeball being called 'kennings'. Its other major medicinal use, reflected in its common names, is to clear warts. The 'yellow spit', which is the sap taken from broken or cut stems, is applied to the wart in order to remove it. On a more general note, we are told that the wearing of celandine will dispel melancholy, lift the spirits and cure depression.

PIPSISSEWA

Botanical Name: *Chimaphila umbellata*
Family: Pyrolaceae
Common Names: Butter winter; false winter green; ground holly; love-in-winter; king's cure; prince's pine; princess pine; rheumatism weed

Pipsissewa is used with other plants, in order to gain wealth. It can be crushed and mixed with rosehips and violet flowers in order to form an incense which is burned in order to draw good spirits and benefit from their magical aid.

CHRYSANTHEMUM

Botanical Name: *Chrysanthemum X grandiflorum* (Synonym: *Dendranthema X grandiflora*)
Family: Compositae
Common Name: Mum
Gender: Masculine
Planet: Sun
Element: Fire
Meanings: Cheerfulness under adversity; a desolate heart; friendship (bronze); 'I love', reciprocated love (red); truth (white); discouragement, slighted love (yellow)

Although highly prized in Japan the chrysanthemum has its origins in China. It is mentioned in the writings of Confucius some 500 years before the birth of Christ. It was introduced into Japan, via Korea, in about AD 386 but it was a further 250 years before it really caught the imagination. The Chinese had not bred the plant, but the Japanese did. It became the national emblem towards the close of the eighth century, when the Order of the Chrysanthemum was the highest award that the state could bestow on any citizen. As the symbol of Imperial Japan the emperor's throne is known as the chrysanthemum throne. In the east, the chrysanthemum is seen as a symbol of purity and long life.

Myth and Legend

One account of the origin of the chrysanthemum tells of a young Chinese girl, Kuku-no-hana, who asked a spirit how long her forthcoming marriage would last. The spirit told her that it would last for as many years as there were petals on the flower she would wear on her wedding dress. Kuku-no-hana searched for the flower that had the most petals and eventually found one which had seventeen. Then, with a pin, she split each petal first into two, and then into four. This then was the first chrysanthemum, and her marriage lasted for sixty-eight years.

A folk tale from Europe suggests a different origin. Hermann, a woodsman from the Black Forest, found a child abandoned in the snow on Christmas Eve. He carefully picked him up and took him to his home, caring for him for the next forty-eight hours. On Christmas Day the child revealed that he was actually the Christ, and shortly afterwards he disappeared. When next Hermann ventured into the forest he saw that where he had found the child there grew a magnificent stand of golden chrysanthemums.

In Japanese legend the god Izanagi and goddess Iznami, were sent across a bridge of clouds to Earth, because there were too many gods in heaven. On arriving on Earth, Iznami created the gods of the wind, the mountains and the sea, but she was killed in the flames that sprang up when she was creating the god of fire. Izanagi missed her and set out to follow her into the place known as the Black Night. However, when he eventually caught a glimpse of her, he was chased from the night by a witch, and fled back to Earth. He decided to wash himself in the river, and undressed on the bank. The articles of clothing he dropped onto the ground turned into twelve gods. His jewels became flowers: his bracelets becoming an iris and a lotus flower, and his necklace becoming the golden chrysanthemum.

A Chinese legend relates how an elderly emperor heard about a magical herb that could grant eternal youth. It grew on Dragonfly Island but could only be picked by young people. The emperor sent twenty-four children on the hazardous journey to collect it for him. When they eventually arrived on Dragonfly Island they found it deserted, with no sign of the magic herb. What they did find was a golden chrysanthemum, which even today symbolizes the Chinese people's connection with their country. Mao Tse Tung changed the colours, replacing the imperial golden yellow with the red of the People's Republic.

Chrysanthemums are known as Christ's flower, because it came into bloom on the morning of his birth.

Magic and Lore

Though not a plant that we might immediately associate with magic the chrysanthemum is said to protect the wearer against the wrath of the gods. Having it growing on your garden will ward off all evil spirits.

Some people consider it unlucky to have chrysanthemums in the house. This is doubtless because they are a favourite funeral flower in some areas of Europe. It is probably in this context that they have become associated with All Soul's Day, a day when the dead are remembered. In Italy, apparently, a gift of chrysanthemums was tantamount to saying 'I wish that you were dead'.

In China the chrysanthemum is considered symbolic of rest and ease. Perhaps because of the story of the elderly Chinese emperor, chrysanthemum has also become a plant of longevity. Placing a single petal at the bottom of your glass when drinking will give you a long and healthy life, and dew collected off the flowers promotes long life. In Japan the orderly unfolding of the chrysanthemum petals was taken to represent perfection.

Medicinal

It was once recommended that you should drink an infusion made from the plant as a cure for drunkenness. I suspect that the plant used was actually a close relative *Tanacetum parthenium*, feverfew, which is commonly used to treat headaches.

VEGETABLE LAMB

Botanical Name: *Cibotium barometz*

Sir John Mandeville brought the legend of the vegetable lamb to England. It was supposed to be a lamb-like organism that shared plant and animal characteristics. It is generally assumed that the thing that Mandeville had spoken of was the cotton plant. A description of the vegetable lamb is given in the *Talmud Ierosolimitanum*, written in AD 436, which says, 'It was in form like a lamb, and from its navel grew a stem, or root, by which this zoophyte or plant animal was fixed.' If you think that this is ridiculous remember that Gerard, who poured scorn on many superstitions associated with plants, describes the 'barnacle goose plant' in his herbal of 1597.

ENDIVE

Botanical Name: *Cichorium endivia*
Family: Compositae
Gender: Masculine
Planet: Jupiter
Element: Air
Meaning: Frugality

Myth and Legend

Legend informs us that the endive had a magical, though tragic, origin. It sprang up from the tears of a young woman who waited in vain for the return of her lover's ship.

Magic and Lore

If the plant is to be used for mystical or magical purposes it should be collected on St Peter's Day (27 June) or St James's Day (25 July). No tools containing iron should be used in gathering it as this will dispel the magical properties of the plant. Various texts recommend the use of either golden tools, or stag's horn. Once gathered, however, endive can be worn as an amulet in order to attract love and affection. To be effective it must be fresh and so should be replaced at least every three days. The more normal use of endive, as an ingredient in salads and other foods, has an added dimension as it is said to induce lust.

CHICORY

Botanical Name: *Cichorium intybus*
Family: Compositae
Common Names: *Barbe de capucin*; blue endive; bunks; hard ewes; hendibeh; monk's beard; strip-for-strip; succory; wild succory
Gender: Masculine
Planet: Sun (Jupiter, according to Culpeper)
Element: Air
Meaning Frugality

Myth and Legend

The origin of chicory is described in a German folk tale. A young woman sat down at the side of a road to weep for her dead lover. None could console her, and so great was her grief that she resolved never to stop weeping unless she was turned into a flower. She was transformed into the wegwort, the chicory.

An alternative tales tells of an attractive maiden whose beauty had captivated the sun. She refused all the sun's advances and, to escape his unwelcome attentions, was transformed into the chicory plant. She still turns her head to watch as the sun passes across the sky during the day.

Magic and Lore

Many of the magical powers attributed to mandrake have been applied to chicory. If it is to be used for magical purposes a certain amount of ceremony must be observed in gathering it. The most auspicious times to collect it are at midday or midnight, at Midsummer, or on St James's Day (25 July). As with endive it is important that no tools containing iron are used in gathering it, as this will dispel its magical powers. In this case a golden knife should be used (though some sources suggest a stag's horn and others a 'sun disc', i.e. a gold coin), and the whole operation must be conducted in total silence. If a word is uttered the person gathering the plant will die.

Once collected it can be used to remove all obstacles that might hinder your progress in life, and when the leaves are applied to locks on doors or boxes it will open them. It has also been claimed that rocks will open to those bearing the plant, exposing entrances into the Underworld. Some writers have said that chicory can only be used to open locks on St James's Day. They suggest it should be done in silence by holding a golden knife and some chicory leaves against the lock.

If it is properly prepared and used, chicory is alleged to be able to render the bearer invisible whenever they wish. The juice can be used to anoint the body, and this should be done in order to win the favours of those in positions of authority. Dyett, in *Dry Dinner*, says: 'It hath bene and yet is a thing which superstition that beleeved, that the body anoynted with the juyce of chicory is very available to obtaine the favour of great persons.'

Generally, the plant is said to bring good fortune, especially on journeys of exploration. Those prospecting for gold in North America may have carried it to ensure their success.

Linnaeus, the 'father' of plant nomenclature, used chicory in his 'Flora's Dial', at his home in Uppsala. At that latitude the flowers will open at approximately 5 o'clock and close again at 10 o'clock. In Britain the flowers will open at approximately 7 o'clock to close again at noon.

On upland slopes the shepherd's mark
The hour when to the dial true
Cichorium to the towering lark
Lifts her soft eyes serenely blue

This is one of the bitter herbs of the Jewish Passover, and is generally considered to be a symbol of the good fortune that is to follow.

Medicinal

It has been suggested that water distilled from the flowers of chicory can be used to aid poor or failing eyesight. This is a reference to the sympathetic magic followed in the Doctrine of Signatures as the flowers of chicory respond to the position of the sun.

BLACK COHOSH

Botanical Name: *Cimicifuga racemosa*
Family: Ranunculaceae
Common Names: Black snake root; bugbane; rattle root; squaw root
Gender: Masculine

Black cohosh is mainly prized in folk magic for its protective properties. An infusion made up from this herb can be sprinkled around a property to ward off evil forces. Adding a little of the infusion to bath water will also protect the bather from evil. It has the added benefit of curing impotence. Carrying a sachet containing black cohosh will attract love to the bearer, and will give courage to the faint hearted.

PERUVIAN BARK

Botanical Name: *Cinchona pubescens* 'Succirubra'
Family: Rubiaceae
Common Names: Cinchona bark; Jesuit's powder; red bark

This plant was named in honour of the Countess de Chinchon, wife of the Viceroy of Peru. There is a story that in 1638 she was cured of malaria with it. In central Java it was once the main source of commercial quinine. It was a part of the Japanese strategy, in the Second World War, to cut off this supply of quinine to the Allied troops. If the pharmaceutical companies of the West had not been able to synthesize the drug the jungle campaigns in the East could probably not have been fought.

The folklore of the plant tells that carrying it will bestow protection, to the bearer, from all bodily harm or evil. It also ensures that whoever carries it enjoys the benefits of great good fortune.

CAMPHOR

Botanical Name: *Cinnamonum camphora*
Family: Lauraceae
Common Names: Gum camphor; laurel camphor
Gender: Feminine
Planet: Moon
Element: Water
Meaning: Fragrance

This 'true' camphor is often unavailable and so there are many materials that are used as substitutes. Camphor has a number of properties of significance in folk magic and folklore. In magic it can be added to incenses burned to induce prophetic states and to encourage powers of divination.

Amongst the indigenous tribes of Sarawak it was believed that if women were unfaithful whilst their husbands were away gathering camphor in the jungle, then any camphor that the cuckolded husband had collected would magically evaporate. The men, it was claimed, could discover whether their wives had been unfaithful from certain knots in the tree. In former times it needed no further evidence of the woman's guilt to justify the husband killing her. Furthermore, the women would not dare to touch combs whilst their husbands were away as the fibres of the tree would not be filled with the precious resin but would be as empty as the gaps between the teeth of the comb.

The scent of camphor is recommended to reduce sexual desires. If this is desirable a small piece of camphor should be kept at the side of the bed. Wearing a sachet containing camphor, about the neck, is a folk remedy that will prevent the wearer from catching colds and flu.

CINNAMON

Botanical Name: *Cinnamonum zeylanicum*
Family: Lauraceae
Common Names: Ceylon cinnamon; sweet wood
Gender: Masculine
Planet: Sun
Element: Fire
Deities: Aphrodite; Chang-o; Venus

Cinnamon is another herb that has been used for various purposes for centuries. The ancient Egyptians used the essential oil in their mummification processes. The Hebrews used the same oil as an ingredient in their holy anointing oils, and in ancient Rome wreathes of cinnamon leaves were used in temple decorations.

When Cinnamon is added to incense it raises 'spiritual vibrations', stimulating the psychic powers and aiding healing processes. It has the additional benefit of attracting wealth to those who are burning it, as well as conferring on them protection from evil.

MELANCHOLY THISTLE

Botanical Name: *Cirsium helenioides* (Synonym: *Cirsium heterophyllum*)
Family: Compositae
Common Name: Plumed thistle

Some authors claim that it is the melancholy thistle rather than the cotton thistle (*Onopordon acanthium*), that was the original symbol of the House of Stuart and so of Scotland. Certainly, it is more common north of the border, where it is known as *Cluas au Fleidh*. In his herbal Culpeper reports that Dioscorides recommends it as a treatment for all those diseases causing melancholia, and it is from this that the plant gains its common name.

WATERMELON

Botanical Name: *Citrullus lanatus*
Family: Cucurbitaceae

There are a number of superstitions linked to watermelons, most of which relate to its cultivation. You should, for example, sow watermelon seeds during the first three days of May if you want to be sure of a good crop. Some writers recommend that the seed be sown on 1 May, before sun-up, with plenty of manure. If you do so while you are still wearing your nightclothes you need not fear any insect damage to your crop. Sowing watermelon seed during the sign of Cancer, and in the dark of the moon (a waning moon) will give a large crop and the melons will be especially sweet.

If you cut open a watermelon and find it is unripe it indicates that there will be sickness and sorrow in your house. To find out if the melon is fully ripe simply roll it away from you. If it rolls back by itself, it is ready for eating.

LIME FRUIT

Botanical Name: *Citrus aurantifolia*
Family: Rutaceae
Common Names: *Limetta fructus*
Gender: Masculine
Planet: Sun
Element: Fire

Lime can be used to remove spells and curses. A talisman must be made by piercing a fruit with old pins, needles, nails and spikes. It must then be buried in a deep hole in the garden, where it will remain undisturbed. Merely carrying twigs from the tree will act as a protection, as it will avert the evil eye. In common with other species of citrus fruits the peel of the lime fruit can be added to love potions and incense.

There is a potential for confusion between the citrus lime and the linden tree or European lime. No doubt various traditions relating to each have been transferred in the past. I hope that I have not contributed to this confusion rather than clarifying the situation.

Medicinal

Lime can be used to treat physical pains. Wearing a necklace of lime fruits is recommended as a cure for a sore throat, and may make an interesting fashion statement. To cure toothaches you must drive a nail into the trunk of the lime tree; however, you must thank the tree for its help first.

ORANGE

Botanical Name: *Citrus aurantium*; *Citrus sinensis*
Family: Rutaceae
Common Names: Bitter orange (*C. aurantium*); love fruit; neroli; sweet orange *(c. Sinensis)*
Gender: Masculine
Planet: Sun
Element: Fire
Deities: Juno; Jupiter
Meanings: Bridal festivity; chastity; loveliness; purity; virginity; 'Your purity equals your loveliness' (orange blossom); generosity (orange tree)

Myth and Legend

In mythology, there is a tale that Jupiter gave Juno a 'golden apple', an orange, as a gift at their marriage. This has found is way into folklore, as there is a tradition that where a young man gives an orange to a woman it will induce love between them.

Magic and Lore

Traditionally orange blossom was included in the bride's bouquet at weddings as a symbol of good luck, happiness, chastity and fertility. This practice appears to have originated in the Holy Land amongst the Saracens and was brought to Europe by returning crusader knights. It is considered unlucky for orange blossom to be included in any bouquet other than a bride's. Natural magic takes this one stage further, by recommending the use of orange blossom in sachets worn to ensure wedded bliss. This link between oranges and love can be found elsewhere.

Neroli is the essential oil distilled from the blossoms of the bitter orange. It is considered to be one of the finest of the flower essences, and is an ingredient of eau de Cologne. It can also be added to bath water in order to aid relaxation. Neroli may also be used as an ingredient in love or lust potions. Adding the fresh or dried flowers of the orange to bath water is said to make the bather more attractive to the opposite sex. Even the dried peel of the orange can be used. It is added, with the orange pips, to sachets carried to inspire love, or added to incense burnt to attract wealth.

Oranges can be used to help in decision making. If you are confronted with a difficult choice then take a break and eat an orange. As you enjoy the fruit, concentrate on the problem that faces you, and when you have finished the fruit count the pips. An odd number of pips mean 'yes' and an even number 'no'. You must be careful never to swallow an orange pip, of course, lest an orange tree should grow in your stomach.

Those witches and wizards who drive to their meeting rather than using the more traditional broomstick, will be pleased to learn that orange juice is a suitable substitute for wine in most magical rituals.

There are conflicting interpretations of oranges as they appear in dreams. Some sources state that if you should dream of oranges it is always an unfavourable augury, betokening misfortunes in the future. It may indicate infidelity in a lover and that you should beware of placing implicit trust in casual acquaintances. Alternatively, other sources suggest that it is an excellent omen if you see an orange on a tree in your dreams.

LEMON

Botanical Name: *Citrus limon*
Family: Rutaceae
Common Names: Citron; citronnier; *Citrus medica*; neemo; leemoo; limone; *limoun*
Gender: Feminine
Planet: Moon
Element: Water
Meaning: Discretion; fidelity in love (lemon blossom); zest

Like other citrus plants the lemon can be used in simple charms and spells to win the love and affection of others. Dried lemon flowers and the peel of the fruits are added to love mixtures and sachets. The leaves can be used in potions and teas to induce lust. To provide a suitable gift for a loved one, ensuring their love and as a symbol of your affection, choose a lemon from a tree you have grown from a pip (it might take quite a while, however, to grow a tree big enough to produce fruit). Serving your loved one lemon pie is an alternative way of strengthening their fidelity. If you want to ensure the friendship of a visitor to your home your should conceal a slice of lemon beneath the chair that they will sit on.

Lemon can also be used to make a good luck charm. A small, unripe, fruit should be used, no more than an inch and a half in diameter. It must be as fresh as possible. It must be stuck with as many coloured headed pins as possible (black pins must not be used); this can then be hung up in the house as a talisman.

Lemons are added to healing incense and used in spells for the same purpose. When eaten, lemons help to increase psychic powers and lemon juice mixed with a little water makes a good cleaning agent for magical objects. A wash in lemon juice will remove all negative vibrations, particularly from objects that have been acquired second hand. Lemon juice can also be used in purificatory baths, and is most effective at the time of the full moon.

In the Jewish Feast of the Tabernacles a large lemon, as a substitute for a cone of cedar, may be passed amongst the congregation, so that they might smell it and give thanks to God for the 'sweet odours' that he has given mankind.

A dream of lemons indicates quarrels between husband and wife, or the breaking off of an engagement.

CLEMATIS

Botanical Name: *Clematis vitalba*; *Clematis* spp.
Family: Ranunculaceae
Common Names: Blind love; grandfather's whiskers; hedge vine; maiden hair; old man's beard; traveller's joy; virgin's bower
Meaning: Poverty (evergreen); artifice; mental beauty; intellectuality; 'I pay tribute to your brilliance and cleverness'

The 'virgin's bower' trails over arbours to form a refuge for young ladies. Some say that it was named in honour of Elizabeth I as it was introduced into Britain during her reign. Other sources have stated that the plant gained this name as it sheltered Mary and the infant Jesus during their escape into Egypt.

HOLY THISTLE

Botanical Name: *Cnicus benedictus*
Family: Compositae
Common Name: Blessed thistle

The holy thistle is said to have gained its name because it had a reputation as being a panacea for all ills, including the plague. Whereas most thistles are seen as a blight on the land and on mankind, the holy thistle has been cultivated for its medicinal properties.

The holy thistle can be worn as an amulet to protect against all nature of evil forces. It can be added to purificatory baths, or used in spells created to break curses cast against you.

COCONUT

Botanical Name: *Cocos nucifera*
Family: Palmae
Gender: Feminine
Planet: Moon
Element: Water
Deity: Te Tuna

Myth and Legend

Amongst the Samoan peoples there is a myth that a coconut palm grew at the entrance to Puloto, the world of the spirits. This tree is known as Leosia, 'the watcher'. If a spirit struck against it it would have to return to the mortal realm and relatives would rejoice at their return from the point of death saying that the person had 'come back from the tree of the watcher'.

In China the coconut is called *ye-tsu*, or *yűe-wang-t'ou*, meaning 'the head of Prince of Yűe'. The legend tells that Prince Lin-yi was at war with the Prince of Yűe. He sent an assassin to kill his enemy who beheaded the Prince of Yűe while he was intoxicated. The head was hung from a tree but was magically changed into a coconut, with the place of the eyes still visible in the shell.

The Maori myth relating to the origin of the coconut says that it sprang from where the head of Te Tuna, the eel-god was buried. Te Tuna is a Christ-like deity who was sacrificed to redeem mankind. In a form of transubstantiation the eating of coconuts became akin to partaking of the divine flesh. The unripe fruits were said to represent heaven and the underworld.

Magic and Lore

Although one might expect that coconut would be an aphrodisiac, its use in folk magic has been as a protection. To produce a protective talisman for guarding a property against evil the coconut it is drained of its milk and cut in half. It is then filled with the appropriate mixture of herbs before being sealed up and buried in close proximity to the site to be protected. An alternative is to hang up a whole, fresh, coconut in the building.

The Dyaks of Borneo regard the coconut as the shelter of souls. They specifically 'transfer' the souls of their newborn children into coconut shells to protect them in the first year of their life.

On Fiji and Bali a coconut palm was planted whenever a child was born, the fate of the child being ever after linked to that of the tree. (A similar practice is carried out in parts of Europe, where apple and pear trees are planted.) The tree is referred to as the 'tree of life' and some islanders would never think of eating the coconuts from a tree without first seeking the tree's permission.

The links between childbirth and the coconut can be found elsewhere. In the Philippines, the Baujaus bury the afterbirth in coconut shells and the West African Wanika tribe consider the felling of a coconut palm comparable to matricide, from the belief that each tree embodies a spirit.

The Melanesians believed that someone who broke taboos would be driven to madness, causing them to commit suicide either by starving themselves or by throwing themselves from the top of a coconut palm.

COFFEE

Botanical Name: *Coffea arabica*
Family: Rubiaceae
Common Name: Arabian coffee

Coffee originates in the Middle East. According to an Ethiopian legend, a young goatherd named Kaldi was out minding his herd when he noticed that the goats were energetically cavorting rather than peacefully grazing. Having watched them carefully for a short time he noticed that they were eating the bright red berries off a shrub that grew nearby. His concern for their safety led him to try some of these berries himself and he was amazed at the sensation of alertness they caused. That night he took some of the berries back with him, which he gave to the mullah who lived in his village. The mullah thought that their use might help people to stay awake and alert during the evening religious ceremonies and they are still used to aid wakefulness even today.

JOB'S TEARS

Botanical Name: *Coix lachryma-jobi*
Family: Gramineae
Common Name: Tear grass

In folk medicine Job's tears is used as treatment for sore throats and to aid children when they start teething. The seeds, or 'tears', are strung onto a cord and worn as a necklace. Great care must, of course, be taken if a necklace of this type is to be used to help teething babies. In both cases it is the seeds of Job's tears that are supposed to absorb the pain of the person wearing them.

Job's tears may be carried as amulets in order to ensure good luck. They can be used for making wishes. Seven 'tears' must be held in the hand as the wish is made, and then these must be thrown into running water if the wish is to come true. An alternative suggestion from another source states that you must concentrate on your wish whilst counting out the seven seeds and then carry them with you continuously for a week if you want the wish to succeed.

AUTUMN CROCUS

Botanical Name: *Colchicum autumnale*
Family: Liliaceae
Common Names: Daggers; fog crocus; go-to-sleep-at-noon; kite's legs; meadow crocus; meadow saffron; *mort-au-chein* ('dog killer'); naked boys; naked jacks; naked ladies; naked maidens; naked men; naked nanny; naked virgins; naked

whores; pop-ups; purple crocus; snake flower; son-before-the-father; star-naked-boys; strip-jack-naked; upstarts
Meaning: 'My best days are over'

The name Colchicum gains its name from Colchis, on the Black Sea, where the plant grew freely. This is the place where, according to myth, Jason and the Argonauts found the Golden Fleece. The autumn crocus was much prized by Medea who used it in her magic spells. It was at Colchis that she fell in love with Jason and prepared a potion to restore his youth. Where drops of this potion fell onto the ground, up sprang the autumn crocus.

William Turner, in his herbal of 1568, copies a warning from Dioscorides, that it is important to be able to recognize this plant 'that a man may isschewe it, it will strangell a man and kyll him in the space of one daye, even as some kindes of Tode Stolles do … if any man by chaunce have eaten anye of this, he remedye is to drink a great draught of cowe milke.' Early herbalists were wary of using it with good reason, as it is very poisonous.

MYRRH

Botanical Name: *Commiphora myrrh*
Family: Burseraceae
Common Names: Bal; bola; bowl; didin; didthin; gum myrrh; karan; mirra balsom odendron; morr
Gender: Feminine
Planet: Moon
Element: Water
Deities: Adonis; Astarte; Isis; Mariamne; Marian; Mary; Myrrha; Ra
Meanings: Gladness; 'How bitter and precious is our love'; lessons hard won

It would seem that the myrrh referred to in the early part of the Bible cannot be the same plant as that named in later references. *Commiphora* are natives of Abyssinia and Arabia and not indigenous to Gilead or Israel. It is likely that the early references were to a fragrant resinous gum, collected from the leaves of rockroses, *Cistus salviflorus* or *C. villosus*. This is now usually called ladanum. In New Zealand these rockroses are known as Gallipoli roses as they were first grown from seeds brought back by returning WW1 soldiers.

Myth and Legend

Legends tell the story of a beautiful young maiden called Myrrha who became besotted with her own father. Although he tried to avoid her advances she contrived a situation that allowed her to sleep with him without his being aware that it was his daughter. Filled with anger at her deception he chased her from his bed and sent her into exile, where she spent much of her time weeping. She pleaded for the gods to take pity on her and rescue her from her woes. Their response may not have been quite what she had in mind, because they transformed her into a tree.

> The earth gripped both her ankles as she prayed.
> Roots forced from beneath her toenails, they burrowed
> Among deep stones to the bedrock. She swayed.
>
> Living statuary on a tree's foundations.
> In that moment, her bones became grained wood,
> Their marrow pith,
> Her blood sap, her arms boughs, her fingers twigs,
> Her skin rough bark. And already
> The gnarled crust has coffined her swollen womb.
>
> It swarms over her breasts. It wraps upwards
> Reaching for her eyes as she bows
> Eagerly into it, hurrying the burial
> Of her face and her hair under thick-webbed bark.
> Now all her feeling has gone into wood, with her body.
> Yet she weeps.
>
> The warm drops ooze from her rind.
> These tears are still treasured.
> To this day they are known by her name – Myrrh.
> – Ovid, *Metamorphoses,* translated by Ted Hughes in *Tales from Ovid*

However, she was pregnant by her father and so the tree gave birth to a son. He was named Adonis, and was raised by the woodland nymphs. Myrrha, unable ever to hold her son, continued to weep and her tears of resin are the source of the incense that bears her name.

Magic and Lore

There is a long-established use of myrrh as an ingredient in incense and perfumes. It was also an ingredient in the holy oil that the Jews used to anoint the Ark of the Tabernacles. It is perhaps best known as one of the gifts brought by the Magi to the infant Jesus. The significance of this may stem from its use in the embalming processes of the ancient Egyptians. Plutarch tells us that it was prepared to a magical formula and was used in fumigations and medications as well as embalming. The ancient Egyptians would also have burned myrrh to the god Ra, at noon each day, and in the temples of Isis.

Myrrh is used in incense because of its purifying powers, removing all negativity and bringing peace. The smoke rising from burned myrrh can be used to purify objects such as amulets and talismans. Although it can be burned on its own, it is more often burned with frankincense or other resins. When it is added to other ingredients to make up incense it is supposed to increase the efficacy of the final mixture. It may also be added to a little frankincense and carried in a pouch or sachet in order to heighten the powers of concentration.

BALM OF GILEAD

Botanical Name: *Commiphora opobalsamum*
Family: Burseraceae
Common Names: Balessan; balsam tree; bechan; mecca balsam
Gender: Feminine
Planet: Venus
Element: Water
Meanings: A cure; healing; relief

There appears to be some debate as to the true identify of balm of Gilead. Most authorities say that it was originally a local plant but was later imported from southern Arabia. The early references appear to be to the soapberry tree, *Balanites aegyptica*, whereas the later references are to *Commiphora*.

Myth and Legend

This tree gains its common name from being grown on the slopes of Mount Gilead. Josephus tells us that it originated in Arabia and was taken into Judea by the Queen of Sheba. She presented it as a gift to King Solomon during her visit to his court. Perhaps her intention in so doing was to win the king's affections, as carrying the buds of this plant is said to bring love to the bearer.

Magic and Lore

The buds of balm of Gilead can be steeped in red wine to form a love potion. Carrying the plant is credited with other benefits of which the Queen of Sheba might have been aware. It is another plant that can be used to give protection against the forces of evil, and if her attempt to win over the king failed, she could also use it as a cure for a broken heart.

Medicinal

Balm of Gilead has been used in a number of medicinal applications. However, there is a danger of incorrectly identifying it, as there are several other species called by the same name, including *Abies balsamae, Cedronella triphylla* and *Populus candicans*.

HEMLOCK

Botanical Name: *Conium maculatum*
Family: Umbelliferae
Common Names: Bad man's oatmeal; bad man's whotmeal; beaver poison; break-your-mother's-heart; caise; Californian fern; cart wheel; Devil's blossom; Devil's

flower; gipsy curtains; gipsy flower; hare's parsley; hech-how; herb bennet; hever; Honiton lace; humlock; humly; kaka; kakezia; kecksies; kelk; kesh; kewse; kex; kexies; koushe; koushle; lace flower; lady's lace; lady's needlework; musquash root; Nebraska fern; nosebleed; pick pocket; poison hemlock; scabby hands; spotted corobane; spotted hemlock; stink flower; water parsley; winter fern
Gender: Feminine
Planet: Saturn
Element: Water
Deity: Hecate
Meaning: Perfidy; scandal; 'You will be my death'

Myth and Legend

The generic name *Conium* is derived from the Greek *Konas*, meaning 'to whirl about'. This relates to the sensation of dizziness and vertigo that precedes death if it is ingested. The specific epithet *maculatum* means spotted and refers to the marks on the plant stems. An old English legend tells that these purple streaks are representations of the mark placed on the head of Cain after he had killed his brother Abel.

Magic and Lore

The extremely poisonous nature of hemlock is well known. In ancient Athens it was administered to those criminals found guilty of crimes against the state, as a form of capital punishment. As such it was the method used at the execution of the philosopher Socrates.

The use of hemlock in magic is also well known. It is supposed to be an aid to astral projection and was an ingredient in the witches' flying ointment. The juice of the plant could be used to both purify and empower the knives and swords used during magical rites. The herb itself was used in spells to destroy sex drive. As a virulent poison it could potentially kill more than the sex drive of anyone foolish enough to use it.

LARKSPUR

Botanical Name: *Consolida ambigua* (Synonym: *Delphinium consolida*)
Family: Ranunculaceae
Common Names: Knight's spur; lark's claw; lark's heel; lark's toe
Gender: Feminine
Planet: Venus
Element: Water
Meanings: Flippancy (blue); fickleness (pink); haughtiness (double); agility; ardent attachment; flight of fancy; levity; lightness; 'Read my heart'; swiftness; trifling. The name *Delphinium* derives from the Greek word 'delphis', meaning dolphin

Myth and Legend

In classical myth the plant arose during the battle at Troy: the armour of a brave soldier was given to Ulysses rather than to Ajax. Ajax, in his disappointment, killed himself and from his blood delphiniums sprang up.

Magic and Lore

This pretty blue flower is likened to the lark as it flies so high that it is a mere speck in the blue sky. It is used as a protective herb against various problems. The flowers, it is said, will ward off all venomous beasts, such as snakes and scorpions as well as protecting against ghosts and ghouls. On a potentially more practical level it is supposed to aid eyesight. Looking at a midsummer fire through a bunch of larkspur will ensure that your sight remains sharp, at least until the next midsummer.

LILY OF THE VALLEY

Botanical Name: *Convallaria majalis*
Family: Liliaceae
Common Names: Convallaria; convall-lily; dangle bells; faeries' bells; faeries' cups; Jacob's ladder; innocents; ladder to heaven; lady's tears; lilies and valleys; lily confancy; lily constancy; linen buttons; liriconfancy; little white bells; male lily; may blossoms; may lily; our lady's tears; white bells
Gender: Masculine
Planet: Mercury
Element: Air
Deities: Asclepius; Apollo; Maia; Mercury
Meaning: 'Friendship is precious, talk to me not of love'; 'Let's forgive'; maidenly modesty; return of happiness

Myth and Legend

One old legend, from Sussex, tells how St Leonard fought a great dragon called Malitia in the woods near Horsham. He eventually defeated and beheaded the beast after many hours of combat, but received several grievous wounds during the battle. Throughout the woods, wherever his blood had fallen, lily of the valley sprang up as if to commemorate his epic struggle. The woods, called St Leonard's Forest, were carpeted with this little white flower.

Another legend tells of the love between a lily of the valley and a nightingale. Each year, the nightingale would fly away once the flowers faded on its beloved flower, and would not return until the following May when it bloomed once more. Perhaps this is because the sweet scent of the lily of the valley was supposed to draw the male nightingale from out of the hedge and lead him to his chosen mate. It is from this legend that the sentiment 'return of happiness' was derived.

A legend from the Cherokee of North America tells the story of Little Dawn Bird, who had learned plant lore from her father, Big Tree. One day she set out alone to go into the mountains but her father, worried for her safety, secretly followed her. Believing she was alone Little Dawn Bird marked her way by dropping small white, quartz pebbles onto the leaf litter as she went. After a while, as she grew tired, she settled down and fell asleep. Big Tree continued to watch over her from a distance and when he saw a mountain lion about to attack her, he shot it with an arrow through its heart. Little Dawn Bird awoke and set out to follow the trail she had marked for herself, only to find that the pebbles have turned into lily of the valley ringing their tiny bells to guide her safely home. Since the plant is not native to North America it is more than likely that the story has undergone some changes down the years.

Lily of the valley is, like the hawthorn, a plant of Maia or May, the oldest of the Seven Sisters and the goddess of growth, increase, fields, and spring. The older astrological texts place this plant under the dominion of Mercury, as Maia was the mother of Mercury in mythology. She is also one of the goddesses that Robert Graves identifies with the White Goddess in his book of the same title. As with the 'Christianization' of so many other plants, it became associated with the Virgin Mary and gained the name 'lady's tears. It was supposed to have grown from the tears shed by Mary at the foot of the cross. Its flowering time led to it being linked with Pentecost (Whitsunday) when the Holy Spirit came to the apostles.

Magic and Lore

The whiteness of the flower and its association with the tears of the Madonna, have led to it being seen as a symbol of purity. It is sometimes considered to be the fifth thing a bride should carry on her wedding day. In common with many other white-flowered plants it is usually considered unlucky to bring the flowers into the house. In France, people still exchange gifts of the plant on May Day, in order to ensure that they enjoy good luck throughout the year. In Britain it is said that you should never give a bunch of the plant to a friend, as it will destroy your friendship.

The common name of 'faeries' cups' comes from a resemblance of the flowers to cups that faeries hang up while dancing. The flowers are said to ring when faeries sing and to form ladders for them to use, enabling them to reach reeds from which they can weave their cradles.

Lily of the valley is thought of by some as a sign of death, and it is supposed to be unlucky to accidentally come across the plant in bloom. According to folklore, it blooms on the grave of someone executed for a crime of which they are innocent. In the West Country, it is said to be unlucky to plant a bed of lily of the valley in the garden, as it invites an early death for the gardener, although it is also claimed that planting it protects the home from spectres and evil spirits. The scent is supposed to give people the power to see a 'better world' and to attract nightingales. One piece of advice for the gardener who would have lily of the valley thriving in their garden; it is said only to grow well where its 'husband', Solomon's seal, grows nearby.

Medicinal

The drug convallamarin is derived from lily of the valley. It operates in a similar way to digitalin, in that it slows and strengthens the heart. Although not as poisonous as digitalin, convallamarin can still be toxic, especially to children.

It is a flower to lift the spirits and so will cheer the heart of anyone entering a room in which it is placed. Furthermore its presence is said to be a benefit to the mind and memory of those nearby. Folklore holds that a distillation of the sap, dabbed on the forehead and nape of the neck, restores common sense.

FIELD BINDWEED

Botanical Name: *Convolvulus arvensis*
Family: Convolvulaceae
Common Names: Bearwind; bedwind; bell-billy-clippe; billy-clipper; bine-lily; cornbind; corn lily; devil's garters; Devil's guts; dralyer; earwig; faeries' umbrella; faeries' winecup; field convolvulus; gipsy's hat; granny-jump-out-of-bed; granny's nightcap; hedge bells; hell weed; jack-run-in-the-country; kettle smock; lady's smock; lady's sunshade; laplove; lily; morning glory; old man's nightcap; parasols; robin-run-in-the-field; robin-run-in-the-hedge; ropeweed; ropewind; sheepbine; thunder flower; white smock; willow-wind; withwind; withywind; young-man's-death
Meanings: Repose (blue); worth sustained by judicious and tender affection (Pink); bonds; uncertainty

Many of the common names and, indeed, much of the folklore seems to be common to both the field bindweed and hedge bindweed (*Calystegia sepium*).

The common name of 'young-man's-death' springs from the superstition that if a young woman picked the flowers of the bindweed her boyfriend would surely die. In Shropshire the plant was known as 'thunder flower' as, in common with many other species, it was said that picking the flowers would cause thunderstorms. These tales may have been sufficient to discourage children from collecting the flowers and in doing so damaging the crops amongst which the bindweed grew.

Bindweed may be used to bind spells, in the three days before a new moon. An image formed from clay, dough or any other malleable material would be made and baptized in the name of the person to be bewitched. The image would then be bound nine times, clockwise, with the bindweed stem whilst the following was said, 'I bind (name of person) against (or to) (whatever action is required). So shall it be.' The image would then be buried somewhere that the person targeted would be bound to walk.

TI

Botanical Name: *Cordyline terminalis*
Family: Agavaceae
Common Names: Hawaiian good luck plant; *ki*; Polynesian ti; red dracaena
Gender: Masculine
Planet: Jupiter
Element: Fire
Deities: Kane; Lono; Pele

Ti is grown as a houseplant in Britain, but most people will be unaware of the superstitions that surround it in its native country. Red ti is dedicated to the god Pele, but is usually considered to be the bearer of misfortune. It is said to bring bad luck to the owners of any house where it grows. Green ti, on the other hand, should be grown about the home as a protective plant to ward off evil forces and to bring good fortune. Carrying the leaves of ti, when at sea, will give protection from drowning and keep storms at bay. At home, placing a piece of ti under the bed at night will protect the occupant through the night. Rubbing it against the forehead is also supposed to cure headaches.

CORIANDER

Botanical Name: *Coriandrum sativum*
Family: Umbelliferae
Common Names: Chinese parsley; cilantro
Gender: Masculine
Planet: Mars

Element: Fire
Meanings: Concealed merit; hidden worth

Coriander has been a valued spice for centuries. The Romans introduced it into Britain from the East. Although best known for its culinary uses it has been widely used in magic.

Magic and Lore

On a basic level coriander is thought to have some aphrodisiac properties. When the powdered seeds are added to warmed red wine they make a potion that can be used to inspire love or lust. The spice might also be added into love inducing sachets, or used in love spells. On another level it was believed that it could be eaten by a pregnant woman to ensure that her child would grow up to be a genius. In China, it was thought that coriander was one of the plants that had the power to confer immortality on those who used it.

Medicinal

Coriander also has its place in folk medicine. One folk remedy suggests that wearing it seeds will help to ease the pain of headaches. Turner, in his herbal of 1551, suggests coriander as a treatment for other complaints. He asserts, 'Coriandre layd to wyth breade or barly mele is good for Saynt Anthonyes Fyre.'

DOGWOOD

Botanical Name: *Cornus floridus*
Family: Cornaceae
Common Names: Bitter red berry; box tree; boxwood; budwood; common white dogwood; cornel; *cornouiller à grandes fleurs*; dog tree; eastern flowering dogwood; flowering cornel; flowering dogwood; green osier; *hat-ta-wa-no-min-schi*; *mon-ha-can-ni-min-schi*; Virginia dogwood
Deities: Circe; Cronus; Proteus; Saturn
Meanings: 'I am perfectly indifferent to you' (blossom); durability; 'Our love will endure adversity'

Myth and Legend

In Roman myth Romulus, one of the founding fathers of Rome, marked out the boundaries of the proposed city using his javelin. Having finished he threw the javelin over Mount Palatine. When it struck the earth it took root, growing into a fine tree. This was taken as a good omen for the establishment of the Roman Empire. The hardness of the timber led to it being commonly used for spears and other such weapons.

Magic and Lore

Like so many other plants, the significant time of the year for using dogwood is at midsummer. It is then that you should go out and collect a little of the sap of the plant in a clean handkerchief. Thereafter, so long as you carry that handkerchief with you your wishes will come true. In America the dogwood comes into flower during the third week of May with a reasonable degree of regularity. In consequence it has been used by the Native Americans as an indicator that it is time to start planting their corn.

Dogwood has been accredited with protective powers. The leaves, or small pieces of the wood, may be collected and worn as protective amulets.

CORNEL

Botanical Name: *Cornus mas*
Family: Cornaceae
Common Name: Cornelian cherry
Deitiy: Apollo

The cornel was considered as sacred to the god Apollo in ancient Greece, and was offered on the altar of his temple at times of festival. One such festival was called Cornus, and was held in honour of the god as an appeasement because the Greeks had cut down a consecrated thicket of sacred cornels that grew on Mount Ida.

HAZEL

Botanical Name: *Corylus avellana*
Family: Betulaceae
Common Names: Cobbedy; cobbley-cut; coll; filbeard; filbert; hale; halse; hasketts; hazelnut tree; nuttall; victor nut; witch-halse; woodnut
Gender: Masculine
Planet: Sun (Mercury, according to Culpeper)
Element: Air
Deities: Aengus; Artemis; Diana; Hermes; MacColl; Mercury; Thor
Meaning: Peace; reconciliation

Myth and Legend

The hazel is one of the great trees of mystery and magic. There is a story that it was protected by an elf called Hind Etin. When the May Margaret came to gather the nuts off the tree Hind Etin caught her.

He tied her to the tree by the tresses of her golden hair, and threatened to kill her. At length he relented, allowing her to live so long as she consented to marry him. They had seven children. Even today we can still see the evidence of the May Margaret on the tree as the catkins hang down, like her golden tresses, every spring. In the north of England local tradition has it that the hazel was guarded by two fierce goblins called Churnmilk Peg and Melsh Dick. Perhaps the reason such emphasis was placed on the protection of the tree lay as much in the potential value of the crop of nuts as in its alleged magical properties.

Although there is little written of the hazel in English folklore it was an important tree to the Celts, who regarded it as one of the ancient sacred trees and associated it with beneficial magic and divine blessings. For the druids it was the tree of knowledge and wisdom, of beauty and fertility, of fire and poetry. The hazelnuts, therefore, were viewed as the receptacles of all wisdom, fit only to be eaten by poets and scholars, as they gave skill with words.

Irish folklore contains many references to this tree. In the *Triads of Ireland*, for example, it was recorded that the hazel and the apple were the only two sacred trees whose felling incurred the death penalty. These two trees were also linked with the oak in the *Dinnshenchas*, an early topographical treatise of Ireland. The Great Tree of Munga is recorded as having the qualities of all three within it; possibly to represent wisdom, strength and beauty.

Hazels can be found associated with holy wells across Britain, and pilgrims would hang votive offerings on them. In the treatise of the Great Tree of Munga there is also a description of Connla's Well. This was said to be a beautiful fountain over which grew the nine hazel trees of poetic art, which produced fruit and flowers at the same time, representing wisdom and beauty. When the crimson nuts fell into the well the salmon that lived there ate them, and the number of bright spots on each salmon's back indicated the number of nuts it had consumed. By eating the hazelnuts the salmon knew all things. They are referred to in legend as the 'salmon of knowledge'. Even the gods were forbidden to approach this well. One woman, named Boann, driven by curiosity, broke this taboo. As she neared the well the waters rose up in flood. Baonn escaped the deluge but the waters never returned to the confines of the well. They became the River Boyne and now the sacred salmon swim in the depths of the river searching for their hazelnuts.

In Scottish myth there were two mystical salmon that were the spirits of similar wells and hazel trees. It is no surprise that people wanted to eat the nuts of the tree, which were so highly esteemed that they were called the food of the gods, and their custodians, the salmon. However, just as it was a crime to harm the tree, being their guardians these fish were also sacred, and to harm them was a crime punished by the gods themselves. The old name of Scotland, Caledonia, is said to have derived from *cal dun*, meaning 'the hill of the hazel'. In Scotland a grove of hazel trees was called *calltuin* (or *calltaina*). Various places are called Calton, a derivation of this. The name marks them as entrances into the 'other world'. (There is, perhaps, a link to this in the story that St Joseph of Arimathaea built the original abbey at Glastonbury from hazel branches.)

Other Irish legends relating to the hazel link it to Finn MacColl, one of the last kings of the Tuatha de Dunaan and one of the earliest rulers of ancient Ireland. The name 'MacColl' (MacCuill) literally means 'son of the hazel'. There were a triad of god-kings in the Tuatha de Dunaan, MacCuill, MacCecht ('son of plough') and MacGréine ('son of sun'). This triad split Ireland into three and were set above all else, therefore one third of the country was under the sun, one third under the plough and the final third given over to the hazel. The Tara, the chief seat of kingship in Ireland, was set near to a hazel grove. An old Fenian story tells of how Mear, the wife of Bersa of Berramain, fell passionately in love with Finn MacColl and attempted to seduce him by giving him hazelnuts from the well of Segia that had been bound with love charms. Finn refused to eat them, proclaiming them to be the 'nuts of ignorance' rather than of wisdom and knowledge. He discarded them, burying them deep in the earth.

In other mythologies hazel is also closely associated with wisdom. In Norse mythology the tree is sacred to Thor and called the 'Tree of Wisdom'. Hebrew myth tells us that the staff of Moses was cut from a hazel bush, the same plant that Adam had cut in the Garden of Eden. Wood from the hazel plant was incorporated by Noah into his Ark and was passed, via his son Shem, to Abraham and then Jacob. In some translations of the Bible the tree identified is the almond.

Some legends tell that the staff used by St Patrick, when he banished all snakes and serpents from Ireland, was made of hazelwood. Hazel is still considered by some to be a powerful protection against snakes. One blow of a hazel rod is supposed to be sufficient to kill a snake and a cross made of the wood, laid over the wound, is said to be a sure cure for an adder bite.

Classical myth informs us that Apollo gave the hazel to Mercury, god of intelligence and the messenger of the gods. Mercury used it to calm human passions and improve their virtues. The staff of Mercury is usually pictured as being intertwined by two serpents and topped by wings. It was adopted by Asclepius to become the symbol of the medical profession. In Celtic myth it was Aengus, the god of love, who carried the hazel rod.

In the Christian legend of St Mungo (seventh century) the saint was unable to light the monastery lanterns on the day when it was his duty to do so, as some mischievous or malevolent boys had extinguished all the fires. Mungo walked from the monastery in despair and as he walked he plucked a hazel twig, returning with it into the chapel. There he knelt and prayed for heaven's aid, at which point fire burst from the hazel twig.

Magic and Lore

The hazel has been widely associated with faerie lore, and proper use of the tree was said to allow access into the faerie realm. A seventeenth-century recipe purported to give the power to see faeries. It directed those who wished to use it to:

> First pick wild thyme from the side of a hill where faeries still live. Take a pint of sallet oyle and mix it with rose and marigold water, the flowers of which should be picked in the east. Shake or stir the oyle until it becomes white and then put it in a glass vial, adding buds of hollyhocks, flowers of marigold, the flowers from wild thyme and the buds of young hazels. Then add the grasse of a faerie throne, and allow them to dissolve for three days in the sun.

This mixture could then be stored until it was wanted. The person wanting to see faeries would be anointed with a little of the mixture and would have to drink some. Any failure of this to work is presumably due to the person gathering the ingredients failing to correctly identify a faerie throne when collecting the grass.

Hazel has been valued for centuries as a protective plant. In the legends of Finn MacColl we are told that he had a shield of hazelwood that rendered him invincible in battle. Likewise, nothing harmful is supposed to be able to penetrate a fence or hurdle of hazelwood set around a house. May Day was the traditional time to gather hazelwood that could be used as a protective talisman. It afforded protection against all kinds of different dangers and difficulties, particularly those caused by malicious faeries, which might sneak into the house at night in order to abduct babies or young children. They might also be responsible for the mysterious curdling of the milk in the dairy. The traditional use of hazel for making the hurdles used around a farm may stem in some part from this superstition. In a superstition, that appears to recall the tale of the Hind Etin and May Margaret, it is said that if a girl goes 'a-nutting' on a Sunday she will wed the Devil. It should be pointed out that 'gathering nuts' and 'going a-nutting' have been euphemisms for sexual intercourse.

Another reference to the protective powers of this plant comes from the magical method of protecting people and property. To prevent attack from evil supernatural influences a circle should be drawn around the person or building, using a hazel stick. Nothing supernatural will then be able to cross the line and enter the circle.

In the Highlands of Scotland the hazel is often considered to be unlucky. However, the double hazelnuts, sometimes called St John's nuts or *cnó chomblaich*, were greatly valued as they could be thrown at witches to drive them away or carried as a lucky amulet. Hazel was once commonly put into the bridles of horses as a protective amulet against the mischief of faeries and witches. It was when evil became synonymous with witchcraft that the belief that hazel could protect against the mischief of faeries led to it being used to ward off the malice of witches. In *The Discoverie of Witchcraft* (1584) it is recommended that a hazel wand cut 'upon the Sabbath daie before rising' should be used as a charm against witches and thieves.

Hazelwood was also used as a protection against severe weather, possibly because of its association with Thor, the Norse god of thunder. Hazel twigs could be gathered on Palm Sunday and placed in pots of water around the doors and windows of a house to ward off lightning. These small bunches of twigs, tied with the hair of the householder, might be kept by the hearth to ensure domestic happiness and protection from fire and 'flash'. Three pins made from hazelwood driven into the timberwork of a house were thought sufficient to protect it from fire. Sailors might choose to wear hazel in their hats, when at sea, as it is reputed to protect them against the dangers of the most violent storms.

From ancient times hazel has been the preferred wood for making staffs of power. The obvious example of this is its use for making magic wands, but it was also used for making royal sceptres. In Egyptian and Chaldean records there are references to its use in both, and we have seen that the staffs of Moses and St Patrick and the caudeus of Mercury were made from it. Hazel is still the wood preferred by many dowsers for making their divining rods. For this purpose, it seems, one-year-old wood is the ideal material. Dowsers might also benefit from eating hazelnuts, as they are supposed to improve the powers of divination. Another type of hazel wand can be formed from a twig around which honeysuckle has grown. The resulting intertwined wand is known as a 'honeystick', and the possession of such a wand is supposed to be a sure way to attract women.

It is not only rulers and magicians who may use hazel rods and wands. Pilgrims might also choose to take a hazelwood staff to aid them on the journey. They would retain this all their lives, and it might even be buried with them. In addition to their well-known use as divining rods for locating water, up until the seventeenth century, they were also used for locating treasure, thieves and murderers. The common name 'hazel' is derived from the Anglo-Saxon word *haesl*, meaning 'a baton of office'.

A further use of the wood of this tree is in the 'wishing-cap'. Any wishes made whilst wearing a crown of woven hazel twigs is sure to come true. Another legend relates that wearing such a 'crown' will make the wearer invisible, although this may have arisen from the mixing of two beliefs. It was alleged that a forked hazel twig, the type used by dowsers to locate water, could be used to find fern seeds. It was also believed that fern seeds could be used to render the user invisible. The *Book of St Albans* claims that it is possible to become invisible by carrying a staff 1½ fathoms in length and, in a particular manner, inserting into it a green hazel twig. This takes us back to the magical properties of the hazel staff or wand.

Hazel had long been used in pagan rites; it was a favourite wood for use in the festival fires at the Beltane feasts (30 April), before the coming of Christianity. In country regions farmers would drive their cattle though the fires at Beltane and Midsummer to protect them from bad luck and mischievous faeries. The backs of the cattle would be singed with hazel twigs to protect them against diseases and the evil eye, and the rods that had been used to scorch them were kept and used to drive them in the coming year.

The plant became dedicated to St Philibert, the founder of the abbey at Jumiége on the Seine. It is the *noix de St Philibert*, and hence called a 'filbert'. The feast day of St Philibert is 22 August, when the nuts are ready. Even the ceremonies of collecting the wood to be used for magical purposes were 'Christianized'.

Whether the wood was to be used for dowsing rods, magic wands or for staffs of office it would always be gathered on some auspicious date, such as May Day or Midsummer. The most powerful wands were made from wood gathered when a given day fell on a Wednesday, as that is the day governed by Mercury. Later writers recommended that the hazel be collected on holy days, such as Good Friday or St John's Day, but never on a Sunday as this would attract the attention of the Devil. St John's Day is 24 June and is, therefore, very near to the Midsummer festival. Furthermore, it is not sufficient simply to collect the hazel on one of the great feast days of the year; some specific conditions must be observed in gathering it. The wood should always be collected from 'virgin' branches, that is to say from branches that have no side shoot. When cutting the wood the collector must face the east and only cut wood from the eastern side of the tree. It should be severed from the tree by the single stroke of a magic, preferably golden, sickle. The wood should be gathered on the first night of a full moon, although some authors say it ought to be the last night before the new moon. In either case the freshly cut wood must be 'presented' to the first rays of the rising sun.

Halloween has also been called 'Nutcrack Night' and at this time of year hazelnuts were put to use in love divination. A row of the nuts was placed on the hot embers of a fire, each one being identified with one of the girls present in the room. The girls would then whisper the name of their suitor and if, as the name was said, the appropriate nut 'jumped' then the success of their relationship was assured.

If you love me pop and fly,
If you do not lie and die.

Alternatively a girl might put two or more hazel nuts into the fire and identify each with one of her suitors. Whichever of the nuts 'popped' first or burned brightest indicated which of her suitors loved her the most.

Two hazel nuts I threw into the flame,
And to each nut I gave a sweetheart's name.
This, with the loudest bounce me sore amazed,
That, with a flame of brightest colour blazed.
As blazed the nut, so many passions grow,
For t'was thy nut that did so brightly glow

Another Halloween ritual would have a couple place two hazelnuts onto the hot coals. If the nuts continued to lie together as they burned it indicated that the couple were faithful to each other. However, if one of the nuts moved in the heat it showed that one of the partners was being unfaithful.

In ancient Germany hazel was seen to symbolize immortality. This may be due to its flowering at the end of the winter. At weddings in the Black Forest the pastor conducting the ceremony would carry a wand of hazel wood. It was a medieval symbol of fertility, and in this instance symbolized a happy marriage. It is still occasionally to be seen being carried at weddings or included in the floral decorations. In nineteenth century Devon it was traditional for an old woman to greet a bride with the gift of hazel to ensure fertility, in much the same way as rice (and later confetti) has been used in modern times. Hazel torches might be burned at the wedding festivities to ensure a peaceful, happy and fruitful union.

Hazelnuts can be used in weather forecasting. According to the saying,

If the nut shells are thick, the winter will be bleak.
If the nut shells are thin the winter will be mild.

It is also claimed that a bumper crop of hazelnuts is a sign that the winter will be especially harsh, whereas if the hazelnut crop fails it indicates that rain will follow.

A less pleasant country saying warns that if there is a plentiful crop of hazelnuts then there will be many deaths in the coming winter. It says simply, 'Many nuts, many pits [graves].' In contrast to this dire portent another country saying claims:

Many nuts in the autumn,
Means lots of babies in the spring.

Dreaming of hazelnuts also promises success in love and marriage, with a large and prosperous family. Such a dream is always taken as a good omen as it further signifies good health and long life. Discovering hazelnuts in a dream predicts that you will uncover hidden treasures. Dreaming of cracking and eating hazelnuts indicates that you will achieve riches and contentment following toil.

Medicinal

In Devon a double hazelnut, or two hazelnuts on one stalk, is called a 'loady-nut', and can be used for curing toothache. The crushed kernels of hazelnuts, mixed with a little mead or honeyed water, is an old recipe for a cough medicine. Carrying a hazelnut has been said to be a treatment for rheumatism and lumbago. This relates to the power of the hazel against the mischief of faeries as both of these ailments are supposed to be caused by 'elf-shot'. Eating hazelnuts was believed to have two beneficial effects: firstly it would increase wisdom, and secondly, fertility. Some writers suggest that just carrying the nuts is sufficient to aid fertility.

To treat adder bites a piece of hazelwood was laid across the bite with the words:

Underneath this hazelin mote,
There's a baggoty worm with speckled throat,
Nine double is he,
Now from nine double to eight double,
And from eight double to seven double,
And from seven double to six double.

And so it went on until:

And from one double to no double,
No double hath he.

HAWTHORN

Botanical Name: *Crataegus monogyna*; *Crataegus oxyacantha*
Family: Rosaceae
Common Names: Aglet tree; azzy tree; boojun; bread and cheese tree; cheese and bread tree; cuckoo's bread and cheese tree; faerie thorn; gaxels; hag; hag bush; hagthorn; halves; haw; haw bush; hazels; heg-peg bush; hipperty-haw tree; holy innocents; huath; ladies' meat; mahaw; may; may blossom; may bread and cheese bush; may bush; may flower; moon flower; mother-die; peggall bush; quich thorn; quick; *skayug*; *sgeach*; *skeeog*; thorn; tree of chastity; whitethorn

The berries are variously called: agald; agarve; agasse; aggle; aglet; agog; bird's eegle; butter haw; cat haw; chaw; chucky cheeses; cuckoo's beads; eglet; hag; hagall; haggil; halve; harve; harsy; haw; hawberry haw gaw; hazle; heethen berry; heg-peg; hip-haw; hipperty-haw; hogarve; hogazel; hogberry; hoggaw; hog-gosse; hog-haghes; hog-haw; may fruit; pigall; pig berry; pig-haw; pixie pears
Gender: Masculine
Planet: Mars
Element: Fire
Deities: Blodeuwedd; Cardea; Eris Flora; Hymen; Maia; Olwen
Meanings: 'Despite your answer I shall strive for your love'; hope; marriage

> Mark the fair blooming of the Hawthorn Tree,
> Who, finely clothed in a robe of white,
> Fills full the wanton eye with May's delight.
>
> – William Browne, *Brittania's Pastorals*

Myth and Legend

In various parts of the world the character of the fertile Earth Mother takes on different guises. Examples include the fertility goddesses Cardea and Hymen. The altar in the temple of Hymen was lit with torches of hawthorn as a symbol of 'present happiness and future hope'. Cardea was the goddess who presided over childbirth and protected infants. Her symbol was the hinge of the door of the year. Her consort was the two-headed god Janus, who opened the year and presided over the start of all new things.

Another spring goddess associated with the hawthorn is Olwen. She is identified as the daughter of Yspaddeden Pencanr, a wild man, also called the 'Giant Hawthorn'. What immediately ties Olwen to the cult of the White Goddess is another name used for her, 'She of the White Track'. This is because white trefoils were said to spring up wherever her feet trod. Blodeuwedd, the goddess of spring in Welsh myth, was also linked to the hawthorn. She was created magically by the sorcerer Gwydion as a wife for LLew-LLaw Grywffes, the son of Arianrhod, the sun goddess. She was created from nine spring flowers, and the Queens of the May at the spring festivals are dressed in flowers to represent her.

Hera, one of the great goddesses of classical myth, was the mother of Ares, and his twin sister Eris is another deity linked to the hawthorn. The Greek Ares is better known under his Latin name of Mars, god of war. Hera is supposed to have conceived these children after having touched a hawthorn bush. However, perhaps the most instantly striking Earth Mother goddess associated with the hawthorn is Maia, from whom the name May, for both the plant and the month, is derived. Maia was the mother of Hermes and her month was a period of purification in readiness for the summer rites. Sexual intercourse was forbidden, temples were swept out, literally 'spring cleaned' and old clothes were worn until this unlucky month had passed.

In Scandinavian mythology we are told that the thunder god, Thor, created the hawthorn with a single bolt of lightning. It was therefore one of nine timbers to be used for the funeral pyres of fallen warriors. In German traditions it was used in funeral pyres as the burning of hawthorn was supposed to aid the ascension of spirits into heaven.

As with other plants the stories of pagan and Christian origin have become intermixed. There is, for example, a Judaeo-Christian legend that a species of hawthorn, *Crataegus pyracantha*, was the burning bush from which God spoke to Moses on Mount Horeb. *Pyracantha*, of course, translates as 'fire thorn'. Another tale relates that another species, called *Albespyne*, was used for the crown of thorns placed on the head of Christ prior to the crucifixion, the red haws, presumably, being the drops of the saviour's blood. Later, when pagan imagery was suppressed, the use of hawthorn was considered blasphemous and the plant was said to be under the influence of the Devil.

This Christian myth is not the only connection that the hawthorn has with crowns. Henry VII adopted hawthorn as a heraldic device. Legend has it that during the battle of Bosworth Field he found a small crown, which had come loose from the helmet of Richard III, caught on a hawthorn bush. This was an omen that he would succeed in his quest to be the ruler of the country. It gave rise to the saying:

> Cleave to thy crown,
> though it hangs on a bush.

In Scotland it is still used as a heraldic device as it is the badge of the Ogilvies.

Magic and Lore

The flowering of the hawthorn marks the time of year when winter gives way to spring. For the gardener and farmer it is always thought of as a hedging plant, and this is reflected in some of the common names of the tree, names such as hagthorn and the German *hagedorn*. Another commonly used name is May and it is indeed a May Tree, but not the only one. Sycamore, birch, holly, hazel and others have all been used in the May festivals at one time or another. Maypoles could be made from, or decorated with, branches of hawthorn loaded with blossom. The link with the May Day fertility rites, such as the erection of a Maypole, sets the hawthorn very firmly in place as a tree of the pagan White Goddess. May is the month of love after the cold winter days, and it might be from this that hawthorn is linked with lovemaking.

Hawthorn is traditionally symbolic of love, fertility and betrothals. Its flowers have a heavy musky smell. Some people find the scent unpleasant, whilst others suggest that it is potentially erotic and reminiscent of sexual perfumes. In France there is a tradition of placing hawthorn flowers in the bedroom windows of young girls on May Day. Under the old calendar, 1 May would have been thirteen days earlier than now and the flowers of the hawthorn would have been fully open. This tradition undoubtedly related to the power of the plant to increase fertility.

Another situation in which this attribute is applied is in the inclusion of hawthorn blossom in a bride's wedding bouquet. This, in turn, may be derived from far older practices, as in ancient Athens its timber was used for the torches that were burned at weddings, whilst brides would be crowned with chaplets of the blossoming twigs. Young maids would carry boughs of flowering hawthorn at the wedding of a friend, a role similar to that of the bridesmaids at a modern wedding. The wedding altar could be decked with sprigs of hawthorn as the flowering shoot symbolized the couple's blossoming future, edged with thorns. In Celtic Britain weddings often took place at Beltane, when the hawthorn would have been in full bloom.

Paradoxically, having said that hawthorn flowers are usually thought of as promoting fertility, the plant can also be symbolic of cleansing and chastity. This clearly links to the observation of abstinence during the month of May and the cleansing of the temples of Maia. This seems to have led to the superstition that placing the leaves of the plant beneath a mattress or scattering them about the bedroom, will enforce the chastity of any woman sleeping there. One other connection between the flowering hawthorn and fertility comes in an old country rhyme that says:

> The fair maid who, on the first of May
> Rises at the break of day,
> To wash in the dew of the Hawthorn Tree,
> Shall ever after handsome be.

Before 1752 there was a tradition, in the villages of Northamptonshire, whereby a blossoming hawthorn branch would be planted in the ground outside the home of the prettiest girl in the village on 1 May. However, after the change in the calendar it was difficult to find the plant in flower on 1 May. In Huntingdonshire, on May Eve, a girl might hang a spray of hawthorn on a signpost and just leave it there overnight. The following morning, if the wind had blown it round, it would indicate the direction from which her future husband would come. If it had blown down she would never marry.

Perhaps even the scent of the flowers is not as wholesome as it may first seem. The hawthorn is pollinated by flies, and its flowers smell of carrion to attract them. Some writers have described the plant as smelling of the plague.

It was not only for use in fertility rites that hawthorn was collected; several superstitions identify it as a protection against witches and faeries. It is one of the three 'faerie trees' of Britain and like the other two, the oak and the ash, it is a sacred tree. A grove made up of these three trees is supposed to be the prefect habitat for faeries and, therefore, the perfect place to see them.

This is a tree of mixed fortunes. It brings good luck and prosperity to the owner of land on which it grows but, as a plant of the White Goddess, it will cause a year of misfortune and illness for those foolish enough to bring it into the house. Misfortune is sure to befall anyone who damages the tree, cuts wood from it or cuts it down. Taking any part of the plant into the house will bring misfortune on all who live there especially, according to a Suffolk superstition, if it is taken into the bedroom. The very least problem is that it will bring rain and the worst is that it will cause the death of a child or mother. In the early twentieth century a construction firm ordered the felling of a faerie tree at Down Patrick, in Ulster, but the foreman had to undertake the job himself as all his workers refused. It is claimed that when he dug out the roots lots of white mice, supposedly the faerie folk, ran out. Whilst the foreman was clearing away the spoil from the site a horse shied and crushed him against a wall, resulting in him losing one of his legs. As recently as 1982 workers at the ill-fated DeLorean car factory in Northern Ireland claimed that the company's problems were a result of a faerie tree having been disturbed when the factory was built. Another tree was bought and planted, with due ceremony, but apparently without appeasing the faeries, as the company eventually failed.

When pieces of hawthorn are to be collected it has to be done on a particularly auspicious date, and the person gathering them should be fasting. In Shropshire it was cut on Holy Thursday. Some authors say that the one safe day to collect wood for the tree is on Ascension Day, and others on Palm Sunday, but it would be fair to assume that the original, pre-Christian, day was May Day. If collected on this day a piece of the wood could be safely taken into the home and hidden in the loft to protect the house from being struck by lightning and to safeguard the household from witches, spirits, faeries and storms. If you do not want to risk taking the wood inside, and thereby insulting the White Goddess, you might plant a hawthorn in your garden as this will also serve to protect the house from the worst effects of storms and lightning strikes.

The suggestion that hawthorn can be a protection against lightning can be found in the old rhyme:

Beware the oak,
It draws the stroke,
Avoid the ash,
It courts the flash,
Creep under the thorn,
It will save you from harm.

Sir John Mandeville summed this up in *The Travels of Sir John Mandeville* with the words:

> And ye shall understand, that our lord jesu, in that night that he was taken, he was led into a garden; and there he was first examined right sharply; and there the Jews scorned him, and made him a crown of the branches of albespine, that is white thorn, that grew in that same garden, and set it on his head, so fast and so sore, that the blood ran down by many places of his visage, and of his neck, and of his shoulders. And therefore hath the white thorn many virtues, for he that beareth a branch on him thereof, no thunder ne no manner of tempest may dere him; nor in the house, that it is in, may no evil ghost enter nor come unto the place that it is in. And in that same garden, Saint Peter denied our lord thrice.

When considering the connection between the hawthorn and witches there are some apparent anomalies. The wood was seen as being able to keep evil forces away. It was once believed that on May Day the sun rose extra early and that on May Eve and May Day witches and faeries were especially active. Milk, butter, cheese and other farm produce were very likely to be stolen by them. In Scotland and Ireland the rowan was accepted as the ideal protection against the malice and mischief of these evil forces. In England and France it was the hawthorn. A bough of hawthorn might be hung above the main door into the house or other building to prevent anything evil from entering, and obviously a hawthorn hedge would have more benefits than simply being a physical barrier to keep livestock in a field.

Other sources claim that witches would prefer to meet where the hawthorn grew and even that the trees themselves were really witches, transformed by magic. In Ireland it was believed that the faeries lived in, or under the hawthorn trees, and that the tree might bleed, or scream, if it was cut. Such trees were called 'gentle bushes', 'lone trees', 'faerie trysting trees', or simply 'faerie trees'. They were frequently to be found growing on mounds or barrows, or at crossroads and other favourite places for pagan altars. Just sitting under a thorn tree could, however, be potentially dangerous as it might allow the faeries to gain power over you. It is reasonable to think that these ideas came after the widespread acceptance of Christianity in Britain. Those who chose to follow pagan rites would, thereafter, be seen as witches and so be thought of as evil.

Like the hazel, the hawthorn may often be found growing close to holy wells, the thresholds to other realms. Here they may be festooned with votive offerings, usually ribbons, placed by pilgrims. It is said that such a hawthorn stood over St Patrick's stone on an island in the River Shannon. Dew which collected in a hollow in the tree was supposed to have great healing powers. St Brigit's Well in Cork was also said to collect dew falling from a faerie thorn that grew over it.

In magic the power of a plant used in a spell or as an amulet is increased by weaving it into a shape, as is the case with corn dollies. Hence, the effect of the hawthorn would be greatly increased if it were shaped into a crown. It has been claimed that it is because of being the 'crown of thorns' that the wood of the hawthorn is useful as an amulet against evil. In Herefordshire a hawthorn globe was fashioned each year in farmhouse kitchens. It was made early on New Year's Day and then hung up until the following New Year. At about 5 o'clock in the morning on New Year's Day it was taken down for the ritual of burning the bush. The old globe was taken out into the fields, often the field that was to be sown first in the coming season, and ceremonially burnt by the men, while the women made up the new globe that would ensure a good harvest in the coming year.

There is a traditional saying:

Cast not a clout
'Til May be out.

Many have taken this to mean that you should not put away your winter woollies until the end of the month of May. The 'May' in question however, is the hawthorn, and one explanation of the rhyme suggests that it was believed that hawthorn flowered after the last frost of the winter. Another possibility is that this is a reference to a taboo on wearing new clothes before the hawthorn comes into flower, so as not to insult and risk the fury of the White Goddess. The saying most likely dates from the requirement, amongst the followers of Maia, to wear old clothes until after the end of May and the cleansing of the temples.

There is another rhyme that links the hawthorn to cold weather:

Many haws, many sloes,
Many cold toes.

– *Norfolk Garland*, 1872

The suggestion that a bumper crop of fruits indicates that there will be a severe winter has been applied to many different fruiting plants.

If hawthorn should appear in your dreams it is an indication that you will find a new lover.

Medicinal

With many of the other plants in this book there have been beliefs that the plant, or some part of it, could be used to treat specific ailments. With the hawthorn the opposite is true. Some writers claim that it was once thought that eating the haws, or fruits, of the thorn would cause jaundice. However, infusions made from the leaves were used to remedy insomnia, and decoctions of the flowers prescribed for treating acne and facial blemishes.

William Coles, in *The Art of Simpling* (1656) suggested a treatment for the wounds made by thorns. This consisted of a cloth soaked with the distilled water of hawthorn, which was to be applied as a compress to 'any place where into thorns, splinters etc. have entered and be there abiding, it will notably draw forth, so that the thorn gives a medicine for its own prickling'.

GLASTONBURY THORN

Botanical Name: *Crataegus monogyna biflora*
Family: Rosaceae
Planet: Mars

The legend of the miraculous thorn tree growing in the ruins of the medieval abbey at Glastonbury which flowers on Christmas Day, seems to have first appeared in Hearne's *History and Antiquities of Glastonbury* of 1722. According to the legend, Joseph of Arimathaea travelled widely following the death of Christ, eventually arriving in Britain. He landed on the west coast, on the Isle of Avalon, on 6 January (old Christmas Day), to spread the story and teachings of Christ. His long journeys had wearied him and he feared that his mission in this far corner of the Roman Empire would prove to be a failure. When he reached Wearyall Hill (now called Worrel Hill), he stopped to rest and lent heavily upon his staff, his weight driving the tip into the soft ground. His followers settled around him to rest a little and to pray. While the group rested the staff, which had been cut from a hawthorn tree many years before, began to develop buds and shoot. This was taken as a divine sign that the new faith of Christianity would also take root and flourish in Britain. When the calendar changed in 1752 crowds turned up to see whether the flowering time of the thorn had changed; it had not.

The Isle of Avalon is now a part of mainland England and when Glastonbury Abbey was built the hawthorn tree which had grown from Joseph's staff was carefully dug up and brought into the abbey precincts. The most marvellous attribute of this tree, indicating its miraculous origins, is that it flowers twice in each year. In addition to the normal flowering time in May it flowers again at Christmas. Perhaps as an embellishment to the original story it was said that Joseph of Arimathaea's staff was cut from the same tree that provided the crown of thorns used at the crucifixion. It is possible that the attribution of religious overtones to this tree was an attempt to overcome its pagan significance.

Apparently there was an attempt during the reign of Elizabeth I to cut the tree down. This must have failed as in the reign of James I there was a great demand for cuttings from the tree, and these changed hands for considerable sums of money. In view of the story of its origin the tree is clearly marked as being sacred. Anyone who damages it or tries to cut it down can therefore expect nothing but misfortune. This original tree was said still to be growing at the time of the English Civil War. When a soldier of Cromwell's army went to cut it down splinters of the wood hit him in the eye and blinded him. 'He was well serv'd for his blind Zeale, who going to cut doune an ancient white Hauthorne-tree, which, because she budded before others, might be an occasion of Superstition, had some of the prickles flew into his eye, and made him Monocular.' (James Howell, *Dodona's Grove*)

CLOTH OF GOLD

Botanical Name: *Crocus augustifolius*
Family: Iridaceae

The cloth of gold, if it is gathered correctly, will give those who wear it the ability to understand the languages of the birds and animals. Those who seek to gain these powers must clothe themselves in white and go barefoot. Furthermore an offering of bread and wine must be left when the plants are gathered.

SAFFRON

Botanical Name: *Crocus sativus*
Family: Iridaceae
Common Names: Autumn crocus; crocus; karcom; krokos; kunkuma; saffer; Spanish saffron
Gender: Masculine
Planet: Sun
Element: Fire
Deities: Ashtoreth; Eos; Thoth
Meanings: Beware of excess; cheerfulness; marriage; mirth; smiles; 'I rejoice in you'

Although there are more than 80 species of crocus the early naturalists recognized only two, *Crocus sativus* and *Crocus vernus*; all others were thought to be forms of these two. It must be presumed, therefore, that the folklore relating to the crocus can be applied to almost any species.

Some early writers suggested that the words 'crocodile' and 'crocus' shared the same origin. Thomas Fuller, in *The Worthies of England* said of saffron: 'In a word, the sovereign of Saffron is plainly proved by the antipathy of the crocodiles thereto: for the crocodile's tears are never true, save when he is forced where crocus groweth (whence he hath his name crokodeilos, or saffron bearer) knowing himself to be all poison, and it all antidote.'

The ancient Greeks and Romans knew and valued the spice saffron, and the saffron crocus is still widely cultivated throughout the Mediterranean region. The yellow spice is obtained from the stigmas of the flower. To obtain half a kilogram of the spice requires between 60,000 and 100,000 flowers. In consequence, the spice is incredibly expensive. It was once grown in Essex near Saffron Waldron, which even today bears its name. However, the majority of the saffron used in Britain has always been imported.

Appropriately the 'royal' colour of ancient Greece was yellow. At one time the emperors of ancient Rome would have saffron scattered in the streets before they walked there, presumably because of its sweet fragrance. The Greeks and Romans also adopted the saffron crocus as a symbol of marriage. It was said of brides, that they 'wear the bridal saffron', and their wedding beds might also be bedecked with the flowers. This, perhaps, indicates that the plant was thought of as being able to increase fertility. Certainly it appears in such a role elsewhere. The Phoenicians ate a crescent-shaped cake made with saffron as part of their adoration of Ashtoreth, goddess of fertility and of the moon. A direct comparison can, of course, be drawn between Ashtoreth and the White Goddess described by Graves in his book of the same name. In Persia pregnant women would commonly wear a small ball formed from saffron at the pit of their stomachs. This, they believed, would ensure the speedy delivery of their child.

In Ireland there was a superstition that a little saffron should be added to the water in which bedding was to be washed, as it would strengthen the arms and legs of those who slept on it. In magical usage saffron is still added to the water used for ritual handwashing prior to conducting healing spells.

This is not the only use of saffron in magic, nor its sole use as a medication. Drinking an infusion of it is supposed to give the power of divination, and it is supposed to dispel all melancholy. One early author suggests that care should be taken when using it, however, as over-indulgence might lead to death from 'excessive joy'.

You might choose to keep a little saffron in your kitchen cupboard, even if you never use it in your cooking, as it is claimed that having some saffron in the house will prevent lizards from entering. You might opt to have the saffron crocus rather than the spice, as you could then fashion for yourself a chaplet of the flowers to wear when you go to parties; this was believed to prevent inebriation.

During the reign of Henry VIII laws were passed to prohibit the dyeing of bed linen with saffron, as the dyed sheets were not washed frequently enough.

In some parts of Cornwall there was a superstition amongst fishermen that saffron would be the cause of bad luck. If saffron cakes were taken on board fishing boats it would spoil any chance of a catch.

The juniper has been known as saffron in some parts of England and there has also been confusion between this crocus and the safflower, or false saffron, *Carthamus tinctorius*.

CROCUS

Botanical Name: *Crocus vernus*
Family: Iridaceae
Gender: Feminine
Planet: Venus
Element: Water
Deities: Britomartis; Juno; Jove
Meanings: 'Abuse not'; youthful gladness

Myth and Legend

In classical mythology there is a story of a youth named Crocus, the son of Europa, who was killed whilst playing quoits with the god Mercury. The plant that we know by that name sprang up for where the blood of the young man was spilt. A different tale tells that he loved a beautiful shepherdess called Similax. She, however, did not care for him and so he pined for her and died of a broken heart. The gods, recognizing the cause of his death, dipped him in 'celestial dew' and transformed him into the flower that now bears his name. Similax was ultimately transformed into a yew tree.

The crocus was dedicated to St Valentine, a herbalist and healer who lived under Roman rule in the third century. One day the local jailer brought his blind daughter to him and asked whether she could be healed. Over several weeks, while Valentine applied waxy ointments to the girl's eyes, he befriended her and took her for walks in the countryside. There the girl would pick bunches of crocus for her father. During an uprising in the streets of Rome the Christians were persecuted. Valentine was arrested and jailed. Whilst being prepared for execution on 14 February 270, he wrote a note to the jailer's daughter. When she opened it out fell a crocus, which she was immediately able to see. The note said simply 'from your Valentine'. Thereafter, each year crocuses are supposed to come into flower on this date at dawn.

Magic and Lore

The ancient Egyptians used crocus as an ingredient in incense. It was burned, mixed with alum, in a censer. Where this was done anyone present who had been robbed would be able to see a vision of the thief.

Although we usually cultivate the crocus for its ornamental beauty there are other benefits to be gained from growing it in our gardens. Tradition has it that love and affection will be attracted to those who live where the crocus grows.

A dream of crocus indicates dangers in love and warns that you ought not to trust the dark man to whom you are attracted.

COMMON MELON

Botanical Name: *Cucumis melo*
Family: Cucurbitaceae
Common Name: Muskmelon

There are several species of melon in cultivation today, and many have been used throughout Asia since ancient times. The Egyptians, Greeks and Romans were all familiar with it. Indeed, according to some writers the 'cucumber' referred to in the Bible was probably more correctly a reference to the common melon.

Dreaming of melons has a significant meaning. If a young woman dreams of them it indicates that she will marry a person of foreign origin and live with him in a foreign land. If a young man sees them in his dreams it shows that he will marry a wealthy foreign woman and have a large family. Unfortunately, however, they will die young. When a sick person dreams of melons it is a sure sign that they will make a full recovery.

CUCUMBER

Botanical Name: *Cucumis sativus*
Family: Cucurbitaceae
Common Names: Cowcumber; gherkin
Gender: Feminine
Planet: Moon
Element: Water
Meaning: Extent

The cucumber has been in cultivation for some 3,000 years. Originally it would have been the size of a gherkin, rather than the type that we are used to today. It would have been this type that, according to Pliny, was served daily to the Emperor Tiberius and the same sort that is mentioned in the herbals of the sixteenth century. Its cultivation began in the East.

It may be something of a surprise, considering some modern references, that it has traditionally been thought to hinder lust rather than promoting it. It is true that the seeds were eaten as an aid to fertility, but the modern cucumber has no seed in it. One use of cucumber you might care to try is as a cure for headaches. It has been suggested that binding a section of cucumber peel to the forehead will relieve them.

Dreaming of cucumbers is usually taken as a good omen. For an invalid it denotes that they will make a quick recovery. For lovers it indicates that they will soon marry. For those who have not yet found their true love it means that they will soon fall in love. For the trader or business person it is a sign that their venture will prosper. And for a sailor it shows that they can expect a pleasant journey.

GOURD

Botanical Name: *Cucurbita* spp.
Family: Cucurbitaceae
Gender: Feminine
Planet: Moon
Element: Water
Meanings: Bulk; extent

In various parts of the world gourds are dried, emptied and filled with dried beans to form a rattle; this is not a child's toy, however, but to scare away evil spirits. Alternatively the top may be cut away, the contents removed, and the shell filled with water to form a scrying dish. Pieces of dried gourd can be kept and carried in purses or pockets as amulets to ward off evil forces. Whole gourds might be hung outside the main entrance to a house in order to protect those inside from fascination and the evil eye.

PUMPKIN

Botanical Name: *Cucurbita maxima*
Family: Cucurbitaceae
Gender: Feminine
Planet: Moon
Element: Water
Deities: Chicomecohuatl
Meaning: Bulkiness

Myth and Legend

The Taino people of Puerto Rico blame the pumpkin for making their land an island. They claim that it was once a mountain, which rose above a vast plain. Magic seed was planted on the mountain slopes and produced on verdant forest. From the midst of this grew a vine on which there was a beautiful golden flower. In due time the flower withered and the fruit formed: a pumpkin. A while later two men came across it and began to argue over who should have it, but in their struggle it was dropped and rolled away only to shatter on a rock. Water burst out of the pumpkin and so flooded that area that the mountain, Puerto Rico, became an island.

Magic and Lore

The pumpkin is a plant from the New World. The ancient races of South America would offer human sacrifices to ensure that their pumpkin crop would thrive. Along with peppers, wreaths of corncobs, roses and various types of seed, they were used to decorate the temples of the Aztec maize goddess Chicomecohuatl. In Madagascar, it was believed that any evil that afflicted a person could be transferred into a pumpkin, which could then be carried away and smashed in order to be rid of them.

The pumpkin came to people's attention in the Quest through the story of Cinderella, which was based on a tale from *Histoires ou contes du temps passé* by Charles Perraults (1697). More recently it has been the centre of attention at many horticultural shows as competitions are run to see who can grow the largest specimen.

The widely spread tradition of placing candles inside hollowed out pumpkins at Halloween has its origins in Ireland and was taken to America by Irish immigrants. It came about from the tale of jack-o'-lantern and originally a root vegetable was used (see under Turnip). Usually a face is carved out of the pumpkin so that the light shines out, supposedly to frighten away malicious faeries and witches.

Medicinal

American Indians have eaten pumpkin seeds to expel worms and consumed the whole squash as a treatment for snakebites.

CUMIN

Botanical Name: *Cuminum cyminum*
Family: Umbelliferae
Common Names: Cumino; Cumino Aigro
Planet: Mars
Meaning: Cupidity

Cumin is accredited with the power of retention. Amongst the ancient Greeks it symbolized cupidity and one avaricious character, Marcus Aurelius, was nicknamed 'Cumin'. Misers, it was said, must have eaten cumin. This power of retention extends to objects, and any item that contains the spice will be protected from theft. Unfortunately, for most people this now means that only their spice rack is safe from the attention of burglars! In parts of Germany and Italy cumin would be added to the dough when making bread in the belief that it would prevent it being stolen by wood spirits. An alternative meaning of retention is fidelity, and cumin seeds were used in love spells. However, the seeds could also be added to wine to form a potion to promote lust.

On the whole this is a plant of good fortune. The flowers might be worn or carried by brides, not to promote fidelity or fertility, but to ward off negativity. Carrying the flowers is said to impart peace of mind. An incense of cumin mixed with frankincense can be burned in order to drive evil forces away.

Another use of cumin might be considered to be something of a cheat. The followers of the rhetorician Porticus Latro smoked cumin seeds in order to achieve pallor to their skins. This was supposed to be consistent with many hours shut away studying.

One odd instruction for gardeners is that they must curse the seeds of cumin as they sow it. If they do this, it will ensure a bumper crop.

CYPRESS

Botanical Name: *Cupressus sempervirens*
Family: Cupressaceae
Common Names: Italian cypress; tree of death
Gender: Feminine
Planet: Saturn
Element: Earth
Deities: Aphrodite; Apollo; Artemis; Ashteroth; Astarte; Cranae; Cupid; Diana; Hebe; Heqet; Hercules; Jupiter; Mithras; Osiris; Pluto; Zoroaster
Meanings: Death; despair; eternal sorrow; mourning

Myth and Legend

This is a tree of grieving. It is named after Cyparissos, son of Telephus, who was greatly loved by Apollo. Cyparissos accidentally killed one of Apollo's favourite stags when playing with it on sacred Ceos. Overwhelmed with grief at his mistake, he killed himself, appealing to the gods to make him suffer eternal remorse. He was transformed into the cypress tree, a symbol of mourning.

Another legend says that the daughters of Eteocles were carried away by goddesses in a series of whirlwinds. After a very long time of revolving in seemingly endless spirals they fell back to Earth, landing in a pool of water. Gaia, goddess of the Earth, having pity on them, transformed them into cypress trees.

The cypress tree was held sacred to the Persian magicians and was planted near to their temples as the form of the tree was said to resemble a flame. Persian myth tells of two huge cypress trees that grew in Khorasan, one at Kashmar near Turshiz and the other at Farmad near Tuz. It was believed that these plants grew from shoots brought from Paradise by Zoroaster. It was widely planted and venerated as a reminder of the loss of Paradise. One tree, planted near to the tomb of the Persian Cyrus the Great, was said to leak blood every Friday. This was the object of particular attention as Friday is the Muslim holy day.

The imperishable nature of the wood led to it being used in the construction of the Phoenician ships, and for the costly chests used in the Temple of Diana at Ephesus. The Greeks carved the figures of the gods that are found in some temples from cypress wood. It has also been suggested that the pillars at the Temple of King Solomon, Noah's Ark, Cupid's darts, the club of Hercules and the cross on which Christ was crucified were all made of cypress wood.

Magic and Lore

In many parts of Europe the cypress is the traditional graveyard tree, much as the yew is in Britain. Like the yew it was once considered to be sacred and was worshipped by the Minoans. Its use migrated westwards from the Indus valley to Lebanon and eventually to Cyprus, the island that bears its name.

> Dark trees, still sad when other's grief is fled,
> The only constant mourner o'er the dead.
>
> – Byron, *The Giaour*

It was traditional to plant a cypress on or near to graves and to place one at the front of a house in the event of a death. The Greeks and Romans called it 'the mournful tree' and made it sacred to the gods of the underworld. In ancient Egypt the timber was used for making coffins and in Greece the coffins of Athenian heroes were manufactured from the wood as it was considered to be a symbol of immortality. A grove of cypress trees was preserved on the Acropolis and another, at Phlius in the Peleponnese, was a place where fugitives from justice sought sanctuary. Escaped prisoners would hang their chains on the trees, as they had no further use for them. The Chinese chose to grow both cypress and pine on or near graves in order to strengthen the souls of the dead and to preserve the bodies from decay.

In much the same way as rosemary used to be used, sprigs of cypress might be worn at funerals, especially those of close friends or relatives, in order to allay the grief felt. As with rosemary the pieces of cypress would be tossed into the grave in order to give the deceased luck and love.

Another characteristic that the cypress shares with the yew is that, in addition to being a tree of death and grief, it is emblematic of rebirth and life, owing to its evergreen nature and the fact that the timber retains its red colour long after being cut. It was considered to be imperishable and indestructible. As a symbol of eternity it was thought that carrying small pieces of the wood would increase longevity. In Persia it was dedicated to Mithras, the god of light and wisdom. It thus became associated with fire and a symbol of the sun.

Its properties also made it a symbol of the resurrection. From this it was not a big step to its use in healing magic. A suitable branch was cut very slowly taking about three months to remove it from the parent tree. The resulting wand would then be passed over an invalid and the afflicted areas of their body touched with it. The tip of the wand had to be plunged into a hot fire after each area of infection was touched, in order to cleanse it. Pieces of foliage and cone, or wax cones containing the essential oil, could also be burned at healing rituals. The essential oil of the cypress is still valued for its invigorating qualities.

Another rather odd belief was that a mallet made from the timber could be used to discover thieves, although how this was done is now lost. It was once believed that growing it near to the home would bring blessings on the whole household and protect all inside.

Dreaming of cypress is usually taken to indicate that obstructions and afflictions can be expected in business undertakings.

ZEDOARY

Botanical Name: *Curcuma zedoaria*
Family: Zingiberaceae
Common Names: Olena; tumeric

Zedoary is used in Hawaiian magic for protection and purification. When used for protection it is scattered on the floor, or around a magic circle, to prevent the approach of evil spirits. To purify an area before the practice of some rite, an infusion of zedoary, mixed with salt, was sprinkled over the whole site.

DODDER

Botanical Name: *Cuscuta epithymum*
Family: Convolvulaceae
Common Names: Adder's cotton; beggar weed; clover devil; devil's guts; devil's net; devil's thread; epiphany; faerie hair; fireweed; hairweed; hairy bind; hell bind; hellweed; hellwind; lady's laces; love vine; maiden's hair; mulberry; red tangle; scald; scald-weed; strangle tare; strangle weed; witches' hair

Gender: Feminine
Planet: Saturn
Element: Water
Meaning: (Dodder of thyme) business

Clover is a plant of good in Christian myth: St Patrick even used it to explain the concept of the trinity. Dodder, according to folklore, was created by the Devil to be its counterpart. It is said that he spun dodder through the night so that it might strangle clover. The plant is a parasite on other vegetation, and this characteristic is used for divination purposes. If you want to know whether a person who claims to love you is being entirely honest then pick a piece of dodder, turn around, and throw it over your shoulder back onto its host plant. If, when you return to the same plant on the following day, the dodder has reattached itself then the person's love will be true. If it has not then the would-be lover is being less than honest.

The twisting threads of dodder have been used in magic to form knots. However, it must never be tied too tightly, as the strands will tend to break.

CYCLAMEN

Botanical Name: *Cyclamen* spp.
Family: Primulaceae
Common Names: Groundbread; *pain-de-porceau*; sowbread; swine bread
Gender: Feminine
Planet: Venus
Element: Water
Deities: Hecate
Meaning: Diffidence; distrust; goodbye; indifference; 'I choose not to hear your protestations'; resignation

The names by which the cyclamen is sometimes known indicate that it has been common to feed it to pigs. In southern France, Italy and Sicily, the corms are used for this purpose because they give a particular flavour to the pork.

Magic and Lore

When the plant is kept in a bedroom it will protect those who sleep in the room through the night.

> St John's Wort and fresh cyclamen
> She in her chamber kept,
> From the power of evil angels
> To guard him while he slept.
> – *Plant Lore, Legends, and Lynes*

Whether grown indoors or out, where the cyclamen flourishes it will render any spells cast to cause harm ineffectual.

Cyclamen is a magical and medicinal plant that has been used in various aspects of love magic including love charms and aphrodisiacs. According to the Doctrine of Signatures, cyclamen was once thought to be an aid to fertility, probably owing to the heart-shaped leaves or the shape of the corms. It was also taken in an effort to relieve the heartache caused by lost or unrequited love.

Medicinal

Some claim that the cyclamen is a drinker's plant because it is said to love the vine, and where a little piece of cyclamen is added to wine it is supposed to cause drunkenness more quickly. Another of its alleged properties is to prevent poisoning, including snakebites.

Hanging a cyclamen about the neck of a woman during labour is supposed to aid childbirth. This can have a disadvantage, as, according to some superstitions, it is very dangerous for a pregnant woman to tread on a cyclamen, as it will cause her to have a miscarriage.

LEMON GRASS

Botanical Name: *Cymbopogon citratus*
Family: Gramineae
Gender: Masculine
Planet: Mercury
Element: Air

Lemon grass is usually associated with oriental cookery, but it also has uses in folk magic. It is included in infusions to increase psychic powers and put into potions to induce lustfulness. Furthermore, if it is planted around the house, it will keep snakes away.

HOUND'S TONGUE

Botanical Name: *Cynoglossum officinale*
Family: Boraginaceae
Common Names: Dog-bur; dog's tongue; gypsy flower; little burdock; navelwort; rats and mice; rose-noble; scald-head; sticky buds; tongue of dog; woolmat
Gender: Masculine
Planet: Mars
Element: Fire

Hound's tongue gains its name from the shape, and texture, of its leaves. It may be from the Doctrine of Signatures that the idea came that it could be used to prevent dogs from barking at anyone bearing the plant. John Gerard, in his herbal of 1597, says of it, 'It will tye the tongues of Houndes so that they shall not bark at you, if it be laid under the bottom of your feet.' More modern authors simply suggest that you should hide a piece of the leaf in your shoes for the same result.

PAPYRUS

Botanical Name: *Cyperus papyrus*
Family: Cyperaceae
Common Names: Egyptian paper reed; Egyptian papyrus
Gender: Masculine
Planet: Mercury
Element: Air
Deities: Baalat; Hapy; Hathor

The best known use of papyrus must be for making 'paper' in ancient Egypt. In fact it was widely used for this purpose, from ancient times until the reign of the Emperor Charlemagne, when it was superseded by parchment. However, this was not the sole reason for its cultivation. It was also used for fuel, cordage, sails, and baskets, as well as for the manufacture of canoes and punts. Egyptian myth says that, following the death of Osiris, the goddess Isis recovered her dead husband's body. Set, who had killed Osiris, found it and cut it into fourteen pieces. These he scattered across Egypt. Isis searched out the various pieces of her husband, sailing up and down the Nile in a vessel made from papyrus. It is for this reason that it is believed that papyrus should be carried in a boat to prevent crocodiles from attacking those on board.

LADY'S SLIPPER

Botanical Name: *Cypripedium calceolus*
Family: Orchidaceae
Common Names: American valerian; nerve-root; Noah's ark; our lady's shoe; sleeve of the virgin; venus's slipper; whippoorwill shoes

Planet: Saturn
Element: Water
Deities: Aphrodite; Venus
Meaning: Capricious love; fickleness; 'Win me and wear me'

Lady's slipper is used as a magical protection against evil. It is added to sachets made up to counter curses and spells, and to ward off the evil eye.

BROOM

Botanical Name: *Cytisus scoparius*
Family: Leguminosae
Common Names: Banadle; banal; banathal; bannal; basam; basom; beesom; besom; bisom; bizzon; bream; broom tops; browme; brum; brushes; cat's peas; genista; golden chair; green basom; green broom; greenwood; Irish broom; Irish tops; lady's slipper; link; scobe; Scotch broom
Gender: Masculine
Planet: Mars
Element: Air
Meaning: Ardour: devotion; humility; neatness; severity

Myth and Legend

The Bible tells us that after the birth of Christ, the holy family had to escape into Egypt in order to evade King Herod. Legend has it that during this journey Mary cursed the broom, as she feared that the cracking sound made by the ripe pods as they split when they were touched would alert Herod's soldiers. The consequence of ths curse is that broom is used for menial tasks like sweeping the floor.

Strangely another line from the Bible comes to mind. Amongst the teachings of Christ is 'The last shall be first' and the humble broom was elevated to the status of royal emblem when it gave its name to the English royal house the Plantagenets. Geoffrey of Anjou picked a piece of broom from the bank where it grew, to wear in his helmet as he led his troops into battle so that they could identify him easily on the battlefield. An alternative story claims that he killed his brother in order to seize his kingdom. He then made a pilgrimage to the Holy Land, in remorse for this act, during which he scourged himself with 'genets' *(Genista)* and adopted the humble broom as his symbol rather than the haughty feather. It thus became an emblem of Brittany and Henry II, his son, used it as his family symbol.

> With a great sprig of broom which he bore as a badge in it,
> He was named from this circumstance, Henry Plantagenet.
>
> – Thomas Ingoldsby, *The Ingoldsby Legends*

The medieval name of the plant was *planta genista*, and the first use of broom in English heraldry is on the great seal of Richard I. In France King Louis continued the use of broom as a royal symbol. He founded an order of chivalry called the

Colle de Genet in 1234 on the occasion of his marriage. The collar of the order is marked with the broom flower and the royal fleur-de-lis. Richard II of England was a member of this order, and his tomb is marked with the open, empty pods of the broom. In Scotland broom is the badge of the Clan Forbes, and it was worn in their bonnets in battle, to inspire heroism.

A Christian myth claims that when Christ was walking in the garden of Gethsemane he stepped on a branch of broom, and it was the hollow rattling of the pods that betrayed his presence to Judas and those who had come to arrest him.

Magic and Lore

Although broom is cursed to be used forever for menial tasks it is usually considered to be unlucky to use it when it is in full flower. At this time it will bring misfortune on anyone who brings it into the house and will 'sweep away' any good luck there. An old Sussex saying warns:

> If you sweep the house with blossoming broom in May,
> You are sure to sweep the head of the house away.

It is also often claimed that broom must never be pruned in May, the Roman month of death, as ill fortune is sure to befall the transgressor.

Broom can be used as a protection against evil. Where magical rituals are to be carried out in the open air the floor should first be purified by sweeping it clean with broom. A sprig of the plant can be hung up around the outside of a house to ward off evil forces. Sprinkling an infusion of broom around the house will help in the exorcism of troublesome poltergeists. Infusions were also drunk at one time, to increase psychic powers, but this is not recommended, as broom is poisonous.

In pagan festivals broom is taken as a symbol of fertility. It would be used in the decorations erected at the Whitsun festival. At Whitsun weddings sprigs of broom, tied with ribbon, were given to the bride in order to ensure good fortune and a large family.

In weather magic, broom can be used to raise and calm the wind. To raise the wind sprigs are thrown into the air, preferably from the top of a mountain, whilst invoking the spirits of the air. Burning broom and burying the ashes is supposed to calm strong winds.

LUCKY HAND

Botanical Name: *Dactylorhiza maculata* (Synonym: *Orchis maculata*)
Family: Orchidaceae
Common Names: Dead men's fingers; hand of power; hand root; helping hand; *Palma Christi*; salap
Gender: Feminine
Planet: Venus
Element: Water

It is easy to see how this species of orchid gained its common name. The tubers of the orchid are divided into several finger-like lobes. Lucky hands have been popularly used in magical spells and practices in the region of New Orleans in the USA. It is not necessary to use them fresh; they can be collected and stored in a jar of rose oil until needed. When required they must be taken out and left to dry thoroughly before use. If seeking love, the 'hand' can be worn over the heart. If you want to travel it should be placed in one of your shoes. To attract wealth put it into your purse or wallet. It can be added to sachets that are kept as lucky amulets to ensure success, help in securing or maintaining employment, and protection against all misfortunes.

DRAGON'S BLOOD

Botanical Name: *Daemonorops draco*
Family: Palmae
Common Names: Blood; blume
Gender: Masculine
Planet: Mars
Element: Fire

Dragon's blood is one of the most famous ingredients of the potions made by witches in faerie tales, but few people would actually expect it to exist in reality. However, there are several sources of 'dragon's blood', of which this plant is probably the best known. The 'blood' is a reddish resinous exudate collected from the fruits of the tree.

This material can be concealed beneath a mattress or pillow to cure the impotence of the person sleeping on it. Burning it as incense will bring a wayward lover back to you. Such an incense is also supposed to remove all evil and negative vibrations from the area. It may also be added to other incenses in order to increase their potency. In noisy homes dragon's blood can be used to bring some welcome calm. It should be mixed with a little sugar and salt, bottled, covered and concealed somewhere in the house.

DAHLIA

Botanical Name: *Dahlia* spp.
Family: Compositae
Meaning: Dignity and elegance; 'Thine forever'; pomp; 'My gratitude exceed your care'; good taste (single); participation; 'I partake of your sentiments' (double); rebuff (red); dismissal (white); distaste (yellow); instability (group)

Dahlia is a native of Central America and was cultivated by the Aztecs for its tuberous roots, which they valued for use in medicinal preparations.

In Aztec mythology the dahlia, known as cocoxochitl, was the emblem of the serpent woman, who conversed daily with the eagles on Serpent Mountain. One day, whilst the eagles were conveying to her the messages of the sky gods, she noticed a rabbit beside an agave. In its mouth was a dahlia with eight red flower rays. The serpent woman took the flower and, on the instructions of the sky gods, impaled it on one of the sharp spiked leaves of the agave. This she then held close to her breast throughout the night. The following morning she gave birth to a fully grown son whom she called Uitzilopochtlithi, god of war. Uitzilopochtlithi was fully armed with a sword-like leaf of the agave. The eight red rays of the dahlia had given him strength for war and a thirst for blood. The Aztecs sacrificed their prisoners to Uitzilopochtlithi every eight years, removing their hearts and placing them on stones surrounded by agaves and dahlias.

THORNAPPLE

Botanical Name: *Datura stramonium*
Family: Solanaceae
Common Names: Angel's trumpet; apple of Peru; applethorn; Devil's apple; Devil's trumpet; ghost flower; jamestown weed; jimson weed; love will; mad apple; madherb; manicon; sorcerer's herb; stinkweed; toloache; witches' thimble; *yerba del diablo* (herb of the devil)
Gender: Feminine
Planet: Saturn (Jupiter, according to Culpeper)
Element: Water
Meaning: Deceitful charms

Magic and Lore

This plant was considered to be sacred by the ancient Aztecs in Central America, and was used in the magical rites of the shamans. Even in early Britain it was considered to be a plant used by witches in their potions and spells. During the period of 'witchmania', when many innocent people were accused of witchcraft and killed, it would be a very unlucky plant to find growing in your garden, as it could lead to the householder being denounced as a witch. Ironically this same plant can be used as a protection against evil and the malice of witches. Sprinkling thornapple around a house will help to break a spell and guard against the entrance of evil.

In 1694 John Pechey maintained that the powdered seeds, added to beer, caused temporary madness, which lasted some twenty-four hours. It, he says, would be given by thieves to those that they intended to rob. He goes on to say, 'Wenches give half a dram of it to their Lovers, in beer or wine. Some are so well skill'd in dosing of it, that they can make men mad for as many hours as they please.'

Medicinal

Other uses for this plant include curing insomnia. To do this you must place the leaves of the plant in your shoes and put the shoes under your bed with the toes pointing towards the wall. To prevent sunstroke and apoplexy, place leaves of the thornapple on the crown of your hat – mind you, there are those who would say that wearing the hat on its own might be equally successful. Great care must be taken when handling this plant, as the thornapple is poisonous, like deadly nightshade and other plants to which it is closely related.

CARROT

Botanical Name: *Daucus carota*
Family: Umbelliferae
Common Names: Bee's nest; bird's nest; crow's nest; curran-petris; eltrot; fiddle; keggas; kex; philtron; pig's parsley; rantipole
Gender: Masculine
Planet: Mars (Mercury, according to Culpeper)
Element: Fire

The wild carrot has a white parsnip-like taproot; our garden carrot is derived from the Mediterranean species *Daucus carota sativus*. Early cultivated carrots were available in a range of colours, notably purple, red and yellow. Dutch growers developed the modern orange carrot in the sixteenth and seventeenth centuries. There is a story, probably apocryphal, that the orange carrot was bred to honour William of Orange.

Magic and Lore

The best-known belief relating to carrots must be that they will help improve eyesight. This was used in propaganda during the Second World War, when it was claimed that the pilots of the RAF were being given extra carrots so that they would have an advantage over their German counterparts. In reality it was the invention of radar that enabled them to see the enemy coming, not carrots.

In dreams carrots signify strength and profit for those who are pursuing an inheritance. Given the shape of the taproot we should not be surprised to discover that they were once thought of as an aphrodisiac.

Medicinal

Including carrots regularly in your diet is said to reduce rheumatism and ease asthma. Eating carrots was supposed to promote lust, cure impotence, and help women to conceive.

SWEET WILLIAM

Botanical Name: *Dianthus barbatus*
Family: Caryophyllaceae
Common Names: Bearded pink; bloomy-downs; Jove's flower; London tufts; toleiners; velvet williams

Meanings: Coquetry; craftiness; flirtation; gallantry; 'Grant me one smile'; 'I was only teasing you'; a smile; treachery

There are several theories about how this plant got its name. Prior, in 1863, suggested that the name was a derivation from French *oeillet* (in Latin *ocellus*) meaning a little eye. This, he says, was corrupted to Willy and thence to William. It is also claimed that it was named to commemorate a wide range of people including William the Conqueror, St William of Rochester, King William III and St William of Aquitaine. One of the best-known of these accounts relates to William, Duke of Cumberland. Following the defeat of the Scots at Culloden in 1746 he became known as the Butcher of Culloden. It is said that when the English named this pretty flower in honour of the duke the Scots immediately named their most obnoxious weed, the ragwort, after him calling it 'Stinking Billy'.

In some parts *Dianthus* plants with broad leaves are known as sweet Williams, whereas those with narrow leaves are called sweet Johns.

The sweet William is the birthday flower of 15 May.

CARNATION

Botanical Name: *Dianthus carophyllus*
Family: Caryophyllaceae
Common Names: Clove carnation; clove gilloflower; gillies; gilloflowers; gillyflower; gillyvors; incarnacyon; jove's flower; nelka; scaffold flower; sops-in-wine
Gender: Masculine
Planet: Sun
Element: Fire
Deities: Jupiter; Zeus
Meanings: Encouragement; 'I will remember you'; 'Your charming token pleased me well' (pink); capriciousness (purple); 'I must see you soon'; passionate love (red); deep love and affection; heartache; heart break; 'Alas for my poor heart' (deep red); admiration (light red); refusal; regret that love cannot be shared (striped); pure affection; woman's love (white); disdain; 'You disappoint me'(yellow); distinction; fascination; pride; pure love

This is one of our oldest garden plants and Pliny informs us that it was to be found growing in Spain during the reign of Augustus Caesar and called Cantabrica. The common name may come from a corruption of 'coronation', as the flowers were once fashioned into ceremonial crowns in ancient Greece, and later worn at weddings. Alternatively it may come from *corone*, meaning flower garlands, or *carnis*, meaning flesh, from the colour (pink) of the original flower.

Myth and Legend

The name 'incarnacyon' is clearly a corruption of 'incarnation', 'God made flesh'. This is a reference to Christ's passion; the red and flesh-coloured flowers represent the flesh of Christ and the drops of blood. An alternative Christian legend tells that pink carnations first appeared when a tear fell from the eye of the Virgin Mary, when she beheld Jesus Christ carrying the cross. The flower sprang up when her tear hit the ground. The pink carnation is therefore representative of a mother's undying love.

Magic and Lore

The name 'carnation' was first used in the 1578 translation by Henry Lyte of *Historie des Plantes* (1557) by Rembert Dodoens. Carnations are, of course, a type of pink. The name 'scaffold flower' comes from the Elizabethan practice of wearing a carnation in the hope that it would prevent an untimely death on the scaffold. It is still used in protective spells, as well as in healing ones. When rituals are being carried out to cause or aid healing a carnation should be placed onto the altar. Carnations are very suitable to give to someone convalescing after an illness as they impart strength and energy to speed the healing process.

In medieval art the young women were depicted wearing a pink carnation to show that they were engaged to be married. In France, 29 June is celebrated as Carnation Day. It is dedicated to St Peter and St Paul. Today carnations are commonly worn in the buttonhole at weddings.

Carnations are worn on Mother's Day in the USA as a symbol of maternal love. The origin of Mother's Day in the USA is very different from that of Mothering Sunday in the UK. During the American Civil War, Ann Jarvis, the wife of a West Virginian Methodist minister, organized women to tend the wounded of both sides. After the war she promoted 'Woman's Work Day', which emphasized pacifism, and she organized meetings of the mothers of soldiers from both the Union and the Confederacy. Following her death, her daughter, Anna Jarvis, campaigned to have Mother's Day recognized as a national holiday. To this end, on 10 May 1908, the third anniversary of Ann's death, a service was held at Andrews Methodist Episcopal Church in Grafton, West Virginia, to launch the observance of a general memorial day for all mothers. Anna Jarvis sent 500 white carnations to the church to be given to all the participating mothers. Over the next few years she sent more than 10,000 carnations, red for a living mother and white to commemorate one who had died. These became symbolic of the purity, strength and endurance of motherhood. Mother's Day was duly declared a recognized holiday by President Woodrow Wilson in 1914.

One odd superstition is that if you dig small holes close to where you intend to sow carnation seeds and into these put a spot of any colour of dye, that is the colour the plants will eventually turn out to be.

Although many carnations are still well scented, there is an urban myth that scent was intentionally bred out of florist carnations at the request of funeral directors.

This was to avoid people being overcome by the fragrance and fainting at funeral services.

A dream of carnations indicates a passionate love affair and a secret admirer.

PINK

Botanical Name: *Dianthus pulmarius*
Family: Caryophyllaceae
Meanings: Admiration; amiability; boldness; divine love; morning light; 'No matter what the years may bring for me your beauty will never die'; timidity; welcome; always lovely (Indian); pure and ardent love (double red); refusal (striped or variegated); talent; ingenuousness (white); woman's love (single red)

To the Athenians this plant was *Di-anthos* (*dios* meaning 'a god' and *anthos* 'a flower'), the flower of Jove. 'Pink' may have been derived from the Celtic *pic*, meaning peak. Thus the 'pink of perfection' is the peak of perfection. This sentiment is carried in *The Country Housewife's Garden* by William Lawson who says, 'I may call them the king of flowers except the rose.'

Another suggestion is that the name is a combination of Dutch and English words meaning a small twinkling or winking eye. The word 'pink', as in pinking shears, also describes the scalloped edge. The plants do not get their name from their pink flowers, rather the colour pink is so called because it is the colour of the pink plant.

In medieval art the pink symbolizes divine love and signifies that a lady was engaged to be married.

BLEEDING HEART

Botanical Names: *Dicentra cucullaria*; *Dicentra spectabilis*
Family: Fumariaceae
Common Names: Dutchman's breeches; lady-in-a-bath; staggerweed
Gender: Feminine
Planet: Venus
Element: Water

The botanical name *Dicentra* is derived from *dis*, meaning 'twice', and *kentron*, meaning 'a spur', a reference to the plant's flower. In the USA it is called 'staggerweed' and is very unpopular with farmers as cattle that eat it could suffer convulsions or even death.

Magic and Lore

Bleeding heart is one of the most important love charms of the Menomini tribe of Native Americans. A young man might throw it toward his intended, or he might chew the root and circle around the girl, breathing out so that its scent will carry to her. When she catches the scent, she will feel compelled follow him wherever he goes.

It is a popular cottage garden plant. The striking flowers resemble a lover's heart that has been split open, or a pair of Dutchman's breeches, hence two of its common names. It is presumably because of the resemblance to a heart that wearing the root was believed to attract love. It will certainly attract some comments from those who see it.

Growing bleeding heart in your garden is one way to attract love and affection to yourself. However, you would be well advised not to pot up the plant and bring it into the house, as it will produce negative vibrations. If you insist on having bleeding heart as a houseplant slip a coin into the compost in the pot, as this will prevent these vibrations. If you want to discover whether your love for someone is reciprocated simply crush a flower; if the sap is red then you can be sure that the person loves you. If the sap is white or clear, they do not.

BURNING BUSH

Botanical Name: *Dictamnus albus*
Family: Rutaceae
Common Names: Dittany; dittany of Crete; gas plant; fraxinella
Deities: Juno; Lucina
Meaning: Birth

On warm summer evenings burning bush emits a volatile oil, as a vapour, which has been known to ignite spontaneously without any damage to the plant. It is from this that the plant gets its name. Some authors go further, suggesting that it was from this plant that God spoke to Moses.

The volatile oil has naturally lent itself to being used in incense. It is used in those that are burned in order to promote astral projection, mixed with equal parts of sandalwood, vanilla and benzoin. The oil can be burned on its own as a base for the manifestation of spirits which, when conjured, appear in the vapours as they rise from the censer.

It is reputed to be able to ward away venomous beasts if the sap is smeared all over the body before venturing out.

FOXGLOVE

Botanical Name: *Digitalis purpurea*
Family: Scrophulariaceae
Common Names: Bee catchers; bee hives; blobs; bloody bells; bloody fingers; bloody man's fingers; bunch of grapes; butcher's fingers; clothes pegs; cottagers; Coventry bells; cow flop; cowslip; claquet; dead men's bellows; dead men's bells; dead men's fingers; dead men's thimbles; digitalis; dog's fingers; dog's lungs;

dragon's mouth; duck's mouth; faerie bells; faerie caps; faerie fingers; faerie gloves; faerie hat; faerie petticoats; faerie thimbles; faerie weed; finger cap; finger hut; finger root; fingers; fingers and thumbs; finger tips; flop-a-dock; flopdock; flop docken; flop-poppy; floppy-dock; flops; floptop; floss docken; flowster-docken; folk's gloves; fox and leaves; fox bells; fox docken; foxes' glofa; fox fingers; fox flops; foxes' glove; foxter; foxy; gap-mouth; goblin's thimbles; goose-flops; granny's bonnet; green pops; green poppies; harebell; hedge poppy; hill poppy; king's elwand; lady's fingers; lady's gloves; lady's slipper; lady's thimble; lion's mouth; little folks' gloves; long purples; lusmore; *lus-na-mbau-side*; *maneg ellylln*; our lady's gloves; pop bell; pop dock; pop glove; pop guns; pop ladders; poppers; poppy; poppy dock; rabbit's flower; scabbit-dock; Scotch mercury; snap dragon; snap jack; snaps; snauper; snomper; snowper; snoxum; the great herb; thimble flower; thimbles; throatwort; tiger's mouth; virgin's fingers; virgin's gloves; wild mercury; witches' bells; witches' fingers; witches' gloves; witches' thimbles

Gender: Feminine
Planet: Venus
Element: Water
Meaning: Insincerity; shallowness; a wish; youth

Myth and Legend

In Hartland, Devon, foxgloves are associated with St Nectan. According to tradition the saint and his sister arrived from Wales. During their journey through the county they were set upon by robbers. The saint was decapitated. This, however, was not enough to slow down their progress as the saint simply picked up his head and continued on his way. Wherever drops of his blood fell foxgloves grew.

Another tale tells of the story of a young physician, Dr Witherington, who took up residence in a small country village. Try as he might, he could not shake the devotion of his potential patients to the herbal remedies dispensed by the wise woman of the village. Motivated by jealousy he tried spreading rumours that she was a witch, but even this had no effect. The young doctor then decided to learn the secret of the old woman's healing powers. One day he saw a man known to be suffering from heart problems enter the old woman's house. Soon after, the woman came out into her garden and picked some leaves off the foxgloves growing there. The young doctor never did establish a successful medical practice in the village, he left soon after this incident and published an account of a wonderful treatment for heart disease under the title *An Account of the Foxglove and some of its Medicinal Uses* (1785). To this day the foxglove is still used for the same purpose. Dr Witherington died in 1799 and a foxglove was marked on his tomb at Edgbaston Old Church.

Magic and Lore

Foxgloves are strongly associated with faerie folk and can be used in all faerie magic or for evoking elves and other earth elementals. The plant was originally the 'folk's gloves', that is the gloves of the 'good folk', the faeries. Faeries were supposed to dwell in the woodland dells where foxgloves thrive, and picking the flowers would offend the 'little people'. One northern legend tells that it was actually the bad faeries that gave the flowers to the fox so that it might wear them on its toes to quieten its footsteps when prowling amongst the chicken roosts. The mottled markings on the foxglove flowers, and those of the cowslip, are said to be the fingerprints where elves have placed their hands. An alternative suggests that foxes wore their 'gloves' so that the eerie ringing of the bell-like flowers would frighten away hunters.

In Ireland it was believed that the foxglove could be used to bring back children who had been carried off by faeries. Similarly, in Scotland it was thought that the juice could be used to get rid of the changeling children left by faeries in place of the children they had stolen. If your child is peevish or bad tempered it may be that you have a changeling, a faerie child, left in place of your own. To make sure, take the juice from a foxglove and place three drips onto the child's tongue and three drips in each of their ears. Then, sit the child on a shovel and hold it at the door of the house. Swing the shovel out of the door three times saying, 'If you're a faerie, away with you.' If it is a faerie child it will die, if not it will recover. Great care must be taken, of course, as foxglove is very toxic.

Usually the plant is used as a protective herb. Planting them is an invitation to faerie folk to inhabit your garden. Growing it in the garden will keep evil away. A black dye made from the leaves was used in Wales to mark crosses on the stone floors of cottages. These were initially to keep faeries away, but later were used to ward off all evil influences. In some areas of Britain it is believed that foxgloves should never be taken into the house, as it will allow devils and witches to enter the property. Wearing a foxglove will attract faerie energy to you and the sap of the plant can be used to break all enchantments.

A slightly more disturbing belief says that the flowering foxgloves could be inhibited by the souls of the dead. It is always thought to be an unlucky plant as to hear them ring means that you will soon die. Foxgloves should also never be taken on board ship, as they will bring misfortune. Picking foxgloves is supposed to offend the faeries that live within the flowers and so result in misfortune or even death for the picker.

In Scotland the foxglove is the clan badge of the Farquharson family.

Medicinal

Early herbalists considered the plant to have little or no medicinal value. It was not until the 1700s that its value for treating heart disease was discovered.

VENUS FLYTRAP

Botanical Name: *Dionaea muscipula*
Family: Droseraceae
Gender: Masculine
Planet: Mars
Element: Fire
Deities: Dione; Venus
Meanings: Deceit; 'Fly with me'

The botanical name is taken from Dione who, in mythology, was the mother of Aphrodite. It was apparently given this name in allusion to its beauty. In mythology Dione can be identified as the goddess Venus.

Growing the Venus flytrap at home has two benefits: first it will attract love into the home, literally trapping it there; and secondly, it acts as a protective plant to ward off evil forces.

EBONY

Botanical Name: *Diospyros ebenum*
Family: Ebenaceae
Common Names: Lama; obeah wood
Deity: Pluto
Meanings: Blackness; 'You are hard'

Ebony had always been highly valued as a timber for making furniture and ornaments. In magic it is used to make protective amulets and magic wands. It is claimed that an ebony wand gives unadulterated power to those who use it. Maybe a combination of these two beliefs can be found in the assertion that Pluto, god of the Underworld, was seated on a throne made of ebony wood.

AMERICAN PERSIMMON

Botanical Name: *Diospyros virginiana*
Family: Ebenaceae
Common Names: Possumwood; tree of chairs of the gods
Gender: Masculine
Planet: Venus
Element: Water

Persimmon, it is said, has the power to cause a sex change in those who eat it. In Alabama, it was believed that if a young girl ate nine unripe persimmons she would become a boy within the following two weeks.

Recurrent chills and colds could be cured by using it. You must tie a knot in a length of string for every chill you have suffered and hang it in a persimmon tree, and you will not suffer another cold. A general good luck spell can be made by burying three green persimmons in a secluded part of the garden: you will enjoy good luck for the whole household so long as the they remain undisturbed.

TEASEL

Botanical Name: *Dipsacus fullonum*
Family: Dipsacaceae
Common Names: Bath of Venus; fuller's teasel
Meanings: Inopportunity; misanthropy

Teasel was used in the woollen industry in fullering, the process of cleaning the wool to eliminate oils and dirt. (In various parts of Britain a fuller might also be known as a 'tucker' or a 'walker'.) This process is reflected in the botanical name of the plant.

Medicinal

It is not the plant that is used in folk medicine, but the droplets of water caught at the bases of the leaves, where they join the plant stem. This water could be used to treat weak eyes or warts. It might be applied as a cosmetic or be used for cleansing the feet.

TONKA

Botanical Name: *Dipteryx odorata*; *Dipteryx oppositifolia* (Synonym: *Coumarouna odorata*)
Family: Leguminosae
Common Names: Amburana; coumaria nut; cumaru; *imburana de cheiro*; tonqua; tonqin bean
Gender: Feminine
Planet: Venus
Element: Water

There are a number of plants called tonka and, as ever, the use of the same common name for several different plants may have resulted in the attributes ascribed to one of these plants being, wrongly, applied to the others.

Wearing or carrying tonka beans will protect the bearer against illness and bestow courage. It also attracts both riches and love. The beans can be added to mixtures and sachets that are used to induce thoughts of love. If you want your dreams to come true, and you happen to have a few tonka beans about your person, then make the wish with a bean held in your hand then throw it into running water.

DRAGON PLANT

Botanical Name: *Dracaena* spp.
Family: Agavaceae
Planet: Mars

The species of the genus *Dracaena* are all dragon plants. The botanical name is derived from the Greek *drakaina*, 'a female dragon'. 'Dragon's blood', a deep red resin that burns brightly when ignited, seeps from cracks in the trunk of the species *Dracaena draco*.

The plants can be used in magic for love spells and in male sex magic. More generally they have been used as protective herbs and in spells to give inner strength.

WINTER'S BARK

Botanical Name: *Drimys winteri*
Family: Winteraceae
Common Names: True Winter's bark; wintera; Winter's cinnamon

This is not 'winter bark' but 'Winter's bark'. It is named after Captain Winter, who accompanied Sir Francis Drake on his circumnavigation of the world. It was he who learned of the medicinal properties of the plant during the voyage. It can be carried or burned to ensure that a venture will be successful.

SUNDEW

Botanical Name: *Drosera* spp.
Family: Droseraceae
Common Names: *Lus na feàrnaich*; *lus ny greih*; *lus ny eiyrts*; red rot
Planet: Sun
Element: Fire

Myth and Legend

A commonly held belief was that plants that grow in wet, acid soils, including sundew, caused illness to cattle and other livestock. In Scotland sundew was sometimes said to be the plant of *earnaich*, a cattle disease that can be identified with murrain.

Magic and Lore

On the Isle of Man sundew was used as a love charm if hidden surreptitiously in the clothing of the person to be attracted.

Medicinal

On Colonsay the plant was mixed with milk to make a skin lotion that could be used to ease sunburn and clear freckles. In folk medicines it was believed that sundews could help cure corns. It has now been found that the plant contains natural antibiotics.

MALE FERN

Botanical Name: *Dryopteris felix-mas*
Family: Dryopteridaceae
Common Name: Male shield fern
Gender: Masculine
Planet: Mercury
Element: Air

It is not surprising that the male fern was considered to be a lucky for men. Carrying or wearing it was supposed to attract good luck and draw women to the wearer.

Occasionally 'lucky hands' amulets were made from the roots. These roots were also known as dead man's hands or St John's hands, as the amulets were made from roots dug up on St John's Eve (23 June) and trimmed into the shape of a hand, all four fingers and a thumb. They were then baked in an oven and afterwards carried as a protection against all kinds of evil and witchcraft. They could also be used to protect people, households and possessions.

CONEFLOWER

Botanical Name: *Echinacea angustifolia*
Family: Compositae
Common Names: Black Sampson; coneflower; niggerhead; rudbeckia; spider flower

This plant holds a particular significance for some tribes of Native Americans, who include it in the offerings they make to the spirits. In magical usage echinacea is beneficial in increasing the potency of spells. Its only other major relevance is as an aphrodisiac, although there is little evidence that it is effective.

Medicinal

Native Americans have used coneflower in the treatment of a number of ailments including toothache, sore throats, mumps, measles and smallpox. Poultices made up from the plant were applied to treat insect stings, snakebites and other wounds.

VIPER'S BUGLOSS

Botanical Name: Echium vulgare
Family: Boraginaceae
Common Names: Blue bottle; blue cat's tail; blue devil; blue thistle; blue weed; bugles; cat's tail; iron weed; our lord's flannel; our saviour's flannel; snake's flower; soldiers and sailors; today and tomorrow; wild borage
Planet: Sun (Jupiter in Leo, according to Culpeper)

The botanical name of *Echium* is derived from *echis*, 'a viper'. In ancient times the viper's bugloss was believed to be a cure for the bites or stings of venomous beasts. It could also be taken as a preventative, by drinking wine in which it had been soaked. Planting bugloss around a property will ward off snakes and therefore reduce the risk of snakebite still further.

CARDAMOM

Botanical Name: *Elettaria cardamomum*
Family: Zingiberaceae
Gender: Feminine
Planet: Venus
Element: Water
Deity: Erzulie

Cardamom has long been valued as a spice and as an aid to digestion. In magical terms it is usually considered to be an aphrodisiac, and the ground seeds can be added to warmed wine to make a simple love potion. It can also be added to apple pies to encourage love or lust in those who eat them. It is also added to incense and sachets used to promote lustfulness.

COUCH GRASS

Botanical Name: *Elytrigia repens* (Synonyms: *Agropyron repens*; *Elymus repens*)
Family: Gramineae
Common Names: Dog grass; quick grass; quitch; scotch quelch; squitch; twitch grass; witch grass
Gender: Masculine
Planet: Jupiter (Mercury, according to Culpeper)

Couch grass is the bane of many gardeners' lives, and it is rare to find a good word being said about it. However, in folk magic it has several benefits. An infusion of couch sprinkled around the home is said to disperse spells and curses, and drive out spirits. Alternatively, wearing a little of the grass will dispel depression and raise the morale, and sprinkling a little beneath the bed will attract new lovers.

Couch grass has a reputation for spellbinding audiences. It must be gathered just before dawn, pulled up by the roots, with the words:

As the birds wake and sing,
So will I wake and sing;
As we listen to them,
So they listen to me.

The roots should then be washed under running water and eaten. This is to be done on three consecutive mornings during a waxing moon, and will give the power of great oratory.

HORSETAILS

Botanical Names: *Equisetum arvense*; *Equisetum hyemale*; *Equisetum telmateia*
Family: Equisetaceae
Common Names: Bottle brush; Dutch rushes; field horsetail; paddock pipes; pewterwort; scouring rush; shave grass
Gender: Feminine
Planet: Saturn
Element: Earth

Horsetails are one of the most ancient groups of plants. Large specimens probably made up the greater part of the Earth's vegetation during the carboniferous period. In the seventeenth century horsetails were known as 'scouring rush' or 'pewterwort'. This was because they were used to scour pewter as the high levels of silica contained in the stem gave a fine finish. It was also occasionally used to put a fine finish to cabinetwork.

Placing horsetails in a bedroom is supposed to increase the fertility of those who sleep there. They can also be added to mixtures concocted for the same purpose. The stems can be used to make whistles, which are used to attract snakes – though why anyone would want to attract snakes is another matter.

WINTER ACONITE

Botanical Name: *Eranthus hymalis*
Family: Ranunculaceae
Common Names: New Year's gift; Romans; Roman soldiers

Aconite shares the genus *Eranthus* with wolfsbane. Indeed Dodoen in the *Newe Herball or Historie of Plants* calls it 'little yellow woolfes-bane'. One explanation of the derivation of the name is that it comes from *akoniton* meaning 'without struggle', possibly a reference to its use as a poison.

It was said that the pretty little flower only grew where the blood of Romans had been shed.

YERBA SANTA

Botanical Name: *Eriodictyon californicum*
Family: Hydrophyllaceae
Common Names: Bear's weed; consumptive weed; gum bush; holy herb; mountain balm; sacred herb

Carrying or wearing yerba santa was said to bestow many benefits. When carried it was said to promote psychic powers and increase spiritual strength. It was claimed that it would help to improve beauty. It can be worn around the neck as a protective amulet against evil forces, to ward off illnesses and to protect against injuries.

A little could also be added to bath water. It might also be added in incense burned to promote healing.

SEA HOLLY

Botanical Names: *Eryngium campestre*; *Eryngium maritinum*
Family: Umbelliferae
Common Names: Eryngo; sea holme; sea hulver
Gender: Feminine
Planet: Venus (Moon, according to Culpeper)
Element: Water

Various explanations have been given of the derivation of the scientific name *Eryngium*. One is that it is a diminutive of *eerungos*, meaning 'the beard of a goat'. Initially this makes little sense, but, it becomes clearer when one bears in mind Plutarch's claim that if one goat from a flock took some sea holly into its mouth it would cause first that goat, then the whole flock, to stand still. They would stay stationary until the goatherd removed the sea holly from the goat's mouth. Another strange attribute credited to it is the power to stop quarrels. If the sea holly is presented to a quarrelling couple it is said that it will bring peace. Strewing it about the home will prevent further quarrels.

The roots of the sea holly are reputed to have some aphrodisiacal properties. In the seventeenth century they were candied and sold as sweetmeats. In general terms, then, this is a plant of good luck. Travellers might choose to wear or carry it during their journey as an amulet in order to ensure their safety and bring good fortune.

WALLFLOWER

Botanical Name: *Erysimum cheiri* (Synonym: *Cheiranthus cheiri*)
Family: Cruciferae
Common Names: *Baton d'or*; bee flower; bloody warrior; cherisaunce; drops-of-blood-of-Christ; gilly flower; hand flower; heartsease; keiri; wall-gilleflower; wallstock-gillofer; wall violet
Planet: Moon
Meanings: Lasting beauty (garden); constancy; fidelity in adversity; misfortune; 'Through sunshine and storm I am true to you'

Myth and Legend

The sentiment 'fidelity in adversity' seems to date from the twelfth century, when troubadours would wear a sprig of wallflower to signify that their love would survive time and misfortune. It is said that this is based on seeing the plant growing on the ruins of fallen towers, the rather romantic image of beauty and fragrance amongst great desolation. An alternative suggestion comes from a Scottish folk tale. In the fourteenth century, the son of a Scottish chieftain fell in love with Elizabeth, the daughter of the Earl of March. The Earl occupied Neidpath Castle on the Tweed, and wallflowers grew along the castle walls. Elizabeth was betrothed, against her will, to the son of Robert III, King of Scotland, but was in love with the chieftain's son. She was confined to her chamber in the castle when her father became aware of her romance. In due time her secret lover came beneath her window, disguised as a minstrel, and sang to her. The words of his song told her to drop a wallflower to him if she was prepared to elope with him. In reaching out to pluck a flower off the wall Elizabeth overbalanced, and fell to her death. Thereafter, whenever the wallflower was in bloom, the young man would wear one on his tunic in remembrance of her. Another variation of this tale says that her lover threw up a rope for her to climb down but she failed to attach it properly and fell to her death. The powers of white magic caused her to be transformed into a wallflower. It was ever after adopted as a symbol of fidelity in Scotland. In truth King Robert III blocked the marriage of his son, Prince David, Duke of Rothesay, to the daughter of the Earl of March for fear that it would give the prince undue influence in setting Scottish policy towards England.

Magic and Lore

In Palestine the wallflower is known as the 'blood drops of Christ'. It was introduced to Britain some 300 years ago, though some sources claim that it was originally introduced by the Romans. At that time it was called 'cherisaunce' or 'chevisaunce', meaning 'comforter', because of the cordial qualities of the warm scent of the flowers. The Normans are said to have planted wallflowers beside the windows of their castles, so that the scent would drift into the rooms on warm days. At festivals it was often carried in the hand, and so was called 'hand flower', or botanically *Cheiranthus*.

Apple growers might do well to underplant their orchards with wallflowers because growing them close to apples is supposed to boost the fruit crop. Young soldiers would wear a sprig in their caps to indicate their constancy to their girlfriends at home.

Dreaming of wallflowers is a sign to those in love that their sweetheart will be true to them. To an invalid it is a promise of a quick recovery. If a woman dreams of picking wallflowers for a bouquet it indicates to her that the most worthy of her admirers has yet to propose to her.

Medicinal

Galen, the Greek physician, recommended wallflower for regulating the menstrual cycle, to relieve pains in childbirth, for liver and kidney problems and to clear cataracts. Modern use of the plant in herbal remedies is rare, owing to its toxicity.

AMERICAN ADDER'S TONGUE

Botanical Name: *Erythronium americanum*
Family: Liliaceae
Common Names Adder's tongue; adder's mouth; amber bell; dog's tooth violet; serpent's mouth; trout lily; yellow adder's tongue; yellow snowdrop
Gender: Feminine
Planet: Moon
Element: Water

American adder's tongue is used in folk medicine to treat wounds and bruises. The leaves are soaked in water, wrapped in cloth and than applied to the injury as a poultice. Once the herb has ben warmed by the body heat, it must be taken away and buried in a muddy place if the wound is to heal.

EUCALYPTUS

Botanical Name: *Eucalyptus globulus*
Family: Myrtaceae
Common Names: Blue gum tree; ironwood; stringy bark tree
Gender: Feminine
Planet: Moon
Element: Water

Eucalyptus is one of the quickest-growing plants. Some species will compete with *Sequoia*, the Californian redwood, for the record as the tallest living thing on the Earth.

The essential oils are quite volatile and have been used as antiseptics. The oil is now readily available for use in steam inhalation, to clear the head in the case of colds and flu. It has been used in folk medicine to treat the same ailments. One way to clear colds is to put a ring of eucalyptus leaves around a green candle, and then burn the candle right down to the base whilst visualizing the sufferer becoming healthy. To prevent colds and flu a sprig of eucalyptus should be hung over the bed, and a pod from the plant placed under the pillow.

To treat aches and pains, place finely chopped leaves into warm water or fat for a few minutes and then pour the resulting mixture into a clean handkerchief. Place it against the area of pain to give relief within two or three minutes. Simply carrying the leaves is supposed act as a protective amulet to prevent illness.

SPINDLE BUSH

Botanical Names: *Euonymus atropurpureus*; *Euonymus europaeus*
Family: Celastraceae
Common Names: Bitchwood; burning bush; cat tree; catty tree; death alder; dog timber; dog tree; dogwood; foulrush; fusanum; fusoria; gadrose; gatten;

gatter; gatteridge; gatter tree; Indian arrow wood; ivy-flower; pegwood; pigwood; pincushion shrub; prick timber; prickwood; skewer wood; skiver; skiver timber; skiver tree; skiver wood; spindle tree; wahoo; witchwood
Meaning: 'Your charms are engraven on my heart'

The botanical name derives from Euoyme, the mother of the Furies in Greek mythology. The timber of this plant was deemed to be fit only for skewers, toothpicks, spindles and cattle goads. It is from the latter usage that such names as 'gatter', 'gatten' and 'gadrose' are derived.

In folklore there are other uses of a less mundane nature. Carrying some of the wood will ensure that you gain success in all your undertakings. It is also said to give the bearer courage. It can be used to remove spells and curses; an infusion made from the bark is rubbed, or traced in the form of a cross, onto the forehead of the cursed person. At the same time the word 'wahoo' must be said seven times. This will lift the curse and bring peace.

HEMP AGRIMONY

Botanical Name: *Eupatorium cannabinum*
Family: Compositae
Common Names: Andurion; black elder; crow rocket; filaera: holy rope; jack-o'-lantern; raspberries and cream; St John's weed; thread flower; virgin Mary; water agrimony

Hemp agrimony has a leaf that resembles that of hemp, and it is from this resemblance that it takes both its common and its botanical name. In some areas it has been called 'holy rope' from a belief that it was used to make the ropes that bound Christ.

BONESET

Botanical Name: *Eupatorium perfoliatum*
Family: Compositae
Common Names: Ague weed; crosswort; feverwort; Indian sage; sweating plant; teasel; thoroughwort; wood boneset
Gender: Feminine
Planet: Saturn
Element: Water

Boneset was a favourite medicinal plant of the Native Americans, who knew it as 'ague-weed'. In addition to its several uses in herbal medicines it was used to ward off evil spirits. An infusion made from the plant was sprinkled around the home.

GRAVELROOT

Botanical Name: *Eupatorium purpureum*
Family: Compositae
Common Names: Gravel weed; hempweed; joe-pie weed; joe-pye; jopi-weed; queen of the meadow root; purple boneset; trumpet weed

The botanical name of this and the previous two plants is derived from Mithridates Eupator, King of Pontus, who is said to have discovered its medicinal benefits. He used a species of *Eupatorium* as an antidote to poisoning. The folk names of 'joe-pye' and 'jopi' come from an American Indian who used the plant as a treatment for typhus. The juice of gravelroot is also said to be beneficial in the treatment of open sores and bruises.

Gravelroot can be used to boost confidence when talking to members of the opposite sex. It is claimed that placing a few of the leaves of this plant in your mouth, before engaging in conversation, will be sufficient to ensure success in your amorous advances. Failing this you might try carrying a few of the leaves in your pocket as this is supposed to cause all who meet you to regard you with favour and respect.

SUN SPURGE

Botanical Name: *Euphorbia helioscopia*
Family: Euphorbiaceae
Common Names: *Lus y Bwoid Mooar*; seven sisters; warty girse

The Manx name of this plant, *lus y bwoid mooar* has been translated as 'plant of the big knobs'. More correctly it means 'plant of the big penis'. The sap was rubbed onto the penis, and was said to cause it to swell greatly. If this caused discomfort, the organ could be dipped in milk – soured milk was most usually recommended!

On Orkney and elsewhere the sap was recommended as a treatment to remove warts. The liquor resulting from boiling up the plant could also be drained and drunk, diluted with equal parts of water, as a remedy for rheumatic pains.

SPURGE

Botanical Name: *Euphorbia* spp.
Family: Euphorbiaceae
Common Names: Crown of thorns (*Euphorbia milli*); darkmous; dermuse; euphorbium bush; gum euphorbium; poisonous gum thistle; wolf's milk
Planet: Mercury
Deitiy: Munsa

All of the spurges are toxic to a greater or lesser extent, as the sap causes skin irritation. They are used as protective plants, whether grown in the garden or kept as houseplants. In Jordan it was common to take a branch of spurge into the house after the birth of a child, in order to cleanse it.

Euphorbia milli is known as the 'crown of thorns plant' as it is supposed to have been the plant used to form the rough crown put on the head of Christ prior to his crucifixion. In India *Euphorbia ligularia* is held to be sacred to Munsa, the god of serpents. The roots of this plant are powdered, and mixed with black pepper as a treatment for snakebites.

The poinsettia, *Euphorbia pulcherrima,* commonly seen at Christmas is named after Joel Roberts Poinsett, the first US ambassador to Mexico and an amateur botanist. A Mexican legend tells of a child, with no means to buy a Christmas gift, who picked a bunch of weeds from the roadside, which he placed on the church altar on Christmas Eve. These weeds were miraculously turned into the red and green flowers of the poinsettia.

EYEBRIGHT

Botanical Name: *Euphrasia officinale*
Family: Scrophulariaceae
Common Names: Bird's eye; euphrosyne; faerie flax; *lus nan leac*; *peeweets*; *rock rue*
Gender: Masculine
Planet: Sun (in Leo)
Element: Air

Myth and Legend

The botanical name is derived from Euphrosyne, who was one of the Graces in Greek mythology; euphrosyne means gladness. It is thought that this name was applied to the plant as the preservation of eyesight 'filled the patient with joy'. It was also applied to the linnet, which according to tradition first made use of eyebright leaves to clear the sight of its young. It passed on this knowledge to mankind, who named the plant in its honour.

In *Paradise Lost* Milton tells of how the Archangel Michael ministered to Adam, following the fall from grace:

> But to nobler sights
> Michael from Adam's eyes the film removed,
> Which that false fruit that promised clearer sight
> Had bred; then purged with euphrasy and rue
> The visual herve, for he had much to see.

Magic and Lore

Eyebright is also said to give those who use it insight or even second sight. It can be made into an infusion and drunk to clear the mind and aid the memory. Carrying the herb is supposed to increase psychic powers and it is claimed that repeated applications of eyebright onto the eyelids will induce clairvoyance.

Medicinal

As the name implies, eyebright was used as a treatment for sore eyes and possibly for blindness. In the Highlands of Scotland an infusion of eyebright in milk would be applied to eyes using a feather.

Despite its reputation for being good for sore eyes, it can do more harm than good. Instances have been reported of people being almost blinded by its usage. In addition to being commonly recommended for eye complaints, infusions of the leaves have also been suggested as a treatment for hay fever, coughs, colds, fevers and sore throats.

BEECH

Botanical Name: Fagus sylvatica
Family: Fagaceae
Common Names: Bog; bok; boke; buche; buk; European beech; faggio; fagos; faya; haya; hetre
Gender: Feminine
Planet: Saturn
Deities: Ammon; Apollo; Athena; Diana; Zeus
Meanings: Grandeur; prosperity; 'The halcyon days of our love are at hand'

Myth and Legend

The legend of St Leonard is set in a beech wood in a larger forest on the Sussex/ Hampshire border. St Leonard was a recluse, a hermit who had given his life over to prayer and enjoyed the tranquillity of the woodland. However, he found it difficult to spend his days in prayer because of the serpents in the woods, and his nights were disturbed by the singing of the nightingales. He prayed that all the nightingales and serpents would leave the woods so that he could have some peace. Thereafter no serpents or nightingales were ever seen there again. Perhaps this is why beech bark is claimed to be deadly to snakes.

Magic and Lore

The botanical name of *Fagus* is derived from the Greek word meaning 'to eat', in reference to the edible beechnuts or mast. For some it is the beech, rather than the birch, that is the queen of the woods beside the oak king. In its role as 'mother' it is protective and nurturing. The common names of the tree can be found throughout the Teutonic dialect either as 'book' or 'beech', as the runic tablets for the early 'books' were written on the wood. In ancient European legend the beech is associated with wisdom and learning.

It does seem slightly odd, considering its stately grandeur, that there is relatively little folklore associated with the beech tree. In the Eifel Mountains of Rhenish Russia it was customary, on the first Sunday in Lent, for young people to gather brushwood and stack it around a young, slender beech tree. A piece of wood was attached to the tree to form a cross, and the result was known as a 'castle' or 'hut'. The brushwood was set alight and the assembled people would march around it carrying burning branches to ensure a good harvest in the coming year, and to be successful the smoke from the fire must drift across the cornfields.

Although a beech grove is usually considered to be an unlucky place to be after dark, small pieces of the wood may be carried as lucky charms, and as a way to enhance your creative powers. The swellings frequently found on the bark of beech trees are, it is claimed, representations of the evil eye.

To ensure that your dreams will come true you can carve or mark your wish onto a beech stick and then bury it. As long as the stick is buried in a place where it will not be disturbed your wish will come true. In sympathetic magic, carving a heart into the bark of the tree and marking it with your own initials and those of your sweetheart will ensure that your love will endure so long as the tree survives.

Beech did have a reputation as a non-conductor of electricity. Native Americans would choose to shelter beneath it during thunderstorms in the belief that it would not be struck by lightning.

Medicinal

In folk remedies a tea made from beech, and using lard, was taken as a treatment for rheumatism. The water caught in a hollow beech tree was also used in folk medicines. It was reputed to be good for the treatment of scab and scurvy in humans and livestock.

ASAFOETIDA

Botanical Name: *Ferula assa-foetida*
Family: Umbelliferae
Common Names: Devil's dung; food of the gods; giant fennel
Gender: Masculine
Planet: Mars
Element: Fire

As might be deduced from the name, this plant has a strong, foul, nauseous odour, which tends to limit its usefulness. It has been used as an amulet to cure colds and fevers, worn about the neck to drive away the germs. Unfortunately it is likely to drive away friends and family as well! Small quantities of asafoetida are added to incense and burned as a protection or during exorcisms. A little of the herb, thrown onto a fire or burned in a censer during magical rites, will destroy any manifestation or spirit that lingers there.

Ferula is the classical Latin name of the great fennel. The word is also applied to a rod used to chastise schoolboys for minor offences. Giant fennel has tall stick like stems and was used by the Titan Prometheus, in classical mythology, to transport fire stolen from the gods.

SUMBUL

Botanical Name: *Ferula sumbul*
Family: Umbelliferae
Common Names: Euryangium musk root; *jatamani*; *ofnokgi*; *ouchi*

Sumbul, like its close relative asafoetida, can be worn as an amulet, about the neck, to ward off diseases. This is also supposed to bring the person who wears it good luck. Carrying sumbul, burning it as an incense or adding a little to bath waters is said to be a sure way to attract the love and affection of others. Burning sumbul is said to have the additional benefit of increasing the psychic powers.

BANYAN TREE

Botanical Name: *Ficus benghalensis*
Family: Moraceae
Common Names: Arched fig; Indian fig tree; Indian god tree; vada tree
Gender: Masculine
Planet: Jupiter
Element: Air
Deities: Maui: Shiva

Myth and Legend

The banyan tree holds a sacred position in many Eastern cultures. While some writers of Indian mythology tell us that there is a banyan tree in heaven, beneath which Vishnu was born, many others point to it as being the place of God's birth. The banyan is regarded as a tree of knowledge, of importance to seers and holy people, and as such is revered by the Hindus and can often be found planted near to their temples. There is a banyan growing near Surat on the River Nerbudda, which is reputed to be the oldest in India, more than 3,000 years old, and to have been visited by the officers of Alexander the Great. It is never trimmed and never touched by metal tools. A figure carved into the bark is painted red and presented with offerings by visiting pilgrims.

It is not only in India that the banyan has a special significance. It has been revered by Chinese Buddhists and connected to the worship if the god Maui on the islands of Hawaii.

Magic and Lore

Just to sit under a banyan tree is said to bring good luck, and to be married beneath its branches will ensure a couple's continued happiness.

FIG

Botanical Name: *Ficus carica*
Family: Moraceae
Common Names: Common fig; Mediterranean fig
Gender: Masculine
Planet: Jupiter
Element: Fire
Deities: The Apsaras; Dionysus; Isis; Juno
Meanings: Argument; longevity; prolific

Myth and Legend

The fig tree is an important plant in Roman and Christian myths. For the Romans it was responsible for the very site on which Rome was built. The cradle containing the infants Romulus and Remus became caught in the roots of a large fig tree as it drifted down the River Tiber. A she wolf was resting under the tree, and it was she that nurtured and raised the children who would eventually found the city. The fig tree was therefore considered sacred in Rome, and was used in religious ceremonies. The Romans associated it with several of their deities. One story of the origin of the plant tells how the Titan Lyceus was being pursued by the god Jupiter. Rhea, the mother of the gods, came to his rescue and transformed him into a fig tree. Another legend claims that the fig was created by Saturn, father of the gods, and in Cyrene the inhabitants wore crowns of fig when sacrificing to him.

The fig was also dedicated to the god Bacchus, and an alternative account of its origin says that Bacchus created it, and it was the source of his virility. During bacchanalian feasts Roman women would wear collarettes of fig leaves as emblems of fecundity. Their male counterparts might carry statuettes of the god Priapus that had been carved from fig wood. Certainly the fig was a symbol of fertility and fecundity. It was associated with Juno, in her guise of goddess of marriages.

Although the apple is usually depicted as the 'forbidden fruit' of the biblical story there are a number of legends that suggest that it was the fig. The fig is the first fruit mentioned in the Bible and the many references to it indicates its significance. On one hand it symbolizes food and security, but on the other it embodies the principle of procreation. It would make an ideal candidate for the symbolic forbidden fruit, as it is supposed to bear a resemblance to female genitalia.

Not all Christian legends relating to the fig are centred on it as a symbol of fertility. There is one that suggests that it was on a fig tree that Judas Iscariot hanged himself after his betrayal of Christ. This, to some, makes it a plant of ill omen.

At St Newlyn's Church in Cornwall a fig which grows out of the south wall is said to have sprung from the staff of St Newlina, a somewhat obscure virgin martyr. She was a Christian princess who is supposed to have pushed her staff into the ground to show where her church was to be built, although unfortunately, the church dates from the fourteenth century and figs were not introduced into Britain until the sixteenth.

Like many other sacred plants there is a tradition that death will follow for anyone who so much as plucks a leaf off the plant.

Magic and Lore

The Apsaras, sometimes called 'sky dancers', are the fig tree faeries of Hindu mythology. They can be compared to the Muses of ancient Greece. They bless humans at significant stages in life but also have a slightly dark side. They will seduce scholars and scientists, sexually exhausting them, to distract them from discovering things better left alone.

The fig is supposed to be an aphrodisiac. Eating fresh figs has been recommended in order to overcome male impotence and infertility. It is also supposed to help female infertility and promote conception. An amulet in the form of a small phallus carved from the wood can be carried for the same purpose. A gift of figs to a member of the opposite sex, will ensure that they find you ideal company.

In ancient Greece the fig was one of the principle articles of sustenance. It was a main part of the diet of their athletes as it was thought to increase their speed and strength. The high importance that was placed on the fruits in Greece is indicated by laws that prohibited the exportation of the best quality fruits.

In folk magic the fig can be used for divination. If you need the answer to a specific question then write it onto the back of a fresh fig leaf and then leave it to dry out. If it dries quickly the answer to your question will be 'no'. If it is slow to dry out then the answer will be positive. Hanging a branch of fig over a doorway before embarking on a long journey will ensure that you will return safely. Growing figs in a house will protect the whole household from harm and bring good fortune to all who live there. It was once believed that to tame a mad bull it must be tethered to a fig tree.

Dreaming of figs indicates that you can expect great happiness and riches, especially in your old age. It promises the fullfilment of your wishes and wealth you have not worked for, possibly through an inheritance or a lottery win.

The saying 'not worth a fig' has nothing to do with figs at all. It comes from the Elizabethan expression '*fico*', which was a gesture of contempt in which the thumb was placed between the first two fingers of the same hand.

BODHI

Botanical Name: *Ficus religiosa*
Family: Moraceae
Common Names: Asvattha; bo-tree; peepul tree; pippala; pipul; sacred tree; tree of wisdom
Gender: Masculine
Planet: Jupiter
Element: Air
Deities: Buddha; Krishna; Vishnu

Myth and Legend

Bodhi is native to India and Sri Lanka, and in myth is said to have sprung from the centre of the universe, at the point from where Buddha was born. It is the tree of knowledge beneath which Buddha sat to meditate for seven years until he attained the knowledge which 'maketh free'. The leaves of the tree still tremble remembering those divine vibrations. His mother, worrying about the health of her son, wished him to return to a normal life so ordered the tree destroyed. This distressed him rather than shaking him from his meditations. He poured 100 jars of milk over the tree stem and vowed that he would die if the tree did not recover. Immediately the tree was rejuvenated, recovering to its original size.

The tree is held sacred throughout the Buddhist and Hindu worlds. It is the cloud tree in which the heavenly flame is stored, guarded by dark demons. It is the 'way of safety' as it first grew on the banks of the river separating heaven from earth and, therefore, the only way in which mortals can pass across between the two. This is also said to be the tree beneath which the Hindu god Vishnu was born. Vishnu is the second in the Hindu triad; Brahma is the creator, Vishnu the preserver and Shiva the destroyer.

In Sri Lanka, the Mahabodhi tree of Anvradapura is believed to date from 245 BC and to have been grown from a branch taken off the original bo-tree.

Magic and Lore

Sacred fires are still fuelled with the timber of the bodhi tree. Burning the leaves as incense is said to help gain wisdom. Another superstition claims that if you feel the presence of evil you should walk several times around a bodhi as this will drive away any evil which is near to you.

Medicinal

If a barren woman walks naked beneath the bodhi she will be cured of her infertility.

MULBERRY FIG

Botanical Name: *Ficus sycamorus*
Family: Moraceae
Common Names: Sycamore; sycamore fig; tree of the virgin
Deities: Hathor; Isis; Nut; Osiris

According to the Egyptian Book of the Dead twin sycamores of turquoise stood at the eastern gate to heaven, from which Ra went forth every morning. The sycamore, or mulberry fig, was regarded as one manifestation of Nut, the sky goddess, who shielded Osiris and rejuvenated his soul amidst her branches spread across the heavens. The tree also acted as a focal point for an ancient tree cult, centred at Memphis, and dedicated to the goddess Hathor, the 'Lady of the Sycamore'.

In ancient Egypt it became identified as a funeral tree to be planted near burial sites as a representation of the eternity of the soul. Many tombs would have had their own sacred trees and each temple dedicated to Osiris, god of the dead, had a grove of the trees consecrated as a place for his spirit to reside. A single tree beside the sarcophagus would symbolize the resurrection after death.

An ancient sycamore stands on the outskirts of Cairo, which is said to be the tree beneath which the holy family found shelter during their escape into Egypt, hence the name the 'tree of the virgin'. It is still revered by Muslims and Coptic Christians.

MEADOWSWEET

Botanical Name: *Filipendula ulmaria* (Synonym: *Spiraea ulmaria*)
Family: Rosaceae
Common Names: Bittersweet; blackin-girse; bride of the meadow; bridewort; courtship and matrimony; dolloff; dropwort; goat's beard; gravel root; hatriff; kiss-me-quick; lady of the meadow; little queen; may of the meadow; meadow maid; meadow queen; meadow soot; meadow wort; new mown hay; old man's pepper; queen of the meadow; queen's feather; steeple bush; summer's farewell; sweet hay; tea flower; trumpet weed; wireweed
Gender: Masculine
Planet: Jupiter (Venus, according to Culpeper)
Element: Air
Deities: Blodeuwedd
Meanings: Adornment; uselessness; 'I seek a lover who is something more than merely decorative'

Meadowsweet is a strewing herb, which was put on the stone floor of old houses so that the scent emitted when it was trodden on would cheer the heart. It would also mask the less pleasant smells that might be in the air. It was also strewn on church floors at weddings not only to sweeten the atmosphere but also to promote love. It is placed on the altar during love magic rituals and the dried herb can be added to mixtures to encourage love. Its use is not new; it was one of the plants that were considered sacred by the druids. It was used to flavour beers and other beverages before the widespread use of hops, and is referred to as an ingredient in save, the drink mentioned in Chaucer's Knight's Tale. Considering its use in love magic this must have made this drink a love potion.

An Icelandic tradition says that, if you have been robbed, meadowsweet can be used to learn the gender of the thief. It must be gathered at Midsummer or on St John's Day and placed on some water in a bowl. If it floats, the thief is a woman, and if it sinks it was a man. The plant used to be associated with death, and was considered unlucky to have in the house. It was thought that the scent of the flowers might induce a sleep from which you would be unable to wake. Likewise it was believed that it was unlucky to fall asleep in a field where meadowsweet was plentiful as the heavy scent might afterwards cause you to suffer fits.

Those who inhale the sweet scent of the plant are supposed to be granted the gift of second sight and be able to converse with faeries. Ancient herbalists say, 'The smell thereof makes the heart merry and joyful and delighteth the senses.'

FENNEL

Botanical Name: *Foeniculum vulgare*
Family: Umbelliferae
Common Names: Fenkel; finhal; finkle; fyukylesede; spingel
Gender: Masculine
Planet: Mercury
Element: Fire
Deities: Dionysus; Prometheus
Meanings: Strength; force; praise worthy

> Above the lowly plants it towers,
> The fennel, with its yellow flowers,
> And in an earlier age than ours,
> Was gifted with these wondrous powers,
> Lost vision to restore.
> It gave new strength and fearless mood,
> And gladiators fierce and rude,
> Mingled it in their daily food;
> And he who battled and subdued
> A wreath of fennel wore.
>
> – H.W. Longfellow, *The Goblet of Life* 1846

Fennel has been in use since ancient times and has always been greatly valued for its aromatic shoots and fruits. Although it was known and used by the Romans, it is unclear whether it was in cultivation in northern Europe at the time.

Myth and Legend

The Greeks prized fennel and used it in the ceremonies of Dionysus, the stalks being adorned with pine cones to form a *thyrsus* or wand. Sophocles tells us that when the Titan Prometheus stole the spark of fire from the gods on Mount Olympus, he carried it to mankind concealed in the stem of a giant fennel. This story may have arisen from a common practice of kindling fires from charcoal kept in the pith of a fennel stem.

Magic and Lore

Fennel has a place in folk magic as a protective plant. In medieval times it was used, with St John's wort and other herbs, against witchcraft and evil influences. It could be grown around the home or hung over doorways and windows, especially at Midsummer Eve, to ward off evil and protect against enchantment. Carrying the seeds of fennel is also supposed to act as a protection for the bearer. Fennel was used in the nine-herb charm, along with chamomile, thyme, betony, watercress, wormwood, nettle, plantain and wild apple. A more mundane use of fennel is to prevent being bitten by ticks when walking through woodlands; put small pieces of stem into your shoes, and rub some of the juice onto your lower leg. However, be warned it is much better to acquire fennel rather than grow it as tradition says, 'Sow fennel; sow trouble.'

Medicinal

Fennel was known to the Greeks as *marathon*, meaning 'to grow thin', perhaps because it dulls the appetite or simply because it grew on the Plain of Marathon outside Athens. They believed that fennel could bestow strength, courage and long life. They also believed that it could be used to improve eyesight and clear misty or inflamed eyes. This belief persisted for centuries afterwards. Pliny recorded fennel as being a remedy for twenty-two ailments. Amongst these is an aid to eyesight saying that when snakes shed their skins they sharpen their vision with fennel juice by rubbing against the plant. An old English herbal also acknowledges this property of the plant.

> Whaune the heddere [adder] is hurt in eye
> Ye red fenel is hys prey,
> And yf he mowe it fynde
> Wonderly he doth hys kynde.
> He schall chow it wonderly,
> And leyn it to hys eye kindlely,
> Ye jows shall sang and hely ye eye
> Yat beforn was sicke et feye.

In addition to being a treatment for the eyes, it is supposed to be a general restorative, but its most unlikely attribute is as a treatment for obesity. Whilst it is fair to assume that by subduing the appetite it will aid in treating obesity the folk remedy suggests that it should be added to bath water. Whilst this is unlikely to make you thin it might help your appearance as fennel oil is used as an anti-wrinkle agent.

STRAWBERRY

Botanical Name: *Fragaria vesca*
Family: Rosaceae
Planet: Venus
Deity: Freya, Friggaa
Meanings: 'Be on the alert'; foresight; perfection; 'I esteem you as a friend but not as a lover'

> Doubtless God Almighty could have made a better berry, but he never did – Dr Botelier, quoted by Izaak Walton

In Norse mythology the strawberry is dedicated to the goddess Friggaa who was, amongst other things, goddess of marriage. It was her gift to children who had died in infancy; enabled them to ascend to heaven, concealed in a strawberry leaf. With the coming of Christianity it became associated with the Virgin Mary and St John the Baptist.

It appears as a motif on the designs of coronets worn by dukes and other members of the nobility. On the coronet of a duke or earl there are eight strawberry leaves, and on that of a marquis there are four. The coronets of junior members of the royal family are marked with strawberry leaves and fleurs-de-lis.

Dreams of strawberries mean that you can expect a new love and a happy marriage. It may suggest that you will pay a visit to the country with someone who loves you.

DEER'S TONGUE

Botanical Name: *Frasera speciosa*
Family: Gentianaceae
Common Names: Vanilla leaf; wild vanilla
Gender: Masculine
Planet: Mars
Element: Fire

Deer's tongue is another of those herbs which is used to attract the opposite sex. In this case it is suggested as being suitable for a woman who may sprinkle a little of it over her bed in order to attract men to it. It is also used in mixtures to aid and improve the development of the psychic powers.

ASH

Botanical Name: *Fraxinus excelsior*
Family: Oleaceae
Common Names: Esh; nion; the husband tree; venus of the forest; whinshag
Gender: Masculine
Planet: Sun
Element: Fire
Deities: Achilles; Andrasteia; Cerridwen; Gwydion; Mars; Nemesis; Neptune; Odin; Poseidon; Thor; Wotan; Uranus
Meanings: Grandeur; 'My love is iron-hearted, high as the mountains, deep as the oceans'

Myth and Legend

The *Snorri's Edda* of Nordic mythology tell us about the great world tree, Ygg'drasil. The universe is inside the skull of the giant Ymir, who was created from the fusion of ice from the north and fire from the south. This skull is held in place by four dwarfs, positioned at the north, east, south and west, to prevent it falling down onto the earth. This skull then represents the sky, and within it the sun and moon. The sky is like an upturned bowl supported above a flat earth which is surrounded by seas. Midgardur is at the centre of the earth disc and is inhabited by mankind. It is cut off from the outside by a high wall that keeps out the giants who live along the shoreline.

In the midst of it all stands Ygg'drasil, a giant multi-stemmed, ash tree which, we are told, represents the life-giving forces of nature and existed even before the beginning of time. Ygg'drasil grew up from the subterranean source of all matter, and stands at the centre of the universe to hold heaven and earth together. Its branches encompass all the celestial regions, and are home to a mighty eagle with a hawk on its head. The eagle is Odin, the sky god. The leaves are the clouds and the fruits are the stars. The centre stem of Ygg'drasil rises up through Midgard. It grows from the mountain, Asgard, where the gods assembled at the base of Valhalla, reached by means of a rainbow bridge, Bifrost. Around the roots of the tree there coils a giant serpent called Nidhogger, and smaller snakes chew at the roots.

The roots of Ygg'drasil extend down to the realm of the gods. There are three main roots, each of which reaches a well or spring. One root extends into Asgardur, the realm of Aesir, and reaches into the well Urdarbrunner. About this well there are three maidens, called the Norns, who control the destinies of men. They are called Urdr (Fate), Verdandi (Being) and Skuld (Necessity). These three water the tree daily and smear clay onto its bark to whiten it and preserve its life. Another root enters the well Hvergelmir in the frozen northern lands of Niflheimar. The third grows into the well of Mimisbrunner in Joetunheimar, the land of the giants.

Just as there are three roots there are also three stems. One, as has already been said, grows up through Midgard. The second rises up in Muspellsheim, where the past, present and future can be found. This is where the gods sit in judgement on mankind. The third stem arises at Nifleheim, the cold north. It is from here that all of the knowledge of mankind comes. It flows from a fountain which belongs to the frost giant Mimir, who personifies wisdom. We are told that one day Odin, who regulated night and day, and the passage of the seasons, came to Mimir to beg a drink from his fountain. He was given some of its miraculous water but had to leave one of his eyes as a payment. This then represents the going down of the sun each evening as the world comes under the influence of the god's blind eye. The mead drunk by Mimir each morning gives the colour of the new dawn. Another tale relates that the god, his side pierced by a spear, was hung upside down on the giant ash tree for nine days. At the end of this time he received enlightenment in the form of the runes. There are obvious parallels to be drawn between this and the story of the crucifixion of Christ.

Ygg'drasil constantly suffers anguish from the life it hosts. Nidhogger, the serpent at the roots, represents the female energies of the earth. The eagle in the canopy represents the male energies of the sky. The tree then links all things together, male and female, underworld, Earth and heaven, the gods, mankind and the dead.

The Vikings were called by some the *aescling* – 'men of the ash'. This was due to their great reliance on their sacred ash tree. In Norse mythology the first man was created from an ash tree. The gods Odin, Lodr and Hoenir were walking along a seashore when they came upon an ash and an elm growing close together. From the ash they formed Aska, the first man, while the elm was formed into Embla, the first woman. It was Odin who gave the couple breath. Hoenir gave them warmth, and Lodr provided them with souls. Elsewhere in Norse myth it is the mighty ash that is the repository and source of all unborn souls.

Norse myth is not alone in having a creation story depicting man being formed from an ash tree. Both Greek and Roman myths contain such stories. One such tale tells how Zeus formed the third, or bronze, race of men from ash trees. In Greek mythology the ash is the image of the clouds, and the nymphs of the ash were cloud goddesses. The ash was dedicated to the sea god Poseidon, symbolizing the power that resides in water, and it was frequently used in sea rituals. In addition to being associated with Poseidon the ash was dedicated to the Meliai, spirits that sprang up from the blood shed by Uranus when he castrated Cronos. Elsewhere it can be found dedicated to Mars as the god of war seized ash poles and placed them into the hands of his warriors. The spears of Achilles, and those of the Amazons, were made from ash poles.

The goddess Nemesis carried an ash branch as the emblem of divine justice. She is the epitome of the female Furies, or Fates, who dispensed justice through the ash tree. It was Nemesis who meted out happiness and misery to mankind, thus ensuring that fortune was shared between all people and was not the reserve of the few. If someone who had benefited from the good graces of the goddess boasted of their riches, or failed to give sacrifice to the gods, or did not relieve the necessities of the less well off then Nemesis was swift to take what had been given. Her ash scourge could use humiliation as the weapon of justice.

Nemesis is also identified with Andrasteia, the daughter of Oceanus, the sea god. The later Greeks knew her as 'Nemesis of the rain-making ash tree', emphasizing the connection between ash and the life-giving properties of water. The ash scourge of Nemesis was now used in the ritual beatings to ensure the fruitfulness of crops and fruit trees. Nemesis also carried an apple branch that was a reward for heroes. It was the passport to the Abode of the Blessed, the Elysian Fields.

The ash may be represented as a tree of rebirth and inspiration. Its seeds, the 'keys', symbolize the power to unlock the future although, clearly, the future will take time to open. In Celtic mythologies, as in Norse, it connects the circles of existence – Abred, Gwynedd and Ceugant, the past, present and future or confusion, balance and creative force.

Magic and Lore

The idea of linking ash with water can be found closer to home. In ancient Ireland and Wales coracle slats and oars were commonly made from ash wood. Those travelling over water would carry a solar cross, one with all arms of the same length, made of ash wood, as protection from drowning whilst at sea. However, the ash is a sacred tree and as such anyone who cuts or damages it without asking the tree's permission first, can expect bad luck.

Nidhogger is not the only snake with which the ash is associated. In folklore, the tree is probably the most powerful weapon against serpents and snakes, especially adders. Wearing ash twigs will protect against being bitten by snakes, whilst placing a circlet of the twigs around the neck of someone who has been bitten will cure them. Pliny, writing in the first century AD tells of the efficacy of ash in the control of snakes. Gerard, in his herbal of 1597, notes that the ash will flower before snakes appear and drops its leaves after the serpents have gone. It was claimed that snakes would even avoid its shadow, and that they would rather die in a fire than cross an ash twig. It was also said that an ash rod could be used to kill a snake outright.

Christian mythology associates the snake with the Devil. Consequently ashwood has been used as a protection against the activities of the Devil and his minions. In Ireland faggots of ash twigs would be burned in order to drive away evil forces. Similarly at Christmas, in Devonshire, ash wood was burned for the same purpose. One suggested origin of this practice comes in the legend that the Virgin Mary first washed and dressed the infant Jesus before a fire made of ash twigs.

In addition to protecting against the works of the Devil, the ash also protects against other forms of evil, particularly the activities of wizards and witches. The keys are said to be particularly potent in this case. Garters made from the green bark of the ash can be worn to protect against the powers of wizards and sorcerers. An ash staff hung above a door wards off all malign influences, and a house or other site can be further protected by scattering the green leaves at the four points of the compass.

The sap has been used to protect babies form evil. John Lightfoot, in the *Flora Scotia* (1777) says: 'In many parts of the highlands, at the birth of a child, the nurse or midwife, from what motive I know not puts one end of a green stick of this tree into the fire, and, while it's burning, receives into a spoon the sap or juice which oozes out at the other end, and administers this as first spoonful of liquor to the newborn babe!' The idea appears to have been that this will give the strength and protection of the ash tree to the child. Ash will also protect against the malice of faeries, who are unable to harm anyone who stands in its shadow. Placing ash seeds into the cradle of a child will prevent the faeries stealing it and leaving a changeling child in its stead.

Paradoxically, at one time ash was identified as a favourite tree of witches. This may date from the time of the spread of Christianity, which replaced the ancient Norse gods; the characters of the ancient deities became identified as witches and the sacred Ygg'drasil the witches' tree. This theme continues in the story of the witches of Oldenburg, who were said to eat all the ash buds when they travelled to their Walpurgis Night festivities, so that on St John's Day the tree appears to have none.

The wood of the ash is supposed to be ideal for use in spells and magical rites of cleansing and purification. It is said to be beneficial in removing mental and emotional strife. It is also claimed that it is a good conductor of power and so ideal for making magic wands.

It is usually considered to be lucky to find an ash leaf with the same number of leaflets on either side of it and the leaflet at the tip missing completely. This is the 'even-ash' which, if carefully picked, can then be worn as a lucky amulet. It will ensure happiness and protect the wearer from misfortune.

Even-ash I do thee pluck,
Hoping thus to meet good luck,
If no luck I get from thee,
I shall wish tee on the tree.

Placing an ash leaf beneath your pillow, before retiring to bed, will cause you to have prophetic dreams, and the tree is a tree of divination. If, for example, it fails to produce seeds in June it is said to be a sure sign that some great calamity can be expected. Some authors suggest that it betokens the death of the reigning monarch. Tradition has it that there was no ash seed produced in England in the year that Charles I was executed. Ash keys are also used in love divination. A young woman might pick a key from an ash tree saying:

Even Ash, even Ash, I pluck thee,
This night my own true love to see,
Neither in bed,
Nor in the bare,
But in the clothes he doth every day wear.

This ensures that she will see her future spouse before the end of the day. An alternative to this is for the young woman to pick an even-ash leaf and say:

The even-ash-leaf in my hand,
The first I meet shall be my man.

She should then place it in her glove and add:

The even-ash leaf in my glove,
The first I meet shall be my love.

Finally she should put the leaf close to her bosom saying:

The even-ash leaf in my bosom,
The first I meet shall be my husband.

Her future husband will be sure to make his appearance.

In country weather forecasting a bumper crop of ash seed is usually seen as an indication that the coming winter will be severe. One old rhyme asserts:

Oak before ash in for a splash
Ash before oak, in for a soak.

In other words, whichever of the trees is the first to come into leaf indicates either a wet or dry summer. Another country superstition is that ash trees particularly should be avoided during thunderstorms, as they attract lightning.

Avoid the ash,
It courts the flash.

However, because it attracts the lightning strikes it will act as a protection to other trees and buildings around it. That is if the rhyme is right, of course. The link between the tree and weather have extended to it being used in European fire and rain making ceremonies.

A children's game developed to be played on Ash Wednesday, much as other games were played on Oak Apple Day. Children would carry a small piece of ash twig, with a black bud clearly visible, to school. Anyone unable to produce a piece of twig would have their feet stamped on. As with many such games it only lasted until noon.

Ash is always a preferred wood for burning on open fires as it burns well whether dry or fresh cut. If it is burned as the Christmas Yule log it will ensure prosperity for the whole of the coming year. In parts of Devon faggots of ashwood held together by bonds from the ash tree were burned instead of a Yule log. As it burned young women would each choose one of the bindings. Whichever binding broke first showed who would be first to marry.

Ashwood has frequently been used to make tool handles. Tradition has it that tools with ashwood handles will do twice as much work as tools with handles made of any other wood.

Medicinal

Another aspect of the ash tree has been its usage in folk remedies. The best recorded remedy is that to heal a child of ruptures. A pollarded ash tree was selected and a hole made in it large enough to pass the child through. The hole was held open with oak wedges. The naked child was passed through the hole three times, always from east to west, before sunrise. The hole was then bound up to heal over, and as the tree healed so did the child. The tree then had to be preserved as if it was cut down the rupture would return to the child, regardless of how far away from it the child was at the time; the tree and the child were ever after linked.

The tree was used in a similar way to cure impotence. A young ash stem was split and fastened to hold it open, forming a symbolic vulva. The reluctant male member was then pushed into the opening and, on withdrawing it, the wound in the tree bound up. Then, as the wound in the tree healed, the impotence would be cured.

In Scotland ash is recognized as the 'proper' wood for herdsmen to use for their walking sticks, because the wood will not ever harm livestock and can be used to heal them. There was once a belief that stiffness in the limbs of cattle was caused by shrews running over them. Shrews were called 'faerie mice', and the cause was also seen, in this case, as being the cure: the affected limbs were rubbed with 'shrew ash'.

A piece of ash-wood was selected and a hole drilled in it. A live shrew or mouse was put into the hole, which was then sealed up. It was the slow death of the shrew, entombed in the ash, which was supposed to ensure the cure. Shrew ash was used as a treatment for various afflictions of livestock, simply being rubbed against the afflicted area. More generally, ash twigs were collected at appropriate phases of the moon and kept as a panacea in cases of illness.

One recorded use of ash as a remedy occurred when James II was at Salisbury in 1688. He had been suffering nosebleeds, which had persisted over two days. None of the remedies tried had any discernible affect until a local surgeon, William Nash, applied a piece of ash to the king's nose. The bleeding stopped.

Ash has also been used as a cure for warts. The warts would be pricked with a pin and then the pin driven into the tree bark. As the growing bark absorbed the pin so the warts would disappear. Alternatively there have been several variations of the following simple spell: push a pin into an ash tree, up to the head, and say:

> Ashy tree, Ashy tree
> Pray buy these warts of me.

Ash leaves were used in preventing illness and in improving magical powers. To prevent ill health a few ash leaves were placed in a bowl of water at the side of the bed each night, changing the water daily.

CROWN IMPERIAL

Botanical Name: *Fritillaria imperialis*
Family: Liliaceae
Common Name: Persian lily
Meaning: Majesty; power

The crown imperial occurs naturally in only two colours, orange and yellow, but Christian myth suggests that it was originally white. The story says that during the period Christ spent in the Garden of Gethsemane this was the only flower not to bow its head. Ever since it blushes and hangs its head in shame. The nectaries at the base of the petals are the tears of repentance in its eyes.

A Persian legend suggests that the 'tears' of the crown imperial are those of a beautiful queen whose fidelity was, unjustly, called into question by her husband. In a moment of rage he banished her from the palace. She was transformed into the flower by an angel so that her beauty could continue to be admired by the world, but the tears will remain until she is reunited with her husband.

From its introduction to Britain the flower has been dedicated to St Edward, the tenth-century king and martyr. Tradition says that the plant comes into flower on 18 March, the saint's feast day.

BLADDERWRACK

Botanical Name: *Fucus vesiculosus*
Family: Fucaceae
Common Names: Black-tang; bladder fucus; blasentang; cutweed; kelp; kelp-ware; meeriche; seasprite; sea wrack; seetang
Gender: Feminine
Planet: Moon
Element: Water

For the most part seaweeds are seen as being a protection for those on the sea or flying over it. Bladderwrack is used in spells to summon up the sea spirits by casting the bladderwrack onto the waters whilst calling on them in order to gain their favour. The wind can also be summoned up using it. A strand of the seaweed is held aloft and whipped around clockwise, whilst whistling. This must be done on the seashore.

Scrubbing floors and doors at a place of business with it will cause good vibrations and so attract customers. It can also be used in spells to attract money. One such spell requires you to pickle a small quantity of bladderwrack in a jar of whisky, which is then kept tightly capped and put on the kitchen windowsill. From there it will draw money into the home. It has also been used in preparations to increase psychic powers and to prevent madness.

FUMITORY

Botanical Name: *Fumaria officinalis*
Family: Fumarinaceae
Common Names: Babe in the cradle; beggary; birds on the bush; dicky birds; earth smoke; fevertory; fuminterry; fumiter; fumiterry; fumus; god's fingers and thumbs; hemitory; jam tarts; lady's lockets; lady's shoe; mother-of-thousands; nidor; *scheiteregi*; snapdragon; *taubenkropp*; vapour; wax dolls
Gender: Feminine
Planet: Saturn
Element: Earth
Meaning: Anger; spleen; 'ill at ease'

The names by which this plant is known, such as vapour, fumus and fumitory, all allude to its smoky appearance. Medieval doctors knew it as *fumus-terre*, the 'smoke of the earth', a reference to gaseous odours from its roots as well as its general appearance. This led people to believe that the smell and look of fumitory came from gases which had risen up through the Earth, or even that the plant grew, not from a seed but from those gases. One of the qualities which fumitory is said to possess is that of protecting against evil. In Germany it was called 'the thunderer's plant' and was burned to drive away evil spirits. Elsewhere it has had a long tradition of use in the incense and mixtures used in exorcisms.

To attract wealth you should sprinkle some fumitory about your home weekly, and rub a little onto your shoes for good measure. Another use of this plant is in an ointment, as a treatment to clear freckles.

If you wish to be pure and holy
Wash your face with fevertory.

This is more of an old wives' remedy than a medical recommendation. A more reasonable use of fumitory is in the treatment of scurvy, a usage for which it was once widely grown.

Fumitory is the birthday flower of 1 September.

SNOWDROP

Botanical Name: *Galanthus nivalis*
Family: Amaryllidaceae
Common Names: Belle of the snow; bulbous violet; Candlemas bells; Candlemas flower; death's flower; dewdrops; dingle bells; drooping bell; drooping heads; drooping lily; Eve's tears; fair maids; fair maids of February; February's fair maids; foolish maids; Mary's tapers; milk flower; naked maidens; pierce snow; purification flower; procession flower; snowbells; snow droppers; snow-piercers; white bells; white queen; winter bells
Meanings: 'Accept my consolation'; consolation; friendship in adversity; hope in sorrow; 'I make a fresh bid for your affections'; 'I wish you well for the future'; purity; renewal

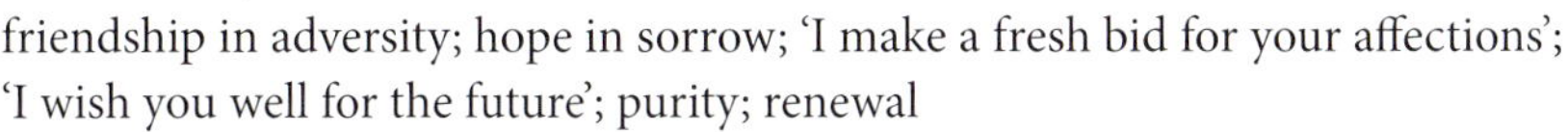

The snowdrop in purest white array,
First rears its head on Candlemas Day.

The botanical name, *Galanthus nivalis*, is derived from the Greek *gala*, meaning 'milk', and *anthos*, 'a flower'. *Nivalis* simply means 'of the snow'. Hence, the snowdrop is the 'milk flower of the snow'.

Myth and Legend

In Homer's Odyssey, the god Mercury gave Ulysses a herb called moly (the snowdrop), which made Ulysses immune to the forgetfulness poison of Circe and counteracted the amnesia that Circe had inflicted on his crew.

Another story tells how a young prince, the son of King Albion, fell deeply in love with Kenna, the daughter of Oberon, King of the Faeries. Oberon was unhappy about this mortal courting his daughter and drove the prince from his lands. The prince returned home and raised an army in order to avenge this indignity. He led his army into battle against the faerie folk, but lost his life in the fighting. Kenna wept for her fallen lover, and then tried to return him to life again using the restorative powers of a herb. However, when the juices of the herb touched her dead prince he was transformed into a snowdrop.

A German legend about the origin of this beautiful little white flower relates that at the creation all things had a colour. The sky was blue, the leaves green, the earth brown and the flowers all the colours of the rainbow. The snow complained to God that there was no colour left for it and it would be as invisible as the wind. God told it to speak to the flowers to see if one would give it a little colour. Every flower refused until, finally, the snowdrop consented to let the snow take some of its whiteness. Consequently the snow is white, and each year thanks the flower that gave it colour by keeping it warm in the late winter.

A Christian myth tells that when Adam and Eve were banished from Eden, Eve was despondent and lonely. After the bliss of Eden she found her new home cold and barren. Summer had turned to winter and driving snow covered the land. An angel saw her and took pity upon her. He breathed onto the falling snowflakes and turned them into snowdrop flowers. The tapering blossoms bowed themselves humbly at Eve's feet and each open bloom revealed a touch of green promise for the return of summer.

The snowdrop is sometimes called the 'Candlemas flower', probably because it comes into flower about the time of Candlemas, 2 February. An alternative explanation comes from another Christian myth which reminds us that candles are blessed at Candlemas as Simeon, the priest, had said that Christ would be the Light of the World. This was the time when Jesus was taken to be presented at the temple at Jerusalem and, according to the myth, snowdrops sprang up in his footsteps.

Magic and Lore

Candlemas, in times past, marked the latest time by which the Christmas greenery should be removed from churches. It could be replaced with images of the Madonna and snowdrops as emblems of her purity. Other writers suggest that at Candlemas, when the image of the Virgin Mary was taken down, a handful of snowdrop blooms were scattered in its place. Candlemas marks the date of the purification of Mary as according to Mosaic Law a purification ritual should be carried out forty days after the birth of a male child. It was once customary for girls dressed in white to walk in procession to the church bearing the flowers, hence 'procession flowers'. Taking a bowl of snowdrops into the house at Candlemas was said to be giving it a 'white purification'. In folklore it was said that if the Christmas greenery was left in the church any later than Candlemas goblins would appear and fill the church. There would be one goblin for every leaf left from the Christmas decorations.

The snowdrop is normally considered symbolic of hope and purity, and the harbinger of spring and rebirth. However, it can also be seen as an omen of death, possibly because it commonly grows in churchyards. Some superstitions state that it must never be taken into a house where someone is ill, as it will result in their death within the year. Other sources simply state that it must never be taken inside at all; if it is there will be a death in the family 'before it blooms again'. The single snowdrop is supposed to be especially symbolic of death, as it resembles a corpse in a shroud. There was a custom by which the first snowdrops to come into bloom in spring would be placed on family graves, rather adding to their reputation as a 'death flower'. Deaths in February, when the snowdrop comes into flower, were often blamed on this pretty flower.

It is better to pick a bunch of snowdrops rather than an individual blossom, and you should, of course, always pick them before St Valentine's Day to have any hopes of being married in that year. Wearing a snowdrop is said to bless you with pure thoughts.

If snowdrops are taken into a farmhouse when the hens are sitting, it will undoubtedly result in the eggs failing to hatch. In the dairy the milk will be thin and pale, and the butter colourless.

A dream of snowdrops indicates that you should not continue to conceal your secret, but will be happier if you confide in someone.

GOAT'S RUE

Botanical Name: *Galega officinalis*
Family: Leguminosae
Common Names: *Herba ruta caprariae*; Italian fitch
Gender: Masculine
Planet: Mercury (in Leo)
Element: Air
Meaning: Reason

Traditionally goat's rue was used to promote the flow of milk in goats and other domestic livestock. It was also used in rituals concerned with healing. One example of this use was as a treatment for rheumatism; putting the leaves of goat's rue into your shoes was said to prevent and cure rheumatic pains.

GOOSE GRASS

Botanical Name: *Galium aparine*
Family: Rubiaceae
Common Names: Barweed; beggar's lice; catchweed; cleavers; clivers; eriffe; everlasting friendship; goosebill; grip grass; hay riffe; hedgeriffe; sticky willie
Planet: Moon

Goose grass was traditionally prescribed by the old wives, as a cure for obesity. It could be added to soups and broths, and was said to ensure weight loss. It was also said to be good for clearing the complexion, removing freckles, and curing sunburn.

WOODRUFF

Botanical Name: *Galium odorata*
Family: Rubiaceae
Common Names: Blood cup; hay plant; herb walter; kiss-me-quick; ladies-in-the-hay; lady's needlework; madder; master of the woods; muge-de-boys; new mown hay; rice flower; rockwood; scented hairhoof; sweet grass; sweet hairhoof; sweethearts; sweet woodruff; witherips; woodrep; wood-rowell; wood-rova; woodruffee; wuderove
Gender: Masculine
Planet: Mars
Element: Fire
Meaning: Modest worth

Among the various powers attributed to this plant is the ability to attract wealth and prosperity. Carried in a leather pouch it will also help to protect against harm, warding off any evil spirits. For soldiers and athletes the carrying of woodruff guarantees success in their endeavours.

FRAGRANT BEDSTRAW

Botanical Name: *Galium trifolium*
Family: Rubiaceae
Common Name: Madder's cousin
Gender: Feminine
Planet: Venus
Element: Water

This bedstraw, if worn or carried, will attract the love and affection of others.

LADY'S BEDSTRAW

Botanical Name: *Galium verum*
Family: Rubiaceae
Common Names: A-hundredfald; broom; cheese rennet; cheese renning; cheese running; creeping jenny; fleaweed; gallion; gold dust; halfsmart; hundredfald; keeslip; lady's bed; lady's golden bedstraw; lady's tresses; maiden's hair; maid's hair; our lady's bedstraw; pettimugget; pretty mugget; rennet; robin-run-the-hedge; strawbed; wild rosemary; yellow bedstraw
Planet: Venus

The common name of lady's bedstraw may be derived from the story that this was one of the 'cradle herbs', in other words that it was among the hay in the manger into which the infant Christ was laid. A German legend goes further, saying that the manger was lined with a mixture of bracken, which at that time had a flower, and bedstraw. The bracken would not recognize the divinity of Christ, and so lost its flower. The lady's bedstraw did acknowledge Christ's divinity and its flower was turn to gold.

During the reign of Henry VIII this herb found popular use as a hair dye, and so became known as maid's hair. The names referring to rennet come from the use of bedstraw to curdle milk for cheese making.

GARDENIA

Botanical Name: *Gardenia augusta*
Family: Rubiaceae
Common Name: Cape jasmine
Gender: Feminine
Planet: Moon
Element: Water
Meanings: Concealed love; 'Like unto this virgin flower you are'; 'My love for you is secret'; sweetness

The gardenia is a plant of good luck and good health. The dried flower petals are added to mixtures and incense to promote healing. The fresh blooms should be put into the rooms of those who are feeling unwell in order to speed recovery. In magical terms these flowers are said to have 'high spiritual vibrations', and as such attract good spirits. Scattering gardenia petals will induce a peaceful atmosphere.

WINTERGREEN

Botanical Name: *Gaultheria procumbens*
Family: Ericaceae
Common Names: Boxberry; checkerberry; deerberry; mountain tea; partridgeberry; spicy wintergreen checkerberry; teaberry; *thé du canada*
Gender: Feminine
Planet: Moon
Element: Water

Wintergreen is used to give infants their best start in life, by placing a little on their pillow to ensure good fortune. This has the added benefit of affording them protection from harm in their immediate future as well. The protective characteristics of wintergreen can be used to remove curses and spells. If it is scattered about the house, for best results it should be mixed with a little mint.

GENTIAN

Botanical Name: *Gentiana lutea*
Family: Gentianaceae
Common Names: Bitter root; bitterwort; felwort; yellow gentian
Gender: Masculine
Planet: Mars
Element: Fire
Meaning: 'I love you best when you are sad'; virgin pride; 'You are unjust'; ingratitude

The gentian derives its name from Gentius, King of Illyria. In legend it was another king, Ladislaus of Hungary, who sought a cure for a plague that was spreading through his people. The yellow gentian was revealed to him as a cure for this pestilence.

In magic gentian can be added to incense in order to increase its potency; particularly the case when trying to break spell and curses. Gentian can also be added as an ingredient of sachets used to induce love and affection.

CRANESBILL

Botanical Name: *Geranium maculatum*
Family: Geraniaceae
Gender: Feminine
Planet: Venus
Element: Water
Meanings: Envy; folly; steadfast piety; stupidity

Myth and Legend

According to Turkish legend the Prophet Mohammed, whilst resting beside a stream, decided to wash his shirt in the water. Having done so he spread it out to dry on a bed of mallows that grew close by and then lay down and slept. As he slept a miraculous transformation occurred. First the mallows blushed with pleasure at having the opportunity to serve the great man. Then, by the time he awoke, the mallows had all been transformed into cranesbills.

Magic and Lore

Traditionally, cranesbills are a potent protection against snakes and serpents. As a country rhyme puts it;

> Snakes will not go
> Where geraniums grow.

To dream of geraniums (or possibly *pelargoniums*) is an indication that you need not worry about a recent quarrel, as it meant nothing and will soon be forgotten.

AVENS

Botanical Name: *Geum urbanum*
Family: Rosaceae
Common Names: Assarabaccara; benet; blessed herb; city avens; clove root; clover root; colewort; golden star; goldy star; hare foot; herb bennet; minarta; pesleporis; ram's foot; star of the earth; way bennet; wild rye; wood avens; yellow avens; yellow strawberry
Gender: Masculine
Planet: Jupiter
Element: Fire

Avens, or herb bennet, has long been thought of as 'blessed'. Indeed there is a school of thought that says that the name 'herb bennet' is a corruption from 'blessed herb', because it affords protection against evil spirits and venomous beasts. The *Ortus Sanitatis*, of 1491, says; 'Where the root is in the house, Satan can do nothing and flies from it, wherefore it is blessed before all other herb, and if a man carries the root about him no venomous beasts can harm him.'

It has been added to incense used in exorcisms and purification rites, and carried or worn as a protective amulet. An alternative suggestion for the derivation of the name is linked to this tradition. Dr Prior, in *Popular Names of English Plants*, suggests that this was originally 'St Benedict's herb', a reference to the story that the saint was presented with a goblet of wine, which contained a poison. Before drinking, he blessed it. The poison, in the form of a demon, flew from the goblet with such force that the goblet was shattered. Another reference to the 'blessed' nature of this plant comes from the symbolism associated with it in medieval times. The trefoil leaf was seen as emblematic of the holy trinity, and the five petals of the flower symbolized the wounds of Christ.

Collecting the plant can present difficulties; for medical use the roots should be collected in the spring, around the 25 of March, when the soil is dry; and although it grows best in shady sites, only those plants that grow in full sunlight should be used in magical spells.

Native American men have used avens to attract the affection of women.

GLADIOLI

Botanical Name: *Gladiolus* spp.
Family: Liliaceae
Common Names: Corn flag; sword lily
Meanings: Generosity; moral integrity; natural grace; pain; sincerity; strength of character; 'Your words have wounded me'

Ceres, the Roman goddess of grain and harvests (Demeter in Greece), loved a sacred grove close to Thessaly, near which lived Erisichthon, a wealthy but evil man. Erisichthon cared nothing for the gods and their followers and freely cut the wood from the trees in this grove. On one occasion, when the followers of Ceres tried to stop him, he decapitated one of them. From the blood, Ceres caused bright red sword-leaved lilies, gladioli, to grow. Her revenge was swift. She ordered famine to enter Erisichthon's body, so that he was never able to consume sufficient food to sate his appetite. So great was the hunger that he sold his own daughter in order to buy more food. She escaped into the forest, where Ceres found her and changed her into a gladioli so that she might watch over the man her father had unjustly killed. As for Erisichthon, his appetite drove him to sell all he had to buy still more food, and when he had nothing more to sell he consumed himself.

In ancient Rome the legionary's sword was called a *gladius* and the short sword a *gladiolus*. The men who lived and died by the sword were 'gladiators'.

GROUND IVY

Botanical Name: *Glechoma hederacea*
Family: Labiatae
Common Names: Alehoof; bird's eye; blue runner; catsfoot; creeping jenny; deceiver; devil's candlestick; fathen; field balm; gill; gill-creep-by-the-ground; gill-go-by-the-ground; gill-go-by-the-hedge; gill-go-over-the-ground; ground ivvins; grundavy; hay-hoa; haymaids; hedgemaids; hen-and-chickens; jenny-run-by-the-ground; lion's mouth; lizzy-run-up-the-hedge; monkey chops; monkey flower; moulds; nep; rabbit's mouths; rat's foot; rat's mouths; reed-hofe; robin-run-in-the-hedge; robin-run-up-dyke; run-away-jack; run-away-robin; running dyke; tun-hoof; wandering Jew
Planet: Venus

Ground ivy is used to discover who might be working magical spells against a person, by placing some around the base of a yellow candle that is then lit – but it must be done on a Tuesday.

During the reign of Henry VIII ground ivy was commonly added to beers as a clearing agent. It also improved its flavour and keeping qualitiy. It was also prescribed as a general cure-all for the insane. 'Reed-hofe' was an Anglo-Saxon remedy for 'ringing in the ears' and 'gill-tea', with added sugar, was an old wives' cure for colds and coughs.

LIQUORICE

Botanical Name: *Glycyrrhiza glabra*
Family: Leguminosae
Common Names: Lacris; licorice; licourice; lycorys; reglisse; sweet root; sweet wood
Gender: Feminine
Planet: Venus (Mercury, according to Culpeper)
Element: Water

We normally think of liquorice as a flavouring in foods. Chewing the roots is said to induce passion, and they can be added to sachets to attract love or induce lust. Paradoxically the other use of liquorice in magic is in spells to ensure the fidelity of a lover.

LIFE EVERLASTING

Botanical Name: *Gnaphalium polycephalum*
Family: Compositae
Common Names: Blunt-leaved everlasting; catsfoot; chafe weed; eternal flowers; everlasting; field balm; Indian posy; none-so-pretty; old field balsam; silver leaf; sweet scented life everlasting; white balsam
Planet: Venus

Life everlasting is used in herbal and magical preparations to ensure a long life. With all the panache of an advertising slogan, the saying advises,

> Chills and ills, pains and banes,
> Do your fasting with Life Everlasting.

Taking life everlasting first thing in the morning, before any other food or drink, is supposed to ensure that you will enjoy a long life free from serious illness. It helps, of course, to say the rhyme when taking it, if only to remind you why you are taking it. Growing the plant in the home, or carrying some part of it, is also said to prevent sickness. It has been recommended for use in those magical spells aimed at restoring lost youthfulness.

COTTON

Botanical Names: *Gossypium barbadense; Gossypium herbaceum*
Family: Malvaceae
Gender: Feminine
Planet: Moon
Element: Earth

Cotton has been used in Europe since prehistoric times and was already in use in the Americas when the Spanish conquistadors landed there. It was once used in almost 90 per cent of all the clothing in the world, but modern synthetic materials have now greatly reduced this.

For magical purposes it is second only to wool as a fabric, and is commonly used in the making of sachets. For example a sachet to bring back a lost love is made by wrapping a pepper in cotton fabric and sewing it into a sachet.

Cotton can be used to bring good luck. There are two suggested methods: first simply throw a piece over your right shoulder at dawn; or alternatively place a piece of cotton plant in a sugar bowl. Growing cotton near to the house, or scattering it around the area will ward off evil spirits, ghosts and ghouls. Placing cotton balls soaked in vinegar on windowsills will also ward off evil. Burning cotton is said to be a sure way to cause it to rain.

WITCH HAZEL

Botanical Name: *Hamamelis virginiana*
Family: Hamamelidaceae
Common Names: Snapping hazel; spotted alder; winterbloom
Gender: Masculine
Planet: Sun
Element: Fire
Meaning A spell; spellbound

The medicinal benefits of witch hazel are well known. It is commonly applied as an anti-inflammatory or as an astringent. In magic it is used as an alternative to common hazel for making wands and divining rods. The bark and twigs of witch hazel can be used as protective amulets to ward off evil influences. Those suffering from a broken heart might carry a small piece of its wood, as it is supposed to cool passions and ease heartaches.

HEBE

Botanical Name: *Hebe* spp.
Family: Scrophulariaceae
Common Name: Hedge veronica

In mythology Hebe was the daughter of Jupiter and Juno. In some stories it is suggested that she was conceived as a result of her mother eating lettuce leaves. She is described as being 'ever fair' and 'ever young', and was given the prestigious role of cup-bearer of the gods. Unfortunately she lost her job when, at one of the great feasts held on Mount Olympus, before handing a cup of nectar to Jupiter she took a taste. Being unused to the heady mixture she fell down 'in an indecent posture'. Her role of cup-bearer was given to Ganymede and she was given the more humble task of harnessing the peacocks that drew Juno's chariot. Her story does, however, end happily as she married the great hero Hercules.

IVY

Botanical Name: *Hedera helix*
Family: Araliaceae
Common Names: Bentwood; bindwood; gort; hibbin; ivin; ivery; lovestone; robin-run-in-the-hedge
Gender: Feminine
Planet: Saturn
Element: Water
Deities: Artemis Tridaria; Attis; Bacchus; Cissia; Dionysus; Hymen; Osiris; Rhea
Meanings: Assiduous to please; bonds; 'Be my bride' (ivy sprig with tendrils); amiability; ambition; elegance; fidelity; friendship; 'I die where I cling'; 'I desire you above all else'; immortality; marriage; absurdity; tenacity; the resurrection; wedded love; 'We shall cling together in the spirit of fidelity and lasting friendship – nothing will separate us'

In the ancient Celtic Ogham alphabet ivy is known as gort, and represents ruthlessness and achievement.

Myth and Legend

Ivy is dedicated to the god Dionysus in Greek mythology (Bacchus in Rome). He is usually depicted wearing a crown of ivy leaves, holding a chalice and carrying a thyrsus (a wand) entwined with ivy and vine leaves. The reason for this dedication is given in the story of Semele, daughter of Cadmus, King of Thebes. Semele was one of several maidens and goddesses who became mistresses of the god Zeus. Ultimately she became pregnant by him, which inflamed the jealousy of Hera, his long-suffering wife. She plotted to kill Semele and visited her disguised as a nurse. During her visit she suggested that Semele should persuade the great god to unveil himself so that she could see him in all his might, so Semele did so the next time the god visited her. The outcome was just as Hera had planned: Semele, being a mere mortal, was consumed by divine fire. The unborn child would also have perished but for an outgrowth of ivy nearby. Zeus rescued it and implanted it in his own thigh until it was ready to be born. The child was named Dionysus.

An alternative account tells that Semele simply abandoned the child on a mound of ivy. A further account tells of Kissos, a nymph who participated at the feast of Dionysus. In the midst of the feast she fell down dead, having literally danced herself to death. The assembled gods transformed her into an ivy plant. Kissos was an early name applied to the plant.

As the ivy leaf was the symbol of Bacchus/Dionysus it symbolized unrestrained drunkenness and feasting. According to Robert Graves the feasts were probably celebrated with spruce ale laced with ivy. Those involved were known to tear animals, children and men to pieces in their frenzied state. In the story of Orpheus we are informed that his ultimate demise, as a sacred victim of the rage of the goddess Ariadne, came about when he was torn to pieces by a group of Bacchanalian women, who were delirious, having been intoxicated with a mixture of ivy and toadstool sacred to Dionysus.

Ivy is also the plant of the Roman god Saturn. It was used in decorations at the festival of Saturnalia, a seven-day festival starting on 17 December. Its use for decorations in the later Christmas festivals is probably a continuation of this.

In a variation of the Cornish legend of Tristran and Iseult, following the death of the lovers the king ordered that their graves be placed far apart. Even in death the two were not to be parted, however, and from each grave grew a long shoot of ivy. These two shoots eventually joined and twisted together to form a true lover's knot.

A legend from Florence tells of a monastery in the city, beside which stood a tall tree. Ivy covered both the tree and a good part of the monastery wall. The traditional belief amongst the brethren was that if the ivy fell from the tree it would also fall from the wall; and if the wall was uncovered the monastery itself was endangered. When plague broke out in the city there were many appeals to the monastery for aid but, although the monastery was prosperous, no assistance was given. The abbot told all who came that the monks were concerned with the ways of heaven not those of men and further added that his order forbade its members from going into the world. They would not, therefore, minister to the sick and dying or help to bury the dead.

After a day or two a family came to the monastery begging for refuge. They were told that the monks were at prayer and could not be interrupted, but were permitted to take shelter beneath the trees to wait. Their fevers of their illness began to worsen and, though they waited for many hours, no help was forthcoming. At sundown the eldest of the family, realizing that no aid would be given them, rose and cursed the monastery and all its inmates. At the same time the youngest child, in a fit of petulance, chopped though the stem of the ivy. By the time the monks had completed their devotions and came into the garden for air the plague had claimed the lives of the family sheltering there. The very next day the ivy on the tree was dead, and the abbot called the brethren to offer prayers for the safety of the monastery. He realized now that his search for heaven was as selfish as love of earthly wealth. He directed the monks to go into the community and give whatever assistance they could. Unfortunately it was too late. The plague had ravaged the city and was now breaking out amongst the brethren of the monastery, and as the ivy died so did the monks who had hidden behind the ivy-covered walls. The ivy never recovered and, eventually, the monastery was deserted and ruined.

Magic and Lore

At the rites and ceremonies of Dionysus, wreaths and a thyrsus would be wound round with ivy. The followers of the god would chew ivy leaves during festivals in order to induce states of frenzy (it was held in high esteem by the ancients and the poet's crown was formed from its leaves). The practice of binding ivy about the head when indulging in bouts of drinking no doubt came from this association. It was believed that ivy could be used to prevent intoxication and cure hangovers. Before a drinking bout a glass of vinegar, in which ivy berries had been steeped, would be consumed, and then a wreath of ivy leaves worn whilst indulging in a drinking spree. To cure hangovers, some ivy leaves might be boiled in a little wine and the resulting concoction drunk. None of these practices is to be recommend, as ivy is slightly poisonous. One that is safe to test however, is the claim that a wife might cure her husband's drinking habit by merely spreading ivy across the pathway he will walk along on his way back from the pub.

It is not only the leaves and berries of ivy that were thought to have special powers over wine, the wood was also thought to have some valuable attributes. It was believed, for example, that if wine diluted with water was put into a bowl made of ivywood the two liquids would separate out. And if wine was left standing, in any container made of ivywood, it was thought that it could prevent infection.

In former times inns and tavern were marked by the sign of an ivy bush above the door. This, in turn, brought about the saying, 'Good wine needs no bush', meaning that it does not need to be advertised.

In addition to its associations with Dionysus, in ancient Greece, ivy was also dedicated to Hymen, the goddess of marriage, and was used to decorate the altar in her temple. It is perhaps from this association that it has gained its reputation as a kindly, feminine, plant. Greek priests would present newlyweds with an ivy wreath, as an emblem of fidelity. It would be included amongst the bride's flowers as it was thought to be particularly lucky for women.

In addition to being a symbol of marriage and resurrection it has also been used to symbolize poetry. In classical times it was used to form the poet's crown, perhaps because it represented inspiration and ecstasy, which will have had connection back to the god Dionysus.

Ivy has long been associated with Christmas. Indeed, it was said by many that the plant should never be taken into the home except as the start of the festive season and should be removed promptly on Twelfth Night. Even then, because it is unlucky, it should always be accompanied by holly to counter the negative forces that might otherwise cause misfortune. Holly is often said to represent the male, and ivy the female. Bringing the two into the house together will ensure that the entire family benefits from their blessing throughout the coming year. And whichever of the two was brought into the house first was taken as an indication of who would rule the household throughout the following twelve months, the man or the woman.

As we have seen, it was used in the decorations at the pagan midwinter festival of Saturnalia, which was replaced by Christmas with the spread of Christianity. As an evergreen plant, it was taken as a symbol of rebirth. At one time the custom of decorating houses and churches with ivy at Christmas was banned by the early Council of Churches because of its use in Bacchanalian rites, but this ban was ultimately overturned because of the usefulness of the plant as a decoration. It was an Oxfordshire custom for the master of the house to provide sufficient ivy to decorate all the rooms at Christmas. If he should fail to do so one of the maidservants might take a pair of his breeches and nail them outside the gate so that all who passed would know his disgrace.

In the Scottish Highlands and islands ivy was used at Christmas to ward off evil. Circlets of the plant, either on its own or with rowan and honeysuckle, would be hung on the lintels at the entrance to the byre to prevent evil or mischievous forces from damaging either the cattle or the dairy products. Growing ivy up a wall deters misfortune and is said to protect those inside from witchcraft and evil. If the ivy dies or falls away from the wall, the household will suffer hardships, especially financially. In Wales this was taken as an indication that the house would change ownership, because of the owner either suffering financial difficulties or failing to produce any heirs.

Ivy can be used in divination. Pick a healthy ivy leaf on New Year's Eve and place it in a bowl of water. On Twelfth Night examine it; if it is still green and healthy your health in the year ahead will be good, but if there are any black spots on it, it indicates that illness should be expected. If the discoloration is in the middle of the leaf it shows that the ailments will affect the head and neck. Spotting at the pointed apex suggests that it will be a problem affecting the feet or legs.

Like many other plants, ivy can also be used in love divination. A young woman could, for example, collect several leaves at Halloween and put them next to her heart whilst saying:

Ivy, Ivy I love you,
In my bosom I put you,
The first young man to speak to me,
My future husband shall he be.

A variation of this custom recommends that the young maid should carry an ivy sprig in her pocket or handbag rather than wearing it next to her bosom. A custom from Warwickshire and Staffordshire was for the young woman to hide a shoot of ivy beneath her pillow, or under her bed on St Thomas's night. She could then find out when they will marry by praying to St Thomas, when they retired to bed, with the words:

Stand by the bed
and say the day that I shall wed.

In Oxfordshire there was a tradition that if a young woman wanted to know whom she would marry she should pick an ivy leaf and hide it in her pocket or handbag. The first man she met when she next went out would be the man she would marry, even if he was already married. A variation on this says that if you wear an ivy leaf over your heart for three days, and then the first person with whom you shake hands after that period is unmarried, you will be married within a year. If a man wishes to learn whom he will marry he should collect ten ivy leaves at Halloween. One must be discarded, and the remainder put under his pillow before he goes to bed. This will cause him to see the image of the woman he will wed in his dreams that night.

Ivy has long been regarded as emblematic of fertility, love and fidelity. Greek priests would present a wreath to newly married persons. Even today it is still associated with weddings, and is often included in the wedding flowers as well as the in the bouquets carried by the bride and bridesmaids.

A dream of ivy is usually considered a good omen as it indicates good fortune, success and happiness, as well as honour and the friendship of others and the aid and comfort of a faithful friend. Some sources suggest that a dream of ivy might also indicate that a romance will break up.

In Scotland the ivy leaf is the badge of the Gordon clan. A sprig of ivy is the birthday 'flower' of 24 September.

Medicinal

A number of folk medicines used ivy. In the seventeenth century a vinegar of ivy berries was recommended as a cure for the plague. A preparation made from the leaves of the plant was used as a treatment for skin rashes, eye disorders, burns and even corns. The juice of the plant was suggested as a cure for colds, but had to be taken through the nose like snuff. Food given to children with whooping cough would be served in bowls made of ivywood that had been cut at the 'appropriate' phase of the moon. Wreaths of fresh ivy would be worn, following illness, to prevent hair loss or placed onto the heads of children to cure rashes on their scalps. Ivy leaves steeped in vinegar were reputed to be a useful disinfectant, and this was used in London at the time of the plague.

Ivy has also been used in the treatment of livestock. Leonardo da Vinci claimed that a wounded wild boar would roll on ivy in order to heal itself. In parts of Norfolk there are tales of ivy being fed to ewes after lambing and those recovering from illness. It was said to be of benefit in the treatment of cattle suffering from symptoms of foot and mouth disease, and it was claimed that the animals would instinctively seek it out themselves. It was also claimed that if an alcoholic drank from a cup of ivywood they to would be cured.

SUNFLOWER

Botanical Name: *Helianthus annuus*
Family: Compositae
Common Names: Corona solis; marigold of Peru; *sola indianus*
Gender: Masculine
Planet: Sun
Element: Fire
Deities: Apollo; Daphne; Mithra
Meanings: Adoration (dwarf); affection; constancy; false riches; glory; gratitude; haughtiness; infatuation; lofty and pure thoughts; ostentation; 'That which glitters is not always precious'; 'Your devoted admirer'

There is some confusion, as the name 'sunflower' appears in English texts before this plant was introduced into the country. The name was probably originally applied to the marigold, which also turns its flower to track the movement of the sun across the sky.

Myth and Legend

Greek mythology tells of Clytie, the daughter of Orchamos, a water nymph who was deeply in love with Apollo, the sun god, having seen him one day on Mount Olympus. However, her love was not returned as Apollo sought the Princess Leukothea. Driven by jealousy Clytie informed the king of his daughter's misconduct. The king, enraged, had Leukothea buried alive. Despite having disposed of her competition Clytie could still not interest Apollo, who returned to the heavens without so much as a backwards glance at her. Clytie fell to the ground, overcome with a mixture of guilt and remorse. She pined for the love of Apollo. She did not eat, nor drink. Her golden hair remained uncombed. She just sat, always watching the chariot of the sun god pass across the heavens, waiting for some look of pity. After nine days, she died of a broken heart and was transformed by the gods into a sunflower. Even now the sunflower will turn its head to always face the sun as it crosses the sky.

It is believed that the people of the Inca Empire worshipped a giant sunflower as images of sunflowers were found in the temples in the Andes. Their priestesses wore large sunflower discs made of gold on their garments and bowls of sunflower seeds were placed on the graves of their dead. It seems that the Incas believed the sunflower's magic came from the perfection of the flower's geometry.

Magic and Lore

The sunflower 'lights up' the area around it. If you plant one in your garden it will bring good fortune to everything there, warding off pests and giving the gardener the best of good fortune. The flowers can be used to ensure that your deepest wish will come true as long as it is not too extravagant. All that you need to do is cut a

sunflower at sunset whilst making your wish. It will be granted before sunset the next day. Sleeping with a sunflower beneath the bed is said to allow you to perceive the truth in any situation.

One drawback of having sunflowers growing in the garden may be found in an old saying, 'Where sunflowers grow, beans won't go.'

The sunflower is the birthday flower of 30 June.

Medicinal

Sunflower seeds are rich in vitamins E and B_1, as well as magnesium, potassium, zinc and calcium, making them a beneficial addition to the diet. They also help lower cholesterol. It is claimed that they are an aphrodisiac and an aid to female fertility. They are supposed to improve sexual potency and increase the likelihood of conception. Ideally the seed should be bruised before being eaten. The seeds, strung as a necklace were once recommended as a protection against small pox.

HELIOTROPE

Botanical Names: *Heliotropium arborescens*; *Heliotropium europaeum*
Family: Boraginaceae
Common Names: Cherry pie; Peru; Peruvian heliotrope; turnsole
Gender: Masculine
Planet: Sun
Element: Fire
Deities: Apollo
Meanings: 'Devoted to you'; devotion; faithfulness; intoxicated with pleasure

The story of Clytie described under Sunflower may well be applied to the heliotrope, as although it is now often associated with the sunflower, that plant is not native to Europe.

The heliotrope is a true sun-loving flower. As the sun moves across the sky from east to west the flower head will move so that its face is always toward the sun. At night it will turn east again ready to meet the sun as it rises on the following day.

This is one of the plants that are supposed to render those who carry it invisible: wearing or carrying a small horn filled with it is said to prevent one's movement and actions from being noticed. Keeping the flower in your pocket will also attract money and riches to you and, it can be use to help you discover the identity of a thief, sleeping with it beneath your pillow will induce prophetic dreams, and in this situation you are sure to dream of the person who has robbed you. A little heliotrope can be added to incense used during exorcism and healing rituals.

HELLEBORE

Botanical Name: *Helleborus niger*
Family: Ranunculaceae
Common Names: Bear's foot; christe herbe; Christmas bloom; Christmas rose; holy night rose; melampode; rose of noel; setterwort
Gender: Feminine
Planet: Saturn
Element: Water
Meanings: Anxiety; calumny; female inconstancy; lying tongues; 'Relieve my anxiety'; scandal; wit

Myth and Legend

Legend tells that hellebore first grew in the gardens of heaven, tended by angels, and was called the 'rose of love'. When man fell from the grace of God all Paradise lay shrouded in a covering of snow. All the beautiful flowers that had been cared for by Adam and Eve died away. The angels appealed to the Almighty to allow them to carry this one flower out into the world as a symbol of God's love and mercy.

Since when this winter rose
Blossoms amid the snows,
A symbol of God's promise care and love.
– Anon.

Better known, perhaps, is the story of a young girl named Madelon. At the birth of Christ in Bethlehem, when the Magi brought their valuable gifts, this poor shepherd girl looked on and wept. She had come, with the other shepherds, to see the place where Jesus was born, but was so poor that she had no gift to present to the child. Even the landscape offered no flower that she might pick. God, seeing her distress sent the angel Gabriel to her. He asked why she wept, and being told he touched the ground with his staff. Immediately, the earth opened and hellebores grew and bloomed. These she took as her gift to the infant Jesus. Some versions of this tale say that the flowers sprang up from where the tears of Madelon touched the ground and were all as white as the snow. It was when she presented the flowers to the Christ child and he touched them, that they blushed, turning pink.

Magic and Lore

The old name of melampode commemorated Melampus, a soothsayer and physician who, in 1400 BC, used the hellebore to treat the daughters of Proetus, King of Argus, for insanity. The plant thereafter gained a reputation as a cure for madness.

One possible derivation of the botanical name is that it comes from the Greek *helein*, meaning to kill, and *borus*, a food – literally a food that kills – as it has a well-deserved reputation as a poison and was one of the four classic poisons, the others being nightshade, aconite and hemlock. King Attalus III, one of history's great fanciers of toxins, seems to have had a particular perverse fondness for hellebore because the toxin 'racked the nerves and caused the victim to swell'.

This plant was associated with witches and witchcraft. It was supposed to be used by witches as only one 'finger' of its leaf was evil, but it took a witch to know which one.

In magic, hellebore is reputed to be able to render the user invisible. A French romance tells how a sorcerer was able to walk unseen through the camp of his enemies by scattering powdered hellebore into the air before him as he went. However, the plant had to be gathered in a particular way. There are great similarities in the gathering of hellebore and the collection of other magic herbs such as mandrake. Pliny instructs that a circle must first be drawn around the plant with the point of a sword. One should then pray for permission to gather the plant, then scan the skies for an eagle. If an eagle were seen it signified that the person gathering the plant would be dead within the year.

Gathered correctly the plant was also used as a protection against evil, being used in exorcisms and as a protective amulet. It could be used to purify homes and drive out malevolent spirits. At one time cattle would be blessed with it to keep them safe from evil spells. When used for these purposes the plant had to be collected whilst observing various ceremonies. It has also been suggested that it could be used to induce astral projection and, when used with borage, dispel melancholy.

Medicinal

Hellebore was once widely grown or collected for use against spots and boils. It was also used as a somewhat drastic and dangerous purge to control worms in children. One writer noted, 'Where it killed not the patient … it would certainly kill the worm; but the worst of it is, it will often kill both.'

DAY LILY

Botanical Name: *Hemerocallis fulva*
Family: Liliaceae
Common Names: Forget-your-troubles; lily-asphodel; plant of forgetfulness
Meanings: Coquetry (yellow); reviving pleasure

The botanical name of this plant comes from the Greek *hemera*, meaning 'day', and *kallos*, meaning 'beauty', a reference to the short time each flower will stay open. The 'plant of forgetfulness' is said to have the power to remove grief by making those who use it forget their problems.

How shall I get the plant of forgetfulness?
I would plant it in the north of my house;
Lovingly I think of my husband
And my heart is made to ache.

The plant is still eaten in some parts of China and Korea.

AMERICAN LIVERWORT

Botanical Name: *Hepatica americana*
Family: Ranunculaceae
Common Names: Edellebere; heart leaf; herb trinity; kidneywort; liverleaf; liver weed; noble liverwort; round-leaved hepatica; trefoil
Gender: Masculine
Planet: Jupiter (in Cancer)
Element: Fire

The name, and the medicinal usage of this plant, derive from its appearance. In the Doctrine of Signatures liverwort is recommended for treating liver disorders, the colour and shape of the leaves being said to resemble the human liver. There is a tradition that if a young woman wishes to secure the love of a man she should carry a small sachet of the plant with her at all times. A more romantic tale is that the flower petals are stolen by faeries as soon as the flowers open to make their clothes.

HIBISCUS

Botanical Name: *Hibiscus* spp.
Family: Malvaceae
Common Names: Kharkady; shoe flower
Gender: Feminine
Planet: Venus
Element: Water
Meaning: Delicate beauty

Hibiscus is native to Asia and the Pacific islands; it is the national flower of Malaysia and the State flower of Hawaii. It is known as the queen of tropical flowers, as it is symbolic of peace and happiness.

Myth and Legend

Long ago, there was a man who was so angry with his wife that he cast her adrift on the ocean on an open raft. The gods, however, had intervened and warned the woman of her husband's plan. They had instructed her to prepare all the supplies she would need to survive, including ashes, a coconut shell and a hibisicus branch. When the raft had drifted far out beyond the reef, the woman scattered the ashes on

the surface of the water, as instructed by the gods. She pushed the hibiscus branch into the seabed and placed the coconut shell on top of it. This created an island on which the woman landed and so was saved. This island is now called Kayangel, in the Pacific island state of Palau, and the sacred hibiscus tree still grows there.

Magic and Lore

In North Africa the flowers of the red hibiscus are used to make a refreshing drink, usually called kharkady. They can also be used to make a strong red tea, which is drunk as an aphrodisiac as it is supposed to induce lustfulness. In Egypt, it was forbidden for women to drink this concoction.

In Dobu, in the western Pacific, a bowl of water on which hibiscus flowers have been floated is used in scrying. There, and elsewhere in the tropics, hibiscus flowers are included in wreaths at wedding ceremonies. They are also added to love inducing incense.

In Hawaii a woman may wear the bright red flowers; worn behind the right ear it indicates that she is married, while worn behind the left it shows she is unwed and seeking a partner. Worn behind both ears it would mean that she is married but still looking for another lover.

Medicinal

Teas made from hibiscus have been recommended to prevent bladder infections and as a treatment for high blood pressure.

BARLEY

Botanical Name: *Hordeum distichon*
Family: Gramineae
Common Names: Perlatum
Gender: Feminine
Planet: Venus (Saturn, according to Culpeper)
Element: Earth

Magic and Lore

There are strong grounds for believing that barley is one of the oldest cereals cultivated by the Aryan race if not *the* oldest. It has been argued that the name of Demeter, the Greek 'corn-mother', was derived from an alleged Cretan word, *deai,* meaning 'barley'. The nutritional and demulcent qualities of barley were well known to the Romans, and it was eaten by the gladiators, who were called the *hordearii,* to increase their strength. Long before the use of hops became common barley was used in the brewing of beers and ales, which the Romans drank as barley wine. The ancient Hindus used barley in their religious ceremonies as it symbolized wealth. The symbol of an ear of 'corn', usually depicted as wheat or barley, is still considered to represent riches or plenty.

There are various rhymes and sayings from around Britain that give recommendations as to when to sow the barley crop. You might, for example, choose to 'Sow when the blackthorn or whitlow grass is in flower', or 'Sow when the elm comes into leaf', or 'Sow when the ground is warm to your bare bottom.' Other suggestions include:

When blackthorn is white,
Sow barley both day and night.

and

When the elm leaf is as big as a mouse's ear,
Then you sow your barley and never fear.
When the elm leaf is as big as an ox's eye,
Then I say hie, boys, hie.

There are even recommendations regarding the preparation of the ground to be sown. One says:

Dry your barley land in October
Or you'll always be sober.

This is derived from the fact that when barley is sown in wetland, yields are poor. Barley, as we have seen, is used in brewing, hence the reference to being sober.

There is a little piece of ritual love divination that can be enacted at St Agnus Eve (20 January). If a young woman wants to discover whom she will wed she must scatter barley seed beneath an apple tree whilst reciting the words:

Barley, barley, I sow thee,
That my true love I might see,
Take thy rake and follow me.

The figure of her future husband should be seen walking along behind her raking up the seed she had sown.

One way to purify an area and clear evil forces from it, possibly before the commencement of some magical rite, is to scatter barley on the ground.

Medicinal

Barley is used in a folk remedy for toothache. Wind a barley straw around a stone, whilst visualizing your pain in the stone. Once the straw is secure throw the stone into running water and, as it sinks, so your pain will go down with it.

WAX PLANT

Botanical Name: *Hoya carnosa*
Family: Asclepiadaceae
Common Names: Pentagram flowers; pentagram plant; wax flower

Gender: Masculine
Planet: Mercury
Element: Air
Meaning: Susceptibility

In magical usage the Pentagram, or five-pointed star, is usually considered to be invaluable as a protective device. The flowers of the wax flower appear as just such a five-pointed star. Consequently the plant and its flowers have been recommended as protective amulets to ward off evil forces. It is said that the plants should be grown throughout the whole house, especially in bedrooms. The flowers can be collected and dried to be worn or carried as protective amulets.

HOP

Botanical Name: *Humulus lupulus*
Family: Cannabidaceae
Common Names: Beer flower; bine; *flores de cerveza*
Gender: Masculine
Planet: Mars
Element: Air
Meaning: Apathy; injustice; passion and pride

We now associate hops with the brewing industry but it has not always been so. In the fifteen and sixteenth centuries there were a number of attempts made to maintain the distinction between ale, which was un-hopped, and beer. Beer was considered an inferior drink. To this end, various authorities forbade ale brewers from using hops as flavouring.

In a document dating from January 1530, Henry VIII's ale brewer at Eltham Palace, near London, was instructed not to use hops or brimstone (sulphur, used for fumigating casks) when brewing. These moves to prevent the use of hops in ale brewing led to suggestions that there had been attempts to completely outlaw the use of the herb. Walter Blinth, in *The English Impower Improved*, 1653, notes that 'It is not many years since the famous City of London petitioned against two nuisances, and these were Newcastle coals, in regard of their stench, &c, and hops, in regard they would spoil the taste of drink and endanger the people.'

In his book, *The History of theWothies of England*, published in 1662, Thomas Fuller wrote that hops were 'not so bitter in themselves, as others have been against them, accusing hops for noxious; preserving beer, but destroying those who drink it. These plead the petition presented in parliament in the reign of King Henry the Sixth, against the wicked weed called hops.' The unfounded assertion that Henry VI had completely outlawed the use of hops by brewers seems to arise from this publication.

It seems that hops were probably introduced into Britain from Flanders. An often repeated, if somewhat inaccurate, rhyme claims:

> Hops, reformation, bays and beer,
> Came to England all in one year.

Alternatively:

> Turkey, carp, hops, pickarel and beer,
> Came into England all in a year.

One rhyme, relating to the cropping of hops, warns:

> Rain on Good Friday and Easter Day,
> A good crop of hops but a bad one of hay.

Hops are used in an old wives' cure for insomnia; stuffing one's pillow with hops assures a sound night's sleep.

One custom, which I am sure the hop growers loved, was that any visitor to the hop fields of Kent should contribute 'foot money' to prevent luck leaving the fields.

Hop is the birthday flower of 7 April.

BLUEBELL

Botanical Name: *Hyacinthoides non-scriptus*
Family: Liliaceae
Common Names: Adder bell; auld man's bell; bell bottle; bloody man's fingers; blue bonnets; blue bottle; blue goggles; blue ganfer-greygles; blue rocket; blue trumpets; bummack; bummuck; calverkeys; crake-feet; crawfeet; craw-taes; cross flower; crow bells; crow picker; crow's legs; crowstoes; cuckoo; cuckoo flower; cuckoo's boots; cuckoo's stockings; culver bells; culverkeys; deadmen's bells; faerie bells; ganfer-gregors; goosey gander; gowk's hose; grammer-greygles; granfer-griggles; granfer-grigglesticks; greygles; griggles; harebell; jacinth; link; locks and keys; pride of the woods; ring o' bells; ring of death; rook's flower; single gussies; snake's flower; snapgrass; wild hyacinth; wood bells
Meanings: Constancy; 'I am faithful'; kindness; solitude and regret

A good deal of confusion seems to arise when identifying the English bluebell and the harebell, probably because the plant known in most of England as the harebell, *Campanula rotundifolia*, is called a bluebell in Scotland.

Myth and Legend

The botanical name for the bluebell was originally *Endymion non-scriptus*, the name being taken from Endymion, a youth in Greek myth with whom the moon goddess Selene fell in love. The bluebell is related to the hyacinth, but unlike the hyacinth was not marked by Apollo, and so has the specific epithet *non-scriptus* – 'not written on'.

Magic and Lore

One Scottish name for this plant is 'death bell'. Tradition has it that if you hear the bluebell ring it is ringing your death knell, and you will die soon after. It is also very unlucky to pick the flowers for use in the house, although you can make a wish when you see the first bluebell open in the spring.

The bluebell is often portrayed as one of the most potent of the faerie flowers, and woods filled with them are dangerous places of faerie spells and enchantments. Some writers say that you should never walk through bluebells, as the enchantment will lead you to the faerie folk, who will spirit you away, or at the very least keep you captive until rescued by someone else. At daybreak, however, the bluebells are said to ring, to call the faeries back from the woods. The bells of the flowers are held upright through the night, and then nod forward at the first rays of the rising sun. This rings in the start of each new day. Children who pick them, when walking alone in woodland, might never be seen again, presumably spirited off to the faerie realm.

They are not always considered to be plants of ill omen however, on the contrary, some traditions see them as luck flowers. If, for example, you are able to turn a bluebell flower completely inside out without tearing it, it indicates that you will win the heart of the one you love. To ensure good luck you should pick a bluebell and conceal it in your shoe, saying:

> Bluebell, bluebell, bring me some luck,
> Before tomorrow night.

There is another belief which was once associated with the bluebell, namely that if someone wore the flower (or a wreath of flowers) they would be incapable of telling lies. This may have made life interesting on 23 April, St George's Day, as it was once common to wear a bluebell then as the flowers were reputed to come into bloom on that date.

To dream of bluebells is an ill omen as it indicates that you will have a nagging spouse. It may also indicate that you will enjoy a passionate but stormy relationship.

Bluebell is the birthday flower of 30 September.

HYACINTH

Botanical Name: *Hyacinthus orientalis*
Family: Liliaceae
Gender: Feminine
Planet: Venus
Element: Water
Deities: Hyakinthos; Zeus
Meanings: Admiration; 'I esteem you highly' (white); devotion; constancy; fidelity; 'I shall devote my life to you' (blue); sorrow (purple); cheerfulness; games; play; sorrow; sport

Myth and Legend

Hyakinthos, the son of Amyclas, King of Sparta, was a youth of great beauty. He was a favourite of the sun god Apollo, but there were others who were attracted to him, including Zephyrus, the west wind, who was jealous of the time Hyakinthos spent with Apollo. One day, when Apollo and Hyakinthos were playing quoits on the banks of the Eurotas, Zephyrus came upon them. In a blind rage he caused a quoit thrown by Apollo to veer off course and strike Hyakinthos on the temple, killing him. From his blood Apollo caused the hyacinth to grow as a symbol of his sorrow, and marked its leaves with the mournful exclamation *Ai*, meaning 'alas'. Hyakinthos

was worshipped in ancient Crete as a god of flowers and in Sparta an annual festival, called Hyacinthia, commemorated his death.

> The hyacinth be wary the doeful 'Ai',
> And calls the tribute of Apollo's sigh,
> Still on its bloom the mournful flower retains
> The lovely blue that dyed the stripling's veins
> – *Camoens Lusiads Canto IX*

Another legend tells how Ajax was driven to madness following his defeat by Ulysses. Believing them to be his enemies, Ajax slaughtered the sheep belonging to the Greek army. When he discovered his mistake, he killed himself and it was from his blood that the hyacinth grew.

However it seems that the name is derived from an earlier, non-Greek, language, (Thraco-Pelasgian) and is apparently connected with the blue flowers being the colour of water. The plant marked *Ai* by Apollo is more likely to have been a gladiolus rather than a hyacinth. *Gladiolus italicus* has marking on the lower leaves that could be taken to represent '*Ai, Ai*', Apollo's mournful exclamation.

Magic and Lore

There was a belief that eating hyacinth bulbs would delay the onset of maturity and retard the signs of puberty such as hair growth – a link with the story of the early death of Hyakinthos.

Growing hyacinth in the home, as a pot plant can prove beneficial. Putting it in a bedroom is said to prevent nightmares. Generally, the scent of the fresh flowers is supposed to raise the spirits and relieve depression and grief.

ORPINE

Botanical Name: *Hylotelephium telephium* (Synonym: *Sedum telephium*)
Family: Crassulaceae
Common Names: Arpent; arpent-weed; harping johnny; heal-all; healing leaf; Jacob's ladder; life everlasting; live long; long live; midsummer men; orphan John; orpies; orpy-leaf; Solomon's puzzle; stonecrop

The virtues of this plant are said to have been discovered by Telephus, the son of Hercules. In Christian myth it is dedicated to St Giles, the patron saint of beggars and cripples. It used to be popular in love divination. John Aubrey, writing in the seventeenth century, tells how young maids would collect slips, or leaves, on Midsummer Eve, and set them in pairs into cracks in the roof joists. Each pair of leaves was to represent a young couple. As the leaves dried and bent toward, or away from, each other it indicated whether the relationship in question was blooming, or doomed to fail. If, however, one of the leaves withered up it was taken as an indication that the person it represented would soon die. In a similar way the bending of the plant leaves could be used to indicate the fidelity or faithlessness of a lover.

Orpine was collected at Midsummer and used as a talisman in the house to protect against thunder and lightning. In France it was collected before St John's Eve and 'purified' over a bonfire so that it might be used as a protection against evil. In the Gironde there was a belief that if a witch, male or female, should enter a room where an orpine has been placed it will immediately wither and die.

The plant has acquired names such as 'live long' and 'life everlasting' because it remains green long after it has been cut.

HENBANE

Botanical Name: *Hyosycamus niger*
Family: Solanaceae
Common Names: Black henbane; black nightshade; cassilago; cassilata; *deus caballinus*; devil's eye; hebenon; henbells; henpen; henpenny; hogsbean; isana; jupiter's bean; jusquiame; poison tobacco; stinking nightshade; stinking roger; symphonica
Gender: Feminine
Planet: Saturn
Element: Water
Meanings: Fault; imperfection

Myth and Legend

In mythology, henbane is associated with the souls of the dead. We are told that those who wander beside the River Styx, in the Underworld, are crowned with it.

Magic and Lore

The common name of 'hogs bean' is a direct translation of the generic name, *Hys*, from the Greek meaning 'a pig' and *kyamos* meaning 'a bean'. The plant may be poisonous to people but many animals eat it without apparent ill effect. It is a close relative of deadly nightshade and, like that plant, has been used to dilate ladies' pupils to make them large and lustrous. It was also used in magic rituals to attract love, and was added to love potions. Where it is to be used by a man to gain the affections of a woman it must be collected with some ceremony. It should be gathered in the early morning by a naked man standing on one leg. It can then be worn as an amulet to attract love.

Burning henbane in the open air is said to cause rain, although, as the fumes are poisonous, there are many other plants that can be more safely used for this purpose. It has also been burnt in the past by those trying to conjure up spirits, possibly with fatal results. Inhaling the fumes from burning henbane is said to be an aid to clairvoyance, and the plant can be formed into an amulet to counter the harm caused by witches.

Medicinal

One medicinal use for the plant is as a cure for insomnia; one should bathe one's feet in henbane and water before retiring to bed. It is also is an ancient narcotic, used as a painkiller. Unfortunately, because of its extremely toxic nature, the effect sometimes proved rather more permanent than the patient would have liked. Henbane was also an ingredient in the early painkiller laudanum, which consisted of opium, henbane and wine. Its major drawbacks were that it was addictive and hallucinogenic.

TUTSAN

Botanical Name: *Hypericum androsaemum*
Family: Guttiferae (Clusiaceae)
Common Names: Amber; sweet amber; touch and heal
Gender: Masculine
Planet: Saturn
Element: Fire

Tutsan has bright, glossy berries that turn from green to red and then to black. They are supposed to have been touched by the blood of the Danes, which is the cause of their colouring. The leaves are dried and placed between the pages of books, especially Bibles, as lucky amulets.

ST JOHN'S WORT

Botanical Name: *Hypericum perforatum*
Family: Guttiferae (Clusiaceae)
Common Names: Amber; balm of the warrior's wound; devil chaser; *fuga daemonum* ('the flight of devils'); goat weed; herba John; John's wort; klamath weed; penny John; perforated St John's wort; rosin rose; *sol terrestris* ('sun of the earth'); tipton weed; touch and heal; *xiao lianqiao* (chinese – 'little forsythia')
Gender: Masculine
Planet: Sun (in Leo)
Element: Fire
Deity: Baldur
Meanings: Animosity; oblivion to life's troubles; superstitions; superstitious sanctity

Myth and Legend

A Scottish name for St John's Wort is *ach larson Cholumcille*, 'The little armpit of Columba'. In the Scottish islands the plant is dedicated to St Columba as it is said that the saint once found a child he employed to tend his cattle crying because he was frightened of the dark and afraid the cattle might stray. St Columba is said to have picked a piece of St John's wort and put it beneath the child's armpit, telling him to sleep as no harm would come to him or the livestock.

Having superimposed the festival of St John onto the pagan Midsummer festival various Christian myths were then woven around the plants associated with this festival. In the case of St John's wort it is claimed that the Devil sought to exact his revenge on the plant that could drive his spirits away. The power of the plant limited what he was able to do, but the Devil took a needle and pricked holes across the leaves, which are visible when the leaf is held up to the light and account for the French name *Mil-Lepertuis*, 'thousand holes'. The spots are actually lots of tiny trans-lucent glands in the leaf.

Another Christian legend says that the Devil sought to rid the world of this plant as it was so widely used in herbal medicines, curing those who otherwise would have died and found their way into hell. He took a dagger and stabbed it repeatedly until it bled. However, it bled the blood of St John the Baptist, strengthening its healing powers rather than weakening them. The dark dots to be seen on the underside of the petals and the topmost leaves are where the Devil stabbed the plant. Elsewhere this story is refined by adding that the plant becomes marked with the blood of John the Baptist on 29 August (the 27th according to other sources). On this date dark red spots appear on the leaves, commemorating the blood of the martyr who was beheaded on that day and the red sap in the stem and leaves of the plant is symbolic of the blood of St John.

Magic and Lore

St John's wort, St John's wort,
My envy whosoever has thee,
I will pluck thee with my right hand,
I will preserve thee with my left hand,
Whoso findeth thee in the cattlefold,
Shall never be without kine.

There are a number of plants that have been called St John's wort, or associated with St John's Eve and St John's Day, but this is the one most commonly recognized by that name. It would be safe to assume that, before the spread of Christianity, those plants now associated with St John's Day were previously held to be important at the pagan Midsummer festival. It is noted as a magical herb in every country in which it is grown.

It is said that it should be gathered, naked, on Midsummer Eve, or at sunrise and hung up on May Day to promote fertility. Paracelsus claimed that it could be used to expel 'phantasms and spectres'. The name *Hypericum* is derived from the Greek and means 'over an apparition', in reference to its power to protect against witches and evil spirits. Where the plant was to be kept as a protective amulet it was collected on 23 June, St John's Eve (some authors suggest the 24th), before sunrise, whilst it is still damp with the morning dew. Dew was taken to be a magical and strengthening substance in its own right. Collecting the plant was never straightforward as it was said to be able to move in order to avoid being picked. On the Isle of Man there was a superstition that treading on the growing plant was hazardous as it would result in a faerie horse rising up to carry you away. You would then be deposited far away from your starting point.

St John's wort was said to have the further advantage of attracting love and, when worn by soldiers, of conferring invisibility. The shoots could be hung up in house windows, on St John's Day, to protect against 'ghosts, devils, imps and thunderbolts'. In Scotland sprigs of the herb were carried to counter enchantments and witchcraft. Pieces of the plant might also be placed in dairies to prevent the milk from being curdled by mischievous faeries.

Other superstitions relating to this plant include its use as a cure for infertility. If a woman wished to conceive she would walk naked into her garden on Midsummer Day and pick the flower. This would ensure that she would be pregnant before the next Midsummer Day. And if a young woman wished to capture the love of a man she would collect a sprig, wet with the dew, on St John's Day. She could then be confident that she could have found a husband before twelve months had past. Some sources say that the maid should pick the sprig on St John's Eve and sleep with it on her pillow that night. If, in the morning, the sprig is unwithered then she will be married before the next St John's Day. To dream of the man she will wed, a young woman should gather St John's wort and sleep with the leaves under her mattress.

The young maid stole from the cottage door
And blushed as she sought the plant of power,
'Thou silver glow-worm, lend me thy light
I must gather the mystic St John this night,
The wonderful herb whose leaf shall decide
If the coming year shall make me a bride'.

– Translated from the German

Medicinal

The flowering shoots could be worn to protect the wearer from evil forces, mental illness, colds and fevers. Medicinally it has been recommend as a sedative and a painkiller. It was used also as a treatment for headaches and rheumatism. King George VI, a firm believer in herbal remedies, was convinced of its efficacy. His confidence in it is marked by him naming one of his racehorses Hypericum. St John's wort has interested herbalists for hundreds of years and it has typically been used in the treatment of wounds. The knights of St John of Jerusalem used it to treat the wounded at the Crusades.

HYSSOP

Botanical Name: *Hyssopus officinalis*
Family: Labiatae
Common Names: Hyssop herb; isopo; ysopo; yssop
Gender: Masculine
Planet: Jupiter (in Cancer)
Element: Fire
Meaning: Cleanliness

Hyssop has gained some renown for its use in purification rites. It can be added to bath water, or sprinkled onto people or objects, in order to cleanse them. This attribute is alluded to in the Bible, where it says:

> Thou shall purge me with Hyssop, O Lord, and I shall be clean:
> Thou shalt wash me and I shall be whiter than snow.
>
> – Psalm 51:7

However, as European hyssop is not indigenous to Palestine it is more likely that the plant referred to was a species of Oregano.

Hyssop has been used to dispel evil from property, with small bunches being hung up in the house. It was also at one time strewn on the floor, as a protection against contagious diseases such as the plague.

A dream of hyssop is a portent of peace and happiness being brought by a friend.

HOLLY

Botanical Name: *Ilex aquifolium*
Family: Aquifoliaceae
Common Names: Aquifolius; Aunt Mary's tree; bat's wings; berry holly; berry holm; Christmas; Christmas tree; Christ's thorn; crocodile; hollen; hollin; holly tree; holme chaste; holy; hulm; hulver bush; killin; prick-bush; prick-hollin; prick holly; prickly Christmas; tinne; white wood
Gender: Masculine
Planet: Mars (Saturn, according to Culpeper)
Element: Fire
Deities: Tannus; Taranis; Thor
Meanings: 'Am I forgiven?'; 'Am I forgotten?'; domestic happiness; foresight; 'I thank God that you are restored to health'; recovery; resurrection

Myth and Legend

In pagan myth the holly is forever linked with the oak. The holly king rules nature during its decline from the midsummer solstice (Litha, 21 Jun) through to the midwinter solstice (Yule, 21 Dec), and the oak king rules over the other half of the year.

At each solstice these two engage in ritual combat for the attentions of the goddess and this combat is the subject of mummer plays. At summer's end the holly king would win and would rule the year, while at winter's close the oak king would emerge victorious.

This story appears as the basis for the Arthurian legend of Sir Gawain and the Green Knight. In this tale the Green Knight, an immortal giant, enters the court of King Arthur at the time of the New Year's feast carrying a holly bush as his insignia and challenges those assembled to a strange contest. Sir Gawain steps forward to accept the challenge as the king's champion. In the contest that follows Gawain beheads the Green Knight but, magically, fails to kill him. After six months the two must meet and do battle again. This story may be taken as an allegory of the passage of the seasons. The Green Knight is the ruler of the winter months, represented by the holly, while Sir Gawain can be likened to the Green Man, represented by the oak tree, ruling the summer months, and the two of them behead each other at midsummer and midwinter. When Christianity swept away the pagan rites and festivals these two characters were replaced by St John the Baptist and Jesus Christ. St John is the oak, and his feast day (24 June) is very near the midsummer festival. Christ is the holly, with a feast day in the midwinter. With that in mind it seems appropriate that the Christmas carol reminds us that the holly is glorified above all other trees as, 'of all the trees that are in the wood the holly bears the crown'. Another way of looking at the relationship of these two opposites is that the holly is the waxing ruler and oak the waning ruler. Robert Graves, in *The White Goddess*, goes further, suggesting that the holly replaced another oak, *Quercus ilex*, in the role of ruler of midwinter. *Quercus ilex* is better known as the holm oak or holly oak.

Holly, like mistletoe, was once banned from being used in decorations by the early Christian church, owing to its connections with pagan rituals. This ban stayed in force until the 1600s by which time a number of legends had sprung up linking holly with the Christian stories. One claimed that holly first grew from where Christ's feet had trod. Another suggested that it was the wood of a holly tree that was used for the cross on which Christ was crucified. Until then the berries of the holly had been yellow, but they were dyed red by the saviour's blood. Holly has been taken as being symbolic of the suffering of Christ, the spiky leaves representing the crown of thorns and the red berries drops of his blood. This led to it being known as 'holy tree', or 'Christ's thorn'.

The holly bears a berry red,
The ivy bears a black 'un,
To show that Christ his blood did shed,
To save our souls from Satan.

Another legend says that originally holly was a deciduous plant. Following the birth of Jesus the holy family was forced to flee into the desert to escape King Herod. Mary and Joseph concealed the infant in a holly bush. Mary prayed for protection for her child and in answer to her prayers the holly leaves grew back to hide and protect the baby Jesus. Thereafter the plant has always been evergreen.

In the Scottish tradition, at springtime the *Cailleach*, or Old Hag that presides over the winter months, threw her magic hammer under a holly bush, having been forced to admit that she could not prevent the growing light. Consequently no grass grows beneath a holly tree.

In Shinto mythology the Japanese holly held a similar position to that of the holly in Europe. A Japanese myth tells that Amaterasu, the sungoddess, withdrew into her cavern and refused to come out. Uzume, the erotic clowngoddess, hung a sacred mirror and sacred jewel in the branches of a holly and began to cavort around the black-fruited tree in a funny and sexy manner in order to attract the attention of Amaterasu in the hope of encouraging her out again so that spring would begin.

In Japan a popular New Year's charm consists of a holly leaf and a skewer, which represents the Buddhist monkgod Daikoku. Once, when Daikoku was in danger of being attacked by an oni devil, the rat, his companion, rushed away into the garden and returned with a holly branch, as oni devils will not go near holly.

In Norse mythology the holly is dedicated to Thor and Freya. Being associated with Thor means that it could be used to protect people from being struck by lightning. Both the Norsemen and the Celts would plant a holly tree near to their homes to take lightning strikes and so protect the house and the inhabitants. It might be the crooked leaf margin of the holly that gave rise to this association with lightning. However, hollies do conduct lightning better than most other trees. Having a smooth bark, water forms a sheet over it and lightning is conducted to ground through this surface water rather then through the sapwood of the tree. The tree, therefore, is rarely injured.

Magic and Lore

Holly is possibly the most important British evergreen plant in folklore. Like other evergreens it has been taken to represent immortality, life and rebirth. With its 'partner' ivy it has become closely associated with Christmas festivities but, like ivy, its use pre-dates the spread of Christianity. It was used at the pagan midwinter festival, and at the festival of Saturnalia celebrated by the Romans. The custom of giving gifts at Christmas may have started with the Roman practice of sending a holly branch, along with gifts, as a greeting at that time. It has been suggested that the early Christians, in order to conceal their celebrations of Christ's birth, used holly in their decorations to give the impression that they were celebrating Saturnalia. With the spread of Christianity holly lost its pagan association, becoming a symbol of Christmas.

It was customary amongst the druids to decorate homes with holly and other evergreen plants at midwinter, as an abode of the sylvan spirits. There are similar traditions amongst many northern countries. Holly is said to be a faerie favourite and would be hung by a doorway to provide a shelter for the wood spirits in the hope that this would ensure good luck for the household in the coming year. This is the origin of our holly wreath tradition.

Prior to the introduction of the Christmas tree tradition into Britain great balls of holly, adorned with ribbons, paper roses, apples and oranges were hung to decorate houses. This might be called a 'kissing ball' as sprigs of mistletoe were frequently hung beneath the holly ball.

Just as Christianity replaced the pagan festivals so the traditions and superstitions associated with the midwinter festival became absorbed into Christian mythology. Our Christmas carols are still rich with the image of holly as an emblem of Christmas.

> Christmastide
> Comes in like a bride,
> With holly and ivy clad.

Whether thought of as a plant of Christian or of pagan tradition Christmas is still the most significant time for using holly. Traditionally it should be collected on Christmas Eve if it to be used in decorations. It can then be kept until Twelfth Night, when it must be taken down; indeed to take it down before this will cause misfortune. Gathering it before Christmas Eve exposes one to evil on both a physical and spiritual level, while keeping it after Twelfth Night is said to bring a day of bad luck for every leaf of holly in the house. In Scotland the tradition was that it should be taken down before Candlemas and the festival of Brigid.

Once removed from the decorations it must not be burned in the house, as it will bring misfortune. Having holly in the house at any other time was sure to bring bad luck and lead to quarrels in the family. Traditionally, burning holly that is still green, or squashing the berries, will cause bad luck. Most unlucky of all was to take the flowering holly into the house during the summer.

Care should be taken if both holly and ivy are to be used in the Christmas decorations. Holly, especially prickly holly, is said to represent the male and ivy is taken to represent the female (in some parts of Britain smooth-leaved holly is taken as representing the female aspect and called 'she-holly'). As we see under Ivy, whichever is brought into the house first will show who will be dominant in the home for the next twelve months. There is also a superstition that holly must be hung up before mistletoe. Not to do so will cause quarrels amongst the family in the coming year.

One rather unpleasant custom from Tenby, in Wales, was that men and boys would go 'holly beating' on Boxing Day. They would run through the streets carrying holly branches and use then to beat the bare arms of any girls they found. There was a German custom to collect holly from the hedgerow on one evening during the midwinter festival, bring it home and keep it in the open air, or potted or planted. At the end of a month the branch, if still in full leaf or flower, would be carried amongst the family and friends. Those who needed to be healed, or wanted extra strength, or wished to have children would be touched by it and would reward the person who carried it with gifts.

One reason for including holly in decorations is that it affords protection against the forces of evil at a time of year when there are few other plants that might be used. As in the case of hawthorn and rowan it is the red berries that make it particularly potent. Twisting or twining plants to be used in magic is supposed to increase their efficacy. In the wreaths used at Christmas, holly would have been used either on its own or with ivy. There are obvious sexual connotations with the two plants representing the two genders.

The used of holly as a protective plant can be dated back at least as far as the Romans. Pliny claims that planting holly near to a property will give protection against witchcraft and lightning; it was also thought to prevent poisoning. Young women were advised to hang a sprig of holly on their bedhead at Christmas to protect them from evil and ensure happy dreams in the coming year. A more recent addition is that having a holly tree growing in your garden will keep the taxman away from your door.

Holly is used in a number of divination methods. If a young woman wishes to know whom she will marry she must collect nine fresh, non-prickly, holly leaves and place them in a clean handkerchief folded into a triangle and secured with three knots. This she must conceal under her pillow before going to bed, and she will be sure to dream of her future spouse. Another method, used variously at Halloween, Midsummer Eve, Christmas Eve or New Year's Eve, requires the young woman to pin three holly leaves to her nightgown, over her heart. She must then place three buckets of water in her bedroom before going to bed. If all is done correctly she should be awakened by three terrible wails, followed a little later by the sound of three hoarse laughs. She will then see the image of her future spouse enter the room. If he is to be greatly in love with her he will rearrange the buckets of water but if not he will leave them undisturbed. Men do not need to go to these lengths, they need only carry a few holly leaves and berries in their pockets and women will find them irresistible.

To discover whether a particular plan will succeed, place small candles onto holly leaves and float them on water. If the leaves continue to float the plan will succeed, if they sink it is bound to fail. A general guide to your health can be obtained by taking a few holly leaves on New Year's Day, and keeping them covered until Twelfth Night. If the leaves are unblemished when you uncover them you will enjoy good health. A few blemishes indicates that you should expect some illness. A considerable number of marks foretell a death.

Children in Warwickshire might also use holly leaves in divination. They would count off the prickles around the edge whilst repeating:

I pluck this holly leaf to see
If my mother does want me.

If there is an abundance of holly berries in a year, it is usually thought to indicate that the coming winter will be especially severe.

The holly is a tree of fire and strength in battle. It guides the just to find the strongest arguments in spiritual battles, and gives courage and vigour to fight against enemies. The wood from the tree was used for making spear shafts.

An old spell to help the garden grow requires a quart of ale to be poured into a silver tankard on the night of each full moon, and nine holly berries added that have first been blessed. The berries are washed in the rays of the moon and the tankard held aloft with the words:

> Fair Selene I drink to thee,
> May this mead a potion be!

The contents of the tankard can then be poured over hollyhocks, foxgloves, and other plants in order to improve their blooms.

Generally, holly is considered to be a lucky tree, affirming life and being symbolic of undiminished vitality. To come across a broken holly twig, or branch, is a sign that you will have a turn of unexpected good luck in the near future. There was a belief that the flowers could be used to cause water to freeze over and that the wood, if thrown at an animal, would cause it to turn and lie down by it.

In Scotland the holly is the badge of the Drummonds.

Medicinal

Holly has been recommended as a cure for chilblains. One method suggests that the berries are powdered and applied as an ointment. This is vastly preferable to the alternative, which recommends whipping the sores with a holly twig to 'let the chilblain out'.

Holly has also been claimed to have value to treat asthma, rheumatism, measles, gout and dropsy. Whooping cough could also be treated with it; children were given a hollywood cup to drink from. An old remedy for worms consisted of sage and holly into a bowl of water. When the patient yawned over the bowl the worm would fall out into the bowl.

MATÉ

Botanical Name: *Ilex paraguariensis*
Family: Aquifoliaceae
Common Names: Brazil tea; *houx maté*; jesuit's tea; Paraguay tea; yerba; *yerba maté*
Gender: Masculine
Planet: Saturn

Maté has been valued as an aphrodisiac, promoting lust and attracting members of the opposite sex. Infusions were used as lust potions. An infusion might also be shared by a couple to ensure that they would always stay together. If, however, at some future period you wished to break off the relationship it would be necessary to spill a little of the infusion on the ground to break the spell.

STAR ANISE

Botanical Names: *Illicum anisatum*; *Illicum verum*
Family: Illiciaceae
Common Names: Aniseed stars; badiana; Chinese anise
Gender: Masculine
Planet: Jupiter
Element: Air

Star anise has been used throughout China and Japan as a substitute for the true anise in medicinal preparations. The seeds are said to be endowed with magical powers. They are worn, strung as beads, or burned as incense in order to improve psychic powers. The plant or flower may be worn, or carried, as an amulet to bring good fortune.

PLOUGHMAN'S SPIKENARD

Botanical Name: *Inula conyzae*
Family: Compositae
Common Names: Cinnamon root; cloron's hard; great fleabane; horse heal; nard
Gender: Feminine
Planet: Venus
Element: Water

The common name of 'fleabane' gives the clue that this was once used to discourage insects. The leaves were burned, as their smell was said to be sufficient to kill fleas. If you could live with the scent of the leaves you might try to benefit from other properties attributed to the plant. Wearing it, hung about the neck, was supposed to protect against ill health and bring good fortune. It would also, of course, prevent flea bites.

ELECAMPANE

Botanical Name: *Inula helenium*
Family: Compositae
Common Names: Allecampane; elf dock; elfwort; elicampane; horse elder; horse heal; scabwort; velvet dock; wild sunflower
Planet: Mercury

Myth and Legend

The species name *helenium* commemorates Helena, the wife of Menelaus in mythology, who is known now as Helen of Troy. Some writers claim that when Paris abducted her she was clutching elecampane in her hand. Other sources say that the plant grew from her tears when she was carried off to Phrygia. Yet another says that Helena was the first to discover the medicinal properties of the plant.

Magic and Lore

The roots can be used in faerie magic. Scattering them about the house will attract faeries to your home, and growing the plant will bring them into the garden.

Medicinal

Some of the medicinal values claimed for this plant are slightly suspect. Pliny states, for example, that consumption of the roots will 'cause mirth' and 'doth fasten the teeth'.

HIGH JOHN THE CONQUEROR

Botanical Names: *Ipomoea jalapa*; *Ipomoea purga*
Family: Convolvulaceae
Gender: Masculine
Planet: Mars
Element: Fire

High John the conqueror is closely related to the more familiar morning glory (see below). Carrying a small piece of the plant is said to be sufficient to prevent depression. It is also supposed to act as a lucky charm, protecting against curses, or spells, and bringing success and love. To attract money it is the root that should be carried; this must first be anointed with oil of mint, and then kept in a green sachet.

MORNING GLORY

Botanical Name: *Ipomoea purpurea*
Family: Convolvulaceae
Gender: Masculine
Planet: Saturn
Element: Water
Meanings: Affectation; repose at night

Myth and Legend

A Japanese legend tells how the sun goddess, who passed the day sitting at her loom weaving brightly coloured clothes for her children, became so angry with her devilish brother that she shut herself up in a cave to get away from him. With the sun hidden away in a cave, Japan, 'the land of many islands', was bathed in perpetual twilight. Gods and mortals pleaded with her to return, but to no avail. In time they turned to the god of thought for suggestions. He commanded the gods to gather 500 coloured jewels and hang them on a tree outside the cave. He then placed a great mirror against the tree and ordered the gods to make the sounds of great merriment. Curiosity led the sun goddess to come out of the cave to discover the course of all the noise. She was then dazzled by her own brilliance sparkling in the jewels and reflected in the mirror. The gods rejoiced at her return, but in their joy the mirror was accidentally knocked over and shattered. Its weight was sufficient to cause the tree to fall, and the gods were buried under splinters of broken glass and jewels. The next morning multicoloured morning glories grew upon the site, and ever since the flower has been called the 'Jewels of Heaven'.

Magic and Lore

Morning glory can be used as a substitute for high John the conqueror in various spells and charms. It is a beneficial plant to have growing in the home or garden. Where it thrives it will bring peace and happiness; if you place the seeds under your pillow before going to sleep you can be sure of a peaceful night.

Planting morning glories in the garden is supposed to be a good way to keep hostile faeries, especially nocturnal ones, away from your home.

IRIS

Botanical Name: *Iris* spp.
Family: Iridaceae
Gender: Feminine
Planet: Venus
Element: Water
Deities: Hera; Iris; Isis; Juno
Meanings: Flame; 'I burn' (German); ardour; 'You have set my heart aglow' (purple); flame of love; passion; 'I mourn with you' (yellow); a message; faith; hope; 'My compliments'; promise; valour; wisdom; 'Your friendship means the world to me'

Myth and Legend

Iris is named after the goddess of the rainbow in classical mythology in allusion to the beauty, and diversity, of its colours. Iris was the daughter of Thaumus and Electra, and a messenger of the gods, and is depicted as a beautiful young woman dressed in multicoloured robes, with the wings of a butterfly, and carrying a herald's staff. She glided along the rainbow bearing messages from the gods to men and to the Underworld. She also guided the souls of women to the Elysian Fields (a role Mercury fulfilled for men); consequently irises were frequently placed on the graves of women.

Magic and Lore

From ancient times the iris has been used as a symbol of majesty and power. It was dedicated to the goddess Juno and, it has been suggested, was the origin of the royal sceptre. Christianity took on this symbolism, using the flower as an emblem of royalty. Paintings of the Madonna and child might show a crown of irises, rather than lilies, held above the head of the Virgin, with another held in the hand of the holy infant. Irises might also be depicted growing outside the stable where Christ was born. All of these would indicate that this was a child born to be king. There would be a further allusion to the trinity, as the flowers have three upright petals, called the standards, and three petal like sepals, called the falls. In ancient Egypt it was placed on the brow of the sphinx and on the sceptres of kings, the three sections of the flower being taken to symbolize faith, wisdom and valour. In France, for centuries, the symbol of an iris, the fleur-de-lis (see p. 260), was used as the royal and national emblem.

Since the time of the Romans, irises have been used in purification rites, fresh flowers being placed throughout the area to be cleansed. The sentiment expressed in giving the flower is 'a message'. Similarly, to see an iris in your dreams indicates that you should expect to receive a letter bearing important news.

Medicinal

In ancient times the iris was credited with healing powers simply by virtue of it being a sacred flower.

ORRIS

Botanical Name: *Iris florentina*
Family: Iridaceae
Common Names: Florentine iris; flower de luce of Florence; Queen Elizabeth root; white fleur de luce

Orris is obtained from several different irises. Veronese orris and mogadore, or Morocco orris, come from *Iris germanica* and some was obtained from *Iris pallida*. But the finest is from *Iris florentina*. Orris is used in the preparation of perfumes. Like the perfumes from which it is made, the plant is used to attract love. The powdered root is sprinkled onto sheets and clothing, or applied to the body as a dusting powder. An alternative method is to carry the whole root as an amulet.

In Japan orris root was used as a protection against evil, and both leaves and roots were hung in the eaves of houses to protect all inside. Individuals could bathe in water to which orris had been added. Another use of the root is in dowsing, with the root strung on twine and used as a pendulum.

FLEUR-DE-LIS

Botanical Name: *Iris pseudacorus*
Family: Iridaceae
Common Names: Butter and eggs; cegg; cheeper; crane bill; cucumbers; daggers; dragon flower; duck's bill; dug's lug; flagger; flaggon; flag lily; flag plant; flags; fleur de lys; fleur de luce; fliggers; gladyne; Jacob's sword; laister; lavers; leavers; levers; levver; livers; mekkins; meklin; myrtle flower; pond lily; queen of the marshes; queen of the meadow; saggon; segg; seggen; seggie flooer; shalders; sheep shearing flower; sheep-shears; skeggs; sword grass; sword lily; swords; trinity plant; water lily; water segg; yellow devils; yellow flag
Planet: Moon
Meanings: Eloquence; flame of love; hope; light; 'My compliments'; power; royalty

Myth and Legend

In general irises were considered to be symbols of majesty and royalty. Whilst the fleur-de-lis, was the heraldic symbol of the French kings it was not adopted just because of this symbolism. Legend has it that, early in the sixth century, Clovis, the Merovingian king of the Franks, was at war with the Goths. The battle had reached a decisive point and the Frankish army faced defeat. Clovis prayed to the God of his Christian wife, Clothilde, for an aid. After rejoining his troops he saw some yellow flag irises growing at the side of a river, which until then had impeded his advance. Realizing that, for the plant to be growing there, the river must be shallow at that point he led his army across, surprised his enemies and won a great victory. Clovis took the iris as being a sign from the divine and so replaced the three toads, which had been his badge, with three irises. He also embraced the religion of his wife.

Louis VI, in the twelfth century, was the first French monarch to have the fleur-de-lis motif on his shield. Later, the same emblem was adopted by Louis VII to appear on his banners when fighting in the Crusades. It is said that this led to it being called the fleur de louis, later corrupted to fleur de luce, or fleur-de-lis. Another explanation for the name is that it was derived, not from 'Louis' but from the River Lys, in Flanders, where the iris grew in abundance. The iris was later incorporated in the royal coat of arms of England as King Edward III laid claim to the French crown. The white banner carried by Joan of Arc showed God blessing the fleur-de-lis, when she led French troops to victory over the English.

Magic and Lore

Many of the older common names of this iris refer to its sharp leaves, likening them to swords. Some of the more obvious ones are 'daggers', 'Jacob's sword' and 'sword lily'. 'Segg', is not derived from 'sedge'; another waterside and marsh plant, but was the name of an Anglo-Saxon short sword. Pliny tells us the fleur-de-lis was dug up with some ceremony to be used in ritual purifications by the Romans.

It is the birthday flower of 6 September.

BLUE FLAG

Botanical Name: *Iris versicolor*
Family: Iridaceae
Common Names: Dagger flower; dragon flower; flag lily; liver lily; poison lily; snake lily; water flag
Gender: Feminine
Planet: Venus (Moon, according to Culpeper)
Element: Water

In Christian mythology we are told that the blue flag iris was originally yellow, but took on its dark colour and went into eternal mourning when Christ was crucified.

The root, or rhizome, can be carried as an amulet to bring riches. It can be kept in a pocket, but is probably safer put into a purse or wallet, as the plant is poisonous. If you want to ensure that your business will prosper try placing a piece of the root into the cash register to attract money to your business.

In Japan the blue flag represents heroism. The blue colour refers to blue blood, and irises play a key role in the Japanese spring festival for boys.

JASMINE

Botanical Name: *Jasminum officinale*
Family: Oleaceae
Common Names: Jessamine; moonlight on the grove
Gender: Feminine
Planet: Moon (Jupiter in Cancer, according to Culpeper)
Element: Water
Deities: Vishnu
Meanings: Amiability (white); grace and elegance (yellow); sensuality; sexy (Spanish); elegance; cheerful and graceful; 'How dainty and elegant you are'

Myth and Legend

In Hindu tradition the jasmine is associated with Vishnu, and quantities of the flowers would be included amongst the offering made at some of the ceremonies to the god. The flowers might also be strung together to form garlands that were presented to honoured guests.

A Christian legend relates that the flowers have not always been white. Before Christ's cricifixion they had a beautiful pink hue. At the time of the crucifixion, however, all sweet flowers in the world also died. The pink flowers of the jasmine closed. When next they opened, presumably at the resurrection, they had bleached to white.

An Indian legend explains why jasmine blooms are most fragrant at night. It says that a princess became hopelessly infatuated with the sun god, Surya-Deva, who was unaffected by her beauty and spurned her love. The devastated girl took her own life.

She was cremated and her ashes scattered throughout the realm. Jasmine flowers grew up everywhere they were scattered and, because it was the sun god who had broken her heart, the flowers refused to open during the day, only blooming during the night.

Magic and Lore

The name 'jasmine' is derived from the Arabic, *jas* meaning despair, and *min*, a lie. Jasmine is a plant greatly favoured by gardeners for the sweetness of its scented flowers. The sweetness of the scent has led to it being associated with love. In the Western tradition of magic the dried flowers are added to mixtures or sachets with the aim of attracting spiritual love. In China, it is used by those seeking to find a more physical kind of love. Chinese women might wear the flowers in their hair so that the buds, which continue to open on warm nights, will scent the air and act as a perfume. The carrying or wearing of jasmine is also said to attract money. Drying the flowers and burning them as part of an incense is also supposed to attract wealth.

Burning the flowers in a bedroom or, according to some authors, just smelling them, can lead to prophetic dreams. For jasmine to appear in a dream, whether it be prophetic or otherwise, is a sign of good luck, especially for those in love.

Medicinal

Jasmine was traditionally used as an aphrodisiac and it is claimed that a few drops of the oil, when combined with almond oil and massaged into the body, will overcome frigidity.

WALNUT

Botanical Names: *Juglans nigra*; *Juglans regia*
Family: Juglandaceae
Common Names: Carya; Jupiter's nuts; tree of evil; walnoot
Gender: Masculine
Planet: Sun
Element: Fire
Deities: Artemis; Car; Carmenta; Carya; Jupiter; Metis
Meanings: Intellect; 'Ours would be a marriage of true minds'; stratagem

Myth and Legend

In the 'golden age' of classical mythology, before Ceres created corn, mankind ate acorns as their staple food and the gods fed on walnuts. The walnut was called *Jovis Glans*, or 'Jupiter's nuts', from which we derive the botanical name of *Juglans*.

Magic and Lore

The walnut is the sometimes known as the royal tree or tree of prophecy. If she fell it was said to be an omen that some great calamity would soon follow. An Italian tradition says that witches would prefer to conduct their magical rites and celebrations beneath its protective canopy rather than almost anywhere else. However, they should be very careful as the tree is supposed to attract lightning. Indeed sitting beneath a walnut is usually thought of as courting misfortune. One warning says:

> He that would eat the fruit must climb the tree;
> He that would eat the kernel must crack the nut;
> He that sleepeth under a walnut doth get fits in the head.

In Sussex it was said that it was unsafe to be beneath the walnut, as effluvium from the tree would cause 'derangement of the mind'. Sleeping under the tree, therefore, could result in madness or even death. Another dire warning relating to being beneath the walnut was that if you should walk beneath the branches you would hear the servants of the Devil whispering up in the branches.

The darker side of this tree can, quite literally, be seen in the black dye produced from the shells. It could be used, amongst other purposes, as an instrument of punishment. The eponymous Mikado of the Gilbert and Sullivan operetta declares that any women who dye their hair will have their skin dyed with walnut juice, thereby depriving them of the fashionable pale appearance.

In weather lore the walnut will warn of severe weather conditions ahead. If the walnut crop is especially good, or the husks on them are especially tight, then a very bleak winter can be expected, followed by a good grain harvest in the coming year.

A woman may discover the nature of her future spouse by the use of a somewhat curious spell using walnuts. She must make up some small pills from a mixture of grated walnut, nutmeg, hazelnut, butter and sugar. Before going to bed she should swallow nine of these tablets, and that will induce a dream that will show her future. If she dreams of silk, gold, or precious gems, then she will marry a wealthy man. If she dreams of white linen, her future husband will be a clergyman. A dream of nothing but blackness indicates that she will wed a lawyer, and if she dreams of lightning her husband will come from the armed forces. A dream of rainstorms shows that she will marry a clerk or domestic servant.

Walnuts can also be used as a form of contraception. If she does not want to have children for several years following her marriage she must hide some roasted walnuts in the bodice of her wedding dress, one walnut for each year she wishes to remain childless. This spell is only effective on her wedding day.

A dream of walnuts indicates a faithless marriage, worry, disappointments, rebukes and general upheavals. On the other hand, to dream of shelling and eating walnuts suggests that you will receive money.

Walnut has a reputation for blighting all vegetation growing nearby. As late as the nineteenth century the advice was to avoid planting them close to houses and important plantings, such as strawberry patches or vegetable plots. It is claimed that there is a particular antagonism between walnut trees and oaks, so that the two should never be planted close to each other.

Although having a black walnut nearby is said to be sufficient to blight all the apple trees in the area, the apple and the walnut share a similar tradition. Just as there is a tradition of wassailing apple trees to ensure a good harvest, so there is a similar ceremony with walnut trees. Some growers will whip their walnut trees with chains, to 'scare' them into fruiting and so ensure a bumper crop. An old country saying runs:

A woman, a whelp, and a walnut tree,
The more you beat them the better they be.

Beating walnut trees has a sound basis. The idea is that if the survival of the tree is threatened it will naturally try to produce a good crop of seeds for the future. It is most likely, however, that the beating of the tree had little to do with increasing yields and more to do with damaging the wood to improve the grain and produce what was called burr walnut, which was more valuable as timber.

Medicinal

Under The Doctrine of Signatures it was thought that the walnut corresponded perfectly with the head, as it had a hard shell with a centre that was ridged like the brain. 'Walnuts bear the whole Signature of the Head, the outwardmost green barke answerable to the thick skin whereunto the head is covered, and a salt made of it is singularly good for wounds in that part, as the Kernell is good for the braines which it resembles.' Wearing the nuts as a necklace was supposed to cure epilepsy and prevent mental illness and wearing the leaves on a hat or about the head was a remedy for headaches and the prevention of sunstroke.

In folk medicine walnuts are recommended as a treatment for rheumatism. Simply carrying them was said to be sufficient to prevent the pains and to strengthen the heart. Nicholas Culpeper also prescribed a mixture of walnut with honey, onion and salt as a cure for the bites of any venomous animal or mad dog. Perhaps the strangest claim in folk medicine is that carrying a spider within a walnut shell will prevent fevers.

RUSHES

Botanical Name: *Juncus* spp.
Family: Juncaceae

Myth and Legend

Many of the legends associated with rushes come from Ireland, where they are linked to St Brigid, or St Bride, whose feast day is 1 February. On St Bride's Eve crosses of rushes would be made as amulets. Tradition has it that St Brigid was an abbess living between 450 and 523, but there are few verifiable facts about her. Legend says that she was passing by an old shed when she heard someone moaning inside. When she stepped through the door she saw a dying man. She went across to offer him some comfort and to tell him of God. The man refused to listen to her, and eventually Brigid went outside and formed a cross from some rushes that grew nearby. When she took it back to the dying man he was so moved that he made his confession and received the final sacraments.

Another legend, which comes from County Galway, explains why the tips of rushes so often become brown and withered. One evening, on retiring to bed, St Patrick called his servant, and directed him to sit and listen in case he spoke in his sleep. After he has slept a little while he said, 'Bad luck to Ireland.' The listening servant said, quickly, 'If so let it be on the tips of the rushes.' A short time later the saint again said, 'Bad luck to Ireland.' The boy immediately exclaimed, 'If it be so let it be on the highest parts of the white cows.' Soon afterwards the saint repeated, 'Bad luck to Ireland.' His servant said, 'If so let it be to the bottom of the furze.' When St Patrick awoke he asked his servant if he had said anything in his sleep. The boy told him what he had said and what he had said in response. Ever since, the tips of rushes have been brown and withered, the tips of the horns on white cows have been blackened and the bottoms of all gorse bushes are withered. What is more, every priest will have a boy assisting him when he says Mass.

A rather different explanation tells us that St Patrick was given a piece of dog meat to eat at a house he was visiting. When he learned what it was he cursed the house. Reflecting on the incident he regretted the curse on the house but could not retract it without placing it onto something else instead. He therefore, lifted it from the house and placed it onto the rushes.

It is, perhaps, because the tips of rushes are usually brown that to find a green-topped one is thought to be lucky.

With a four-leaved clover, a double leaved ash,
And a green tipped seave [rush],
You may go before a queen's daughter,
Without asking leave.

When rushes appear in your dreams, it shows that you must beware of a stranger; someone will try to win your confidence for their own ends, and you must be discreet.

JUNIPER

Botanical Name: *Juniperus communis*
Family: Cupressaceae
Common Names: Bastard killer; enebro; Geneva; *gemeiner wacholder*; gin berry; gin plant; ginepro; saffren; saffron; savin
Gender: Masculine
Planet: Sun
Element: Fire
Meanings: Asylum; comfort; 'Please let me care for you, for ever'; protection; succour

Myth and Legend

It is said that, when the holy family fled from Bethlehem into Egypt, in order to evade the wrath of King Herod, the juniper was one of the plants that concealed and protected the infant Jesus. It may well be from this that it is said have the power to counter evil and demonic forces.

Magic and Lore

There is a hint, from an old Welsh superstition, that this plant is one of the sacred trees. It is said that anyone who cuts down a juniper will die within a year, or will lose a member of their close family. This warning is usually found associated with trees such as apple, ash and oak, all of which have traditionally been important in magical rites and ceremonies. But it is not surprising for the juniper to be seen as having a sacred place as it is one of the three British native coniferous trees (the other two being the Scots pine and the yew).

Another attribute that the juniper shares with the trees usually considered to be sacred is its use as a protective talisman against everyday mishaps and all types of evil forces. This use seems to be mainly European in origin. It is claimed that wearing a sprig of juniper will be sufficient to protect against accidents and attacks by wild beasts. It is even said that it will drive away snakes and serpents. It will also protect against evil and will drive away ghouls and demons, as well as breaking any spells cast against the bearer. If you wish to protect property from evil then you must hang a small sprig of juniper over the doorway to the house. You can even drive out any evil forces already present by adding juniper to the incense burned in exorcism ceremonies. In the Highlands of Scotland it would be burned in the house and byres at New Year to protect them for the coming year.

Ironically, juniper was probably one of the earliest incenses used by witches throughout the Mediterranean region. This, presumably, indicates that not all those who practise witchcraft are of evil intent.

The berries have been used as an aphrodisiac. Merely carrying them is supposed to be sufficient to increase male potency. They can also be added to potions and mixtures used to inspire love. They are, of course, also the fruit used to make gin, which may, therefore, prove to be something of a love potion.

In common with a number of other plants there is supposed to be spirit that resides within the tree. In this case it is a goblin called Frau Wachholder, who can be invoked by those who have had items stolen from them. They should go to a juniper bush and bend one of the branches to the ground, securing the tip beneath a stone while calling upon the thief. He is then bound to present himself before the tree and relinquish his ill-gotten gains. That done he can release the branch and replace the stone.

To dream of junipers is usually considered to be an indication of impending bad luck. However, a dream about the berries signifies good luck as it indicates coming success or the birth of an heir.

Medicinal

The juniper was also burned during outbreaks of the plague and other contagious diseases as a protection. Consuming the berries was sometimes recommended as a treatment for dropsy.

RED HOT POKER

Botanical Name: *Kniphofia uvaria*
Family: Liliaceae
Common Names: Flame flower; torch lily

Red-hot pokers are a favourite cottage garden flower, but these popular summer beauties may prove to be prophets of doom. If they flower twice within the same year it is a sure indication that there will be a death in the household.

GARDEN LETTUCE

Botanical Name: *Lactuca sativa*
Family: Compositae
Common Names: Lattouce; sleepwort
Gender: Feminine
Planet: Moon
Element: Water
Deities: Min; Venus
Meaning: Cold hearted; coldness

Magic and Lore

It may be hard to believe now, but in ancient times the lettuce was held in high esteem. It is mentioned amongst Chinese writings from the seventh century BC and Theophrastus, writing in 320 BC describes both wild and cultivated varieties. Its cooling and refreshing nature led to its use in medical treatments. The Emperor Augustus, on recovering from a serious illness, erected a statue in honour of the plant and dedicated an altar to it as he believed that his recovery was solely due to its qualities. The Greek philosopher Aristoxenus is said to have thought so highly of his lettuces that he watered them with his best wines; there is no evidence that this improved the flavour.

Lettuce is a plant of contradictions. To some writers eating lettuce prevents lustfulness. In mythology we are told that when Adonis died the goddess Venus flung herself onto a bed of lettuce in order to cool her passions. A number of writers have picked up on this idea, Culpeper, for example, says that lettuce 'abates bodily lust, represses venerous dreams, being outwardly applied to the testicles with a little camphire'. However, it has also been credited as a powerful aphrodisiac, arousing love and boosting fertility. To aid childbearing women might eat it in a salad or take a decoction of the leaves or seeds.

Lettuce may also be used in love spells and charms. If you wish to gain the love of someone you might write their name in the soil, using a stick or garden cane, and sow lettuce seeds in the groove. Great care must then be taken in tending these plants because if they germinate and flourish so will the person's love for you. To discover how long it will be before you marry, toss a salad while thinking about marriage, and the number of lettuce leaves which fall out of the bowl will indicate the numbers of years before your wedding day. In some parts of the country it is said to be unlucky for women to plant a large number of lettuces in the garden if they are single as it shows they will never marry, or if married they will never conceive.

Growing lettuces in the garden is a good way to protect the home from evil. However, just as eating the leaves will suppress lustful thoughts, so growing excessive quantities of the plants can cause sterility in members of the household. If you come across a white stalk amongst your plants do not to pull it out, as it will result in a death in your family.

Dreaming of eating lettuce leaves is an omen of misfortunes ahead, indicating difficulties in the management of all your affairs.

Medicinal

Some of the ailments which lettuce has been use to treat include seasickness, and insomnia. Eating the leaves is supposed to be sufficient to prevent seasickness and counter the effects of drinking too much wine. Insomnia sufferers will also benefit from eating the leaves. Those who do not enjoy eating lettuce can try rubbing a little of the sap on their forehead as this is said to have the same effect.

WILD LETTUCE

Botanical Name: *Lactuca virosa*
Family: Compositae
Common Names: Acrid lettuce; green endive; *lactucarium*; *laitue vireuse*; lettuce opium; sleepwort; strong-scented lettuce
Gender: Feminine
Planet: Moon
Element: Water

The wild lettuce, like its cultivated cousin, has been called 'sleepwort' because all lettuces have some narcotic properties, but those of the wild lettuce are the strongest. Water distilled from the sap has been used as a mild sedative. The botanical name carries its own warning of the poisonous nature of this plant. *Lactuca*, the genus, refers to the milky appearance of the sap. The species name, *virosa*, means that the plant is potentially poisonous.

CRAPE MYRTLE

Botanical Name: *Lagerstroemia* spp.
Family: Lythraceae
Common Name: One hundred day red flower

'Crape' is the anglicized form of *crepe,* which in turn, is derived from the Latin *crispus*, referring to the ruffed margins of the flower petals.

Korean legend tells how the inhabitants of a small coastal village made an annual sacrifice to a three-headed sea dragon in order to appease him. The victim was inevitably a young maiden, who would be dressed as if for her wedding. One year the sacrifice was to be the beautiful daughter of the House of Kim. She was suitably clothed in a wedding dress and taken to the shore to await her fate. At the crucial moment, when the dragon arrived to claim his 'bride', a handsome and heroic prince sailed by. He cut off one of the dragon's heads and rescued the maiden.

The prince's father arranged for the wedding of his son to the girl but on the day of the wedding all the King's treasure magically disappeared. The King cancelled the wedding ceremony, and his son promised to find the treasure and return to claim his bride within 100 days. In a twist that has similarities with the classical myth of Theseus, the prince, before taking his leave, promised his bride that he would fly a white flag from his ship when he returned if he had been successful, but a red one if he was not. The maiden waited the hundred days until at last, the prince's ship was sighted. When she saw the ship was flying a red flag she died of grief, knowing that she could never marry the prince. However, when the ship docked, everyone could see that the flag was white, but stained with the blood of the dragon that the prince had slain to recover the treasure. The prince arrived in time to attend the funeral of the girl he was to have married. She was buried and from her grave grew the crape myrtle that will bloom for 100 days each year.

LARCH

Botanical Name: *Larix decidua*
Family: Pinaceae
Gender: Masculine
Meanings: Audacity; boldness; disguise; 'Only he who presses his suit with spirit shall win me'

The larch is a tree of ships and bridges. In Siberian myth if was the tree from which the first man was created. There was a belief that larchwood could not be penetrated by fire, which led to its inclusion in sachets which were kept as amulets to protect against fires. The irony, of course, is that turpentine, which is very flammable, is obtained from the larch. A larchwood amulet is also one of the methods used to protect against enchantments and the evil eye.

It should also be remembered that in much of our folklore reference is made to a fir tree, and whilst this was usually a pine, some of the superstitions surrounding that tree were occasionally applied to the larch.

EVERLASTING PEA

Botanical Name: *Lathyrus latifolius*
Family: Leguminosae
Common Names: Pharaoh's pea

Legend claims that seed of the everlasting pea was brought to the West from Egypt, having been found in a pharaoh's tomb. However, as it has not been recorded in Egypt, or the neighbouring countries, this seems somewhat unlikely.

SWEET PEA

Botanical Name: *Lathyrus odoratus*
Family: Leguminosae
Gender: Feminine
Planet: Venus
Element: Water
Meanings: A meeting; delicate pleasures; departure; 'Only in your departure will I find any pleasure'; 'Remember me'; tenderness; 'Your memory is a lingering fragrance'

There are great rewards to be had from carrying or wearing the flower of the sweet pea. If you carry one in your hand all those you meet will tell you nothing but the truth. Wearing it will attract people to you and cause friendships to develop. It will also boost your strength and courage. A small vase of flowers in a bedroom will ensure the chastity of the person who sleeps in the room.

Various 'secrets' have been given for the successful cultivation of sweet peas. They should be planted in rows running north to south for the best blooms. To get the biggest flowers the seed should be soaked in milk overnight before sowing. If the seed is sown before sunrise, on St Patrick's Day (17 March) it will ensure good growth and abundant, fragrant flowers.

BAY LAUREL

Botanical Name: *Laurus nobilis*
Family: Lauraceae
Common Names: Bay; baie; bayes; bay tree; Daphne; grecian laurel; lorbeer; lorer; *laurier d'apollon*; noble laurel; Roman laurel; sweet bay; sweet laurel; true laurel
Gender: Masculine
Planet: Sun (in Leo)
Element: Fire
Deities: Aesculapius; Apollo; Ceres; Daphne; Eros; Faunus
Meanings: Instruction (berry); glory; 'Your elegance and majesty dazzle me' (branch); 'I change but in death' (leaf); award of merit; 'Your determined suit has won my heart' (wreath)

Myth and Legend

A story from Greek mythology tells of the origins of the bay laurel. It seems Daphne, the beautiful daughter of the river god Peneos, had become a target for the mischief of Cupid. Although we usually think of 'Cupid's arrows' as being the cause of love there were other arrows in his quiver. In this case he struck Daphne with an arrow of lead, so that she would abhor all thoughts of love. She resolved to remain a virgin for her whole life. This, however, was only a part of Cupid's mischief, as he had also fired a golden arrow at the god Apollo. Consequently, Apollo was smitten by Daphne who, because of the leaden arrow, rejected his advances. but he was not to be dissuaded, and became ever more amorous. In an effort to evade him Daphne sought the protection of her father, who transformed her into the plant we know as a bay tree. Even this could not cool Apollo's love for her. He declared that, thereafter, he would wear the bay on his brow, and use it to decorate his lyre, forsaking the oak, which had previously been his emblem. All who sought to win the approval of this god were obliged to do likewise and his oracle was to be found in a grove of bay trees.

In Romanian legend there is a tale of a young nymph, called the daughter of the laurel, who dwelt amidst the branches of a laurel bush. Every night the branches would open to allow her to depart and attend the dance in the Valley of the Flowers. On one such evening, while she danced, she was seen by a young man. He was so overcome he rushed forward, praising her beauty and grace, and embraced her. Being a shy creature, she fled and hid among the flowers, protected by the Star Queen, who dwelt in a palace in the clouds. Undaunted, the handsome young man sought her out and, having found her, he promised never to abandon her; he then sang her to sleep in his arms. When she awoke he was gone. She pleaded with the Star Queen to help her find him, but she refused. When she tried to return to her home in the laurel as daylight approached, it would not open its branches to her, saying 'The wreath of honour has fallen from your brow; there is no longer any place for you here.' Unable to shelter from the rising sun the 'daughter of the laurel', dissolved into the morning dew.

Magic and Lore

Our modern use of the bay laurel to symbolize peace and victory, is based on a very ancient tradition. It was one of the chosen plants of the Greek oracles, such as that of Apollo at Delphi, where the prophetess, or Pythia, would chew laurel leaves or burn them and inhale the smoke in order to induce a prophetic trance. The leaves were believed to communicate the spirit of poetry and prophecy. Laurel was used in brews to give clairvoyance and wisdom, or worn as a crown to aid prophecy. It was always a woman who presided at the oracle at Delphi. Early holders of the office were always attractive virgins from noble families but after Pythia was seduced in the sacred cave the subsequent incumbents were as old and unattractive as possible.

Poets were crowned with wreaths fashioned from the leaves, as were victorious athletes and heroes. Some athletes might even choose to wear a crown of bay laurel at the time of the competition as they believed that it would increase their strength, and so enhance their performance. There are modern echoes of this practice of crowning scholars and heroes with bay laurel wreathes in the titles of the degrees conferred upon graduates. The term 'bachelor', as applied to the holder of a degree, may be derived from the bay, or rather *bacca-laureus* ('laurel berry'), through a corruption of the French word *bachelier*.

Daphnomancy is the name given to the use of bay laurel in fortune telling. If, for example, you want to know whether you can expect a period of good luck you need only toss a few fresh leaves onto an open fire. If they crackle as they burn your luck will be good but if they merely smoulder it is a much less favourable omen. One way in which to make a wish come true is to write it onto a bay laurel leaf, and then burn the leaf. There is an additional benefit in burning bay leaves for those who are separated from their loved ones. Tradition has it that the scent of the smoke will bring back the lover.

On St Valentine's Day a spell can be woven to enable a young woman to identify whom she will eventually marry. She must sprinkle a bay leaf with a little water then place it across her pillow before going to bed. She must wear a clean nightdress then lie down saying:

St Valentine be kind to me,
In my dreams let me my true love see.

She will dream of the person that she will marry, and will be married within the year. Placing a sprig of bay beneath the pillow is one way to ensure that she will only have pleasant dreams. An alternative method recommends that on St Valentine's Eve, she should take five bay leaves and pin them to her pillow. One must be placed at each corner and the fifth in the centre. When she lies down she must repeat the following phrase seven times:

Sweet guardian angel let me have
What I most earnestly do crave,
A valentine endued with love,
Who will both true and constant prove.

Between each time of reciting the phrase she must count from one to seven, seven times. If she does all this she is sure to see her future spouse in her dreams. The simplest charm to ensure that a girl will dream of her future husband, however, merely requires her to tuck two bay leaves beneath her pillow before she goes to bed. Merely seeing laurel in a dream would signify that one will achieve success and fame.

A young person could ensure the fidelity of their partner by breaking a small twig from a bay tree and snapping it in half. So long as each partner keeps a piece they would remain loyal to one another.

A spell to attract rain requires a woman to collect nine bay leaves and sufficient strands of hair from a hairbrush to wrap around her hand nine times. During a full, or new moon she should tie the leaves and hair together to form a garland, all the time saying:

> Fair Selene let this be
> A rope of charms and sorcery;
> I root my art in God's good earth
> To give it secret cunning birth;
> May it bloom like the bride at kirk –
> My soul and nature's handiwork.

She must then bow to the moon and place the garland into the hole prepared for planting a tree or shrub. This will ensure the rain needed for the plant to be established.

Not all the superstitions relating to this plant are favourable. There is an old tradition that if the bay laurel withers it is an omen of impending disaster or even a death in the household. More generally the sudden wilting of the trees has been said to indicate the death of the monarch.

> 'Tis thought the King is dead. We'll not stay –
> The bay-trees in our country are withered.
>
> – Shakespeare, *Richard II*

It is also said that before the death of the emperor Nero all the bay laurels in Rome withered.

Bay laurel has long been considered to be one of the most powerful of protective plants. Sprigs would be hung up around the home, and the leaves burned during exorcisms. One report of its efficacy can be found in the story of Mary Hortado who, in 1683 lived in Salmon Falls, Massachusetts. Her home was plagued by a poltergeist, and she was unable to get any peace until she hung shoots of green bay laurel in her home. Thereafter, so long as the bay remained green her home was quiet.

Nicholas Culpeper, in his herbal, tells us:

> It is a tree of the Sun, and under the celestial sign Leo, and resists the witchcraft very potently as also all the evils old Saturn can do the body of a man, and they are not few: For it is the speech of one

> and I am mistaken if it were not Mizaldus, that neither witch nor devil, thunder nor lightning, will hurt a man where a Bay tree is.

Culpeper refers to another traditional belief once widely held. It was thought that the bay tree would never be struck by lightning. People would place twigs in the windows of their homes to ward off lightning and protect their property. During storms they might choose to carry branches over their heads, not so much to keep off the rain as to protect themselves from lightning.

> Reach the Bays –
> I'll tie a garland here about his head;
> 'Twill keep my boy from lightning.
>
> – Webster, *The White Devil*

It seems that this belief dates from antiquity. It is reported that the Roman emperor Tiberius was considerably worried by storms. Apparently he would cover his head with bay laurel leaves and hide beneath his bed when there was thunder.

In magic it is used in purification rites. A sprig may be used to sprinkle holy water about the area to be cleansed or the leaves added to incense, containing sandalwood, which is burned to remove curses and spells. Another practice is to hold a bay leaf beneath the tongue to guard against all bad luck.

Medicinal

In folk medicine bay was recommended for the treatment of snakebites and diseases such as scrofula, the so-called 'king's evil', and the plague.

LAVENDER

Botanical Name: *Lavendula augustifolia*
(Synonym: *Lavendula officinalis*)
Family: Labiatae
Common Names: Elf leaf; nard; nardus; spike
Gender: Masculine
Planet: Mercury
Element: Air
Deities: Aradia
Meanings: Acknowledgement of love; assiduity; diligence; distrust; 'I can only ever be your friend'; negation; sad refusal; 'Sweets to the sweet'

> Lavender is for lovers true
> Which evermore be fain,
> Desiring always for to have
> Some pleasure for their pain.
>
> – Elizabethan lyric

The name 'lavender' comes from the Latin *lavare*, meaning 'to wash'. It was introduced to England around 1568, however this will have been a reintroduction, as there is little doubt that it was brought to Britain by the Romans, who used it in ritual bathing. The ancient Greeks knew the plant as *nardus*, or simply nard, after Naarda, a city of Syria near the Euphrates.

Myth and Legend

Biblical references to lavender and Christian folklore have become mingled. We are variously informed that lavender was the one plant that Adam and Eve managed to take with them when they were banished from the Garden of Eden, and that it was with lavender oil that Mary Magdalene washed Jesus's feet.

Lavender has been valued for centuries because of its wonderful scent. One Christian legend tells that Mary, the mother of Christ, having washed the holy infant's clothes, looked for a suitable place to hang them. She laid them over the grey foliage of the lavender bush and left them to dry in the sun. When she later gathered the clothes in, the bushes had become fragrant with their well-loved perfume.

It was once believed that the plant needed to be approached with caution as the asp, a particularly dangerous type of viper, resided in its shade. This may be due to a legend relating to the death of the great Queen Cleopatra in Egypt. Whilst many people will know that Cleopatra was killed by the bite of a venomous asp they may be unaware that according to the legend, it was lurking beneath a lavender bush.

Magic and Lore

For the most part lavender is considered to be a plant of love and purification. It is entirely appropriate that it has long been used in perfumes as the scent of lavender is supposed to attract men. It has been claimed that Cleopatra wore lavender-based perfumes to seduce Julius Caesar and Mark Anthony. During the Middle Ages, it had a reputation as an aphrodisiac that attracted a lover. Ladies would wear little bags of lavender close to their bosom in order to lure suitors, and Victorian women would bathe in water to which crushed lavender had been added before meeting their lovers. A girl might hide some lavender beneath her lover's pillow to encourage his thoughts to turn to romance. When married, she might put lavender under the mattress to ensure a night of marital passion. Perhaps this is one explanation for lavender being a favourite for scenting bed linen in the seventeenth century, and why it was once customary to scent a bride's bed and clothes with lavender. Then again, the fragrance of the herb was said to soothe wedding night fears.

To discover whom she would wed a young girl could sip lavender tea before retiring to bed on St Luke's Day, and recite the charm,

St Luke, St Luke, be kind to me,
In my dreams, let me my true love see.

Lavender-scented paper was used for writing love letters. The flowers or sachets of the dried plant used to be given by men as love tokens to their sweethearts. Charles II took sachets of lavender as gifts for Nell Gwynn at their secret meetings. It was taken as an emblem of love and remembrance, as in the rhyme:

Lavender grey, lavender blue,
Perfume wrapt in the sky's own hue;
Lavender blue, lavender grey,
Love in memory lives alway.

Lavender grey, lavender blue,
Sweet is remembrance if love be true:
Lavender blue, lavender grey,
Sweeter methinks, is love of today.

Lavender water or oil was once the preferred perfume used by prostitutes as it advertised their profession and attracted customers. Yet at the time of the renaissance, it was worn with rosemary, as it was believed that it would help keep a woman chaste. Sprinkling lavender water onto one's lover's head was said to keep them faithful, a belief that, no doubt, fuelled a demand for lavender.

Ladies would often place the flowers or sachets of the dried plant amongst their clothes in drawers. Whilst some might claim that it is done to ward away moths, given what has already been said, there might well be other reasons, such as to promote chastity or attract men when the clothes are worn.

On high days and holy days lavender might be included amongst the herbs strewn onto the floors of churches. It might also be strewn on days when it was thought that evil forces were particularly active as it used to be considered as a protective plant. Paradoxically, witches threw lavender into Midsummer fires as a sacrifice to the ancient gods. Carrying or wearing a small bunch would avert the evil eye and a cross of lavender was hung on the outside door to safeguard against evil. It is said that those wishing to see ghosts should carry lavender. It is unclear, however, whether this will cause someone to see ghosts or will prove an invaluable protection if they should happen to appear. It will also protect you from savage beasts as even lions and tigers are supposed to become docile at the scent of lavender. Lavender is also used to work magic against unresolved guilt and, as elf-leaf, is used in faerie magic.

Lavender is another of those plants that can be used in order to gain your heart's desire. If, after placing a sprig beneath your pillow before retiring to bed and thinking about the wish you want to make as you lie down, you dream of anything connected with your wish, you can be sure that it will all come true.

The French Riviera town of Grasse has, since the seventeenth century, been a centre of perfume production. At the height of the plague there, the glove makers scented their leathers with lavender oil as it was claimed that the fragrance would ward off the disease. In the Middle Ages it was one of the ingredients of the 'vinegar of four thieves', used by grave robbers to ward off the plague. These uses have some validity, as lavender is used as an insect repellent and would repel the fleas that transmitted the disease.

Like so many other herbs there are superstitions linking the plant with the 'lady of the house'. In this case it is said that lavender grows best in an 'old maid's garden', or to put that another way, where lavender thrives any daughter in the house will not marry.

Lavender is the birthday flower of 9 January.

Medicinal

The sweet scent of lavender has benefits other than those mentioned above. It should be smelt often as it is said to be conducive to a long life. Scattering the flowers about the home will bring about an atmosphere of peacefulness, and just looking at the plant is supposed to be sufficient to dispel melancholy and bring joyfulness. It should be remembered that melancholy was once thought to be a disorder brought about by an excess of black bile within the patient's system.

Being a plant of love, lavender was recommended for 'tremblings of the heart' and used to ease emotional pain. When someone had lost a loved one they would use lavender to heal the heart. Other cures using lavender include a treatment for insomnia, which involves burning a little of the flowering shoots in the bedroom. The smell of smouldering lavender has been claimed to induce sleep. The scent of the flowers has also been used as a relief for headaches. English farmers wore spikes of lavender flowers under their hats to prevent sunstroke.

LESSER DUCKWEED

Botanical Name: *Lemna minor*
Family: Lemnaceae
Common Names: Boggart; creed; digmeat; duck's meat; duck pond weed; groves; grozens; Jenny Green Teeth; mardlens; toad spit
Planet: Moon (in Cancer)

In folklore there are many malicious water faeries that delight in drowning and devouring any small children they can catch. Just such a boggart, or evil elf, is Jenny Green Teeth, who inhabits the ponds and waterways of Yorkshire waiting for the opportunity to attack an unsuspecting child. Wiser children would be able to tell when Jenny Green Teeth was lurking in wait for them by the telltale spread of duck-weed across the surface of the water. Perhaps the stories of Jenny Green Teeth and her cousins, such as Peg Prowler, who could be found in the River Tees, were just created by parents to keep their offspring away from the water's edge.

EDELWEISS

Botanical Name: *Leontopodium alpinum*
Family: Compositae

Edelweiss pulled out by the roots at full moon on a Friday is ideal for making a bulletproof vest. It must be wrapped in fresh, clean, white linen and can then be worn to give protection against bullets and daggers. It can also be formed into a wreath and worn, which is said to make the wearer invisible. Cultivating and nurturing it will, it is said, lead you to gaining your heart's desire.

OX-EYE DAISY

Botanical Name: *Leucanthemum vulgare* (Synonym: *Chrysanthemum leucanthemum*)
Family: Compositae
Common Names: Billy button; bishop's posy; bozzom; bull daisy; butter daisy; cow's eyes; crazy bett; devil's daisy; dog daisy; dog flower; drummer daisy; dun daisy; dunder daisy; dundle daisy; Dutch morgan; espibawn; Fair-Maids-of-France; field daisy; fried eggs; gadjevraws; goldens; golland; gorse daisy; gowan; gowlan; grandmothers; great ox-eye; gypsy daisy; harvest daisy; herb Margaret; horse blob; horse gollan; horse gowan; horse gowlan; horse pennies; large dicky daisy; London daisy; maise; maithen; Margaret; Marguerite; mather; maudlin; maudlin weed; maudlinwort; may weed; midsummer daisy; monnies; moon daisy; moon flower; moon penny; moons; moon's eye; mother daisy; mowing daisy; muckle kokkeluri; open star; poor land daisy; poverty weed; rising sun; sun daisy; thunder daisy; thunder flower; white golds; white gowlan; white gull; white weed
Planet: Venus
Deities: Artemis; Thor
Meanings: Hope; obstacle; a token; temporization

In ancient times the ox-eye daisy was dedicated to the goddess Artemis and was considered to be efficacious in the treatment of 'women's complaints'. An old name by which the plant was known in northern Europe was 'Baldur's brow', although it is more usually associated with the Norse god of thunder, Thor, and thought to be a powerful protection against lightning, hence the old English name of 'thunder flower'. The flowers were hung on haylofts, hayricks and barns as a charm to protect against rick fires. Before the Industrial Revolution a rick fire would have resulted in severe winter food shortages for livestock. With the spread of Christianity the plant became identified with St Mary Magdalene and so became known as maudlin daisy or maudlinwort. In common with other species of chrysanthemum it has also been dedicated to St John. It was used, with other flowers, in the making of 'St John's garlands'.

In common with other white-flowered plants it is usually considered to be unlucky to take the ox-eye daisy into the house. In this case the superstition is somewhat more specific as, it is claimed, when it is taken into a home in which there is an unmarried woman then she will never marry.

It is also associated with the feast day of St Margaret, 20 July and in Christian tradition is symbolic of Christ and the Virgin.

LOVAGE

Botanical Name: *Levisticum officinale*
Family: Umbelliferae
Common Names: Chinese lovage; Cornish lovage; Italian lovage; Italian parsley; lavose; love herb; love rod; love root; loving herb; lubestico; mountain hemlock; sea parsley; Scotch lovage
Gender: Masculine
Planet: Sun (under Taurus)
Element: Fire

Placing a sachet of lovage in bath water is said to make one more attractive to those one meets. This may or may not be connected to another use of the herb; the juices were recommended as a folk remedy to remove freckles and skin blemishes. There is a further benefit to be gained from adding lovage to the water in which you wash; it is said to be a good protection against contagious diseases.

LILY

Botanical Name: *Lilium candidum*
Family: Liliaceae
Common Names: Madonna lily; white lily
Gender: Feminine
Planet: Moon
Element: Water
Deities: Astarte; Britomartis; Hera; Juno; Kwan Yin; Nepthys; Venus
Meanings: 'A tribute to your beauty and spirituality'; innocence; modesty; purity; spirituality; virginity; youth

Myth and Legend

Even among early civilizations the lily was adopted as a symbol of motherhood and fruitfulness. The Assyrians and other Mesopotamian empires held the lily in reverence long before it passed into the religious imagery of the Jews and early Christians. It can be seen in the decoration on pottery from ancient Crete, where it was sacred to Britomartis, the Sweet Virgin pursued by Minos. To the Greeks it was a flower of Hera, goddess of motherhood and marriage. In Rome it was dedicated

to Juno, the Roman counterpart of Hera. It was said to have grown from one of two drops of milk that had fallen from the breast of the goddess as she suckled the infant Heracles (Hercules). Zeus, the infant's father, had wanted his illegitimate son to be an immortal but Hera, his ever-jealous wife, objected. Zeus ordered Somnus, god of sleep, to give Hera a potion that would cause her to fall into a deep slumber. As she slept he set the infant Heracles at her breast to suckle and, thereby, become immortal. One drop of milk that was lost as the baby suckled became the lily, the other spread across the heavens to form the Milky Way. In Rome, lilies were known as *rosa junonis*, or Juno's rose. Wreaths formed from the flowers were used to crown brides on their wedding day as an emblem of their virginity, and fertility.

Christian legend has a somewhat different account of the origins of the plant. It is said to have sprung from the tears of Eve who, having been expelled from the Garden of Eden, discovered that she was pregnant. The early church dedicated the flower to the Holy Virgin. The whiteness of the flower led to it being taken as a symbol of purity in addition to the other sentiments of which it was considered to be emblematic. In most Catholic countries lilies have been included in pictures depicting the Virgin Mary since the twelfth century. The flowers are used on 2 July at the celebrations of the Annunciation. Pictures of the Annunciation were few and far between before the fourteenth century, but even in the earliest of these lilies are rarely absent. Either the Madonna is shown holding a lily and a rose or one, or both, of these flowers appears within the picture. Sometimes Joseph is pictured holding lilies as his staff is said to have 'brought forth' lilies. The works of later artists show the Archangel Gabriel holding a bunch of white lilies. In 1618 a papal edict stated that lilies, roses and palms were suitable plants to be shown being scattered by angels in pictures.

The Catholic Church associates the lily with eight different saints, including St Catherine. Until her time it was scentless. She is said to have prayed that it would become fragrant as a proof to her father of the correctness of her faith. Of course, it became beautifully perfumed, and her father, the Emperor Costis, was convinced and converted to Christianity. Another legend says that St Thomas, having been absent at the death of Christ's mother, would not believe that she had ascended into

Heaven. In order to convince him her tomb was opened, and was found to contain nothing but lilies and roses. Some stories tell that lilies sprang up on the graves of those wrongfully executed to proclaim their innocence.

The combination of love and death come together in a Normandy folk tale about a knight and the lily maiden. The knight had a reputation for coldness, as he had resisted all the 'pleasures of the flesh'. He was often to be found walking in graveyards, where he appeared to be listening for some message that would point the way to his future happiness. On his stroll between the tombs one morning he was surprised to meet a woman of incredible beauty. She was sleeping on one of the marble tombs and was dressed in expensive cloths with jewels about her waist. Her skin was fair and her hair was as golden as lily pollen. In her hand she held a perfect lily. Filled with awe and admiration the knight knelt and kissed her hand, which woke her. She smiled at him and asked whether he would take her to his castle, as she was the one he had sought for so long. But before they left she exacted from him a promise: that in her presence none would speak of death, as she represented life, love and youth. This promise he gladly gave and then he swept her up onto his horse and rode with her back to his home, where they were married and lived very happily. If, occasionally, the knight was troubled again by the melancholic moods that had previously filled him, his wife had only to touch a lily against his cheek to dispel them.

Christmas Eve came and a great banquet was arranged. The tables were adorned with flowers and the guests wore their finest, brightest clothes. The minstrels sang songs of love and adventure, of wars and noble deeds. As the evening progressed they took a more sombre note and sang of heaven and of death. At the mention of the word the knight's wife turned pale and started to fade. Her husband caught her up and cried in anguish as she visibly shrank. In a very few moments he was filled with grief, horror and bewilderment as clasped in his arms was a fading lily, its petals falling to the floor. A great sigh seemed to pass through the banqueting hall and the air was filled with the sweet scent of the lily flower. The knight turned away and went out into the darkness, never to be seen or heard of again. In the world outside a change had taken place. Winter had come with a vengeance. It was cold and snow was falling; white as the scattered lily petals.

Magic and Lore

The sacred association of the lily has led to it being used as a protection against witches and witchcraft, particularly true in midsummer, when the flowers can be used in protective rites on the eve of St John's Day. Planting it in a garden will protect the home against ghosts, evil spirits and the evil eye. It is also claimed that it will keep unwelcome visitors away.

The other aspect with which lilies have long been associated is death, the whiteness of the flowers being taken to represent the removal of sins from the soul of the deceased person. This connection with death and funerals has led to it being thought to be unlucky to bring the flowers into the home. However, not all is doom and gloom as it is usually thought to be an omen of good luck to find the first lily of the year, as it conveys extra strength to the finder. Dreaming of lilies is also a sign of good fortune. If they are flowering it signifies marriage, happiness and prosperity, but if they are flowering out of season or are withered, it indicates the frustration of all your hopes and the severe illness, or possibly death, of a loved one.

There are some rather odd superstitions relating to lilies. One says that if a man steps on one he endangers the purity of his wife and daughters. Another claims that they can be used in the fight against crime. To reveal clues to a crime which has been committed within the previous twelve months, a piece of old leather should be buried in a flowerbed filled with lilies. This will cause vital evidence to come to light. In Worcestershire there is a superstition that if there is prolific growth of lilies in a year then bread will be cheap. Carrying a fresh lily, we are told, is one way in which to break any love spell that might have been cast against you. And the scent was once said to be a cause of freckles.

Perhaps because it is a feminine flower the lily is supposed to grow best in a household where the woman is dominant or, putting it another way, it grows best for a good woman.

Medicinal

In folk remedies, to treat boils, steep forty lily petals in brandy and lay them rough side upwards over the boils. The roots can be stewed to make poultices.

In order to be truly virginal the anthers and pistil must be removed from the flower.

TOADFLAX

Botanical Name: *Linaria vulgaris*
Family: Scrophulariaceae
Common Names: Bacon and eggs; brandy-snap; bread and butter; bread and cheese; brideweed; bridewort; bunny rabbits; butter and eggs; buttered haycocks; calf's snout; chopped eggs; churnstaff; dead men's bones; Devil's head; Devil's ribbon; doggies; dog's mouth; dragon brushes; eggs and bacon; eggs and butter; eggs and collops; faeries' lanterns; fingers and thumbs; flax weed; fluellin; fox and hounds; gallwort; gaping jack; gap mouth; impudent lawyers; larkspur; lion's mouth; monkey flower; patten and clogs; pig's chops; pig's mouth; puppy dog's mouths; rabbit flowers; rabbits; rabbit's chops; rabbit's mouths; ramsted; searchlight; shoes and stockings; snapdragon; snap jacks; snaps; squeeze-jaws; strike; toad; wild gap-mouth; wild snapdragon; weasel snout; yellow rod
Gender: Masculine
Planet: Mars
Element: Fire
Meanings: 'Be more gentle in your wooing'; reluctant lips

Toadflax earns its usual common name, and a few of the more unusual ones, from the simple fact that it resembles flax. And, according to some sources, toads are occasionally given to sheltering underneath it. Toadflax is in the same plant family as the snapdragon, which accounts for even more of the many names of the plant. In folklore and magic the plant is used as a protective herb. It will ward off evil or can be used to break spells that have been cast against someone.

FLAX

Botanical Name: *Linum usitatissimum*
Family: Linaceae
Common Names: Linseed; lint
Gender: Masculine
Planet: Mercury
Element: Fire
Deities: Hulda; Isis; Linda
Meaning: Gratitude (blue); fervour (red); utility (dried); domestic industry; fate; 'I feel your kindness'

Myth and Legend

According to legend it was the goddess Hulda who first taught mankind the cultivation and use of flax. It was therefore used each year at the ceremonies and rites of the goddess.

Magic and Lore

In the Middle Ages flax was used as a protection against witchcraft. Blue flax would be worn as an amulet to protect the wearer during the day. At night a bowl of flax seeds, mixed with mustard seed, would be placed on one side of the bed, and a bowl of water on the other, as a protection whilst sleeping. To prevent any evil from entering the home a box containing flax seed and red pepper should be hidden somewhere in the house.

There are various traditional rites from northern Europe that were held to promote good harvests. An old Prussian ritual required the tallest girl in the village to stand on one leg, on a chair, her lap filled with small cakes whilst holding a cup of brandy in her right hand and a piece of linden or elm bark in her left. She would then pray to Waizganthos, asking that the flax might grow as high as she was standing. She would drink the brandy, the cup would be refilled and she would pour it onto the earth as an offering to the god. The cakes would be thrown down for the spirits that attended on the god. So long as the girl remained balanced on one leg throughout the ritual it was believed that there would be a good flax crop. If, however, she lost her balance and had to stand on both legs, no matter how briefly, it was feared that the crop might fail.

In Switzerland there was a tradition of lighting bonfires high on the mountains on the first Sunday in Lent, which was known as Spark Sunday. When the fire had burned down low people would jump over it in the hope that the higher they jumped the higher the flax would grow that year. Similar rituals could be found in Bavaria into the mid nineteenth century. Midsummer fires were lit, often on high ground but also on lower land. The height of the flames would indicate the height that the flax would grow. Once again people would jump over the fires, as they burned down, in the belief that if they did so they would not suffer backache during the harvest. Another tradition, probably with the same origin, was that three part-burned sticks should be collected from a Midsummer bonfire and pushed into the ground in the flax field to ensure that the flax would grow tall.

A Yorkshire tradition required the farmer to sit on the seed bag, facing the east, before sowing the seeds. It would be even better if there were a few stolen seeds amongst those in the bag. In Thüringen the sower would draw the seed from a long bag that stretched from his shoulder down to his knees. As he sowed he would walk in long strides so that the bag swayed to and fro on his back, so as to encourage the flax to grow long and sway in the breeze.

To discover who she was going to marry, a young woman would go out with a handful of flax seeds on Halloween night and walk across the ridges of the field sowing them. As she went she would say:

Lint seed I sow ye,
Lint seed I sow ye,
Let him it's to be my lad,
Come after and pull me.

If she looked back over her left shoulder she would see the image of her future love crossing the ridges making the actions of pulling up the flax. There are clear parallels between this and other similar rituals such as the hempseed charm.

In Bohemia there was a belief that if a seven-year-old child danced amongst the flax in the field they would grow up to be beautiful. Flax was also used to attract money and ward off poverty. A few seeds would be put into the purse or wallet to bring money, whilst hiding a piece in one's shoe was a sure way to prevent poverty.

Medicinal

Folk medicine attributes a number of cures to flax. To relieve lumbago you could try tying a hank to your loin. For dizziness you must run naked three times around a field of flax after sunset; the dizziness will leave you and move into the plants.

STORAX

Botanical Name: *Liquidambar orientalis*
Family: Hamamelidaceae
Common Names: Styrax; sweet gum; voodoo witch burr; witch burr
Gender: Masculine
Planet: Sun
Element: Fire

Some species of storax are occasionally grown in Britain as ornamental plantings, and the seeds can be used in magical rites: either the pods can be used as protective amulets to safeguard those taking part in magical rituals, or the seed heads can be held or placed on the altar during the ceremony, to ward off evil forces.

TULIP TREE

Botanical Name: *Liriodendron tulipifera*
Family: Magnoliaceae
Common Names: Canary whitewood; tulip poplar; yellow poplar
Meaning: Fame

There is a legend that the tulip tree was Eve's favourite tree in the Garden of Eden. Following the fall from grace, when she and Adam had to leave the Eden, she wanted to take with her a leaf from the tree as a remembrance. She reached out and caught the tip of a leaf, but she was allowed to take nothing from the garden, so she only succeeded in pulling the tip out. From then on the leaves grew which their classic shape, their tip missing.

LOBELIA

Botanical Name: *Lobelia inflata*
Family: Campanulaceae
Common Names: Asthma weed; bladder pod; eyebright; gagroot; Indian tobacco; puke weed; vomitwort
Gender: Feminine
Planet: Saturn
Element: Water
Meanings: Dislike (blue); rebuff (white)

It seems somewhat ironic, given the sentiments that lobelia is taken to symbolize in the language of the flowers, that wearing the flowers is supposed to attract the love and affection of others. Another use is in the control of the weather. Throwing powdered lobelia directly into the face of an oncoming storm is supposed to be a sure way of halting its advance.

SWEET ALYSSUM

Botanical Name: *Lobularia maritima* (Synonym: *Alyssum maritimum*)
Family: Cruciferae
Common Names: Alison; madwort; sweet Alison
Meaning: 'I admire your nobility of character'; virtue and worthiness; 'Worth beyond beauty'

Alyssum is said to be a cooling and calming herb. When it is placed in the hands or on the body of an angry person it averts their wrath and brings peace. Some authors have alleged that it can be used in the treatment of hydrophobia (rabies), but there seems little evidence to substantiate this. Dioscorides recommended it for breaking charms and lifting curses. When it is hung up in the home it will prevent fascination.

DARNEL

Botanical Name: *Lolium temulentum*
Family: Gramineae
Common Names: Cheat; cokil; drake; jura; ray-grass; rye grass; wray
Planet: Saturn
Meaning: Vice

In ancient times there was a popular belief that eating darnel could cause blindness. In France, where this grass is known as *ivraie*, there was a belief that it would cause drunkenness. It is now widely thought that these symptoms were a result of the flour made from darnel being contaminated with the ergot fungus that attacks the grains of rye.

In some areas darnel became known as 'cheat' from the practice of using it to adulterate and bulk up the malt sold for use in distilling liquors. It has been claimed that the country folk of Cheshire believed it was degenerated wheat, although why it should be only the inhabitants of that county is somewhat unclear. There is a tradition, which seems to date back to the fourteenth century, that cokkilmeal (darnel flour) could be used to improve the complexion. It was supposed to remove freckles and other blemishes, leaving the skin soft and white.

HONEYSUCKLE

Botanical Names: *Lonicera caprifolium*; *Lonicera periclymenum*
Family: Caprifoliaceae
Common Names: Dutch honeysuckle; goat's leaf; hold-me-tight; love-bind; woodbine
Gender: Masculine
Planet: Jupiter
Element: Earth
Meanings: Rustic beauty (French honeysuckle); bonds of love; domestic happiness; 'I will not answer hastily'; love's ties (woodbine); inconstancy in love (wild); 'The colour of my fate' (coral); affection; fidelity; generous and devoted; 'My affection for you is boundless'; plighted troth; sweetness of disposition; 'This is a token of my love'

Honeysuckle is a good plant to have growing close to your home. In addition to being fragrant and attractive when in flower, it will also bring you luck. If you allow it to grow up over the doorway it will prevent fevers from affecting anyone in the house. The flowering shoots can be used to attract money by setting rings of them about the bases of green candles. Some writers suggest that if you cut lengths of flowering honeysuckle and place them in a vase in the house you will bring money into the home. In Wales it was considered to be unlucky to bring honeysuckle into the house, whereas in Somerset it is said that a wedding will soon occur if it is brought indoors. Placing the flower in a woman's bedroom will reputedly encourage erotic dreams. Another source recommends that it should never go inside the house, as it will cause all who live there to suffer sore throats.

Like any other twisting climbing plants it has often been used as a love token, the intertwining stems being likened to the fond embrace of lovers. Chaplets of honeysuckle would make a suitable gift to a sweetheart, ensuring their fidelity.

> And those that wear chaplets on their head
> Of fresh woodbine, be such as never were
> To love untrue, in word, thought, nor deed,
> But are steadfast

To find, or be given, a spray of fresh honeysuckle is usually considered to be lucky. However, if you see honeysuckle in your dreams it is an ill omen as it indicates that you should expect domestic strife in the near future, which will cause you sorrow. In magical use it is said that gently crushing the fresh flowers against the forehead will heighten psychic powers.

Honeysuckle is the birthday flower of 20 November.

BIRD'S FOOT TREFOIL

Botanical Name: *Lotus corniculatus*
Family: Leguminosae
Common Names: Bellies and bums; cockies and hennies; eggs and bacon; fingers and thumbs; fisherman's baskets; granny's toenails; horse yakkels; pigtoes; shoes and stockings; tom thumbs
Meanings: Retribution; revenge

In parts of the south of Ireland it was believed that possession of this plant would be a protection against punishment. Children would travel great distances to gather the plant so that they might take it to school with them.

HONESTY

Botanical Name: *Lunaria annua*
(Synonym: *Lunaria biennis*)
Family: Cruciferae
Common Names: Devil's ha'pence; lunary; money-in-both-pockets; money plant; penny flower; satin flower; silver dollar
Gender: Feminine
Planet: Moon
Element: Earth
Meaning: Honesty; frankness; 'I hide nothing from you'

Honesty
Lunaria.

Some of the common names of honesty are derived from the appearance of the seed heads, which are supposed to resemble coins. It's not surprising, therefore, that they are used in money spells and, at the new moon, may be put into a purse or pocket to attract money. Another way in which they have been used, to the same end, is by placing them beneath the base of a green candle that is then allowed to burn right down to its holder. A bunch of the herb can be placed in a wardrobe to bring good luck to the household.

Scattering seed heads as you walk out at night will drive all monsters away from you. Placing a few of them in your shoe at the time of the full moon will enable you to see faeries.

Picturing honesty in a dream always means good luck in matters relating to money.

LUPIN

Botanical Name: *Lupinus* spp.
Family: Leguminosae
Common Names: Old maid's bonnets; sundial; wild pea
Planet: Mars (in Aries)
Meaning: Dejection; 'I conquer all'; over-assertiveness; voraciousness; over-boldness; 'Who goes softly goes far'

Lupins have been cultivated since the times of the ancient Egyptians. The *Herbal Simples* of 1897 tells us that Roman actors would use the seeds as coins in their plays. Roman women, we are told, made a paste from the roots as a face cream. The name is an allusion to the supposed wolf-like nature of the plant in impoverishing the soil.

In reality, in common with other legumes, it fixes nitrogen in the soil and so aids fertility. The old folk name of 'sundial' is derived from the plant folding up its leaves at the close of day as if mourning the absence of the sun.

BACHELOR'S BUTTONS

Botanical Names: *Lychnis flos-cuculi*; *Lychnis* spp.
Family: Caryophyllaceae
Common Names: Billy buttons; catchfly; cock robin; cock's caim; cuckoo; cuckoo flower; Devil's flower; drunkards; evening campion; gipsy flower; Indian pink; indy; Maltese cross; meadow pink; polly baker; rag-a-tag; ragged Jack; ragged Robin; ragged Willie; red campion; red Robin; Robin Hood; rose campion; rough Robin; shaggy jacks; thunder-flower; wild williams
Gender: Feminine
Planet: Venus
Element: Water
Deity: Robin Goodfellow
Meanings: A snare; betrayal; hope in love; single blessedness; youthful love

In his herbal of 1597 John Gerard tells us that the common name of 'bachelor's buttons' is derived from 'the similitude of these flowers to the jagged cloth buttons anciently worne ...'. An alternative explanation is that it was the custom of men in rural areas to carry the flowers in their pockets as an indicator of how they fared with their sweethearts. If the flower retained its colour and freshness they could hope for the best. However, if it faded or died then their romance was doomed. Women might choose to wear it in the hope of attracting a husband. If a man is said to wear bachelor's buttons then he remains unmarried.

There are several plants that are all commonly called bachelor's buttons. Some of the others are the upright coltsfoot, white campion, white ranunculus and the double red campion.

CLUB MOSS

Botanical Name: *Lycopodium clavatum*
Family: Lycopodiaceae
Common Names: Fox tail; ground pine; lycopod; running pine; selago; vegetable sulphur; wolf claw
Gender: Feminine
Planet: Moon
Element: Water

The club moss, when properly gathered, will confer power, protection and the blessing of the gods on the person who bears it. Not surprisingly then, it must be collected with a certain amount of ritual. Before setting out you should take a purification bath, in a running stream. When you have located the plant an offering of wine and bread must be made. Then, like mistletoe and other plants used in mystical rites, iron and steel must not be used to cut the plant. If you do not have a blade of silver with which to cut the piece you want it is quite acceptable to snap sprigs off with your fingers.

CREEPING JENNY

Botanical Name: *Lysimachia nummularia*
Family: Primulaceae
Common Names: Catsfood; creeping Charlie; creeping Jane; creeping Joan; creeping John; ground ivy; Gill-over-the-ground; herb tuppence; herb twopence; Jenny creeper; loosestrife; meadow runagates; moneywort; motherwort; roving sailor; running Jenny; serpentaria; star; string of sovereigns; two penigrasse; twopenny grass; wandering Jenny; wandering Sally; wandering sailor; wandering tailor
Planet: Venus
Meaning: Release from strife

Creeping Jenny is not related to the plant we commonly call 'loosestrife', but it would seem from the botanical name that there should be some connection. The Greek *lysimachos* means 'ending strife', hence 'loosestrife'. Laying the plant on the shoulders of yoked oxen was supposed to be sufficient to end any 'quarrelling' between them. According to several sources it was named to commemorate King Lysimachus of Thrace. The smoke from burning the plant is said to drive away snakes.

In folk medicine it is supposed to be a good cure for headaches. You do not need to take it internally, or even rub it on the skin, merely to smell the scent of the plant.

LOOSESTRIFE

Botanical Name: *Lythrum salicaria*
Family: Lythraceae
Common Names: Blooming Sally; Brian braw; camal buidhe; emmet stalk; flowering Sally; foxtail; grass polly; long purples; lythrum; partyke; purple willowherb; ragged Robin; rainbow weed; red Sally; sage willow; salicaire; soldiers; spiked loosestrife; stray-by-the-lough; willowstrife
Gender: Feminine
Planet: Moon
Element: Earth
Meanings: Forgiveness; peace; 'Please take this plant as a peace offering'

Loosestrife is supposed to literally remove strife and ill feeling from those who come into contact with it. Therefore, if you have had an argument with a friend and wish to mend fences, send them some loosestrife. It will dispense peaceful vibrations and remove negativity so that your quarrel will soon be over. You might also choose to use this plant to dispel evil from a building; strew a little about the house, and peacefulness and calm will be restored.

It is said to soothe the savage beasts and so might be given to difficult horses or oxen which were to be used for drawing ploughs. In addition it was supposedly good for deterring flies and gnats.

MAGNOLIA

Botanical Name: *Magnolia grandifolia*
Family: Magnoliaceae
Common Names: Blue magnolia; cucumber tree; swamp sassafras; sweet bay
Gender: Feminine
Planet: Venus
Element: Earth
Meanings: 'Although you have broken my heart I shall persevere with my dignity'; 'Be not discouraged, better days are coming'; fortitude; love of nature; magnificence

Magnolia is used to guard chastity and fidelity; placing a shoot near or under the bed will ensure that your partner stays true to you, and you to them.

OREGON GRAPE

Botanical Name: *Mahonia aquifolium*
Family: Berberidaceae
Common Names: Californian barberry; holly leaved barberry; Rocky Mountain grape; training grape
Gender: Feminine
Planet: Earth

It is, perhaps, due to sympathetic magic, with the golden yellow colouring of the Oregon grape, that it has been used as an amulet to attract riches. The flowers are very obviously yellow, but this colouring can also be found in the stem and root. It is the root that is carried in the hope of attracting money, securing financial security and increasing popularity.

APPLE

Botanical Name: *Malus communis, Malus domestica*
Family: Rosaceae
Common Names: Fruit of the gods; fruit of the underworld; silver branch; silver bough; tree of love
Gender: Feminine
Planet: Venus
Element: Water
Deities: Aphrodite; Apollo; Athena; Diana; Dionysus; Hera; Hercules; Idun; Olwen; Venus; Vertumnus; Zeus
Meanings: Temptation (fruit); beauty and goodness; preference; 'Fame speak of him great and good'; 'You are as great as your are lovely' (blossom)

Myth and Legend

In Christian mythology the apple is erroneously seen as the 'forbidden fruit' of the Tree of Life in Genesis: it is simply the fruit of this tree that was forbidden to Adam and Eve; nowhere is the fruit identified as an apple. However, because of its place in pagan lore, it has become associated with the Tree of Life.

Our modern apple has been developed over the past 200 years from wild stock, sometimes referred to as Paradise stock, found in the Caucasus Mountains. Apples grown in biblical times would have been small, hard, acidic crab apples rather than the types with which we are now familiar. A further confusion may arise as the Hebrew word *dudim* refers to the physical pleasures of love. In English translations *dudim* is translated as 'apple' which may be another reason for the identification of the forbidden fruit with the apple: the apple might be a euphemism for the physical pleasures of love. To confuse matters still further, before the seventeenth century there was little in the way of classification and many fruits were referred to as 'apples'. Dates were finger apples, tomatoes were love apples, apricots were referred

to as Armenian apples and quince was called a cydonian apple. Pomegranates were known as the apple of Carthage and pine cones would have been pineapples.

One of the best-known Celtic legends is the tale of King Arthur. According to the stories kings and heroes travelled to the paradise of Avalon at their death. The Isle of Avalon is described as being an orchard garden, or an island of apples. It was the land over which Morgan Le Fey reigned, and where the cauldron of rebirth could be found. Morgan is often pictured carrying a branch of apples as a mark of her sovereignty. This connection of the apple with life, rebirth and images of paradise can be found in ancient and modern religious beliefs, and in the Celtic tales of the *Tuatha de Danann*. The gods of Norse mythology retained their youth by eating the apples of Idun, who was the goddess of youth and spring. Thus the apple has become a symbol of immortality, eternal youth, and happiness in the life after death.

This symbolism of an apple as the fruit imparting immortality can also be found in Persian myth. In one tale we are told of Anasindhu and Parvati, a holy man and his wife, who lived as hermits in a wood. Anasindhu was renowned for his wisdom and virtue. He dedicated all his time to meditation and only spoke three times a year. As a reward for his goodness the gods bestowed on him an apple that would make the person who ate it immortal. Anasindhu put the fruit to his lips, but before eating it he thought of his wife. Although she was often overlooked and discontented, she had shared all his hardships. Should she not now share his rewards? He resolved to give her the apple.

To his great surprise Parvati refused his gift. She explained that she would not want to live forever if it meant always being unhappy and living in a wood, where she met no one but passing pilgrims. After a moment's thought, however, she asked her husband if he could not be just as virtuous, and just as useful to the gods, if they lived in the city and had a house with servants rather than begging for their existence. Anasindhu was stunned. How could he, a poor beggar, own such things? His wife's answer was simple. He must sell the apple of immortality. She reasoned that there was no proof that the apple had any such attributes. If it had no special power they would be none the better for eating it and none the worse for selling it. If it did have great powers then the gods could be glorified by the sale of it and Anasindhu could continue in his good works for the remainder of his life. Touched by his wife's pleading Anasindhu went to the city and sold the apple to the king.

However, the king also aspired to holiness. He considered the piety of the old hermit who would, no doubt, pass on the money paid him for his apple to various charitable causes. He decided that he was unworthy of being immortal and so he went to find the queen. He gave her the apple, reasoning that the world could forever delight in her beauty and virtue. The queen did not eat the apple but stole out in the middle of the night to meet her lover, the captain of the palace guard, and she gave him the apple as a love token. The captain of the guard was not a happy man. Although he loved the queen there was a serving wench he loved even more. He resolved that he would pass on the queen's gift to her.

The next day the lowly serving maid came before the king and offered him the apple of immortality. She said that she was unworthy of it but the king could live forever to bestow peace and justice on his people. Taking the apple the king demanded to know how she came by it, only to be informed that it was a gift from her betrothed, his captain of guards. The captain was sent for and quickly grasped the difficult position he was now in. He had no option but to confess that the apple was give to him by the queen. In his rage the king ordered that the captain be executed immediately and the queen be burnt to death in the city square. He reflected on the speed with which he had gone from happiness to misery overnight. He called for his chief priests and ordered them to take all his wealth and divide it amongst the poor of the kingdom. Then, dressed in old clothes, he left his kingdom to become a beggar.

As he left the palace he saw Anasindhu passing by, dressed in the finest silks, riding a gilded litter and attended by many servants. The king gave him the now-withered apple saying that none in the kingdom were worthy of immortality. Anasindhu took this as a sign that the gods really did want him to live for ever and so he accepted it gladly. However, as he put it to his lips, the litter on which he rode was jolted and he dropped it, whereupon a passing dog quickly snatched the apple up and swallowed it whole. Thus was immortality denied to men and in the East there is a dog which wanders from town to town, never happy but unable to die.

Apples are an important image in classical mythology. In ancient Greece, the Festival of Diana was celebrated on 13 August. The ritual meal included apples still attached to their branches. Arguably the most famous reference to the apple in classical myth comes in the story of the judgement of Paris. An apple was sent down from Mount Olympus marked 'for the fairest'. Zeus gave Paris, the most handsome man in the world, the task of deciding between Hera, Aphrodite and Athena which deserved the title of 'fairest'. He chose Aphrodite, not least because she bribed him by promising that, if he did so, he should have the most beautiful mortal woman. That woman is best known as Helen of Troy, and when Paris 'collected' his reward it started the Trojan Wars in which many gods and heroes perished.

The Gardens of the Hesperides, mentioned in Greek mythology, are described as the orchards of Paradise. In them grew an especially sacred apple tree, the fruit of which conferred immortality. It was a wedding present from Gaia to Hera on her marriage to Zeus. Hera planted it in a divine garden on the slopes of Mount Atlas. It was tended, and guarded, by the Hesperides, the nine daughters of the Titan Atlas and the representatives of the goddess of love. These maidens joined hands about the tree to form its outer protection. It was further protected by a serpent called Ladon, which wound itself about its roots, as Hera had discovered that the Hesperides were not above stealing the fruit of the tree themselves. The red of the ripe fruit was likened to the colours of the setting sun and sunset was followed by the rise of the 'star' Hesperus (Venus), which was sacred to Aphrodite, the goddess of love.

The eleventh of the twelve tasks set Hercules by Eurystheus was to collect fruits from this tree. The river god Nereus had told Hercules where Hera's garden was, and had advised him that he should not steal the fruits himself but persuade Atlas to get them for him. Atlas had to bear the weight of the celestial globe on his back, and immediately saw an opportunity to be relieved from this burden, so he was only too happy to do as Hercules asked him. However, he said that first the serpent Ladon must be slain and, secondly, Hercules should shoulder the weight of the globe while he did so. Hercules did as Atlas asked, but when Atlas returned with the fruits he informed Hercules that, rather than handing them over, he would take them to Eurystheus, who had set the task, leaving Hercules supporting the weight of the celestial sphere.

Nereus had advised Hercules not to trust Atlas, however, and the hero had already a plan in mind to deal with this eventuality. He said that he was content to continue holding the globe if Atlas would just lift the weight from his back for a moment so that he might put a pad against his head. Without thinking Atlas laid the apples on the ground and lifted the globe from Hercules. Hercules thanked him, picked up the apples and left.

The theft did not go unnoticed, however, nor unpunished. Hercules was eventually obliged to give the apples up to Athena who, in turn, returned them to the Hesperides. Hera, distressed by the death of Ladon, fixed the image of the serpent among the constellations in the night sky and caused Hercules to become insane as a punishment for killing him.

In Norse mythology the gods retained their youth by eating the sacred apples of Idun. At one point the mischievous Loki stole them, resulting in the aging of the gods until he was forced to return them.

Magic and Lore

The apple tree has long been thought of as a tree of birth and eternal youth, as can be seen in the myths and legends with which it is associated. It has also been seen as emblematic of wisdom and choice, especially the choice between options that appear equally attractive. In common with the hazel, it is said that the fruits should never be eaten by the ignorant.

In Celtic myth there was a matriarchal trinity: the maiden, the fruitful woman and the crone. Symbolically the maiden was the birch tree, the crone the elder tree and the fruitful woman the rowan or hawthorn. However, this symbolism was, at some point, transferred to the apple, making it the emblem of fertility. Through sympathetic magic the fertility of the tree and that of people became linked. In parts of Europe there was a tradition that the first fruit from a tree should be eaten by a 'fruitful', i.e. pregnant, woman so that the tree would always be fruitful. Elsewhere a barren woman might roll on the ground beneath a solitary apple tree in the hope that it would enable her to conceive.

In Somerset the oldest apple tree in the orchard was called the 'Apple-Tree-Man' as it was in this tree that the spirit of the orchard was supposed to live. This tree was singled out for special attention during the wassailing ceremony. Known in Sussex as 'howling', this is probably the oldest and best-known tradition associated with apple orchards. Usually carried out between Christmas Day and Twelfth Night, it is a celebration of the trees, to waken their spirit and prepare them for the coming year. Pieces of cake or toast are placed into topmost fork of the oldest tree in the orchard. Some sources say that it should be the tallest or most prolifically fruiting tree that is singled out, rather than the oldest. A toast to the tree is then drunk in cider. The traditional wording of the toast is:

> Here's to thee, Old Apple tree,
> Whence thou may'st bud, and whence thou may'st blow,
> Hats full! Caps full!
> Bushel-bushel-bags full!
> And my pockets full too. Hurrah!!

In some counties it was:

> Stand fast, root. Bear well, top,
> Pray, good God, send us a howling crop.
> Every twig, apples big. Every bough, apples now;
> Hats full, caps full, fine bushel sacks full,
> And a little heap under the stairs,
> Hulloa boys! Hulloa! and blow the horn.

The dregs of the cider would then be tossed over the tree and gunshots fired into the branches to 'wake the tree up'.

Whilst traditionally associated with apples, wassailing has been carried out for other fruit trees. In Herrick's *Ceremonies at Christmas Eve*, there is the following instruction:

> Wassaile trees that they may beare
> You many a plum and many a peare,
> For more or lesse fruits they will bring
> As you do give them Wassailing.

There are many variations on this theme, all of which aim to ensure that there will be a good harvest in the coming year. One such is that, at the end of the year, thirteen leaves from an apple tree should be buried under its roots to ensure a good crop in the following year. Another says simply that if the sun shines through the branches of an apple tree on Old Christmas Day (6 January) the tree will produce a bumper crop at the next harvest.

> If wold Christmas Day be fair and bright,
> Ye'd have apples to your heart's delight.

Another tradition, with obvious similarities, instructs gardeners to pour cider onto the ground when it is freshly dug if they hope to have a good crop from the soil. Cider would also be poured onto the soil at harvest, before any was drunk, in thanks for the crop and to ensure good luck.

In addition to the 'Apple-Tree-Man' there were other faerie spirits that were said to dwell in old apple trees, amongst these are Awd Goggie (Yorkshire) and Lazy Lawrence (Somerset). These may be one and the same, although Lazy Lawrence would seem to have some reference to St Lawrence's Day (10 August) when it is said that 'Lazy Lawrence guards the ripening fruits'. Then again, this may simply be a confusion of the saint with the faerie spirit. The role of these spirits of the trees may be to protect the fruits. Cynics might suggest that they were the creation of adults to deter children from scrumping the apples.

In faerie magic, apple bark was added to incense burned at Midsummer Eve as an offering, and apples could be left as suitable offerings for faerie folk. As a tree of fertility, the fruits and bark would also be used in all types of love magic.

In Germany, there was a tradition that, on Christmas Eve, the boughs of adjoining fruit trees would by tied together using ropes made of straw to signify that the trees were 'married', so that they would bear more fruit in the autumn.

When harvesting the apples, some of the smaller fruits should be left on the tree for the faeries, and the birds. It is also considered to be unlucky to return to a tree to look for apples that were missed at the first picking. There are many superstitions that focus on Midsummer or St John's Day. An apple that is ripe on St John's Day is called 'Apple-John' and should be carefully picked and stored. These apples are supposed to last for two years and are said to taste best when shrivelled up.

Traditionally, apple trees were 'christened' between the end of June and the end of July, often on St Swithin's Day (15 July) – and although in some places it might be done on St Peter's Day (29 June) or St James's Day (25 July) – the day was known as 'Apple Christening Day'. Rain on St Swithin's Day would ensure that the apple crop would be sufficient to last the winter through. Eating apples before they were christened could result in illness as the apple was bound to be unripe.

When apples blossomed in April it was taken as a sign that there would be a good harvest, but if they flowered in May the harvest would be poor.

There are many love charms that use apples. An Austrian tradition tells that a girl should cut an apple in half on St Thomas's Night, 21 December. She should then count the number of seeds. An even number indicates that she will soon be married, and an odd number that she will never marry. If one seed is cut through it shows that any marriage will be filled with quarrels and strife. Where two, or more, seeds have been cut through it indicates that she will end up a widow.

There is a traditional rhyme that relates to St Thomas's Day but has more to do with wassailing than with love spells. It says:

> Bud well. bear well,
> God send farewell,
> A bushel of apples to give,
> On St Thomas's morning.

Both of these traditions probably pre-date any association with St Thomas, as 21 December is the winter solstice.

St Thomas's Day is not the only date associated with love divination. In Italy there was a tradition that apples might be thrown into the street from balconies on St Andrew's Day. If a young woman should pick one up she would be married within the year. In Germany a girl might sleep with an apple beneath her pillow on St Andrew's Eve. She must take the apple with her to the next church festival, keeping it in her hand, and the first man she sees on leaving the church will be her future husband. Another method of love divination from Europe suggests that a young maid should toss apple pips into a hot fire whilst saying the names of her suitors. If the pips pop then the person she has named is bursting with love for her; if there is no noise then there is no love. Alternatively, to discover whether the man of her choice has any amorous feelings for her a girl could toss just one apple pip into the fire, with the words:

> If you love me pop and fly,
> If you do not lay and die.

If the pip pops, her lover shares her feelings.

To find out whom she will wed the young woman might try peeling an apple, but she must take great care not to break the peel so as to end up with one long, continuous piece. This she must toss over her left shoulder and when it lands it will form the initial of the man she will marry.

> If your hubby's name you wish to know,
> Over your left shoulder an apple peel throw,
> It will wriggle and coil, and you will see
> The first initial of who it will be.
> For the witches plot, and the hexes scheme,
> On the mystic night of Halloween.

One love charm recommends that the young woman should pick an apple from the tree and mark 'aleo + deleo + delato' on it with a sharp knife. She must then present it to the man of her choice with the words:

> I conjure thee, Apple,
> By these names which are written on thee,
> That what man toucheth and tasteth thee
> May love me and burn in my love,
> As fire melteth wax.

A similar old spell says:

> Take an apple in thy hande as it hangeth on the tree and wryte in it these names followinge; Anaell, Satnell, Asiell and then saye, 'I conjure thee apple of apples by the name of cheefe devel, which decetpfully deceaved Eve in Padyce, that what woman soever it be that doe eate or tast of this apple that she may burne in love of me.' Saye this four tymes upon the apple, and then geve the apple to what woman you will.

In Scotland there was a belief that if a girl brushed her hair three times while eating an apple in front of a mirror at midnight on Halloween, she would see the image of her future husband reflected there.

Wee Jenny to her granny says,
'Will ye gae wi' me, granny?
I'll eat the apple at the glass,
I got frae uncle Johnny.'
– Robert Burns, *Hallowe'en*

Another Halloween tradition relates to 'apple bobbing', when apple are floated in a bowl of water and children try to lift them out using only their teeth. The first girl to bite into a apple will be the first to be married.

A rhyme from Lancashire and Cumbria also hints at a charm used to discover the identity of a future spouse:

Pippin, pippin Paradise,
Tell me where my true love lies,
East, West, North, South,
Pilling Brig or Cockermouth.

An apple pip is then flicked into the air to see in which direction it will fall. An alternative to the above can be found in the West Country, where the following rhyme was used:

North, South, East, West,
Tell me where my love doth rest.

Divination, by counting the pips, can by done in the following way:

1 – I love; 2 – I love; 3 – I love, I say; 4 – I love with all my heart;
5 – I cast away; 6 – He loves; 7 – She loves; 8 – Both love;
9 – He comes; 10 – She tarries; 11 – He courts; 12 – They marry;
13 – A happy life; 14 – A happy wife; 15 – A lot of fun; 16 – A little one.

By far the simplest of the love charms requires a young maid to do no more than hold an apple in her hand until it is warm. She should then give it to the man of her choice because if he eats of it he will fall in love with her.

One tradition claims that if a girl can eat a crab apple, without pulling a face because of its sourness, she will marry whomever she wants. Another says that a young woman who is strong enough to break an apple in two will never marry. There is a variation of this, however, that suggests that if the apple is 'named' for her with the name of her suitor and she then breaks it open, he will be her lover. If a girl repeats the name of her true love whilst cutting an apple open, and then finds that it contains twelve pips, she is sure to marry him.

It is a very good omen to dream of apples, being an indication of success, long life and a faithful lover. If the apples that you see in your dream are red then you can expect a gift of money. If they are green then the money will still come to you but it will be some time before it arrives. To see a golden apple means that you should expect great wealth, but there will be some domestic unhappiness. Dreaming of picking and eating ripe apples signifies that you can expect good luck to follow you. Dreaming of an apple on the tree indicates prosperity, whilst dreaming of an orchard suggests that you will receive a large amount of money. To dream of apple blossom is a prediction of a birth.

In the West Country girls might place a large apple, sometimes referred to as an 'Allen apple', under their pillows on Halloween night so that they would dream of their sweethearts.

There are many superstitions that centre on the apple. For example, to eat an apple, without first rubbing it, is to challenge the Devil, and to damage or cut down an apple tree is to court misfortune. If you accept an apple offered to you and, on biting into it, it breaks open, you should return it or else suffer bad luck. In many parts of Europe it is considered to be a sign of great good luck if an apple tree flowers out of season, but in Britain, it has usually been thought to be an ill omen.

Flower out of season
Trouble without reason.

If a crab apple, hanging over a well, flowers out of season it is traditionally a sign that there will be more marriages and births than deaths in the locality in the coming year. It was also thought unlucky to discover blossoms and ripe fruits at the same time on an apple tree. This was given as an indication that there would be a death in the family.

A bloom on the tree when apples are ripe
Is a sure termination of somebody's life.

If an apple remained on the tree through the winter it was thought to be a sign that there would be a death.

There is an exception to this, however, in the legend of the Silver Bough, a branch, from an apple tree, which has bud, blossom and ripe fruit on it all at the same time. According to Celtic myth a person who possesses the Silver Bough could enter the land of the gods, the Underworld. This is the bough carried as a mark of office by Morgan Le Fey in the legends of King Arthur. Hence apples were associated with the Underworld, and so were the food of the dead. In the old ballad of Thomas the Rhymer the Faerie Queen warns against eating the apples or pears off the trees growing in her garden. Traditionally there could be no return to the land of the living for those who had eaten of the food of the Underworld. In some Wiccan traditions an apple is seen as an emblem of the soul and so, at Samhain, one is put onto the altar or buried, so that those who are reborn in spring will have food through the cold winter months. It is not only in Celtic myth that the apple is a passport into the Underworld. In Greek mythology Nemesis bears a bough of apples as a reward for heroes. This fruitful branch was what allowed entry into the Elysian Fields, the abode of the gods.

At one time it was traditional to give apples as gifts on some special days in the year as a token of friendship and a wish for good health. An example of this could be found in the Forest of Dean where, on New Year's Day, apple gifts were mounted on tripods, trimmed with nuts and yew to symbolize sweetness, immortality and fertility. In Glamorgan children would carry decorated apples and oranges from door to door at New Year, and in Cornwall, up until the late nineteenth century, greengrocers would stock up on late apples as traditional gifts for children at All Hallows (1 November).

There are other snippets of folklore relating to the apple harvest. It is said, for example, that if there is sleet in February it will ensure that there is a bumper apple crop that year. In Sussex, there was a belief that if the cuckoo was heard before Spring Day, 14 April, all the apple blossom would be blighted by frosts. When harvesting, it was claimed that apples gathered by the light of the moon would not rot. In Herefordshire it was claimed that hops would never flourish on land where apple trees have been felled. There is also a comment on the lifespan of apple trees in the rhyme:

> Who plants a pear tree
> May set it for a friend.
> Who plants an apple tree
> May live to see it end.

Finally, according to legend, unicorns make their homes beneath apple trees and in orchards. If you visit an ancient orchard early on a misty spring morning, tread softly and you may be lucky enough to see this fabulous beast.

Medicinal

The traditional belief that eating apples will help prevent serious illness is well known. As children most people learn sayings such as 'An apple a day keeps the doctor away', and 'Eat an apple going to bed, make the doctor beg his bread', or 'An apple in the mornin', is the doctor's warnin'.

Apples do have a part to play in folk remedies. To treat an ache or cure an ailment, it was suggested that one should cut an apple into three pieces, rub the affected area with each of the pieces of apple and then take them out and bury them during the waning moon. This should remove all symptoms.

To cure warts the offending blemish must be rubbed with the two halves of a cut apple, the pieces of which should then be buried in the garden. In due time, as the halves of apple rot away, the wart will disappear. A slightly harsher version of this requires that the wart be first 'cut to the quick' until it bleeds. It should then be rubbed with the cut surface of a sour apple, and the apple then buried.

Poultices of rotten apples may be applied to the body to relieve the pains caused by rheumatism.

MALLOW

Botanical Names: *Malva neglecta*; *Malva sylvestris*
Family: Malvaceae
Common Names: Billy Buttons; biscuits; bread and cheese; bread and cheese and cider; butter and cheese; cheese-cake flowers; cheese flowers; cheeses; chucky cheese; custard cheeses; faerie cheeses; flibberty gibbet; French mallow; goodnight-at-noon; horse buttons; lady's cheese; loaves of bread; mallace; mallow-hock; marsh-mallice; maws; old man's bread and cheese; pancake plant; rags and tatters; round dock; tall mallow; truckles of cheese
Gender: Feminine
Planet: Moon
Element: Water
Meanings: Mildness; sweet disposition

Magic and Lore

The mallow may be an unlikely plant of love, but that is certainly one of the roles it can play in folk magic. If the one you love has left you, then place a vase full of mallow flowers in your window, or on your doorstep. This will cause them to be thinking of you, and may be sufficient to make them return. If you would rather find someone new try wearing or carrying a mallow flower, as this will attract love to you.

Medicinal

Mallow was used as a protective plant although it tended to be used only for short-term results. The roots were to be boiled with raisins and the resulting liquor bottled. A small quantity of this liquid taken each morning would be a protection against disease for the remainder of the day. It could also be rendered down to form an ointment which, when rubbed onto the body, would drive out all demons and guard against black magic.

MANDRAKE

Botanical Name: *Mandragora officinarum*
Family: Solanaceae
Common Names: Alraun; anthropomorthon; apples of jan; baaras; brain thief; circeium; circoea; Devil's apples; *galgenmannchen*; gallows; herb of Circe; *hexenmannchen*; ladykins; mandragen; mandragor; mannikin; racoon berry; satan's apple; semi-homo; sorcerer's root; wild lemon; woman drake; *zauberwurzel* (sorcerer's root)
Gender: Masculine

Planet: Mercury
Element: Fire
Deities: Aphrodite; Circe; Hathor; Hecate; Selene; Venus
Meanings: Horror; rarity

Myth and Legend

Mandrake has been considered to be a potent plant in magic and myth for many centuries. Although it is not native to Egypt it was known to the ancient Egyptians, and is depicted on the walls of tombs dating to 1300 BC. Fruits of the plant were found in the tomb of Tutankhamen, placed at regular intervals in the sixth row of the floral collarette; the significance of this is unknown. In ancient Greece it was used in the brews of Circe, daughter of Hecate, which were supposed to turn men into swine.

The thick branching roots of the plant were thought to resemble a human form, either male or female, and were the abode of small familiar spirits. In ancient times the 'human' appearance of the root might be improved by a little bit of judicious trimming and shaping and grains of millet inserted as eyes. These mannequins would be carefully tended, perhaps being kept in small coffins and hidden in cupboards.

It is clear that this is no ordinary plant. It was thought that it grew beneath a gallows, from sperm emitted by hanged men during their death throes: 'It is supposed to be a creature having life, engendered under the earth of the seed of some dead person put to death for murder.' (Thomas Newton, *Herball to the Bible)*

Great care had to be taken in gathering mandrake, as it would not come willingly out of the ground. It was said that, on being pulled from the earth, it (or a demon within it) would emit a powerful scream and the person collecting it would go mad or die. To avoid this fate those who collected mandrake tied a cord (preferably black) around the plant and then secured it to the collar of a dog (preferably white), which had been starved for several days. The gatherer could then plug their ears and lure the dog away with food, pulling the plant out of the ground. So it was the dog that was doomed to suffer any ill effects of pulling up the plant rather than its owner.

Some writers suggest other precautions that should be taken. Josephus warns that, except in quite specific circumstances, there are dangers in even touching the plant. He instructs those who want to gather it to stand with their backs to the wind and then draw three concentric circles around the plant with the point of a sword. They must then pour a 'libation' on the ground and turn to the west. The mandrake can be dug up, using the sword point, without any fear of it 'fleeing away'. This thinking stemmed from a belief that the leaves would shine with a magical luminescence, at night but if any attempt was made to pick them the plant could fly away.

Such lank and deadly lustre dwells
As in the Hell's fire that light
The mandrake's charnel leaves at night.

Bartholomew echoes this warning of Josephus and adds that the plant should only be collected after sunset.

Not everybody believed in the need for any elaborate practices. Theophrastus, who lived between 370 and 255 BC, criticized the 'humbug' of the magic rites practised in the collecting of the plant. Even John Gerard, who subscribed to several odd beliefs, says, There have been many ridiculous tales brought up of this plant, whether of old wives or runnegate surgeons, or physickmongers, I know not, all which dreames and old wives' tales you shall hence forth cast out your bookes of memorie.'

Mandrake was known from ancient times primarily as an aphrodisiac and aid to conception, even being able to overcome sterility. For this it has the best of recommendations, as in the Bible (Genesis 30) we are told that Reuben stumbled upon a patch of mandrakes whilst harvesting wheat. Knowing their significance he gathered some to take home for Leah, his mother. Rachel, Leah's sister wanted some but Leah was reluctant to give them to her. Rachel then offered Leah one night with her husband Jacob, in exchange for them. Nine months later, Leah gave Jacob his fourth son. The Hebrew word *dudaim* refers to a fruit with a sweet and agreeable odour, particularly one that men would find appealing. In English translations of the Bible this word is often rendered as 'mandrake'.

In the story of the crucifixion of Christ it is thought that the sponge used to give him drink, as he hung upon the cross, would have been steeped in mandrake as a painkiller. In stabbing him with a spear the Roman soldier was ensuring that he was actually dead rather than merely under deep sedation from the mandrake.

Magic and Lore

Mandrake was greatly sought after and commanded a high price; the more human-like the root, the higher its value. The roots could be made into dolls or mannequins. which were bathed and dressed, tucked in at night and consulted on important questions. Italian ladies were known to pay as much as thirty golden ducats for mandrakes that were especially lifelike. During the reign of Henry VIII quantities were imported into England to satisfy the demand. What made it so important were the various attributes it was believed to have, not least its power as an aid to conception.

Despite its scriptural support, mandrake was considered by some to be a plant of the Devil and, if collected with the correct incantations, made the Devil do one's bidding. The human-like root tubers were said to mean that it was possessed by demons and evil sprits. It was therefore widely used by witches. By the middle of the sixteenth century tales were being told of witches gathering mandrake roots by night. They would then wash them in stolen sacramental wine and use them in a variety of love potions. Even the witches' flying was said to require mandrakes. In 1630, three women from Hamburg were executed merely for possessing mandrake.

Other authors claimed that devils and demons could not abide the smell or presence of the plant. It was, therefore, used in exorcisms and in the control of ailments thought to be caused by devils. 'For witlessness, that is Devil sickness or demonical possession, take from the body of this said wort mandrake by the weight of three pennies, administer to drink in warm water as he may find most convenient – soon he will be healed.' (*Herbarium* of Apuleius)

Just as rituals had to be observed when gathering the plant, so some authors suggested that when dried root was used it first needed to be 'activated'. It had to be left in a prominent place in the home for three days, and remain undisturbed. The root was then soaked overnight in warm water, after which it was ready to use. The water in which it had been soaked could be sprinkled about the property, or over people, to purify them and protect them from demonic forces. The whole herb was, at one time, used in the rites of exorcism.

This is another herb that has been used as a protective amulet. The whole plant could be placed on the chimney mantle, to protect the household. Not only did it act as a guard against evil forces, but would also ensure prosperity and fertility for those who lived there. A mandrake hung over the bed would protect the person who slept there through the night, and the scent was supposed to help people sleep.

In other rites mandrake was said to have the power to open locks and reveal hidden treasures. It could be used to attract wealth; money, especially silver, which was placed beside a root was supposed to double. Putting a small piece of the root into a locked box with containing money was also supposed to cause the money to double. Some people believed that the plant could be used to tell the future: questions would be put to the 'human' figure of the root that it would answer by nodding and shaking its 'head'.

Medicinal

Mandrake was used as a treatment for epilepsy and eye disorders, presumably because these were thought to have been caused by demonic influences. Doses were administered to lunatics in order to suppress their violent tendencies. Unfortunately, if too large a dose were given, it could be the cause of delirium and madness.

MANGO

Botanical Name: *Mangifera indica*
Family: Anacardiaceae

Indian folklore tells of the origins of the mango. According to the legend the sun had a daughter, who was a particularly beautiful and accomplished girl, more lovely than any other girl on the Earth. She was married to a powerful king and was very happy. However, an evil witch, jealous of her happiness, vowed to make her miserable. She chased the poor girl, screeching and uttering all kinds of threats. In an effort to escape her, the sun's daughter dived into a pool of water and was transformed into a lotus flower. Her husband, finding the exotic lotus, took it home to his water gardens where it was greatly admired.

The witch was furious that the girl should still be admired for her loveliness, even as a flower, and set about destroying the flower with fire. From the ashes of the burned lotus grew the mango. The king, although saddened by the loss of his lotus, thought that the mango flowers were beautiful. When the plant set fruit he could hardly wait until they would be ready to pick, but when he plucked the first mango from the plant he dropped it and it fell at his feet. From the broken fruit burst forth his wife, as beautiful as when he had last seen her, and they returned to their happy life together.

Another tale tells that Kamadeva (Cupid) has five arrows with which he pierces each of the senses of his victims. One day a young girl, who was in love with a particular man, gave him a mango flower and asked him to place it on the tip of one of his arrows when he shot at her beloved. Kamadeva was so pleased with the results that this became his favourite arrow, the arrow of love.

HOREHOUND

Botanical Name: *Marrubium vulgare*
Family: Labiatae
Common Names: Bull's blood; eye of the star; haran; hoarhound; huran; *llwyd y cwn*; maruil; seed of Horus; soldier's tea; white horehound
Gender: Masculine
Planet: Mercury
Element: Earth
Deities: Horus
Meaning: Fire; frozen kindness

Horehound was considered by the Romans to be a good, and sometimes magical, herb. For the most part it is thought to be 'anti-magical', that is to say useful as a protection against sorcery. It could also be ground down and used as snuff or mixed with salt and applied as a treatment for dog bites.

The other main claim to fame of horehound is that it is one of the five bitter herbs to be eaten during the Jewish Passover.

ALFALFA

Botanical Name: *Medicago sativa*
Family: Leguminosae
Common Names: Buffalo herb; purple medic; lucerne
Gender: Feminine
Planet: Venus
Element: Earth
Meaning: Life

Alfalfa, in common with many other leguminous plants, enriches the soil by fixing nitrogen that is needed for plant growth. In folk magic it is used to enrich, or at least prevent impoverishment of the home. A small quantity should be kept in a jar in a cupboard in which food is stored. This will protect the household from poverty and hunger. An alternative, to achieve the same end, is to burn some of the herb and scatter the ashes around the house.

BALM

Botanical Name: *Melissa officinalis*
Family: Labiatae
Common Names: Bee balm; balsam; baulm; bawme; cure all; honey plant; lemon balm; melittena; pentarie; sweet balm
Gender: Feminine
Planet: Jupiter
Element: Water
Meanings: Fun; 'I was but jesting'; pleasantry; social intercourse; sympathy

Myth and Legend

The botanical name of balm, *Melissa*, was the name of the Cretan nymph who, in mythology, first discovered its medicinal benefits. It is also the Greek name for a honey bee. Bees were said to love the scent, and Pliny informs us that when the herb is rubbed on bee hives it will prevent the bees straying, and attract others to the hive.

In Staffordshire there is a legend about Ahasuerus, a wandering Jew, who called at a cottage to find that the owner was ill. Despite his illness and the low esteem in which Jews were usually held, the cottager welcomed the visitor to his home and gave him a cup of ale. Ahasuerus thanked him for his kind hospitality and instructed him to gather three balm leaves, steep them in a cup of ale and then drink it. He should repeat the process, refilling the cup whenever it was empty, and replace the balm leaves with fresh ones every fourth day. After twelve days the cottager was completely cured.

It was even claimed that balm could be used to heal sword wounds. This was not, as one might expect, by applying the herb to the wound but by tying a sprig of balm onto the sword.

Magic and Lore

Wine in which balm has been left to soak was used as a love potion, which was recommended to be shared by a couple to induce love and lust. To attract love, one could simply carry a piece of balm at all times.

Medicinal

The best-known use of balm is as a medicinal herb. John Evelyn says of it, 'Balm is sov-ereign for the brain, strengthening the memory and powerfully chasing away melancholy.' Culpeper adds, 'It causeth the mind and heart to become merry … and driveth away all troublesome cares and thoughts.'

Essence of balm taken every morning in a glass of canary was said to renew youthfulness and prevent baldness: 'An essence of balm, given in canary wine, every morning will renew youth, strengthen the brain, relieve languishing nature and prevent baldness.' (*London Dispensary*, 1696)

It is claimed that Llewelyn, Prince of Glamorgan, who lived to be 108, had herb teas containing balm each morning at breakfast. Balm is also a major constituent of 'Carmelite Water', which is still made in France and is reputed to help in ensuring long life as well as relieving nervous headaches and neuralgia.

MINT

Botanical Name: *Mentha* spp.
Family: Labiatae
Common Name: Sage of Bethlehem
Gender: Masculine
Planet: Mercury (Venus, according to Culpeper)
Element: Air
Deities: Hecate; Minthe; Pluto
Meanings: 'Find someone your own age'; homeliness; virtue; 'You are over-reacting to a small thing'

Myth and Legend

In mythology Minthe was the name of a beautiful nymph who caught the eye of Hades, ruler of the Underworld. Persephone, who had originally been kidnapped by Hades and forced into marriage, was jealous and transformed Minthe into the lowliest of plants that she might be trodden underfoot. Unable to save her Hades gave her a sweet aroma and now the sweet-scented mint plant grows at the mouth of the entrance into the Underworld.

Magic and Lore

Mint has been used in spells and charms to attract wealth, the simplest of which merely requires that a few mint leaves should be placed in or rubbed on a purse or wallet. This will ensure that money finds its way into the purse.

In ancient Greece mint was seen as symbolic of strength and after bathing it would be rubbed onto the arms in order to strengthen them. Parkinson informs us that Aristotle, amongst others in ancient times, forbade the use of mint by soldiers during time of war. This stemmed from a belief that, as it was supposed to increase lovemaking, it would incite venery and reduce the will to fight.

Medicinal

Like many other herbs in common use today mint was introduced to Britain by the Romans. In addition to its well-known values as a culinary herb it was used in medications. Gerard asserted that it was affective as a treatment against 'Beare-worms, Sea scorpions, Serpents and (mixed with salt) the biting of mad dogs'. To cure headaches you could rub mint against your forehead and, if it is worn wound round the wrist it will prevent all illnesses.

Pliny advised scholars to wear crowns of mint to aid concentration; he claimed that it would stimulate their brains.

PEPPERMINT

Botanical Name: *Mentha X piperita*
Family: Labiatae
Common Names: Brandy mint; lamb mint; lammint
Gender: Masculine
Planet: Mercury
Element: Fire
Deity: Pluto
Meaning: Cordiality; warmth of feelings

Magic and Lore

Pliny tells us that at feasts people would crown themselves with wreaths of peppermint and would place sprays of the herb on their tables; this may be because they believed that it had some aphrodisiacal properties. The sap or essential oil would be rubbed into floorboards and furniture to removal all negativity, and drive away any evil forces.

The scent of peppermint is supposed to be an aid to sleep, and a sprig placed beneath the pillow will cause prophetic dreams.

Medicinal

Peppermint has sometimes been regarded as the oldest medicine: archaeological evidence dates it back at least 10,000 years, and the ancient Greeks and Romans valued it. It acts as a local anaesthetic, disinfectant and vascular stimulant and has also been used as a purifying agent and to prevent seasickness.

PENNYROYAL

Botanical Name: *Mentha pulegium*
Family: Labiatae
Common Names: Brotherwort; bishopwort; churchwort; lurk-in-the-ditch; mosquito plant; organ broth; organ tea; piliolerian; pudding grass; run-by-the-ground; squaw mint; tickweed
Gender: Masculine
Planet: Mars
Element: Fire
Deity: Demeter
Meanings: 'Flee away'; 'Leave me quickly'

Magic and Lore

> Peniriall is to print your love
> So deep within my heart.
> – Clement Robinson, *A Nosegay* 1584

Pennyroyal was a popular herb for strewing in churches and banqueting halls because of the fresh, pungent fragrance that is released when it is bruised or crushed, hence the name 'churchwort'. It is said that if it is collected at daybreak on 24 June, St John's Day and kept until Christmas it will naturally dry up, but, if it is then put on the altar when the first Mass is sung it will revive.

At one time it was believed that if pennyroyal was mixed with stone taken from the nest of a lapwing or black plover, and rubbed onto the stomach of an animal, it would cause the animal to give birth to black offspring.

Like balm, pennyroyal was supposed to be particularly effective for those dealing with bees. Placing a little in a container with bees would prevent them from flying away and cause any bees that had almost drowned to recover.

There was a belief that it promoted peace and, therefore, should be given to quarrelling couples. Wearing it was also thought to be beneficial as it gave protection against the evil eye. Those who were involved in business could find that wearing a sprig of the plant would assure their continued success.

Medicinal

The herb has, at various times, been valued as an abortifacient, an aphrodisiac and a contraceptive. In addition to getting rid of an unwanted child and the afterbirth, it was supposed to increase the 'seed' in both sexes and promote lustfulness.

When travelling you should conceal a small piece of it in your shoe, as this will give you added strength and stamina during your journey, and if you are travelling by sea it will prevent seasickness.

SPEARMINT

Botanical Name: *Mentha spicata*
Family: Labiatae
Common Names: Brown mint; garden mint; green mint; green spine; lamb mint; mackerel mint; mismin; our lady's mint; spire mint; *yerba buena*
Gender: Feminine
Planet: Venus
Element: Water

Like other species of mint spearmint has been used as a medicinal herb as well as in cooking. It can be applied in many situations but particularly in treating disorders of the lungs. The scent is said to improve the mental powers. Sleeping on a pillow or mattress stuffed with it will protect you through the night and ensure that you sleep well.

BLACK MULBERRY

Botanical Name: *Morus nigra*
Family: Moraceae
Common Names: Sycamine
Gender: Masculine
Planet: Mercury
Element: Air
Deities: Brahma; Diana; Minerva; San Ku Fu Jen
Meaning: 'I will not survive you'

Myth and Legend

One account of the origins of the black mulberry is in the story of Pyramus and Thisbe, two young lovers who, like Romeo and Juliet, were forbidden to meet by their parents. However, they planned a meeting. Thisbe, arriving first at the rendezvous, saw a lion that had just killed an ox. She ran away in terror, and in her flight dropped her cloak. The lion, with bloody claws, caught it up and ripped it. When Pyramus arrived he saw the lion and the bloodstained cloak and presumed that the great beast had killed Thisbe. In his grief he killed himself beneath a mulberry tree nearby. Thisbe returned to find the body of her dead lover, and killed herself. The blood of the young couple mingled as it soaked into the earth and was taken up by the tree. Thereafter its fruit were changed from white to black.

> Dark in the rising tide the berries grew,
> And white no longer, took a sable hue;
> But brighter crimson springing from the root,
> Shot through the black, and purpled all the fruit.

Although the Romans may first have brought it to Britain, it is King James I who is credited with its introduction, and he certainly instigated its widespread cultivation, with the aim of establishing a silk industry, mulberry leaves being the preferred food of the silk-worm. Mulberry trees were planted in prison yards, and the prisoners were exercised by walking around them each morning. This gives us the children's rhyme of:

> Here we go round the mulberry bush,
> The mulberry bush, the mulberry bush,
> Here we go round the mulberry bush,
> At five o'clock in the morning.

Unfortunately the scheme failed, not least because it should have been the white, and not the black, mulberry that was planted.

Magic and Lore

Mulberry has been seen as a powerful protection against evil, and has also been used for making magic wands. Where a tree is planted in the garden it is said to be sufficient to protect the building from being struck by lightning. For those looking to see when winter has finally passed it can prove to be a useful indicator as it is said never to come into leaf until after the last frost has passed.

A dream of mulberries shows that, through weakness and indecision, you will lose a friend.

BANANA

Botanical Name: *Musa X paradisiaca*
Family: Musaceae
Common Names: *Maia*; plantain; tree of paradise
Gender: Feminine
Planet: Venus
Element: Water
Deity: Kanaloa

The banana is one of the most nutritious foods known to us and it has been an important foodstuff in the tropics from earliest times.

Myth and Legend

According to Burmese mythology, when man was created he went in search of a good food for him and his family. He came across a beautiful, tall, green tree on which were numerous fruit, which were being eaten by the birds. Seeing this the man knew that they must be safe for him to eat as well. He therefore gathered up the fruits and took them back for his family. Ever since, the Burmese have called the banana *Hnget Pyaw*, meaning 'the birds told'.

The Hawaiian people call the banana *Maià*. They say that the fruit was brought to the islands by the brother of the god Pele from Tahiti. They also have a myth that the fruit is the body of Kanaloa, their god of healing.

Magic and Lore

The superstitions and beliefs relating to the banana tend to come to us from Hawaii and the surrounding area. For example the Kapu, or code of taboos, from Hawaii (abolished in 1819) lists various species of banana amongst those foods which women were not allowed to eat, upon pain of death. In Tahiti and Hawaii the tree was occasionally used as a substitute for a human when offering sacrifices to the gods. The fruits are still used in contemporary voodoo rites.

The fruits are supposed to be an aphrodisiac, and have been consumed as a treatment for impotence or infertility. Couples who are married beneath a banana are sure to have the best of good fortune and, presumably, large families. If a woman wished to check that the man in her life was still faithful she could cut the lower tip off a banana with a sharp knife. Where she could see a 'Y' in the banana flesh it showed that he was faithful to her; if she saw an 'O' it indicated that he was not.

The fruit are used in spells to attract money, perhaps because the trees on which they grow are fruitful. Slipping on a banana skin is said to indicate that bad luck will follow. One old tradition is that, when eating a banana, it should be broken into pieces, not cut.

FORGET-ME-NOT

Botanical Names: *Myosotis alpestris*; *Myosotis scorpioides*
Family: Boraginaceae
Common Names: Herbs clavorumin; mouse ear; scorpion grass; think-on-me
Meanings: Constancy; faithful love; forget-me-not; 'Here is the key to my heart'; memories; remembrance; true love; undying love

In ancient times *Myosotis* was the name for madwort, not forget-me-not.

Myth and Legend

In Christian myth, when God was walking through Eden after the creation, he noticed a small blue flower blooming. He stopped to ask it its name and the flower, overawed, whispered, 'I'm sorry, Lord, I've forgotten.' The Almighty answered, 'Forget-me-not, I will not forget thee.'

> When to the flowers so beautiful
> The Father gave a name,
> Back came a little blue-eyed one
> (All timidly it came);
> And standing at its Father's feet,
> And gazing at His face,
> It said in low and trembling tones,
> 'Dear God, the name Thou gavest me,
> Alas! I have forgot.'
> Then kindly looked the Father down
> And said, 'Forget-Me-Not.'

A variation on this story tells that, after Adam had completed his task of naming all beasts, birds, fish and plants God pointed to the little blue-flowered plant, which

Adam had missed. Adam, therefore, called it 'forget-me-not' so that it would never again be forgotten.

A German legend tells of a young couple who were walking along the banks of the River Danube on the eve of their wedding. They saw a small blue-flowered plant about to be submerged and the young man reached down to pick it. Unfortunately he slipped on the muddy surface, fell into the river and was swept away. As he disappeared from sight beneath the water he call back to his love, *Vergils mich nicht* – 'forget-me-not'. A variation of this says that it was a medieval knight and his damsel who were walking beside a river, the knight carrying a bouquet of flowers for his lady love. However, due to the weight of his armour the knight slipped into the swirling waters and as he fell he threw the bouquet to his love shouting 'Forget-me-not.'

In another legend a wayfarer walking through a lonely valley noticed a flower he did not recognize growing at his feet. When he bent down to pick it the mountainside opened. The man entered and found a hoard of gold and precious gems. He began to gather them but, in doing so, he dropped the little flower, which murmured as it fell, 'forget-me-not'. So intent was the man on the riches he was collecting that he ignored the little flower's plea. The rift in the mountain began to close, leaving him barely time to escape, and the little flower that had opened it was gone forever.

Magic and Lore

The pretty blue flower has for centuries been a symbol of eternal love and friendship in countries across Europe and has frequently been incorporated into the seals of lovers. Those leaving on long journeys would often give their lady loves a posy of the flowers before departing. Henry of Lancaster, who became Henry IV, adopted it as his emblem whilst in exile in the belief that if he did so he would never be forgotten. During the Second World War, in Germany, the forget-me-not became a symbol of freemasonry following the organization's banning by Hitler.

Forget-me-not flowers provide protection from mischievous faeries. They are also supposed to be able to unlock the secrets of faerie folk and, as in the tale above, to enable the bearer to find the way to faerie treasures.

If you should see forget-me-nots in your dreams it suggests that you must be firm and break up with your love, as he or she is unsuitable.

Finally, forget-me-not is the best plant to give to those who are setting out on a journey on 29 February.

Medicinal

In folk medicine the forget-me-not was recommended for the treatment of the bites of dogs and snakes. It is also supposed to be a cure for the stings of scorpions, which may be a reference to sympathetic magic as the shoots curl at the end like a scorpion's tail. Steel tempered in the juice of the plant is said to be hard enough to cut stone.

BOG MYRTLE

Botanical Name: *Myrica gale*
Family: Myricaceae
Common Names: Bwrle; fleawood; gale; sweet gale; sweet willow

The botanical name *Myrica* comes from the Greek name for the tamarisk, *Myrike*. Names such as 'fleawood' come from its use as a vermifuge.

Bog myrtle, like willow, was used as a substitute for palms on Palm Sunday in some parts of Britain. It may be because of this religious association that it was considered to be unlucky to use it as a switch for driving cattle with.

NUTMEG

Botanical Name: *Myristica fragrans*
Family: Myristicaceae
Common Name: Mace
Gender: Masculine
Planet: Jupiter
Element: Fire
Deity: Myrrha

Both mace and nutmeg are well-known spices, and both come from the same plant; mace is obtained from the outer covering. It was common, at one time, to carry a little nutmeg in order to flavour foods.

Magic and Lore

To ensure the fidelity of a lover take a nutmeg and cut it into four equal pieces. One piece should be burned, the second buried, the third thrown off a cliff, and the last boiled in water. You must then drink a sip of the water and keep this last piece of nutmeg with you always, sleeping with it under your pillow at night. Your lover will then never be tempted away from you.

If you see nutmeg in your dreams it is an indication that you should expect many changes in your life in the near future, and that you will ultimately rise to high status.

Medicinal

Nutmeg was believed to prevent rheumatism and boils, and was also used as a fumigant against the plague. A necklace of tonka beans, star anise and nutmeg could be made and worn as a powerful amulet. Alternatively nutmeg might be strung on its own, and placed about the neck of a young child as an aid to teething.

MYRTLE

Botanical Name: *Myrtus communis*
Family: Myrtaceae
Gender: Feminine
Planet: Venus
Element: Water
Deities: Aphrodite; Artemis; Ashtoreth; Astarte; Hathor; Mariamne, Marian, Marina; Miriam, Myrrha, Myrto, Venus
Meanings: Affection; 'Be mine for ever'; discipline; duty; fertility; home; instruction; 'Joyously I do return your love'; love; love's fragrance

Myth and Legend

According to one Christian legend, when Adam and Eve had to leave Eden following the fall from grace, they were allowed to take three things with them. They took the date, as the best of all fruits, wheat as the best of all grains, and myrtle as the best of all fragrances. In Muslin tradition Adam only took the myrtle, and this was taken from the bower in which he had first declared his love for Eve.

Two large, old myrtle bushes grew in ancient Rome, one called 'Patrician' and the other 'Plebeian'. The political future of the nobles and the common people could be gauged by comparing the relative health of these two plants. Myrtle was sacred to the goddess Astarte, also known as Aphrodite, and in some tales from classical mythology it is said to be created by her; she was crowned with a wreath formed from it after being declared the fairest of the goddesses at the judgement of Paris. In some parts of Greece Aphrodite was called Myrtilla.

Other myths tell of the nymph Daphne, who turned herself into a myrtle to escape being raped by Apollo, or more vaguely that Myrtle was the loveliest virgin in Greece, who was transformed into the plant to protect her from persecution. In consequence of this the plant has been used as a protective herb for guarding the purity of virgins. In another myth, Venus punished Psyche, a young girl, by beating her with a rod of myrtle for ensnaring the heart of her son Cupid. The nymphs associated with myrtle were usually considered to be beneficial and they were credited with having bestowed the gifts of growing olive trees, keeping bees and making cheese on mankind.

In biblical times myrtle was associated with the peace of the Lord and more specifically with the Feast of the Tabernacles. In Jewish mythology, Myrtle was a woman who was transformed into a myrtle tree after the townspeople accused her of being a witch and then murdered her. This corresponds with a Greek tale which tells that the myrtle was a once human speared to death by barbarous villagers. It is in consequence of this that the leaves have tiny holes in them.

In common with many other evergreen plants myrtle was also taken to be a symbol of resurrection. Greek emigrants would take boughs with them when they set out to found new colonies, as it represented the old life they had left behind and their new life ahead. In Roman myths myrtle was originally seen as a plant of Mars but became associated with Venus, the equivalent of the Greek goddess Aphrodite, after she gained shelter in a grove of the plants when trying to escape from some satyrs. In the crown of Venus myrtle is interwoven with vervain.

Magic and Lore

Being the plant of Venus or Aphrodite, myrtle has long been associated with love and romance and was considered to be a suitable plant to give as a love token. In the language of flowers it represents variously love, marriage and love in absence. The tradition of including a sprig of myrtle in the flowers carried by a bride at her wedding seems to have begun during the time of the Jews' captivity in Babylon, and has spread westwards across Europe. It is common in Germany, where the myrtle is taken to symbolize purity and fertility.

During the Victorian period myrtle was a favourite item to include in bridal wreaths and small bouquets called 'tussie mussies'. It was a symbol of fidelity in marriage, and was used in Queen Victoria's wedding bouquet when she married Prince Albert. She had the myrtle from her wedding bouquet planted in the gardens of Osborne House and the progeny from this original planting provided the myrtle that was included in the wedding bouquets of Queen Elizabeth II, Princess Margaret, Princess Anne and the Princess of Wales. It is traditional that myrtle from the bride's bouquet is planted in her garden after the wedding, not by the bride but by one of her bridesmaids. A chaplet of myrtle might be worn by brides at their weddings to help them avoid becoming pregnant too soon afterwards.

In Tuscany there was a custom that a couple who were engaged would exchange sprigs of the plant, which had to be produced when the couple next met, or else the engagement would be broken. There is a German superstition that a woman who is engaged to be married should never plant a myrtle, or her wedding will be cancelled. On the other hand, when a myrtle bush is seen to flourish and produce a profusion of flowers it is a sign that there will soon be a wedding in the household.

In Somerset the flowering of the myrtle is considered to be an omen of good fortune. Indeed myrtle is generally considered to be a plant of good luck, mainly linked to love, fertility and marriage. When it is grown either side of the main doorway to a house it will ensure that the people living there will enjoy the best of luck, love and peace. However it must never be dug up, or allowed to wither and die, as this will result in misfortune befalling the family. It is also a lucky plant to grow in a window box, bringing luck and money to the household, although it is said that it will not flourish there unless planted by a good woman. However, some sources do claim that growing myrtle in the garden will bring those who cultivate it nothing but sickness and grief.

If a young woman wanted to know whether or not she would marry her sweetheart she had to pick a sprig of myrtle on St John's Eve and lay it across the pages of her prayer-book with the words, 'Wilt thou take me [mentioning her own name] to be thy wedded wife?'

She should then close her book and sleep with it under her pillow. If the myrtle was gone in the morning her sweetheart would be her husband.

Myrtle has a protective use and is considered a good ritual remedy when one is threatened. It was once believed that eating the leaves would enable the eater to detect witches. It was also claimed that if the leaves crackled when they were crushed in the hand, the person's lover would be faithful.

Seeing myrtle in your dreams is an indication that you will receive a legacy and gain the attention of many lovers. If you are already married it is a sign that there will be a second marriage.

Medicinal

Myrtle tea, drunk every three days, is said to be a good way of staying young looking. Wearing or carrying myrtle is supposed to be equally beneficial.

PRIMROSE PEERLESS

Botanical Name: *Narcissus X medioluteus*
Family: Amyrillidaceae

Primrose peerless was a much loved and widely cultivated ornamental that is now naturalized across much of Britain. It is said to have been brought back to Britain by a crusader knight returning from the Holy Land. He had been shocked by the heat and bloodshed he had seen, and had lost much of his fortune. All that he brought back for his wife were two bulbs of primrose peerless, as she had always loved rare flowers. Unfortunately, he discovered that she had been dead for four years. In his despair he threw the bulbs over the churchyard wall and died of a broken heart on top of his wife's grave.

DAFFODIL

Botanical Name: *Narcissus pseudonarcissus*
Family: Amyrillidaceae
Common Names: Affodil; affrodil; asphodel; averell; bacon and eggs; belle-blome; bell flowers; bell rose; butter and eggs; chalice flower; churn; cincliffe; codlins and cream; cowslip; cuckoo-rose; daffadilly; daffodilly; daffy dilly; daffy-down-dilly; dilly daffs; down-dilly; Easter lily; Easter rose; faerie bells; false narcissus; *fleur d'aspholéle*; *fleur de coucou*; giggary; glens; gold bells; golden narcissus; golden trumpets; goose-flop; goose herb; goose leek; gracie-daisies; gracie day; gregories; hen and chickens; hoop petticoats; jonquils; julians; king's spear; lady's ruffles; lent cocks; lent lily; lent pitchers; lent rose; lent rosen; lentils; lents; lenty cups; lenty lily; lide lily; lily; *lus-ny-guiy*; narcissus; *pauvres filles de sainte claire*; porillion; Queen Anne's flowers; St Peter's bell; sun bonnets; Whit Sunday; wild jonquil; yellow crowbells; yellow maidens

Gender: Feminine
Planets: Venus (All non-yellow forms); Mars (All yellow forms)
Element: Water
Deities: Hades; Persephone; Pluto (associated with the Fates – Clotho, Lachesis and Atropos)
Meanings: Chivalry; conceit; deceitful hope; disdain; egotism; new beginnings; regard; regret; refusal; respect; self esteem; self-obsession; 'The sun shines when I'm with you'; unrequited love; 'You love none save yourself'; 'You're the only one'.

Myth and Legend

Narcissus, in classical myth, was a shepherd boy, the son of Liriope, a naiad who had been ravished by Cephisus the river god. He was adored by many young maidens on account of his great beauty (some say that he was blessed with eternal youth and beauty provided he never looked at his own reflection), but to none did he return love. Perhaps the best known of those who sought his love was a nymph called Echo. Hera had condemned her only to be able to repeat the words of others. She followed Narcissus and, turning around, he called out, 'Is anyone there? Let us come together.' Echo replied, 'Let us come together', and ran to him. He rejected her, saying, 'I will never let you lie with me!' and ignored her impassioned reply of 'Lie with me.' She was so smitten that when he rejected her she stopped eating until she faded almost away. The gods transformed her into a pile of stones but even then could not heal her broken heart. She still laments her fate amongst those remote places where she had followed the shepherd she loved, only allowed to repeat the last words which are spoken.

One day Narcissus stopped to rest, and stooping to drink from a small pool of water caught sight of his own reflection. Nemesis, the avenging god, caused him to fall in love with his reflection and so he was transfixed by its beauty. He fell into the pool, as he tried to reach his reflection, and drowned. (An alternative ending to the tale says that he stabbed himself with a sword, knowing that he could never obtain his desire, and from his blood sprang a white daffodil with a red centre.) He was carried down to the banks of the River Styx that flowed through the Underworld. The nymphs had prepared a funeral pyre for him, but his body had changed into the flower that carries his name. The cup in the middle of the flower is said to hold his tears. In consequence of this sad story the daffodil symbolizes unrequited love, vanity and excessive self-love.

A less romantic notion of how the name *Narcissus* came to be applied to this plant suggests that it is derived from the Greek *narkao*, a reference to its narcotic properties. Apparently this led to it being called the 'chaplet of the infernal gods' by the philosopher Socrates. Names such as 'daffodil' and 'daffy-down-dilly' are corruptions of asphodel, a name used by the ancient Greeks. According to myth asphodels were originally all white and were a favourite flower of Persephone, the daughter of Demeter. In springtime Persephone would spend a good deal of her time walking through the Sicilian meadows where they were in bloom. She was unaware that Pluto, god of the Underworld, had seen her during these walks and become besotted by her beauty. One spring morning, as she lay dozing on a grassy bank, surrounded by the white flowers, Pluto reached out from his Underworld kingdom and carried her away. Where he touched the asphodels growing around her they were turned to sulphurous yellow. Persephone had made some of the flowers into a chaplet for her head, which fell onto the banks of the River Acheron, which passed through the Underworld. There they took root and grew in profusion. Being the last flower gathered by Persephone before being taken into the Underworld the daffodil was seen as emblematic of deceit and of imminent death, and as such it was often planted on or near graves.

Chinese legend tells of an old woman from Fukien province who was so touched by the plight of a beggar that she gave him her last bowl of rice which she had been saving for her lazy son. The beggar thanked her for her generosity and greedily ate the rice, spitting the last few rice grains onto the ground, and disappeared. The next morning the old woman was amazed to find daffodils growing on the spot, all of them having white petals and yellow centres. Sale of the flowers made her wealthy and Fukien province famous. Thereafter, in China, the flower came to symbolize benevolence and propriety.

Christian mythology tells that the daffodil first appeared in the Garden of Gethsemane to comfort Jesus in his hour of sorrow prior to his arrest and eventual crucifixion.

Magic and Lore

The daffodil was one of the earliest of cultivated plants. The Prophet Mohammed said, 'He that has two cakes of bread, let him sell one of them for some flowers of narcissus, for bread is the food of the body, but narcissus is the food of the soul.'

Daffodils are associated with St David, patron saint of Wales and are said to bloom on his feast day, 1 March. It is the floral symbol of Wales, and George V wore one at his investiture as Prince of Wales. The current Prince of Wales, as Duke of Cornwall, receives payment of one daffodil bulb as rent for the uninhabited parts of the Isles of Scilly. Welsh tradition claims that whoever finds the first flowering daffodil of the spring will have more gold than silver in the coming twelve months. Wearing a daffodil next to your heart is also said to bring you good luck.

As with many other flowers, it is usually considered to be unlucky to take a single daffodil into the house; you must take a bunch of the flowers or none at all (there are some cynical people who would suggest that flower sellers instigated this superstition). On the Isle of Man the plant is called *lus-ny-guiy*, or goose herb. Farmers are warned that they should never take daffodils into the house when chickens or geese are sitting on eggs, as it will prevent them from hatching. Elsewhere there is a superstition that the number of goslings hatched, and moved, in the year will be governed by the number of wild daffodil blooms brought into the home in the first bunch of the season.

Others say that it is unlucky and so should not be taken into the home. As well as being associated with death, the hanging heads of the flowers are said to bring tears and sorrows.

Placing a vase of daffodils in a bedroom is supposed to boost the fertility of the people who sleep there. If you dream of daffodils, it is an indication that you have treated a friend unfairly and that you should now seek a reconciliation with them.

Medieval ladies of high status occasionally cultivated daffodils in order to collect the juice of the flowers to tint their hair and eyebrows.

At the fall of Singapore in the Second World War, the Japanese referred to the Australians as 'daffodils', because they were beautiful but yellow, alleging cowardice.

The occurrence of daffodils in the wild can be an indication of a former religious settlement. It was recorded in 1797 that at Frithelstock, near Torrington in Devon, the villagers referred to daffodils as 'gregories', as the local priory of Augustine canons, founded by Sir Robert Beauchamp, was dedicated to the Virgin Mary, St Gregory, and St Edmund. In London, the only sizable population of daffodils is at Abbey Woods, the site of Lesnes Abbey.

Daffodil is the birthday flower of 28 January.

Medicinal

Parkinson recommends daffodils for treating everything from coughs and colic to sunburn and splinters. In a mixture with nettle seeds and vinegar, they were also suggested as a treatment for acne. The juice was even considered as a treatment for baldness.

It has been suggested that the toxic alkaloid contained in the bulbs led to them being carried by Roman soldiers so that, if they were mortally wounded, they could eat them and the narcotic effect would mean that they would die painlessly.

In 1990 the flower was adopted as the emblem of the Marie Currie Cancer Care charity.

WATERCRESS

Botanical Name: *Nasturtium officinale* (synonym: *Rorippa nasturtium*)
Family: Cruciferae
Common Names: Brooklime; brown cress; cress; cresson; nasturtium; water cresses; true watercress

The botanical name is derived from the Latin *nasus tortus* ('twisted nose') on account of its pungency. It is said that, in ancient times, Xerxes, the Persian King, ordered his soldiers to eat watercress in order to keep them healthy. The ancient Greeks also believed that eating it would make them 'witty'.

According to tradition watercress is edible in any month that has an 'R' in it, so the summer months from May to August are the 'off-season'. Eating watercress is said to have a number of benefits, not least as an aphrodisiac. However, an Egyptian saying states, 'You do not have to eat the *gargir* [watercress] to increase masculinity; it is enough to put a bundle of it under your bed, and its intoxicating scent will reach your nose.' It is also said that rubbing freckles with it will diminish them. The odour of burning watercress was thought to drive away snakes and help treat the mentally ill.

SACRED LOTUS

Botanical Name: *Nelumbo nucifera*
Family: Nelumbonaceae
Common Names: Pythagorean beans; sacred bean
Deities: Brahma; Shiva; Vishnu

The sacred lotus can be found from Iran to Australia. It is a name applied to several different plant species. The lotus of the ancient world is the *Nelumbo* waterlily, although that of ancient Egypt may well be a species of *Nymphaea* waterlily that shares the habit of lifting its flowers clear of the water.

Myth and Legend

In Ancient Egypt the lotus is associated with the birth of the creator god. This god might be Atum or Ra depending on the locality. The image of the creator springing from the heart of the lotus flower has many similarities with the Hindu story of creation by Brahma. In Hinduism the lotus is the cradle of the universe. It represents the yoni or female generative organ. The spirit of the Supreme Being was pictured as a golden lotus on a great sea. From Vishnu, the lotus-navelled god, issued a lotus that floated on the waters of the flood that had destroyed creation. On the lotus flower sat Brahma, who re-created the world. The lotus expanded into the universe. From its petals rose the mountains, hills and valleys. Shiva is also often pictured seated on a lotus blossom. These deities can be seen as Brahma the creator, Vishnu the preserver and Shiva the destroyer.

The lotus is also a symbol of Buddha. It symbolizes the progress of the individual through the various levels of consciousness. The leaves represent the ability of the intellect to rise above the mud of the common life to reach the domain of the divine.

Magic and Lore

As the flower of the lotus is held up proud of the water surface it is seen as symbolic of chastity and purity. The plant is said to have 'anti-aphrodisiacal' properties.

CATMINT

Botanical Name: *Nepeta cataria*
Family: Labiatae
Common Names: Cat; catnep; catnip; catrup; cat's wort; dog mint; field balm; nep-in-a-hedge; nip; nepte
Gender: Feminine
Planet: Venus
Element: Water
Deity: Bast

Anyone who has tried introducing a young catmint plant into a garden where there is a mature cat will have discovered that the plant is lucky if it survives. The cat will usually lie on it, chew it, rub against it or otherwise find some way in which to do it harm. Surprisingly the old adage,

> If you set it, the cats will eat it,
> If you sow it, the cats don't know it

seems to have some element of truth. Plants grown from seed *in-situ* do seem to survive better. It is said to make a cat 'frolicsome, amorous and full of battle'.

Giving catmint to a cat it is supposed to strengthen the psychic bond with it. And this does not apply only to cats. If you hold a piece of catmint in your hand until it is warm and then hold hands with someone they will be your friend for ever, or at least for as long as you keep the piece of catmint. Chewing the root, it is claimed, will calm even the most quarrelsome and fierce person. It has been suggested that this is the reason that rats, which will normally eat anything, will not touch it even if starving.

It may be because witches' familiars were often portrayed as black cats that growing catmint was thought to attract good and helpful spirits, and bring good luck. Bunches could be hung in the house to achieve the same result. The scent is said to be an aid to a peaceful sleep. The larger leaves of the plant should be collected and pressed as bookmarks for use in books of magic.

GUERNSEY LILY

Botanical Name: *Nerine sarniensis*
Family: Amaryllidaceae

This plant has long been naturalized in Guernsey and has become the island's floral emblem, although it originates from South Africa. It gains its botanical name from the sea nymph Nerine.

A legend from the island has it that a beautiful maiden named Michele De Garis fell in love with the faerie king and was persuaded by him to go with him to his far-away kingdom. Before she left, however, she thought about her family and how they would miss her. She asked her faerie lover to let her leave some token by which they would remember her. He gave her a flower bulb, which she planted in the sands above Vazon Bay before leaving her island home. Later her parents, searching for their missing daughter, found the plant now in flower. The bloom was a beautiful scarlet colour, dusted with faerie gold, but scentless.

TOBACCO

Botanical Name: *Nicotiana tabacum*
Family: Solanaceae
Common Name: Tabacca
Gender: Masculine
Planet: Mars
Element: Fire

Tobacco was once used in the religious ceremonies of the Native Americans. Smoking it was believed, in some cases, to enable one to communicate with the spirits. Indeed, some shamanistic systems required initiates to drink tobacco juice in order to induce visions.

Tobacco's bad press is nothing new. It was introduced to Britain in 1586, and King James I said that smoking was, 'A custom loathsome to the eye, hateful to the nose, harmful to the brain, dangerous to the lungs, and in the black, stinking, fume thereof, nearest resembling the horrible Stygian smoke of the pit that is bottomless'. Likening tobacco smoke to that from the pits of Hell is quite appropriate as in magic it has been used as a substitute for sulphur. It is also occasionally substituted for nightshade and thornapple, to both of which it is closely related. It is included in incense used during purification rites to remove evil spirits.

Not all writers have damned tobacco; some thought that it had many redeeming attributes, if taken in moderation.

> Tobacco, divine, rare, superexcellent tobacco, which goes far
> beyond all the panaceas, potable gold, and philosophers' stones,
> a sovereign remedy to all diseases but, as it is commonly abused
> by most men, which take it as tinkers do ale, 'Tis a plague, a mischief,
> a violent purger of goods, lands, health; hellish, devilish and damned
> tobacco, the ruin and overthrow of body and soul.
>
> – Robert Burton, *Anatomy of Melancholy*

Blowing tobacco smoke into an ear has been suggested as a cure for earache.

WATERLILY

Botanical Name: *Nymphaea* spp.
Family: Nymphaeaceae
Gender: Feminine
Planet: Moon
Element: Water
Deity: Coventia
Meanings: Eloquence; purity of heart; silence; 'The light of your spirit ever shines through'

Myth and Legend

Among the Dakota tribe of Native Americans there is a story of Chief Red Strawberry Man, who was visited in a dream, by the star maiden of the night sky. She told him that she was tired of her life in the skies and wished to come to Earth and live amongst the Dakota. Red Strawberry Man wanted advice from the tribe's adviser, who lived at the other side of a lake, so he sent his son to find him. The son paddled swiftly over the water in the darkness, and the sky maiden sat in the canoe with him. In his haste, and unable to see clearly, the boy's canoe struck a log in the water. The star maiden was thrown out and sank. The next morning, at the point where the star maiden's light had been extinguished in the waters, there grew a waterlily with bright yellow flowers.

A similar story, from the Chippewa tribe, tells of a star maiden who became very attached to the tribe over whom she sparkled in the night sky. She wanted to get closer to them but could not think how to, so she appeared in a vision to one of the young men of the tribe and asked him to put the problem before the elders of his people. This he did, and he was able to go back and suggest to her that she might become the heart of a flower. Delighted at the idea the star maiden first tried to become a mountain rose, but this meant that she was still too far away from the people. Then she became a prairie flower, but found that she was in constant fear of being eaten or trampled by the buffalo that roamed there. Finally she chose to float down the lake beside which the tribe lived, and there spread out her wings onto the waters to become the first waterlily.

The story of the plant's origin is similar in classical mythology. In Greek myths Lotus was a beautiful water nymph who was smitten when she saw the virile demigod Hercules. However, he did not return her affections. She pined for his love and eventually died. Hebe, the goddess of youth and spring, transformed her body into the first waterlily with purple flowers. Later Dryope and Iole, sister nymphs, were out walking and picking flowers with Dryope's son, and came across the beautiful waterlily. When Dryope plucked the flower blood dripped from the broken stalk onto her hands. Immediately leaves sprang out of the hands and her feet took root. Her skin became rough bark and she was transformed into the lotus shrub. Her fruits were the fruits of forgetfulness sought by the lotus-eaters.

SACRED LOTUS

Botanical Name: *Nymphaea edulis*; *Nymphaea lotus*
Family: Nymphaeaceae
Common Names: Bride of the Nile; lotus; white Egyptian lotus
Gender: Feminine
Planet: Moon
Element: Water
Deities: Horus; Osiris
Meaning: Estranged love

In Greece the name 'lotus' was given to a shrub in the buckthorn family, *Zizphyhus lotus*. Hindus applied the name to a species of waterlily, *Nelumbo*. For the Egyptians it was a waterlily that grew on the Nile and held its leaves and flowers proud of the water level. The blue waterlily was the symbol of Upper Egypt, just as the papyrus was the symbol of Lower Egypt, and was the ancient Egyptian's most sacred plant.

Myth and Legend

In Egyptian mythology Osiris, the god of the dead, is sometimes shown crowned by lotus petals and his son Horus, god of silence, is shown seated on a lotus flower.

Magic and Lore

The habit of the lotus lends itself perfectly to mysticism and magic. Rooted in the deep slime of the river the flowers grow up through the muddy water to open to the heavens proud of the surface. The root symbolized indissolubility and the stem the umbilical cord that linked man with his origins. The flowering was symbolic of the rays of the sun and the unfurling of its petals was emblematic of a spiritual unfolding. The seed head was seen as the emblem of the fecundity of the creation.

The first waterlily to achieve religious status was the white Egyptian lotus, *Nymphaea lotus*, the 'bride of the Nile'. This was superseded by the red-flowered *Nymphaea rubra*, the red waterlily of the Old Kingdom, the colourful flowers of which were seen as symbolic of the fire of the sun, the divine light of creation. The habit of these plants linked them to the lunar powers of water, darkness and chaos, and thus associated them with the cycle of birth and death. Each evening the water-lily flowers close their petals and withdraw beneath the water's surface so that they cannot be touched by hand. Every morning, at break of day, they turn to the East to face the rising sun, just as the dead were faced towards the east so that they might metaphorically face the new dawning life of the sun.

By the time of the New Kingdom *c.* 1567 BC) the red lotus had, in turn, been superseded by *Nymphaea caerulea*, the blue lotus. The deceased are depicted in tombs refreshing themselves with the fragrance of the flowers. In ancient Egyptian art the lotus is shown in party scenes, sex scenes and with wine. The body of

Tutankhamen was covered with the flowers. There may be a further significance, as researchers have identified that the flowers contain a sexual stimulant.

It was the lotus that gave rise in Greek art to the pattern of lines known as the 'meander' or 'fret', and thus to the double fret. The symbol of the double fret was used in temples as an emblem of opposites such as good and evil, light and dark, life and death and male and female. The double fret is now better known as the swastika.

BASIL

Botanical Name: *Ocimum basilicum*
Family: Labiatae
Common Names: Albahaca; American dittany; herb royal; our herb; St Joseph's wort; sweet basil; witches' herb
Gender: Masculine
Planet: Mars (in Scorpio)
Element: Fire
Deities: Erzulie; Krishna; Vishnu
Meanings: Animosity; hatred; 'I cannot like you'; poverty; sympathy

Myth and Legend

The ancient Greeks saw basil as symbolic of hatred and misfortune. Poverty was depicted by a sad woman, dressed in rags, with a basil plant growing close by.

One of the best-known and most gruesome tales involving basil can be found in the *Decameron* by Boccaccio. It tells of a woman named Isabella, a maid of Messina, whose brothers were both wealthy businessmen. She developed a romantic relationship with Lorenzo, the manager of their enterprises. Initially, to avoid scandal, the brothers pretended that nothing was happening. Eventually, however, they decided drastic action was required. They called Lorenzo to attend a festival outside the city, and there they killed him. They told their sister that he had been called away on a long journey, but as the days of his absence became weeks she became increasingly concerned. One day, when she was asleep, the ghost of her lover appeared to her and said that he could never return to her as her brothers had killed him. He told her where the body was to be found, so on waking she made her way there, where she found the well-preserved corpse of Lorenzo.

Unwilling to risk discovery by moving the body to consecrated ground she took a knife and removed the head, which she wrapped in a cloth and placed in a large pot. She covered it with earth and planted a basil plant. She tended the basil with great care, often watering it with her tears. The plant flourished, but Isabella's brothers took away the pot in order to try to cure her of her grief. When Isabella wailed for its return the brothers broke open the pot to see if there was anything concealed in its roots, and there they found the still recognizable head of Lorenzo. Knowing that their crime was discovered they buried the gory relic afresh and then fled to Naples, leaving Isabella to die of a broken heart.

This story has its roots in a number of much older beliefs about this herb, one of which was that it would grow best if set on the brains of a murdered man. It was also said that it could cause scorpions to grow in a man's head. There is a suggestion that the name is derived from the Greek *basleus*, meaning 'king', as this is a herb 'fit for a king'. Another possibility is that it is a corruption of '*basilisk*', the 'King of Serpents'. The basilisk was a mythical animal that could kill with a single look, and the association of basil with it comes to light elsewhere, with the tales that link it with scorpions. Tradition has it that scorpions prefer to lie under pots of basil, and in some places it was even believed that if a sprig of basil was left in a pot it would turn into a scorpion. Parkinson tells us that if the leaves of the herb are handled gently they will release a pleasant fragrance, but if roughly handled and the leaves bruised the plant will breed scorpions, and if smelt too closely it will release scorpions in the brain. Culpeper says, 'Mizaldus affirms, that being laid to rot in horse dung, it will breed venomous beasts. Hilarius, a French physician, affirms upon his knowledge, that an acquaintance of his, by common smelling, had a scorpion in his brain …'

Magic and Lore

There was a tradition that a young woman should place a pot of basil in her window if she was expecting a visit from her lover (some writers say that it showed that she was seeking a new suitor). In Crete the herb symbolized 'love washed with tears', and in Italy it might be given as a love token. In Moldavia there was belief that a man would fall in love with a woman if he accepted a piece of basil from her hand. The scent of basil is said to cause sympathy between people and because of this it would be given to quarrelling couples to soothe their tempers. It was also added to incense which was burned to induce love, or rubbed directly onto the skin to act as a natural perfume.

The herb, it was said, could also be used to ensure the chastity of one's lover. All one had to do was to sprinkle a little fresh basil over them, especially over the region of their heart, as they sleep and they would remain chaste. And to discover if a lover is promiscuous, place a basil leaf in their hand. It will immediately wither if they are 'light of love'. Another way to test the chastity of a woman was to have her walk through a swarm of bees while carrying a basil plant. If the plant withered, or the bees attacked her, she was unchaste.

Basil has been regarded both as a plant of great good luck and a herb of ill omen, and in some situations it is a symbol of grief and death. As a plant of good fortune it should be given to those who are moving home to bring them good luck in their new house. Carrying a sprig of the herb, or placing it in a cash register, will attract money into the business at which you work. Placing a small piece of basil in each of the four corners of the house at the start of each season will bring wealth.

In Persia basil, called *rayhan*, was used at funerals. Hindus would refer to it as the 'holy' or 'sacred herb', and in India it is held in high regard and is dedicated to the gods Vishnu and Krishna. It is valued for its purifying properties and is seen as a protective influence within the home as it purifies the atmosphere, cleansing the air. It has also been used as a herb to protect against the activities of witches and for exorcising evil influences from properties. It was strewn on the floors as, where basil lies, no evil can pass. A little of the fresh herb could be added to purification baths, incense and mixtures. Placing a sprig of the herb in each room of the house would keep evil from entering.

Paradoxically, witches might use the plant; they would drink half a cup of the juice before flying on their broomsticks to their meetings. The juice would prevent them becoming inebriated or dizzy.

Some very sage advice has been given regarding the cultivation of basil. The Greeks and Romans asserted that you should curse the seeds as you sow them to ensure that they will germinate. There is also a suggestion that, whereas most herbs should be watered in the early morning or early evening to prevent them drying out, basil should be watered at noon because it is said to grow fastest when watered with warm water. Theophrastus warns that basil will turn pale, through the summer, following the rising of the Dog Star, and that the coriander in the garden will become mildewed. Moreover when laying out a herb garden one should consider the nature of the plants. Basil is a sweet herb and should not be placed near one which is bitter, like rue. Anyone keeping goats near to the garden might find that basil can prove helpful; Francis Barrett, in *The Magus*, asserts that '... goats hate garden basil, as if there was nothing more pernicious'.

Medicinal

For the purposes of folk medicine basil is used like a modern inoculation, to draw out poisons as, like the scorpion, it also has a poisonous side. Another use of this 'king of herbs' is as an aid to dieting; tradition has it that a woman will be unable to eat anything from a plate if, unknown to them, there is a basil leaf hidden under it. Snuff made from basil was once recommended to cure headaches, and Dodoens, the Belgian herbalist, claimed that, if a pregnant woman in labour held a root in one hand together with the feather of a swallow, she would give birth without any pain.

BUSH BASIL

Botanical Name: *Ocimum basilicum* 'Minimum'
Family: Labiatae
Common Name: Greek basil
Gender: Masculine
Planet: Mars
Element: Fire

From the botanical names it is clear that bush basil is very closely allied to sweet basil and much of what has been said about the latter can be applied to this plant.

Bush basil is associated with death in Malaysia and Persia and was planted on graves. In Egypt the flowers of the plant would be scattered onto the graves of friends and family members. For the ancient Greeks it represented misfortune and hatred.

Amongst ancient herbalists there was disagreement about the benefits of this plant. Galen and Dioscorides stated that it was unfit to be taken internally, but its use was defended by Pliny and Arab herbalists.

EVENING PRIMROSE

Botanical Name: *Oenothera biennis*
Family: Oragraceae
Common Names: German rampion; onagra (ass food); tree primrose; war poison
Meanings: Inconstancy; 'Humbly I adore you'; mute devotion; silent love; uncertainty

The name *Oenothera* was given to the plant by Theophrastus. One explanation of its derivation is that it comes from the Greek *oinos*, meaning 'wine', and *thera*, relating to 'hunting' or 'pursuing', an allusion to a Roman belief that eating the roots could produce a desire for drink. It is also claimed by some that the roots smell of wine.

Magic and Lore

Although the plant's name seems to indicate that it might increase the desire to drink, in folklore it is suggested that it will counter the effects of heavy drinking.

In the USA the Native Americans would rub the plant against their moccasins before setting out on a hunt to ensure success. The oil would also act as a protection against snakes and snakebites.

Medicinal

In folk medicine, evening primrose has been used in the treatment of asthma, whooping cough, gastrointestinal disorders and symptoms associated with premenstrual syndrome.

OLIVE

Botanical Name: *Olea europaea*
Family: Oleaceae
Common Name: Olivier
Gender: Masculine
Planet: Sun
Element: Fire
Deities: Allah; Apollo; Athena; Hercules; Irene; Minerva; Ra; Zeus
Meanings: Peace; goodwill

Myth and Legend

It is thought that the olive has been in cultivation for at least 5,000 years; the earliest known live groves can be found in Crete and Egypt. In classical mythology it was a gift to mankind from Athena, the goddess of wisdom. She and Poseidon, the god of the seas, were competing for possession of Attica, which developed into a contest to see who could provide the most beneficial gift to mankind. Poseidon struck the ground with his trident causing salt water to gush forth from the smitten rock. (Other accounts say that he created the horse.) Athena created the olive tree, and that was declared to be the more valuable of the gifts. In consequence the city of Athens was named in her honour. The followers of Athena would punish anyone who dared to harm their olive trees. Goats were often sacrificed to the goddess, not least because they often damaged these sacred trees. In 480 BC, when Xerxes, King of the Persians, sacked Athens, the olive tree on the Acropolis was burned but reappeared as if by magic. Today there is still an olive tree there.

The thyrsus or caduceus, the wand of the god Bacchus, was formed from twisted olive branches with three leaves on them (later it is seen with two snakes curled around it). Kissing this thyrsus, or being touched by it, bestowed the gift of eloquence.

A German legend says that when Adam died the angel who guarded the gates of Eden gave seeds of the olive, the cedar and the cypress to Seth, his son. He placed these seeds into the mouth of his father's corpse before it was buried. The seed germinated and grew to form one multi-stemmed tree. It was a shoot from this tree that the dove carried back to Noah when the waters of the great flood began to recede, and it was under the branches of this same tree that King David wept. Wood from it was used by the Queen of Sheba when crossing the marshes and it was also the wood of this tree that was used to form the cross on which Christ was crucified.

Magic and Lore

Holy oil is a mixture of the oils of several plants, with olive as the base, which represents the *logos* or wisdom of God. To it is added oil of myrrh, which is attributed to Binah, the Great Mother, who encompasses the understanding of the magical and the compassion that comes as a result of the contemplation of the universe. Cinnamon oil is included to represent Tiphereth, the sun. Galangal, representing the first and last, the alpha and omega, is the final ingredient. Olive oil was used for the lamps which lit the temples, and it was also used as an anointing oil especially where this was done as an aid to healing.

In addition to having been considered an emblem of peace the olive is also the symbol of the Greek harvest. The profusion of the fruit produced by the olive tree has led to it being a symbol of prosperity and fecundity. Greek brides would wear a garland of olive leaves to ensure their fertility. The fruits are also said to be something of an aphrodisiac, improving male potency and inducing lustfulness. Paradoxically, however, it was also a symbol of chastity and purity. It may be because it is symbolic of purity that Francis Barrett in *The Magus* says of it, 'There is, also, a total antipathy of the olive-tree and the harlot; that, if she plant it, it will neither thrive nor prosper, but wither.'

Just as a crown of oak leaves was the highest award that could be conferred on a citizen of Rome so a crown of olive twigs was the greatest tribute to be given to a Greek citizen. Crowns of olive leaves were presented to the victors at the Olympic games.

The olive can be a plant of good luck and protection. Wearing the leaves will bring good fortune, and when a branch is hung over a doorway no evil can enter. Moreover if the branch is hung on the chimneybreast it will protect the house from being struck by lightning.

Dreaming of gathering olives, or seeing them on the tree, is a sign of peace and happiness. It is especially favourable for sick people. To dream of eating olives indicates a rise in position and the receipt of a favourable present.

Medicinal

A simple cure for headaches is to press an olive leaf, with the name of Athena written on it, against your forehead.

SAINFOIN

Botanical Name: *Onobrychis viciifolia*
Family: Leguminosae

When Christ was born, all that could be found for him to sleep on was some fresh, sun-dried hay that was placed in the manger. The flowers of the sainfoin began to open as a pink wreath to encircle his head, which is how it obtained its other name, holy hay.

SCOTTISH THISTLE

Botanical Name: *Onopordon acanthium*
Family: Compositae
Common Names: Cotton thistle; woolly thistle
Element: Fire
Meaning: Retaliation

Myth and Legend

The legend of how the thistle became the Scottish emblem tells that it dates from a time when Danish armies where raiding Scottish coastal settlements. On learning that the Scottish king would be staying overnight at a castle on the coast the Danes planned to attack, under cover of darkness, and kidnap him. Having landed on the beach the raiding party made its way toward the castle, and in order not to be heard they removed their heavy footwear. All went well until they reached the castle gardens, which were full of thistles. Their cries of pain and surprise alerted the guards and the Scots drove them back into the sea.

Magic and Lore

The plant was probably introduced as an ornamental species, although it may be native to some areas of East Anglia. It is unlikely, in any event, to have been found in fifteenth-century Scotland. It is unclear exactly which species of thistle is the 'true' Scottish thistle but it is now generally considered to be this species that was the emblem of the House of Stuart. Other strong contenders for the role of Scottish national emblem include the common spear thistle and the cotton thistle.

The use of the thistle as a heraldic device dates back to at least 1458, when wall hangings embroidered with a thistle motif are mentioned among the inventory of possessions made at the time of the death of King James III. It had certainly been adopted as the national emblem of Scotland by 1503, the year that Dunbar wrote *The Thrissill and the Rose*, a poetic allegory of the marriage of James IV of Scotland with Princess Margaret of England. The Order of the Thistle is among the most ancient of British orders of chivalry, second only to the Order of the Garter. It was instituted by James V of Scotland in 1540 and revived by James VII (James II of England), who created eight knights of the order in 1687. It is certainly appropriate to the Scottish national motto, *Nemo me impune lacessit* ('No one irritates me unscathed').

A thistle may be worn as a protective amulet and the emblem was woven into the train of the dress worn by Sarah Ferguson at her marriage to Prince Andrew, the Duke of York.

CACTUS

Botanical Name: *Opuntia* spp. (and other genera)
Family: Cactaceae
Common Name: Prickly pear
Planet: Mars
Element: Fire
Meaning: Ardour; endurance; 'I burn'; 'Our love shall endure'; warmth

In some parts of the world the cactus is viewed as having protective properties. Grown inside it is supposed to guard against unwanted intrusions into the home. As a further protection they may be planted outside, one at each of the four cardinal points of the compass. In various parts of Britain, however, there appears to be a superstition that developed in the mid-1960s that bringing any species of cactus into the house would bring bad luck. This belief certainly exists in Hungary, where they say that, as more spines develop on the cactus, so the number of your troubles increases.

ORCHID

Botanical Name: *Orchis* spp.
Family: Orchidaceae
Common Names: Adam and Eve root; Levant salap; sahlab; salap; salep; saloop; Satyrion
Gender: Feminine
Planet: Venus
Element: Water
Meaning: A belle; beauty; love; luxury; 'You are beautiful'

Myth and Legend

Orchis, in mythology, was a lascivious youth who presided over the feasts of Priapus and Bacchus. He was the son of Patellanus, a satyr, and Acolasia, a nymph. He was ripped to pieces by the followers of Bacchus because in one unbridled moment he laid violent hands on a priestess in the temple of the god. After his death Patellanus prayed fervently to the gods, who transformed Orchis into the flower that bears his name. An alternative version of the tale says that the satyrian orchid grew from semen spilt by Patellanus during his various copulations. Gerard, in his herbal, calls the plant satyrion and suggests that it was the food of the satyrs.

Magic and Lore

One of the folk tales about the plant is that they grow spontaneously on ground wherever animals have mated. In Victorian England it was recommended that women should not own orchids as they sometimes vividly displayed their reproductive organs.

Orchids have long been valued additives to love potions and love spells. The name is derived from the Latin word for testicle, and the plant was thought to be a powerful aphrodisiac and love attractant. Pliny claims that merely holding an orchid bulb would be sufficient to inflame the desires, especially if had been soaked in wine. The twin tubers could also be carried to attract love. The fresh swollen tuber was said to promote true love, whilst the withered tuber would suppress 'wrong passions'. Orchids are sometimes given to a couple as a talisman to ensure their continued happiness. Witches were thought to use the roots in their philtres as some orchid species are supposed to be useful for strengthening psychic powers and inducing prophetic dreams.

EARLY PURPLE ORCHID

Botanical Name: *Orchis mascula*
Family: Orchidaceae
Common Names: Aaron's beard; Adam and Eve; adder's flower; adder's grass; adder's mouths; adder's tongues; baldeeri; boldeeri; beldairy; bloody bones;

bloody butcher; bloody fingers; bloody man's fingers; blue butcher; bog hyacinth; bull's bat; bull-segg; butcher flowers; butchers; butcher's boys; Cain and Abel; candlesticks; clothes pegs; cock-flower; cock kame; cowslip; crake-feet; crawfeet; cross-flower; crowfoot; cuckoo; cuckoo bud; cuckoo cock; cuckoo flower; cuckoo pint; curlie-doddie; dandy-goslings; dead man's finger; dead man's hand; De'il foot; Devil's fingers; dog stones; ducks and drakes; fox stones; fried candlesticks; frog's mouth; gander-guase; gandigoslings; Gethsemane; giddy-gander; goose and goslings; goosey-ganders; gosling; gossips; gramfer-griddle-goosey-gander; grammer griggles; granfer-goslings; granfer griggles; granfer-grigglesticks; hens; hens' kames; herb of enticement; jessamine; johnny cocks; jolly soldiers; keek legs; keet legs; kettle-cap; kettle-case; king's fingers; kite's legs; kite pan; lady's fingers; locks and keys; long purples; lords and ladies; *lus-an-taliadh*; mogramyra; poison more; poor man's blood; priest's pintle; puddock's spindles; ram's horn; red butcher; red granfer-gregors; red robin; regals; sammy gussets; single castles; single guss; snake flower; soldiers; soldier's cap; soldier's jacket; spotted dog; standing gussets; Tom thumb; underground shepherd; wake robin
Gender: Feminine
Planet: Venus
Element: Water

Many of the common names of the plant are derived from the two long tubers it produces below ground. One is swollen ready for next season's growth while the other is shrivelled, having been exhausted in providing the energy for the present season's growth.

Myth and Legend

In various parts of the country this plant is known as 'cuckoo flower' or something similar, as it is said to begin to blossom at the time that the cuckoo can first be heard. In Dorset it is known as 'granfer griggles' and 'granny griggles'; the wild hyacinth can usually be found growing close by. In Cheshire the plant was called 'Gethsemane' because it is said to have grown at the foot of the cross; splashes of the saviour's blood fell onto the orchid's leaves, leaving distinctive dark patches on them.

> Those deep unwrought marks,
> The villager will tell you,
> Are the flower's portion from the atoning blood
> On Calvary shed. Beneath the cross it grew.

Magic and Lore

Early purple orchid has been used to promote conception, increase lustfulness and strengthen genitalia. Some sources suggest that if you wished to be sure of having a male child you would use a large tuber, while for a girl you must find a small tuber. An alternative to this says that when the man eats the large, swollen tuber there will be male children, whereas if the woman eats it there will be girls.

The tubers can also be used in divination. To discover whether a lover will be become a future spouse they must be gathered before sunrise when facing south, and whichever is to be used should immediately be put in spring water. If it sinks it shows that the person in question would be the future husband or wife. Other writers recommend that the tubers should be put into water to see whether they sink or float. Those that float represent hatred, whilst the ones that sink are symbolic of love as nothing goes deeper than true love. The 'love' tuber could be dried and ground down to be added to love potions. The swollen tuber was said to promote lust whereas the flaccid one would dampening ardour. When the powdered orchid is put under pillows it induces dreams of a future partner. The tuber may alternatively be carried as a talisman to attract love.

MARJORAM

Botanical Name: *Origanum vulgare*
Family: Labiatae
Common Names: Bastard marierome; joy of the mountain; knotted marjorane; marjorlaine; mountain mint; oregano; organ; organy; sweet marjoram; wintersweet
Gender: Masculine
Planet: Mercury
Element: Air
Deities: Aphrodite; Venus
Meanings: Blushes; maidenly modesty; maidenly innocence; 'your passion brings blushes to my cheeks'

Myth and Legend

In ancient Greece, marjoram was first called *amarako*, or *amaracon*, after Amaracus, a youth in the service of Cinyres, King of Cyprus, who accidentally broke a vase of expensive perfume. His fear of the consequences was so great that he fell unconscious to the ground. The gods, in their mercy, transformed him into the herb, and over the centuries the name has been corrupted first to majorana, then eventually to marjoram.

Magic and Lore

Theophrastus called the plant 'joy of the mountain' 300 years ago and the botanical name is derived from this, *oros* meaning 'joy' and *ganos* 'a mountain'. It was cultivated by the ancient Greeks, who thought it had been created by Aphrodite. Although we usually now think of it as a culinary herb, all sorts of qualities are attributed to marjoram. It strengthens love, and has also been used in spells and potions to induce love.

Growing marjoram in your garden will protect your home from evil. Carrying a sprig of the fresh herb, or placing a little in each room of your house, will have much the same benefit. It must be kept fresh to be effective and should be changed at least monthly. In magic, it is used in spells and incantations to conjure spirits. It has also been used in spells to attract wealth.

Medicinal

Wearing a mixed bunch of marjoram with violets during the winter months will keep away colds or flu. The essential oil can be rubbed on as a treatment for sprains. The name 'joy of the mountain' implies another use: it has been given to people suffering from depression or melancholia, in order to lift their spirits and boost their energy. The Greeks and Romans saw it as a herb of happiness, and would form crowns of it for young lovers to wear. The Romans believed that where it grew on graves it was an indication that the departed were blissfully happy in their afterlife.

In ancient Greece the herb was used in preparations as an antidote to narcotic poisons.

RICE

Botanical Name: *Oryza sativa*
Family: Gramineae
Common Names: Bras; dhan; nirvara; paddy
Gender: Masculine
Planet: Sun
Element: Air

Rice is the foundation of the diet of many nations. Similar traditions and superstitions to those that we have seen with flax, barley, wheat and corn can be found relating to rice. For example, in Sumatra, rice would be sown by women who had very long hair so that the rice might also grow to be long and luxuriant. In Amboyna, rice was treated like a pregnant woman. When it came into flower no guns would be fired and there would be no loud noises near the fields in case it might cause the plant to 'miscarry', resulting in a crop of straw but no grain.

In the West rice is taken as a symbol of fertility, which is why it was, and sometimes still is, thrown over the bride and groom at a wedding. However, throwing rice into the air is also supposed to cause rain. Brahmins would carry rice as a protective amulet and it may be used to guard a home from evil by keeping a small jar filled with rice grains near to the main door of the house or hidden in the roof space.

ROYAL FERN

Botanical Name: *Osmunda regalis*
Family: Osmundaceae
Common Names: Bog onion; flowering fern; heart of Osmund; Osmund the waterman
Planet: Saturn
Meaning: Contemplation

Myth and Legend

Legend tells us that this fern gains its name from Osmund, the waterman of Loch Tyne. During the various raids by the Danes his daughter would take refuge amongst the ferns that grew in the damp soil at the side of the loch.

Magic and Lore

Like other ferns the royal fern is supposed to flower at odd times, which explains why it is never seen flowering; it is said to flower in the middle of the night during June, and at the first hint of sunlight the flowers disappear, the only evidence that it has flowers being the young plants that grow from the set 'seeds'.

Medicinal

In folk medicine the root is sliced and pounded to form a mash to which a little water is added. The resulting liquid is then bottled, left to set, and used as a rub to ease rheumatism and sciatica.

WOOD SORREL

Botanical Name: *Oxalis acetosella*
Family: Oxalidaceae
Common Names: Alleluia; alleluya flower; cuckoo bread; cuckoo sorrel; cuckowe's meat; faerie bells; hallelujah; sourgrass; sour trefoil; stickwort; three-leaved grass; wood sour
Gender: Feminine
Planet: Venus
Element: Earth
Meanings: Joys; maternal tenderness

Myth and Legend

Some authors suggest that it was a wood sorrel with which St Patrick illustrated his explanation of the trinity, but, it is now more generally accepted that it was a species of clover that is the 'true' shamrock. Wood sorrel was a druidic symbol of ancient Ireland long before the arrival of St Patrick in 432, and was considered lucky, as it was associated with the Celtic sun wheel.

Magic and Lore

John Gerard explains the application of the various common names: 'The apothecaries and herbalists call it Alleluya and Paniscuculi, or Cuckowe's meat, because either the Cuckoo feedeth thereon, or by reason when it springeth forth and flowereth the Cuckoo singeth most, at which time also Alleluya is wont to be sung in Churches.'

It does indeed flower between Easter and Whitsuntide, at much the same time as the cuckoo can be heard throughout the countryside. Another suggestion claims that it is called 'cuckoo flower' because the bird will use it in order to clear its voice. Wood sorrel gets the name 'faerie bells' because the ringing of the flowers is supposed to summon the faeries to their moonlight revelries. It is one of the herbs that can be used in faerie magic when evoking elves.

Medicinal

Wood sorrel is a healing herb in folk medicine. Carrying the dried leaves will protect against heart disease. The fresh herb, put in a room where someone is ill, will aid their recovery, regardless of their ailment.

PAEONY

Botanical Name: *Paeonia officinalis*
Family: Ranunculaceae
Common Names: Peony; sheep shearing rose
Gender: Masculine
Planet: Sun (Moon, according to Culpeper)
Element: Fire
Deities: Apollo; Paeon
Meaning: Bashful shame; 'Beauty is in the heart not in the face'; contrition; 'Please forgive my brusqueness'

Myth and Legend

In Chinese mythology there is a tale of a young scholar who grew various plants, including paeony. One day he was visited by a young woman, who made admiring comments about his garden. He initially employed her as a servant but she quickly became first his companion and then his lover. One day, when he was expecting a visit from a noted moralist, he was unable to find his lover. He searched throughout the house and, eventually, in a dark gallery, he found the ghost of his lover fading into a wall. The spectre told him that she was the spirit of the paeony, who had been brought into human form because of his tender care. She knew that the moralist would never approve of their relationship and so she would return to her place amongst the flowers. From that day the scholar went into mourning for his lost love. He continued to tend his garden with care but never saw the maiden again. In Chinese poetry the paeony has frequently been used as a metaphor for a blushing young girl, a white paeony is symbolic of a young girl of wit and beauty, and the word for paeonies is *sho yu*, meaning 'most beautiful'.

In classical mythology Paeon, the son of Endymion, was a pupil of Aesculapius and learnt the use of plants for healing wounds and ailments. During the Trojan wars he treated the wounds of Apollo and Hades amongst others. In gratitude for his help Hades immortalized him by transforming him into the plant known as 'Paeony'. An alternative telling of the tale says that when Aesculapius heard that Paeon had treated the wounds of Hades and the rest he was jealous and plotted to kill him. Hades learned of the plot and intervened, protecting Paeon by transforming him into the flower.

Another variation claims that Aesculapius, the son of Apollo, was raised and trained as a healer by the centaur Chiron. By the time he had reached manhood he was the physician to the gods and, because of his great knowledge, he was called Paeon, meaning 'helper'. The early doctors were called *paeoni* because of their affinity to this early healer.

Yet another version of this story tells of Paeonia, a nymph, with whom Apollo flirted. One day the goddess Aphrodite caught the two together. The shy Paeonia blushed and the colour never left her face for at that moment the goddess transformed her into a flower.

Another classical tale relating to Paeon tells of how he lost a race on Mount Olympus. He was so disappointed by his defeat that he went into a self-imposed exile in Macedonia where he founded a race of people called the Paeonians.

A myth relating to the gathering of paeony links it to the moon goddess, Silene. Because of this connection it must only be gathered at night as, if it is collected by day there is a risk that Picus, the woodpecker that guards it, will swoop down and peck out the eyes of the gatherer.

There are two Japanese legends that involve paeony. In the tale of the Spirit of the Paeony, Princess Aya is betrothed to the second son of Lord Ako. One night, when walking in the garden, as she approaches her favourite paeony bed, she slips and almost falls into the pond but is saved by a handsome samurai clad in a paeony robe. She immediately falls in love with him but he disappears. The next time she sees him is when she is ill, and then she learns that he is the spirit of the paeony, so she always keeps the flower close by her. However, when she recovers from her sickness her wedding proceeds, and at the hour of her marriage the paeony dies.

The second tale, the Peony Lantern, tells of a maiden named Tsuyu who is visited by a physician and a young samurai, with whom she instantly falls in love. She tells him of her love and says that if he does not call on her again soon she would surely die. The physician, for his own reasons, contrives to ensure that the lovers are kept apart and, as a result, Tsuyu dies of a broken heart. Her servant, ever faithful to her mistress, also dies soon after. The young samurai is grief-stricken and cannot forget Tsuyu. At the Festival of the Dead, when lanterns are hung to guide the spirits, the spirits of Tsuyu and her servant return carrying a paeony lantern. This tale has no happy ending, however, as the samurai is found dead with the bones of Tsuyu about his neck.

Magic and Lore

At one time the paeony was revered as being of divine origin, an emanation of the moon. By day it might be overlooked, hidden amongst the foliage of other plants, but through the night it shone like a star, with its own radiance. Its brightness afforded protection for shepherds and their flocks as it would ward off evil spirits and avert storms.

Some writers recommend that similar care should be taken in gathering paeony as with mandrake: a dog should be used to pull the plant from the ground, as digging it up endangers the collector. Gerard repeats the warning given by Josephus, that paeony should never be collected by day lest the person gathering it is seen by a woodpecker and their eyes plucked out. Having obtained the paeony root it can be carried as a cure for lunacy or epilepsy, or used during exorcism. Pieces of the root can be carved into beads, called 'piney beads', and strung as a necklace. Wearing this necklace will afford protection against imps, faeries and the forces of evil. Where the necklace is strung with coral as well as paeony it acts as a protection against the incubus. Paeony seeds, rather than piney beads, have also been used to form such necklaces.

The number of flowers on a paeony plant has significance in divination. Where there is an odd number it is a sure indication that there will be a death in the household before another year passes.

Dreaming of paeonies indicated that your excessive modesty might cause you sorrow in the future. You should be more forward and grasp your opportunities.

Medicinal

Necklaces made of paeony beads can be hung about the necks of children to prevent convulsions and ease the pain of dentition.

To prevent recurrent nightmares, steep fifteen paeony seeds in wine or mead, strain the liquor and drink it. One warning, however: paeony is poisonous and so the result might prove more permanent than you would like.

GINSENG

Botanical Name: *Panax ginseng*
Family: Araliaceae
Common Names: Five fingers; *jin chen*; man's health; *nin-sin*; red berry; tartar root; wonder-of-the-world
Gender: Masculine
Planet: Sun
Element: Fire

Myth and Legend

Legend says that 2,300 years ago the great Chinese sage Lao-Tse had discovered a drug that gave those who used it long life, and he called it Jen-shun. Another legend says that it was the omnipotent mountain ghost who sent a 'rescuer' to mankind. This saviour was a boy in the form of the human-like turnip called ginseng.

Magic and Lore

The Dutch brought ginseng to Europe in 1610 under the name of 'pentao'. It was valued as an aphrodisiac at the court of Louis XIV, and has maintained this reputation ever since: a tea made from, or containing, ginseng can be used to promote lustfulness. Carrying the root will attract love and ensure sexual potency. It will also bring wealth and act as a protection to good health. Burning the root will ward off any evil nearby and break any curses or spells that have been cast against you. If you have a wish you fervently want to come true, carve it into a ginseng root and then throw the root into running water. Where the wish is impossible to write on the root you should hold the root tightly in your hand and visualize your wish before tossing it into running water.

Medicinal

Ginseng has been held in high esteem in China, Korea and Japan for over 4,000 years. It is regarded as an inexhaustible elixir, the treatment for endless ailments. In the 52 volume Chinese work *Pent-ts'ao kang-mu*, a classification of roots and herbs published in 1597, the details of experiences spanning several thousand years were given. It recommends ginseng as a treatment for disorders of the lungs and a tonic to invigorate the spleen. It claims that it will 'cool the fires', reducing high temperatures, open the heart and cause the blood to circulate, strengthen mental capacities, enrich knowledge, soothe shocks, remove constipation and diarrhoea, and generally cure a whole host of disorders.

The botanical name *Panax* is derived from the Greek *panakos* meaning a panacea. Ingesting a boiled ginseng root was said to be sufficient to renew the vital spark of life, even if it had almost been extinguished. The herb rejuvenates the main organs of the body and so extends life, and it has been seen as a herb of external youth.

POPPY

Botanical Name: *Papaver* spp.
Family: Papaveraceae
Common Names: Blind buff; blind eyes; blind man; blindy buff; bull's eyes; butterfly ladies; canker; canker rose; cheese bowls; cockeno; cock-rose; cock's comb; cock's head; collinhood; cop-rose; corn flower; corn poppy; corn rose; cup rose; Devil's tongue; earaches; fireflout; gollywogs; gye; headaches; headwaak; hogweed; lightnings; old woman's petticoat; paradise lily; pepper boxes; poison poppy; poppet; popple; red cap; redcup; red dolly; red huntsman; red nap; red petticoat; red rags; red soldiers; redweed; sleepy heads; soldiers; thunder ball; thunder bolt; thunder flower; wart flower; wild maws
Gender: Feminine
Planet: Moon
Element: Water
Deities: Aphrodite; Ceres; Demeter; Hypnos; Somnus
Meanings: Silence; 'My heart aspires in silence to thee' (oriental poppy); sleep; 'May sweet sleep attend you, sweetest dreams beguile your slumbers' (pink); 'Be cheered, you still have me'; consolation to the sick; moderation; oblivion (red); fantastic extravagance (scarlet); antidote;

dreams and fantasies; 'I need time to consider'; my bane; sleep; sleep of the heart; temporization; time (white); evanescent pleasure

Myth and Legend

Poppies can be seen growing amongst the wheat and barley in fields throughout Britain, and it has always been thus. Even the Assyrians called them the 'daughters of the fields'. It is no surprise, therefore, that this flower is dedicated to Demeter, the goddess of agriculture, who taught the secrets of cultivation to mankind. In mythology, following the abduction of her daughter Persephone (Kore) by Dis (Pluto), Demeter was inconsolable. She searched for her throughout the length and breadth of Sicily but found no trace of her. When darkness began to fall she climbed Mount Etna and, from its mighty flames, lit two torches to light her way. Eventually, however, overcome by weariness and grief she sat down on a rock and wept. She sat weeping for nine days and nights until the other gods took pity on her and caused poppies to spring up at her feet. She stooped, breathed the heavy scent of the flowers and tasted the bitter seeds. Her sorrow was forgotten as she was overtaken by the sweet oblivion of sleep. Persephone was ultimately restored to her mother, but she had to spend half the year with Dis in the Underworld.

Persephone was the personification of the corn, or wheat, seed that is 'concealed' in the earth during the winter only to put out shoots and appear again in the spring. Bread made from the wheat was dressed with poppy seeds in remembrance of this story. At the festivals of the goddess Demeter poppy seeds would be offered to ensure a successful harvest in the coming year. Figures of the goddess were garlanded with wheat or corn interwoven with poppies to symbolize life and death, summer and winter. The poppy, probably because of the use of opium, was seen as symbolic of death, sleep, forgetfulness and the easing of pain. The god of sleep, Somnus, is always shown crowned by poppies or lying down surrounded by them.

The Greeks also dedicated the poppy to the twin gods Hypnos and Thanatos, the deities of sleep and death, who are often shown crowned with poppies or with poppies in their hands. This perhaps demonstrates an awareness that opium-induced sleep could lead to death. The son of Hypnos was Morpheus, god of dreams; hence we get morphine from the poppy. Morpheus is usually depicted as a winged youth carrying poppies and scattering their seeds on the wind.

Magic and Lore

It is presumably due to the number of seeds that poppies produce that the ancient Greeks considered them to be a sign of fertility. Greek athletes were given mixtures of poppy seeds, honey and wine as the seeds were thought to improve strength and general health.

The poppy being associated with death, it is hardly surprising that it is a plant of ill omen. In Ireland people would dread to touch it because of the misfortune that might befall them. The scent was said to cause headaches, especially in women and most severely in unmarried women. This may have reference, because of the red flower, to menstruation. Children were told that if they looked into the centre of the poppy flower, or placed it too close to their eyes, they would be struck blind, and if they put it near to their ears it would cause earaches. Sniffing the scent would result in nosebleeds and handling the flowers too much would cause warts.

The flowers must never be brought into the house as this will cause illness amongst the members of the household. Furthermore, picking poppies will bring thunder storms and leave the gatherer susceptible to being struck by lightning if petals fall off as the flower is picked. In some parts of Britain there is a superstition that a few poppy flowers can be placed under the roof timbers in order to ward off lightning strikes and protect the home.

Although it was believed that there were certain risks in gathering poppies there were also several benefits in having them. In love divination, for example, a single petal from a poppy flower could be used to indicate whether your love for another is returned. The petal was placed in the palm of the left hand, and then struck with the palm of the right hand. If there was no sound then there was no love.

By a prophetic poppy leaf [petal] I found
Your changed affection, for it gave no sound,
Though in my hand struck hollow as it lay
But quickly withered like your love away.

Poppy seeds and seed heads also have their uses. If the gilded seed heads are used in decorative displays in the home they will attract wealth to the people who live there. To gain the answer to a question which has been troubling you, write it with blue ink on white paper, fold up the paper and place it in a poppy seed head. If you then lay the seed head on your pillow before you go to sleep, you will dream the answer to your question.

Carrying or eating the seed is supposed to promote fertility, induce love and bring good luck. If you would like to become invisible soak some poppy seeds in wine for fifteen days, then fast for the next five days and drink a little of the wine each day. At the end of this time you will be able to make yourself invisible whenever you wish.

Rosemary may have the longer traditional use as a symbol of remembrance, but it is the poppy that we now commonly associate with remembrance of those who died in wartime, serving their country. In 1915 John McRae, a doctor serving in a field hospital near Ypres, saw the poppies which grew in profusion on the disturbed ground between the lines of trenches. Recalling that the poppy was sacred to Somnus, god of sleep, he thought that it was a suitable symbol for the long sleep of the fallen, and asked that poppy seed be sown on his grave. Although it is from the First World War that we date the used of poppies as an emblem of remembrance, there are those who would trace the link back still further. In the nineteenth century poppies were seen growing freely on the battlefield following the Battle of Waterloo; they were said to have sprung from the blood of the fallen soldiers.

A dream of poppies indicates a message bringing you great disappointments.

The poppy is the birthday flower of 10 May.

Medicinal

Poppies were used by the Egyptians, the Greeks, the Romans and many other cultures all over the world. As Poppies symbolize sleep, in folk medicine staring into the black centre of a red poppy was a cure for insomnia, and elsewhere the seeds and the flowers were used in mixtures to the same end. Opium extracted from poppies has, for thousands of years, been used to induce sleep and numb pain.

TRUELOVE

Botanical Name: *Paris quadrifolia*
Family: Liliaceae
Common Names: Devil-in-a-bush; four-leaved-grass; herba Paris; one berry; true-lover's knot
Planet: Venus

The four prominent leaves of truelove are said to be in the form of a lover's knot, hence some of the various common names which this plant has been given. It is a herb of equality, as the parts of the plant are even and harmonious in number: it has four leaves, there are four stigmas and twice four stamens in the flower, there are four inner segments and four outer segments to the perianth, and within the ovary of the flower there are four cells. This harmony led to it being recommended by early herbalists. Moreover it was seen as being a powerful weapon against witches and witchcraft.

The plant can be used by young women seeking to discover whom they will wed. On the night of Beltane, the last day of April, two young women must sit in silence from midnight until 1 o'clock on the morning of May Day. During this hour they should pull as many hairs from their heads as they are old in years. These hairs are then wound around the stem of truelove and wrapped in a clean, white, linen cloth. When the clock strikes the hour at 1 o'clock the linen bundle is unwrapped and each hair burnt separately with the words,

I offer this my sacrifice,
To him most precious in my eyes,
I charge thee now come forth to me:
That I this minute may thee see.

The images of the men they will marry will then appear and walk around the room. Each woman will only be able to see the figure of her own future husband, not both.

PASSION FLOWER

Botanical Name: *Passiflora* spp.
Family: Passifloraceae
Common Names: *Flor-de-las-cinco llagas*; *flos passionis*; flower of the five wounds; grandilla; maracoc; maypops; murucuia
Gender: Feminine
Planet: Venus
Element: Water
Meaning: Belief; consecration; holy love; 'I am pledged to another'; religious fervour; religious superstition; susceptibility

Myth and Legend

It is said that Jesuit missionaries in South America found the plant and named it *flos passionis,* 'passion flower', or *flor-de-las-cinco llagas*, 'flower of the five wounds' because of the symbolism identified in the flower that relates to the passion of Christ (see below). They believed that they had found the plant that, according to one Christian tradition, could have been seen growing on the cross. It was one of the visions of St Francis of Assisi.

Magic and Lore

The parts of the flower are given the following relevancies:

The ten coloured petals	Ten apostles present at the crucifixion (Peter denied and Judas betrayed)
The many filaments	The crown of thorns
The five anthers	The wounds of Christ
The stamen	The hammer which drove in the nails
The three divisions of the pistil	The round headed nails of the cross
The flower's ovary	The hammer used to fix the nails
The column of the stigma	The pillar of the cross
The blade of the leaf	The spear which pierced Christ's side
The tendrils	The whips which cut Christ's flesh and the cords that bound him
The calyx	The glory or nimbus
The white tint of the flower	Christ's purity
The blue tint of the flower	Heaven
The three-day lifespan of the flower	The three years of Christ's ministry on earth

Growing a passion flower in the home will bring benefits as well as being very decorative; it is said to bring peace and calm all troubles. Sprinkling a little of the plant, fresh or dried, on the doorstep of a house will prevent any harm from entering. If you place a flower under your pillow it will help you to sleep soundly; carrying a flower will attract people, increase your popularity and help you to make friends. Burning it as an incense promotes understanding and in spells it attracts love.

PARSNIP

Botanical Name: *Pastinaca sativa*
Family: Umbelliferae

An old adage, which was supposed to ensure the best timing for preparing ground for parsnip seeds, was to start digging whilst still eating the bread baked at Christmas. In common with parsley the parsnip is said to go three times to the Devil before the seed germinates.

GERANIUM

Botanical Name: *Pelargonium* spp.
Family: Geraniaceae
Gender: Feminine
Planet: Venus
Element: Water

Myth and Legend

Geraniums are supposed to be especially good as protective plants. Legend tells of a witch who planted a bed full of magic geraniums near to her cottage. They would warn her of the approach of strangers, and even indicate the direction from which they were coming.

Magic and Lore

All geraniums are protective plants regardless of whether they are grown in your garden, kept as houseplants or used as cut flowers. They protect against snakes, for example:

> Snakes will not go
> Where Geraniums grow.

The scented forms of geranium are supposed to share the protective properties of the plants they smell like, and red geraniums are said to be especially effective as a protection against ill health. Pink geraniums can be used in spells to attract the love of others, and white-flowered forms will aid fertility.

See also Cranesbill (*Geranium*)

AVOCADO

Botanical Name: *Persea americana*
Family: Lauraceae
Common Names: *Ahhocotl*; aguacate; alligator; alligator pear; avigato; avogato; palta; persea
Gender: Feminine
Planet: Venus
Element: Water

Myth and Legend

Legend tells that the avocado was a favourite food of Seriokai, who inhabited the wilder regions of Guyana. He would roam through the forests of the Orinoco gathering the fruits. During one of these excursions the tapir saw Seriokai's wife and fell in love with her and, after a little while, she fell in love with him. When next Seriokai went on his journey to collect fruits his wife went with him to cut wood for fuel. As he was descending from the avocado tree she struck him a violent blow to the right leg with the stone axe she was carrying. The blow was so severe that it severed the leg and left Seriokai helpless. His wife quickly picked up all the fruits he had dropped and rushed off to find the tapir. The pair then hurried away through the forests.

Seriokai was found by one of his neighbours who staunched the wound, carried him home, and nursed him back to health. His leg was replaced with a wooden stump and, armed with a bow and arrow, Seriokai started to track down the runaways. Their trail had long been covered over but he found he could trace it by following the trail of avocado trees which had grown from the seeds dropped by his faithless wife. His journey took him over mountains and across rivers, always following the line of avocado trees. As time went on the trees grew smaller. Then he came across saplings, then seedlings, then seeds. Finally he was following a trail of footprints. So, at last, he caught up with the tapir and his wife. He shot at the tapir as it sprang off the edge of the world and hit it in the eye. On seeing her lover shot the woman also leapt off the world in a vain effort to escape. Seriokai would not allow them to evade him, however, and followed; he now hunts them through the heavens. He is the constellation of Orion, the hunter, his wife is the Pleiades and the tapir is Hyades, with its eye still bleeding.

Magic and Lore

Ahhocotl is the Aztec name for the avocado and means 'testicle tree'; it was believed to be an aphrodisiac and was eaten to induce lust. Having a plant growing in the house will bring love, and carrying the pip is supposed to increase beauty.

REDSHANK

Botanical Name: *Persicaria maculosa*
Family: Polygonaceae
Common Names: *Am boinnefola*; arse smart; ass smart; blood spot; *lus chrann ceusaidh*; persicaria

Myth and Legend

Legends from various parts of Britain tell that this plant was to be found growing at the base of the cross when Christ was crucified. Drops of his blood fell onto the leaves and left permanent marks on the foliage.

In Oxfordshire there is a tale that the Virgin used the leaves to make an ointment. On one occasion, however, she searched in vain for the plant, and only found it when she no longer needed it. She cursed it, giving it the value only of a common weed, and leaving, on the leaves, the marks of her fingerprints, a dark spot in the middle of each leaf.

> She could not find in time of need,
> And so she pinched it for a weed.

Magic and Lore

The alternative common name of 'arse-smart' is an indication of the use of redshank as a laxative or purgative.

BUTTERBUR

Botanical Name: *Petasites hybridus*
Family: Compositae
Common Names: Blatterdock; bog rhubarb; bogshorn; bog's horns; burblek; burn-blade; butterburn; butter-dock; capdockin; cleat; clots; clouts; cluts; dunnies; early mushroom; elden; eldin-docken; ell-docken; flapperdock; gallon; gypsy's rhubarb; kettle dock; langwort; plaguewort; poison rhubarb; rat leaf; rhubarb; snake's food; snake's rhubarb; son-before-the-father; turkey rhubarb; umbrella leaves; umbrella plant; umbrellas; water-docken; wild rhubarb

Magic and Lore

Butterbur has commonly been used in love divination. When an unmarried young woman wished to know whom she would marry she would scatter the seeds in an isolated spot, half an hour before sunrise on a Friday morning, saying the following incantation:

I sow, I sow!
Then, my own dear,
Come here, come here,
And mow, and mow.

An image of her future husband would appear a short distance ahead of her, cutting the grass with a scythe. If she became frightened by the apparition she could say 'have mercy on me' and the figure would disappear. This method was said to be infallible but should only be used as a last resort.

Medicinal

This herb has also found use as a treatment for the plague, although there does not appear to be any great evidence of its efficacy.

PARSLEY

Botanical Name: *Petroselinum crispum*; *Petroselinum sativum* (Synonym: *Carum petroselum*)
Family: Umbelliferae
Common Names: Devil's oatmeal; persele; persely; persil; petersylinge; rock parsley
Gender: Masculine
Planet: Mercury
Element: Air
Deity: Persephone
Meanings: Entertainment; feast or banquet; festivity; honour; to win; useful knowledge

Myth and Legend

In myth parsley sprang up from the blood of the Greek hero Archemorus and was dedicated to Persephone, the Queen of the Underworld. Various Greek myths also claim that it would grow anywhere that the blood of a hero was spilt. Christian myth has the plant dedicated to St Peter who, as guardian at the gates of Heaven, could be seen as the counterpart to Charon.

Magic and Lore

Like the leaves of the olive and bay trees, parsley was used to crown victorious athletes in ancient Greece. Wreathes of the dried herb were used at the Isthmian Games, and fresh wreaths at the Nemean Games. The Nemean Games were held every four years in honour of Archemorus, the 'bringer of doom'.

> Isthmian victory with horses
> Poseidon granted to Xenokrates
> sending a wreath of Dorian parsley
> to bind on his hair as a token of triumph.
>
> – Pindar, *Second Isthmian Ode*

In ancient times parsley was not greatly used as a culinary herb, possibly because it was associated with death. It was used in funeral wreaths and might be strewn on graves or planted on fresh graves. The Romans used it to line graves in order to ward off evil.

Parsley was also supposed to be able to impart or improve strength and stamina. It was eaten by Greek soldiers, and was fed to chariot horses before races.

Romans would carry a little parsley in their togas as a protection against evil. Equally it could be added to a purification bath to prevent all misfortunes. Chaplets would be worn by the ancient Greeks and the Romans, when feasting, as it was believed that parsley would absorb wine fumes and prevent, or at least delay, inebriation. Theocritian lovers, in ancient Greece, would wear the herb at banquets as they felt that it would make them more attractive as well as counteracting the effects of the alcohol.

Its cultivation is clouded in numerous superstitions, some going so far as to say that it is unlucky to grow it at all. It should always be grown from seed and never transplanted; this belief may originate in the superstition that digging it up opened the garden to the Devil. It is sometimes said that the risk is to the person who digs it up, so the person receiving it should dig it up themselves. Then again, parsley should never be given away as to do so is to give away luck and invite misfortune, the result being that someone from where the plants are set will die within the year.

It should be sown on Good Friday because it is one of the Devil's plants and it is only on that day that the Devil has no power over the living things on earth. Planting at other times will result in a death in the family. The seeds are notoriously slow to germinate and this is explained by the saying that 'Parsley goes seven times to the Devil before it grows'. In different parts of Britain the number of times that parsley will travel to the Devil varies from as little as three times in Devon to as many as nine or eleven times in Yorkshire and Herefordshire. In Sussex it is said that when it is sown on Good Friday the herb will sprout quickly and grow to be curly.

The herb is said only to grow outside the home of an honest man, and only grow at its best when sown by the dominant person in the household or, in common with many other herbs, where the 'mistress is master'. To put it another way, 'Parsley is poison to men, and salvation to women.' It is also said that it grows best either where the man is honest or where the owner is wicked, and it grows rankest where the man is a cuckold.

You should never cut parsley if you are in love as, in doing so, you will also be 'cutting' your love. If you cut it and give it to a friend it will only cause you bad luck and them even worse. It is all right, however, if the person you want to give it to helps themselves from your garden. If you pick some from your garden and, as you do so, say a person's name, then they will be dead within seven days. Young lovers should be wary of picking it, as this will result in their sweetheart's death. Where parsley naturally grows near to a house it ensures good luck. Only healthy children will be born there so long as the plant remains.

Even eating parsley has its risks. One warning is expressed in the saying, 'Fried parsley brings a man to his saddle, and a woman to her grave.' In Cambridgeshire there was a superstition that if a pregnant woman ate parsley three times a day, over a period of time it would result in her aborting her child. More generally it was thought that if parsley was eaten it would promote lust and increase fertility.

A British superstition claimed that sowing parsley would encourage conception. If a woman were given a parsley plant it would ensure that she would be pregnant within twelve months. Children might be told that babies are found in the parsley patch, much as some would say under the gooseberry bush.

Small pieces of parsley are used as a decoration or garnish on food; this may stem from the belief that where the herb is placed onto food it will prevent any contamination. Curiously, it is supposed to weaken any glass with which it comes into contact. Tournefoot claimed that if a glass were to be rinsed in water in which parsley had been soaked, it would break at the slightest touch.

Medicinal

Parsley has been used as a medicinal herb for treating human ailments and sheep disorders. Infused in rainwater, collected during a thunderstorm, it becomes a lotion to strengthen eyesight. Sprinkling a few seeds on the head is said to be a good treatment for baldness. It is also supposed to be a good treatment for kidney stones because the plant grows well on stony ground, breaking open the soil, and it was believed to do the same to obstructions in the body.

The benefits of parsley are not restricted to humans as, scattered on fishponds, it will reputedly heal sick fish too.

BEANS

Botanical Name: *Phaseolus* spp.
Family: Leguminosae
Gender: Masculine
Planet: Mercury
Element: Air
Deities: Cardea; Demeter

In parts of Buckinghamshire it was traditional to sow runner beans on the first Tuesday in May, which was the Ashendon Feast Day. French beans, according to a Devonshire tradition must be planted as soon a first cuckoo of the year is heard.

See also *Vicia faba*.

MOSS PINK

Botanical Name: *Phlox subulata*
Family: Polemoniaceae
Meanings: Unanimity; unity

A legend from the United States relates that the Chickasaw tribe trespassed onto the hunting grounds of the Creek tribe, near the Savannah River. After a battle lasting for three days, the Chickasaw tried to retreat but they were prevented by a firewall lit by the Creek to cut them off. A young Chickasaw boy named Chuhla, 'Blackbird', called upon his animal friends to help by stamping out the fire. While they were doing so, the fur of the squirrels became grey with the ash and smoke, the tails of the raccoons became ringed with stained pine resin and the deer lost its tail completely. The Great Spirit, seeing the animals fighting this fire, transformed the fire into a great blaze of fire-coloured flowers. The Chickasaw made good their retreat and the Creek arrived to find the smouldering stumps of the trees and a wide path of flaming moss pinks.

DATE PALM

Botanical Name: *Phoenix dactylifera*
Family: Palmae
Common Names: Arabian tree; bread of the desert; finger apples
Gender: Masculine
Planet: Sun
Element: Air
Deities: Allah; Apollo; Artemis; Ashtoreth; Christ; Damuzi; Enlil; Hecate; Ishtar; Isis; Lat; Latona; Mariamne, Marian, Marina, Myrrha, Ra; That; Tamar; Tammuz
Meaning: Martyrdom; victory

Myth and Legend

> There is among tree one tree which is blessed,
> as is the Muslim among men; it is the Palm.
>
> – Mohammed

The palm was considered to be the loveliest tree in the world. When the phoenix was near to death, after some 500 years, it would build its nest in it. In this nest it would deposit the 'principles of life' from which the new phoenix would arise. The tree and the bird have become symbols of immortality. In legend they are forever linked together and acceptance of one is therefore a belief in both. 'In the South Countree is a manner Palme, that is alone in that Kinde, none other springeth nor commeth thereof: but when this Palme is so olde, that it fayleth all for age: then oft it quickeneth and spingeth again of it selfe. Therefore men suppose that, *Phoenix* that is a bird of *Arabia*, hath the name of this Palme in Arabia.' (Stephen Batman, *Batman vppon Bartholome.)*

The Minoans represented Artemis as standing next to the date palm, dressed in a skirt of palm leaves, holding a small palm tree in her hand and observing the bull calf of the New Year being born from a group of date palms.

At her oracle the prophetess Deborah gave her responses to questioning from beneath a palm tree near Bethel. This tree was said to mark the location of the graves of David and Rachel.

Magic and Lore

The date palm is the oldest tree in cultivation. It has become a symbol of fertility because of the profusion of the fruits it bears. Pieces of the tree would be worn or carried to increase fertility, and the pips might be carried by men as amulets to increase their sexual potency. A dream of a palm tree, especially one in full bloom, indicates that you can expect great success and good fortune. It may be said to predict a journey to foreign lands, although this journey may prove a hazardous one if the palm is withered.

If the tree grows near property it will act as a protection against severe weather. Keeping a palm frond near to the main doorway into the house will prevent anything evil from entering.

The Bible tells us that it was common to carry palm leaves on special occasions, such as occurred when Christ entered Jerusalem. At the time of Judas Maccabeus it was traditional to carry palm leaves in festival processions. Palm fronds, intertwined with myrtle, were carried on the right and citron on the left. Following their use at the Feast of the Tabernacles the fronds would be kept in the house to bring good luck and protection from all ills.

The palm is also a symbol of martyrdom. Early pilgrims were known as 'palmers', as they would often select palmwood from which to make their staff. In early Christian legend angels used palm fronds to carry the soul of a martyr to heaven.

On All Souls' Day (2 November) palm fronds were burned in the belief that the smoke would help the souls in purgatory to find their way to heaven. In Britain the willow was frequently used as a substitute for palm and was sometimes known as the English palm.

REEDS

Botanical Name: *Phragmites communis* (and other species)
Family: Gramineae
Deities: Coventia; Pan
Meanings: Indiscretion (feathery); confidence in heaven (flowering); music; spirit of music; 'You are the sweet inspiration which invests my life with harmony' (bundle with panicles); 'Be careful, others are aware of our indiscretion' (split)

Myth and Legend

The god Pan was besotted with the fair nymph Syrinx and chased her through Arcadia along the banks of the River Ladon. Syrinx, who was not in love with him, called upon the river to help her escape his clutches. She was received into its cold bosom and transformed into reeds. Other tales claim that it was Diana, the goddess of the moon, who transformed her into reeds. Pan was still unwilling to let her escape, and cut several reeds of differing lengths and formed them into the first set of shepherd's 'Pan pipes'.

An alternative story tells that Pan and Apollo were engaged in a competition to establish which of them was the greater musician. Midas, King of Phrygia, was appointed to be the judge and he duly presented Pan with the victor's palm. Apollo, however, was not a good loser. He considered that Midas clearly had no ear for music and decided to improve his hearing by giving him a set of donkey's ears. Midas was forced to wear a cap in order to conceal his disgrace, and only his barber knew. The barber found it increasingly difficult to keep the secret, but feared the wrath of Midas if he were to tell anyone. He therefore dug a deep hole and whispered into it, 'Midas has ass's ears.' He then filled in the hole in the hope of burying the secret, but on the freshly disturbed earth sprang up a reed which, as it waved in the breeze, continually repeated, 'Midas has ass's ears', betraying the king's secret to all who would listen.

Magic and Lore

Reeds are supposed to have great preservative powers. Pens were made from them, the written word being used to preserve knowledge and memory. In the Ogham tree alphabet it is given the symbol Ng, signifying communication. It is also symbolizes the desire to seek out basic truths.

HART'S TONGUE FERN

Botanical Name: *Phyllitis scolopendrium*
Family: Polypodiaceae
Common Names: Adder's tongue; bullock's tongue; horse tongue; lamb's tongue; snake's tongue
Planet: Jupiter
Meaning: Gossip

In Scotland the hart's tongue fern is said to bear the imprint of the Devil's foot. Paradoxically, perhaps, it is also said to be hated by witches because when it is cut through the letter 'X' can be seen, the initial of the Greek word for Christ.

Irish tradition has it that the fern remains flowerless because it was cursed by St Patrick. In spite of this it is still seen as being a gift from God, as when the stem is cut through a 'G' can be found in the first section, an 'O' in the second and a 'D' in the third. A Devonshire tradition says that the fern was a pillow for Christ when there was nowhere else for him to lay his head. In return for this service he left two hairs pulled from his head that the plant treasures inside its stem.

In Surrey it is said that when the stem is cut through on the slant the image of an oak tree can be seen. The more perfect the image the better the luck of the finder.

POKE ROOT

Botanical Name: *Phytolacca americana*
Family: Phytolaccaceae
Common Names: American nightshade; bear's grape; caner-root; coakum; cocan; crowberry; garget; inkberry; jalap; pigeon berry; pocan; pokeberry root; polk root; scoke; Virginia poke
Gender: Masculine
Planet: Mars
Element: Fire

Adding an infusion of poke root to bath water or sprinkling it about the home when the moon is new will break any curse cast against the occupant. Carrying the root will give courage. To help locate any item that you have lost, prepare a mixture of poke root, hydrangea, violet and galangal. Sprinkle this mixture in the vicinity of where you last saw the missing object and it will soon appear.

CHRISTMAS TREE

Botanical Name: *Picea abies*
Family: Pinaceae
Common Names: Norway spruce
There are now many species of tree that are used as Christmas trees, but this is the most common.

Myth and Legend

A legend explaining why a fir tree is used at Christmas says that in 725 St Boniface, an English monk and missionary of the early Church in France and Germany, came across a group of people gathered around an oak tree (sometimes identified as the sacred Donar Oak near Geismar), about to sacrifice a child. To save the child the saint felled the tree, and found a tiny fir growing between the roots. As the shape of the tree pointed the way to heaven he declared that it must be named the 'Tree of the Christ Child'.

A French legend from the seventeenth century tells of a poor but virtuous young woman who fell asleep beneath a tree at Christmas. While she slept spiders wove their webs all over the tree. All their threads were turned into silver as a reward to her for her goodness.

Another tale tells of a woodcutter who found a poor, hungry child in the forest on Christmas Eve. He gave him his own food and the child, having eaten, disappeared. The following day, Christmas Day, he returned and revealed that he was the Christ Child. Breaking a branch from a fir tree he gave it to the couple instructing them to plant it and look after it as at Christmas it would bear fruit. The following year, at Christmas, it was covered with apples of gold and nuts of silver.

Christian legend says that the origin of dressing a tree at Christmas dates from Martin Luther, the sixteenth-century church reformer. It is said that he decorated a tree with candles to represent the stars that filled the clear winter skies.

Magic and Lore

In northern countries, where most trees shed their leaves each winter, evergreen species became symbolic of immortality. They contained good spirits and had the power to overcome the evil forces that caused the 'death' of deciduous trees. Taking pieces of greenery into the house in winter therefore introduced a degree of protection to the home.

The Christmas tree is actually a fertility symbol which long pre-dates Christianity. Decorated trees were used in the Roman Saturnalia festivals. The pagan tradition of decorating whole trees at midwinter is often said to have originated in the Upper Rhine region of Germany. The Church could not ignore the practice and so it was 'Christianized'. By around the fifteenth century it became customary to decorate trees with apples to symbolize Adam's and Eve's expulsion from the Garden of Eden. These trees were erected in time for 24 December, the feast day of Adam and Eve and were called 'Paradise trees'. At that time, few people were literate or had access to books, and these trees became central to Paradise plays, which were used to teach the stories of Genesis.

The first record of a decorated Christmas tree, as we would recognize it, is from Strasbourg in 1605. The description is of fir trees festooned with 'paper, apples, gold foil and sweets'. The tree was topped by a star to represent the guiding star on the first Christmas night. Later, candles were added to the decorations to symbolize the returning sun and, in the Christian tradition, Christ as the 'Light of the World'.

Christmas trees were introduced to Britain in the mid-1800s. Prince Albert, Queen Victoria's husband, popularized the practice amongst the fashionable people of Britain by erecting a large Christmas tree at Windsor Castle in 1841. Initially Christmas trees were a luxury only found in the homes of the relatively wealthy. Later, as goods became cheaper and people became more affluent they became a common feature in every household at Christmas.

In Victorian times the trees would be decorated with paper flowers, gilded nuts and sweets. Stripping the tree on Twelfth Night was as important as decorating it, not least because it gave the children the opportunity to eat the sweets and nuts.

PIMPERNEL

Botanical Name: *Pimpinella* spp.
Family: Umbelliferae
Common Names: Blessed herb; greater pimpernel; herb of Mary; *luib na muc*; poor man's weatherglass; shepherd's weatherglass
Gender: Masculine
Planet: Mercury
Element: Air

> No heart can think,
> No tongue can tell,
> The virtues of the Pimpernel.
>
> – Gerard

Pimpernel should not be confused with the scarlet pimpernel, *Anagalis*. It is a protective herb, and when it is kept in the home it will ward off illnesses and prevent accidents. Carried it will ensure that nobody will be able to deceive you; indeed its possession is supposed to give the bearer second sight. It is said to be so powerful that if it is dropped into running water it will float upstream, against the flow of the current.

In magic this powerful herb is used as a cleanser, to purify the blades of the swords and daggers used during magical rites.

ANISE

Botanical Name: *Pimpinella anisum*
Family: Umbelliferae
Common Names: Anece; aniseed; anneys
Gender: Masculine
Planet: Jupiter
Element: Air

Magic and Lore

Anise was greatly valued in the ancient world. The Romans could use it in the payment of taxes, as the Bible says: 'Woe unto you, scribes and Pharisees, hypocrites! For ye pay tithe of mint, anise and cumin, and have omitted the weightier matters of the law.' (Matthew 23: 23)

It is used protectively in magical rites. The fresh leaves, placed in a room, will keep all evil away and ward off the evil eye. During magical rites the leaves might be placed within the magic circle to aid the sorcerer, especially when calling forth spirits. It will also keep the magician safe from any evil spirits that might appear.

At the first assembly of the American colony of Virginia in 1619 it was decided that each man to whom a division of land was granted must plant thereon six anise seeds.

Medicinal

The seeds of anise have been used in foods and medicines for over 5,000 years. Its known history starts with the tablets of Nineveh and references to it can be found in the Egyptian Ebers papyrus. Pythagoras, in the sixth century BC, claimed that a spray of anise held against the head would prevent epileptic fits. In folk remedies it is claimed that placing a sprig of anise on the bedpost will help those who sleep in the bed to regain their lost youth.

Gerard says, 'Aniseed helpeth the yoexing [belching] and hicket [hiccup].'

BUTTERWORT

Botanical Name: *Pinguicula vulgaris*
Family: Lentibulariaceae
Common Names: Bean weed; bog violet; butter plant; butter root; clowns; earnings grass; ekkel-grise; fly-catcher; marsh violet; rot grass; St Patrick's spit; St Patrick's staff; sheep root; sheep rot; steep grass; steepweed; steepwort; thickening grass; white sincles; yirnin-grise

Myth and Legend

The butterwort is associated with the early missionaries. On the Scottish islands there is a belief that the plant grew wherever St Moalrudha had touched the ground with his staff. It therefore served to show the extent of his travels. In Ireland it was said that, whilst crossing a bog, St Patrick lost his staff. While looking for a tree from which to cut another he came across a butterwort on which the stem had grown so long that he was able to make use of that instead. The flowers on this stem never faded and, as with St Moalrudha, wherever this new staff touched the ground up sprang butterwort plants.

Magic and Lore

Butterwort was considered to be a magical plant throughout the Scottish Highlands. Mixed with juniper and whin it was used to form a charm against witchcraft. Any livestock that ate the plants were safe from the arrows or mischief of malicious elves and faeries. One story tells of how two women, whilst watching over a newborn child, overheard a conversation between faeries that had come to steal the infant. One faerie, seeing the child, said, 'We will steal it.' The other replied, 'We will not. We cannot, because its mother ate the butter of the cow which ate the butterwort.'

Like many other marshland plants, it has been falsely accused of causing illnesses in sheep, cattle and other livestock.

PINE

Botanical Name: *Pinus* spp.
Family: Pinaceae
Gender: Masculine
Planet: Mars (Jupiter, according to Culpeper)
Element: Air
Deities: Artemis; Astarte; Attis; Bacchus; Cybele; Diana; Dionysus; Neptune; Pan; Rhea; Sylvanus; Venus
Meanings: Time; philosophy

The references in mythology to a 'fir tree' may be applied to almost any cone-bearing tree, much as those to 'corn' can be applied to any one of several different cereal crops. There are, however, some that seem to apply specifically to the pine.

Myth and Legend

In ancient Greece and Rome the pine was sacred to the gods of the sea and of fertility, and was widely used in shipbuilding. The tree is described in the story of Attis or Atys, a young shepherd beloved of the fertility goddess Cybele. He was the son of Nana, who had become pregnant after eating the fruits of the almond tree. He was charged with tending the temple of Cybele and therefore had taken a vow of chastity. When he broke his vow, by having an affair with a nymph named Sangaris, the wrath of the goddess was terrible. Driven to madness, Attis, in remorse for breaking his vow, castrated himself and bled to death. In one version of this tale Attis died beneath the pine tree. When the observance of Cybele was adopted by the Romans another version of this myth came about. It says that Attis was turned into a pine tree by Rhea, and the goddess was consoled by Jupiter's promise that the tree would be forever green.

> To Rhea grateful still the pine remains,
> For Atys still some favour she retains;
> He once in human shape her breast had warmed,
> And now is cherished to a tree transformed.
>
> – Ovid, *Metamorphoses*

The Romans enacted the death of Attis each spring. Following a procession to a pine tree a young man would castrate himself. Then there was a period of fasting, which lasted eight days and was ended by the festival of blood, during which the participants flagellated themselves and some of the priests castrated themselves. After the festival came a period of somewhat noisy mourning which ended with jubilation at a symbolic resurrection.

An alternative account is given in the story of Pan, the Greek god of shepherds, pastures, and fertility. He loved a nymph named Pitys (meaning 'pine') and pursued her relentlessly. She shunned his attentions and eventually the gods transformed her into a pine tree in order to escape him.

The pine tree was also dedicated to Bacchus (Dionysus) the god of vinticulture. In most illustrations of the god he is shown with a thyrsus, a phallic wand, topped by a pine cone. In mythology Dionysus gave the first vines to Oeneus, but entrusted the making of wine from the grapes to Icarius. Icarius gave some of his first trial batch of wine to shepherds in the Marathonian woods under Mount Pentelicon, who enjoyed it to the point of intoxication. They began to see double and believed that they had been poisoned or bewitched. They killed Icarius and buried his body beneath a pine tree. However, the whole episode was witnessed by Icarius's faithful dog Maera, which immediately ran off to find Erigone, Icarius's daughter. It led her back to the place where the body had been concealed and dug it up. On seeing the body of her dead father Erigone hanged herself on the pine tree but before she did so she prayed to the gods that all the daughters of Athens might do likewise until the death of her father was avenged.

Icarius was the mortal heir to the Dionysian cult of ecstasy. His death greatly disturbed the ancient world and, although only the gods had heard Erigone's prayer, maidens were found to be hanging themselves on pines throughout the region. This continued until the Oracle at Delphi was consulted and it was explained that the only way to end the death of the innocents was for the guilty shepherds to be found and punished. This was done in due course and in celebration a vintage wine festival was introduced. At this, libations were poured to Icarius and Erigone. Young girls would suspend themselves by ropes from pine trees supporting their weight on small wooden platforms. Masks of the god Dionysus would also be hung from pine trees in the middle of the vineyards, in memory of the hanged maidens. When these turned in the wind it was believed that they fructified the vines that they faced. For its part in the tale, the image of the dog, Maera, was set amongst the constellations as the lesser Dog Star. Icarius has been identified with the constellation of Bootës and Erigone with the constellation of the Virgin.

Other ancient Greek deities are associated with the pine tree, including the goddess Pitthea and the god Pittheus. Pittheus was the father of Aethra who, in turn, was the mother of Theseus.

In many different forests throughout the world there are tales of a genie, or a forest spirit, which would dwelt within an old gnarled tree. His voice could be heard in the rustling of the leaves. In many of the north European countries a pine is identified as being the spirit's favourite abode, known as the 'king of the forest'. The spirit would raise great objection if his tree was to be felled and would beg that it be permitted to live. He is usually depicted as holding an uprooted fir tree as his staff.

A Romanian song tells the story of a young couple who died of love and were buried together. They were transformed into a pine tree and a vine, which grew from their graves. The vine grew around the tree so that the two would continue to embrace each other even after death.

Magic and Lore

Although the pine is a tree of the sun it has generally been considered to be a plant of misfortune, although it has also been heralded as the tree of friendship, especially friendship in adversity. It is also one of the four primary trees of the gods of ancient China.

In ancient Egypt it was closely linked to the myths of Osiris who, in his primitive spirit, can be identified as a tree deity. A pine tree was felled and hollowed out. An image of the god was fashioned from the wood extracted and laid back in the hollow tree trunk. This was kept for a year to watch over the vines and at the season's end was burnt. The ashes were scattered across the ground to symbolically improve the soil's fertility. There are obvious comparisons to be drawn between this and the tale of the body of Osiris being found inside a growing tamarisk tree. There are few pines to be found in Egypt and this tale suggests either that in antiquity there were far more pines or that the plant was mis-identified. In addition to being the god of the dead Osiris is credited with having taught mankind the cultivation of vines and wine making.

Traditionally pine cones which are to be used as talismans must be gathered at Midsummer, when they still contain pine nuts. Eating one of the pine nuts from the cone each day makes one bulletproof. In some parts of Britain pine cones are known as 'dealies' or 'deal apples'.

It is the needles of the pine tree that are generally used in magic, for purification and protection, although the sawdust could be used as a base of incense. Burning pine needles in the winter will cleanse the house of all negativity and will return any spells cast against the occupants back onto the person who cast them. They may also be added to purification baths or scattered on floors to drive away evil. A cross on a hearth formed from pine needles will prevent any evil force from entering the house via the fireplace.

In Japan the pine is considered to be a symbol of constancy and long life. A pine branch placed over a doorway will ensure the continued joy of the people who live there. If a branch is placed over a bed, those who sleep in it will be free from illnesses, although if it is put there too late it can actually make the illness worse. The scent of pine is supposed to be a cure for various ailments. Should a pine tree in a garden be struck by lightning it is an ill omen, indicating that either the master or the mistress of the house will die within the year. Planting a row of pines would leave you open to losing the property on which they stood, or the land might pass to a younger branch of the family rather than to the rightful heir. Likewise to dream of a pine tree portends misfortune, indicating dissolution, and if you see yourself in a forest of fir trees you should expect to undergo great suffering.

Medicinal

Carrying the cones is supposed to increase the bearer's fertility, and give a vigorous old age.

See also fir (*Abies*), Christmas tree (*Picea abies*), Scots Pine (*Pinus sylvestris*) and Larch (*Larix decidua*)

SCOTS PINE

Botanical Name: *Pinus sylvestris*
Family: Pinaceae
Gender: Masculine
Planet: Mars (Jupiter, according to Culpeper)
Element: Air
Deities: Astarte; Attis; Cybele; Dionysus; Neptune; Pan; Rhea; Sylvanus; Venus
Meanings: Elevation; 'You are the sun of my life, which makes all things real'

The Scots pine is prolific in Britain and is the sole northern European pine to have survived the Ice Age. Being conspicuous, they were set as landmarks for drovers' routes as well as indicating those farms in which drovers could expect hospitality.

At the winter solstice the druids would light bonfires of Scots pine to celebrate the passing of the seasons and to draw back the sun. Glades of the trees were decorated with lights and shiny objects as a representation of the Divine Light. In these practices it is possible to see the origins of the Yule log and Christmas tree traditions.

In the Cotswolds, Scots pines were supposed to have been planted as an expression of Jacobite sympathies. It showed fugitives at the time of the Jacobite rebellion where they might find a safe refuge.

See also Fir (*Abies*), Christmas tree (*Picea abies*), pine (*Pinus*) and larch (*Larix decidua*).

MASTIC

Botanical Name: *Pistacia lentiscus*
Family: Anacardiaceae
Common Names: Chios-mastic; gum mastic; lentisco; mastick
Gender: Masculine
Planet: Sun
Element: Air

Mastic is used for various purposes in magic. When it is added to incense it increases its potency and burning it will aid the psychic powers of all those present. Mastic is also burned during the ceremonies of conjuring up spirits. It can be used in potions to induce love or lustfulness.

TREE OF ABRAHAM

Botanical Name: *Pistacia terebinthus*
Family: Anacardiaceae
Common Names: Oak of Mamre; turpentine tree
Gender: Masculine

This is one of the more famous trees of historical tradition. It is supposed to have been the tree beneath which Abraham dwelt and where he was visited by various angelic emissaries. In other myths it is said to have sprouted from the staff of one of his angelic visitors.

The writer Josephus claimed that the tree of Abraham had grown since the creation of the world and continued to survive in his own time. As with many of these plants of ancient myth the 'true' identity of the species is impossible to identify with absolute confidence but this is the plant most usually accepted as the tree of Abraham.

PISTACHIO

Botanical Name: *Pistacia vera*
Family: Anacardiaceae
Common Names: Eustuq; green almond
Gender: Masculine
Planet: Mercury
Element: Air

In parts of Arabia it is believed that the pistachio should be eaten to counteract love spells. Elsewhere it is said that, if given to a zombie, it will release them from their trance and allow them the sweet peace of death. Some writers state that if pistachio nuts are to be used for this purpose they must be dyed red.

PEA

Botanical Name: *Pisum sativum*
Family: Leguminosae
Gender: Feminine
Planet: Venus
Element: Earth

Myth and Legend

Peas are, on the whole, considered to be plants of good fortune. Shelling peas will, it is claimed, bring money into a business and dried peas can be used in mixtures to attract money. When you are shelling peas a pod with a single pea in it is supposed to be a particularly lucky find. It is also very lucky to find a pod that has nine peas in it, the more so when it is the first pod that you open. When you find one you should toss one of the peas over your shoulder and make a wish. The whole pod can be used to discover who it is that you will wed. Hang the pod, still containing the peas, from a single white thread over a doorway. The next person who comes through the door who is unmarried and not a member of the family, will be your future spouse. In some of the English counties, when a young woman finds such a full pod she must also slip a piece of paper under the doormat on which is written:

> Come in my dear,
> And do not fear.

Again, the first man through the door would be her future husband. An alternative superstition says only that the next person through the doorway will share the same name as your future spouse. And if that were not enough, you will be married within the year.

There are various directions for sowing pea seeds to ensure a good harvest. For example:

> Sow beans and peas in the wane of the moon,
> Who soweth them sooner, he soweth too soon.

And:

> Sow beans and peas on St David and St Chad
> Whether the weather be good or bad.

There are also various indicators as to whether a good harvest of peas can be expected. Most of these require that the early part of February must be fairly wet. One states that a good crop should be expected if 'the hedge is dripping on St Valentine's morning'. Another says:

> On Candlemas Day, if the thorns hang a drop,
> Then you are sure of a good pea crop.

Dreaming of eating peas is unlucky, especially if they are hard, as it denotes straitened circumstances and faithlessness in friends. If they are seen growing you may expect your enterprises to succeed.

Medicinal

Peas could be used as a cure for warts. Someone with several warts should touch each separately with a green pea. The peas must then be buried separately in the garden, and as they decay so will the warts.

RIBWORT PLANTAIN

Botanical Name: *Plantago lanceolata*
Family: Plantaginaceae
Common Names: Johnmas pairs; Johnmas flowers; fire grass
Gender: Feminine
Planet: Venus
Element: Earth

In the Shetland Isles and other parts of Scotland, plantain was used in love divination. At Johnmas (St John's Day) two flowering stems, called scrapes, would be collected. These represented the couple, one being the boy and the other the girl. All the visible anthers of the flowers were removed and the scrapes were hidden, wrapped in a dock leaf, under a stone. If more anthers had appeared next day then the love between the couple was certain to flourish.

The name fire grass comes from the problems that arise if too much of the herb is gathered in a hayrick. The rick may overheat and catch fire because of the amount of moisture in the leaves.

Widespread across the north of Britain are various traditional children's games that make use of the flower spikes.

BROAD-LEAVED PLANTAIN

Botanical Name: *Plantago major*
Family: Plantaginaceae
Common Names: Bird seed; bird's meat; broad leaf; canary flower; canary food; canary seed; cow grass; cuckoo's bread; Englishman's foot; great waybrede; hard heads; healing blade; Johnmas flowers; Johnmas pairs; lamb's foot; lark seed; mother-die; pony's tails; poverty grass; rat's tails; ratten tails; ripple gris; slanlis; *slàn-lus*; snakeweed; traveller's foot; waavern-leaf; wabran leaf; way berry; way bread; waybroad; waybroad leaf; wayburn leaf; way fron; weybroed; white man's foot; wibrow-worrou
Gender: Feminine
Planet: Venus
Element: Earth

Magic and Lore

To the gardener and the groundsman a broad-leaved plantain is a pernicious weed to be eradicated. It has been carried across the length of the globe and established itself alongside white settlers in new colonies. Hence, in New Zealand, it is known by the name 'white man's foot' as it seems to appear wherever the white man's foot has trodden.

A Harleian manuscript from the Anglo-Saxon period claims that plantain 'withstands evil and vile things and all the loathsome ones that through the land roved'. In parts of Scotland wearing a plantain root strung as a necklace is said to prevent abduction by the faerie folk. The tradition of using it as a protective herb has been brought up to date with a recommendation that it be hung from the rear-view mirror of a motor car to ward off evil spirits.

The divination practice on Johnmas (St John's Day) in the Shetlands, described under Ribwort Plantain, has also been done with the flower stalks of this species. A young woman could also use it to discover whom she would marry. At noon on Midsummer's Day she must collect a dead old root from beneath a plantain. These roots, because of their appearance, are known as 'coals'. If she sleeps with the coal beneath her pillow, she will see her future spouse in her dreams.

In common with other plants called 'mother-die' it must never be taken into the house, as it will only bring misfortune on the household.

Medicinal

In the Highlands the plantain was called *slàn-lus* or 'plant of healing'. It is supposed to be particularly good as an aid to the healing of broken limbs, probably owing to 'sympathetic magic'; where the plant is trodden on, or crushed, it survives and will soon grow again. It was also recommended as a treatment for the bites of venomous snakes and mad dogs. It was placed beneath the feet to overcome weariness, especially when travelling.

It was one of the nine sacred herbs of the Anglo-Saxons, who called it *weybroed*. As a treatment for headaches it was bound to the forehead using a red thread, and to treat cuts the edge of the leaf was bitten and it was laid over the wound.

PLANE TREE

Botanical Name: *Platanus* spp.
Family: Platanaceae
Deities: Apollo; Xerxes

According to Herodotus, a great plane tree became associated with Xerxes. After he had declared war on the Greeks he was travelling to Sardis when he came across the tree and was so awed by its grandeur that he stopped his entire army so that he might admire it more fully. He referred to it as his goddess, his mistress and his minion, and when he eventually left it he had the trunk encircled with a gold band

and hung the branches with bracelets. A guard was left to watch over it and he had a medal struck with a picture of the tree on it so that he might be able to continue to adore it thereafter.

PATCHOULI

Botanical Name: *Pogostemon patchouli*
Family: Labiatae
Common Names: Pucha-pot
Gender: Feminine
Planet: Saturn
Element: Earth

The fresh essential oil of patchouli is rather unpleasant but it improves as it ages. Its naturally earthy smell led to it being used in fertility amulets and added to sachets used to attract love. It may also be added to bath water for much the same result. It was also included in mixtures and incense used to bring wealth.

MILKWORT

Botanical Names: *Polygala vulgaris*; *Polygala serpyllifolia*
Common Names: Mountain flax; Rogation flower

The botanical name *Polygala* (from the Greek *poly*, meaning 'much' and *gala,* meaning 'milk') is a reference to the copious amount of milky white sap that this plant produces if it is damaged. The colour of the sap has given rise to the common name 'milkwort' and it was believed that, in sympathetic magic, its use would be an aid to milk secretion.

Rogation Sunday is the fifth Sunday after Easter and the Rogation days are the Monday, Tuesday and Wednesday before Ascension Day. Milkwort would be made into garlands and carried at Rogation, hence another of its common names.

SOLOMON'S SEAL

Botanical Names: *Polygonatum multiflorum*; *Polygonatum odoratum*
Family: Liliaceae
Common Names: David's harp; dropberry; Jacob's ladder; Job's tears; ladder to heaven; lady's locket; lady's seal; lily of the mountain; our lady's belfry; St Mary's seal; *sigillum Sanctae Mariae*; Solomon's seal; sow's tits
Gender: Feminine
Planet: Saturn
Element: Water
Meaning: Concealment; discretion

Magic and Lore

It is said that the common name of this plant is derived from marks on the stem, which appear to have been made with a seal. Other writers say that it is the depressions in the root, which look like Hebrew characters, and that they were made by the seal of King Solomon, who was the discoverer of its virtues. In addition to being used as a treatment for wounds and bruising, it has been recommended as a remedy for baldness. The fumes emitted from the brewed flowers of the plant will give inspiration to painters and poets.

When it is used as a protective herb the root must be cut into four sections, one piece being placed at each corner of the house. It may also be used in various exorcism and protection spells. Sprinkling an infusion made from the plant about the house will drive away any evil spirits that might be lurking there.

Medicinal

In his herbal of 1597 John Gerard tells us that Solomon's seal earns its name from an ability to seal up wounds. He goes on to suggest that it can be used to treat bruises. 'The root of Solomon's seale stamped while it is fresh and greene, and applied, taketh away in one night, or two at the most, any bruise, blacke or blew spots gotten by fals or womens wilfulnesse, in stumbling upon their hasty husbands fists, or such like.'

KNOTGRASS

Botanical Name: *Polygonum aviculare*
Family: Polygonaceae
Common Names: All seed; armstrong; arsesmart; beggar's weed; bird's tongue; black strap; centinode; clutch; cowgrass; crab-grass; crab-weed; devil's lingels; finzach; hogweed; iron-grass; man-tie; nine-joints; ninety knots; pig grass; pig rush; pigweed; red legs; red robin; red weed; snakeweed; sparrow's tongue; stone weed; swine grass; swynel grass; tacker grass; way-grass; willow-grass; wire-grass; wireweed.
Gender: Feminine
Planet: Mars
Element: Water

Magic and Lore

There was a traditional belief that knotgrass could be used to stunt the growth of children and animals. Reference to this can be found in Shakespeare's *A Midsummer Night's Dream*:

> Get you gone you dwarf;
> You minimus of hindering knot grass made.

It can be used to rid yourself of all your woes. Simply visualize all the things that trouble you being absorbed into some knotgrass, and then take the plant and burn it.

Medicinal

Carrying a piece of the plant is supposed to be an aid to eyesight, strengthening and preserving it. The name 'arsesmart' is usually associated with the water pepper, *Polygonum hydropiper,* which was used to treat sores, ulcers, swellings and jaundice as well as being applied as a flea repellent in bed straw.

BISTORT

Botanical Name: *Polygonum bistorta*
Family: Polygonaceae
Common Names: Adderwort; dragon's wort; Easter giant; Easter ledger; Easter ledges; Easter magiants; Easter man-giant; Easter mentgions; Easter serpentary; gentle dock; goose-grass; great bistort; meeks; oderwort; osterick; oysterloit; passion dock; passions; patience dock; patient dock; pencuir kale; pink pokers; poor man's cabbage; pudding grass; pudding dock; red legs; silver weed; snakeweed; sweet dock; twice writhen; water ledges
Gender: Feminine
Planet: Saturn
Element: Earth

Magic and Lore

The name is derived from *bis*, meaning 'twice', and *torta* meaning 'twisted', from the appearance of the root. Many of the common names relate to its use in making a bitter pudding eaten at Lent.

Merely for a woman to carry bistort is said to be sufficient to aid her in conceiving. Carrying a sprig is also supposed to attract wealth. Another way to attract money is to burn a little of the plant. Burning it with frankincense is said to heighten the psychic powers and to be especially valuable to those indulging in some form of divination. An infusion made from it can be sprinkled about the home to drive away poltergeists and evil spirits.

Medicinal

This herb was said to give protection against infectious diseases. The eighteenth-century herbalist described the root as the best astringent, as it was 'not violent, but sure'.

BUCKWHEAT

Botanical Name: *Polygonum fagopyrum* (Synonym: *Fagopyrum esculentum*)
Family: Polygonaceae
Common Names: Beechwheat; brank; French wheat; Saracen corn
Gender: Feminine
Planet: Venus
Element: Earth

Buckwheat is cultivated extensively in the Himalayas and is one of the foods that may lawfully be eaten on the Hindu *bart* or fast days. The grain is said to attract wealth and so small amounts are added to money incense. Traditionally some should always be kept in the kitchen to ward off poverty.

When flour is made from buckwheat, it can be used to drive away evil. It may either be sprinkled about the house or used for marking out magic circles when performing magical ceremonies.

POPLAR

Botanical Name: *Populus* spp.
Family: Salicaceae
Common Names: Singing tree; talking tree; the bride's tree
Planet: Saturn
Element: Water
Deities: Apollo; Leuce

Magic and Lore

There are some beliefs that are applicable to all poplars, although they have, over the years, become more associated with particular species. All poplars seem to have a connection with communication and protection against injury and death, for example. They are said to represent the powers of mind over matter, the will over destiny. The association with communication skills leads to them being referred to as 'whispering' or 'talking' trees. It is, after all, the wind, the messenger of the gods, which causes the trees to whisper as their leaves rustle.

The buds and leaves were recommended for use in spells and charms to attract wealth. They might also be used in witches' 'flying' ointments. The modern view of this is that they are used to ensure suitable vibrations for astral projection.

The belief that poplar timber was used to form the cross at the crucifixion of Christ has largely been applied to the aspen tree but was once applied to all poplars. The pagan goddesses associated with poplars are all 'moon goddesses', the various poplars being emblematic of different aspects of the White Goddess and showing different characteristics of the moon.

Poplars are supposed to be unlucky for lovers. It is particularly unfortunate if they have such trees growing in their gardens.

Medicinal

The leaves of the poplar tend to shake even in the slightest breeze, giving the appearance that they are always moving. Their constant shaking led to them being used for the treatments of fevers and agues under the Doctrine of Signatures.

WHITE POPLAR

Botanical Name: *Populus alba*
Family: Salicaceae
Common Name: Abele
Planet: Saturn
Deities: Juno; Persephone
Meanings: 'Our love is timeless'; time

Myth and Legend

In mythology Hercules made himself a crown of white poplar leaves after he had killed the evil giant Cacus, the son of Vulcan and a famous robber in classical mythology, who had stolen some cattle. He was usually depicted as having three heads and vomiting fire. Hercules wore this crown for his journey into the Underworld, where the tops of the leaves were scorched whilst the underside was bleached by his radiance.

In the *Odyssey* the white poplar is one of the three trees of resurrection, the others being the cypress and the alder. It is sacred to Persephone as the goddess of regeneration. She was said to have had a grove of white poplars planted in the west, in the land of the sunset.

Just as the aspen is supposed to tremble from having been used for the cross at Christ's crucifixion, so the white poplar is said to shiver because it grew in the Garden of Gethsemane. Beneath its branches Christ prayed during his night-time vigil before his arrest. Ever since, the tree has been said to shake and whisper its sympathy.

Magic and Lore

The white poplar is emblematic of the fertile goddess. It represents the full moon or the light of the moon, that period between new and full. The leaves are light on one side and dark on the other, representing day and night.

The five-pointed leaves were especially revered by French witches in the Middle Ages. They were supposedly good for 'correcting atmosphere' during flying, and were included in a witch's 'flying' ointment, which was smeared on the body.

White poplar also makes a good weather indicator. The white undersides of the leaves tend to be easily seen when moved in the light summer breezes that often accompany the rain.

BLACK POPLAR

Botanical Name: *Populus nigra*
Family: Salicaceae
Planet: Saturn
Element: Air
Deities: Egeria; Hecate
Meaning: Courage; 'Let not your heart be troubled, all will be well'; loss of hope

Myth and Legend

The black poplar is linked to the tale of the white poplar. When Hercules eventually emerged from the Underworld, the fires of Hades had blackened the leaves.

In mythology, Helios drove his chariot of the sun across the sky. Phaeton, the son of Helios and Clymene, pleaded with his father to be allowed to drive it, even if only for one day. His pleadings were supported by his sisters, the seven Heliades. Helios, against his better judgement, gave in, but very soon after setting off Phaeton realized that it was not as easy as it looked and that he was unable to control the great horses that drew the chariot. The chariot careered off course, coming so close to the Earth that it nearly caught fire. Zeus, annoyed by Phaeton's stupidity, knocked him from the chariot with a bolt of lightning. Phaeton fell, head first, from heaven with his hair blazing and came to earth in the River Eridanus. The naiads built a tomb for him on which was inscribed:

> Driver of Helios' chariot, Phaeton
> Struck by Jove's thunder, rests beneath this stone,
> He could not rule his father's car of fire,
> Yet it was much nobly to aspire.

His sisters, the Heliades, were turned into the black poplars that lined the river-banks, their tears becoming amber as they fell into the waters.

Magic and Lore

The black poplar, with its resinous exudate, represents the crone and symbolizes the dark of the moon, that period from full to new.

There is evidence that, in some places, it was revered as a deity. In some areas of Britain and northern Europe it was traditional to dress a black poplar with coloured ribbons on certain auspicious dates each year. For example in Aston-on-Clun, on 29 May, the black poplar in the centre of the village was decorated, supposedly to commemorate the wedding in 1786 of Mary Carter, a local heiress, to John Marston who was then lord of the manor. This particular tree was locally referred to as 'the bride's tree'. Couples would be given twigs from it on their wedding day as tokens of good luck, but a local vicar stopped the practice when it become too popular. There are other tree-oriented rites observed on, or around, this date in May and it seems more likely that the wedding was timed to coincide with an already established celebration on that date.

The poplar is dedicated to the dark goddess Hecate. In some rural regions of Britain, during the season when lamb's tails are docked, a lamb's tail would be placed under each newly planted black poplar as a sacrifice to the goddess.

Medicinal

In the Doctrine of Signatures the 'shivering' leaves of the poplar have identified it as a treatment for fevers. In Lincolnshire a person suffering from fever must pin a lock of their hair onto a black poplar with the words:

> When Christ, our lord, was on the cross,
> Then didst thou sadly shiver and toss,
> My aches and pains thou now must take,
> Instead of me I bid thee shake.

They should then return home without speaking to anyone. Some writers suggest that, ideally, the person should have been fasting prior to trying this cure. There are obvious similarities between this charm and those that made use of an aspen.

Black poplar could also be used in the treatment of warts. All the sufferer need do was let the balsam-enriched water fall off the leaves onto the blemishes for them to be removed.

ASPEN

Botanical Name: *Populus tremula, Populus tremuloides* (American aspen)
Family: Salicaceae
Common Names: Aps; apsen-tree; old wives' tongues; pipple; quakin' ash; quakin' esp; quaking aspen; shaking asp; shiver tree; shivering tree; snapsen; trembling poplar; whispering tree; woman's tongue
Gender: Masculine
Planet: Mercury
Element: Air
Deities: Hercules; Leuce; Ua Ildak

Myth and Legend

Although the belief that poplar wood was used to make the cross on which Christ died seems to have been applied to almost every species of poplar at one time or another it is now particularly identified with the aspen; it is said to be because of its memory of this usage that the tree continues to tremble still.

> Ah! Tremble, tremble aspen tree,
> We need not ask thee why thou shakest,
> For if, as holy legend saith,

On thee the saviour bled to death,
No wonder aspen that thou quakest
And, till in judgement all assembled,
Thy leaves accursed shall wail and tremble.

In Scotland it is said that the tree was cursed because it failed to bow its head, as all other trees did, during Christ's procession to the place of crucifixion. German legend states that Jesus cursed it during his flight into Egypt for similarly failing to acknowledge him. In Russia it is identified as the tree on which Judas Iscariot hanged himself after he had betrayed Christ.

It is quite appropriate that it is dedicated to Mercury who was the messenger of the gods. In ancient times the wind was also considered to be the messenger of the gods and anything which was particularly in tune with the wind was seen as sacred.

Magic and Lore

The aspen is the smallest member of the poplar group. It represents the virginal goddess and is emblematic of the new moon, filled with potential.

An alternative reason given for the trembling leaves is that the aspen has the most acute hearing of any tree species. Standing between the realms of faerie and man it trembles because of all it hears. Its powers of communication can be transferred to mankind. To receive the gift of eloquence in speech simply place an aspen leaf beneath your tongue.

Its connection with the crucifixion has led to it being considered unlucky. In the Hebrides it would never be used as a building material or when making tools. Although it is generally considered unlucky, there is also a superstition that planting it will keep thieves away from land on which it is planted.

Medicinal

This shivering tree can be used to cure fevers, according to the Doctrine of Signatures. A lock of the patient's hair should be pinned to the tree, or their nail clippings put into a hole in the tree, with the words:

Aspen tree, aspen tree,
I prithee to shake and shiver instead of me.

The person must then return home in silence. This is very similar to the spell using the black poplar.

Another fever cure is to drill a small hole in the tree at the dead of night, and place a few nail clippings in it. The hole must then be plugged to prevent the fever escaping. This should all be carried out at dead of night to be successful.

PURSLANE

Botanical Name: *Portulaca oleracea var. sativa*
Family: Portulacaceae
Gender: Feminine
Planet: Moon
Element: Water

Myth and Legend

Purslane is a potherb of great antiquity and a herb of good fortune with good protective attributes. It may be strewn on the floor around a bed to protect those who sleep in it from magic or placed on the bed to prevent the sleeper having nightmares. Carrying the herb is supposed to bring good luck and ward off evil. It protects the bearer from being struck by lightning or injured by gunpowder. It is due to the latter that soldiers would carry it as they went into battle. Strewing the herb about the home was supposed to bring happiness into the house.

It can also be used to treat excessive lustfulness, it 'doth extynct the arbor of lassyvyousnes, and doth mytygate great heate in all the inwarde partes of man. (Andrew Boorde, *Dyetary)*

Country weather watchers claim that when purslane comes into flower it is sure to rain.

Medicinal

Perhaps the oddest belief relating to this plant is that it can be used to help fix loose teeth.

SILVERWEED

Botanical Name: *Potentilla anserina*
Family: Rosaceae
Common Names: Bread and butter; bread and cheese; goose grey; goose tansy; goosewort; prince's feather; silvery cinquefoil; trailing tansy; wild agrimony; wild tansy
Planet: Venus

Myth and Legend

Silverweed is one of the most widespread of species of *Potentilla*. In times past it was considered a hardship food. Children particularly would dig up the roots to eat them. It was not only a food for people but was also fed to livestock. The name 'anserina' comes from the Latin *anser* meaning a 'goose' because geese were fond of it.

Medicinal

Distillations of silverweed roots were used to aid the complexion. It was supposed to be good for the removal of freckles, pimples and spots. In Leicestershire formulations of the herb were recommended to prevent pitting of the skin following small-pox. Infusions of the herb were also suggested for easing sore throats.

TORMENTIL

Botanical Name: *Potentilla erecta* (Synonym: *Potentilla tormentilla*)
Family: Rosaceae
Common Names: Aert-bark; biscuits; bloodroot; earthbank; English sarsaparilla; esquinancee; ewe daisy; five fingers; flesh and blood; *herb de paralysie*; septfoil; shepherd's knapperty; shepherd's knot; thormantle
Gender: Masculine
Planet: Sun
Element: Fire
Deitiy: Thor

Magic and Lore

Drinking infusions of the herb is said to be a good protection against evil. It may be taken by mediums in order to prevent permanent possession by evil spirits. The roots can be hung up in the house to guard against evil.

Giving infusions to a lover will ensure that they are always faithful, and carrying it will attract love. However, if your lover has left you and you want them back, burn a bunch of tormentil at midnight on a Friday – this will cause them to be so fretful that they will soon return to you.

Medicinal

In folk medicine lotions containing tormentil are used in the treatment of sores, ulcers and sunburn. Boiled in milk it can also be taken as a remedy for diarrhoea, and it is recommended for use in the treatment of rheumatic fevers and paralysis.

CINQUEFOIL

Botanical Name: *Potentilla reptans*
Family: Rosaceae
Common Names: Cramp weed; five finger blossom; five-finger grass; five fingers; goose grass; goose tansy; moon grass; pentaphyllon; silver cinquefoil; silver weed; sunkfield; synkefoyle
Gender: Masculine
Planet: Jupiter
Element: Fire
Meanings: Beloved daughter; maternal affections; sisterly affections

Potentilla means 'little powerful one', which may be an allusion to the power of the plant in magic or as a herbal remedy.

The five parts of the leaf are said to represent love, health, power, wisdom and wealth; carrying it is said to grant all of these. It has been used as a protective herb since the times of the ancient Egyptians. It may be hung on doors and doorframes, or placed on a bed, as a protection against all the forces of evil. Washing the forehead and hands nine times with an infusion of cinquefoil will clean away any curses that have been cast against you. A small pouch of the herb hung at the end of the bed will ensure a good night's sleep.

To find a cinquefoil which has seven leaflets is thought to be every bit as fortunate as finding a four-leafed clover. If this lucky amulet is put at the side of your pillow before going to bed it will ensure that you dream of your future true love.

In the Middle Ages cinquefoil was much used in witches' spells, especially in love magic. One old recipe for witches' ointment calls for 'the juice of five finger grass, smallage and wolfsbane is mixed with the fat of children dug up from their graves and added to fine wheat flour'.

Carrying cinquefoil is supposed to bestow the gift of eloquence on the bearer when he or she is seeking favours from those in authority. It will usually ensure that the favour sought is granted. It was once also used in a special bait for anglers: 'corn boiled in thyme and marjoram water, mixed with nettles, cinquefoil and the juice of houseleek'.

AURICULA

Botanical Name: *Primula auricula*
Family: Primulaceae
Common Names: Bear's ears; dusty miller; mountain cowslip; wild auricula
Meaning: avarice (scarlet); 'Impune not me' (green edged); painting; pride; 'Wealth is not always happiness'

Auricula was known to the Romans and was introduced into Britain in 1570 by Flemish wool weavers. The botanical name is derived from *primus*, meaning 'first', and *auris*, meaning 'ears'. It was formerly called *Auricula ursi*, translating as 'bear's ears', because of the shape of the leaves.

Gerard, in his herbal of 1597, claims that it is native to Switzerland and advises that mountaineers should use it as a treatment for vertigo. He says, 'It preventeth the losse of their best joints (I meane their neckes) if they take a root hereof before they ascend the rocks or other high places.'

COWSLIP

Botanical Name: *Primula veris*
Family: Primulaceae
Common Names: Arthritica; artetyke; buckles; bunch of keys; cars lope; cooslop; cove-keys; cower-slop; cowflop; cow-paigle; cow slap; cow slop; cow slup; cow's

mouth; cow strippling; cow stropple; cow strupple; creivel; crewel; cuckoo; culver keys; drelip; faerie bells; faerie cup; faeries' basins; faeries' flowers; freckle face; golden drops; herb Peter; hodrod; holrod; horse buckle; key flower; key of heaven; lady's bunch of keys; lady's finger; lady's keys; lippe cuy; long legs; May flower; milk maidens; odd rod; our lady's keys; paigle; palsy wort; paralysio; password; peagles; peggle; plumrocks; pretty mulliens; racconals; St Peter's herb; St Peter's keys; tisty-tosty; tosty
Gender: Feminine
Planet: Venus (in Aries, according to Culpeper)
Element: Water
Deity: Freya
Meaning: Charm; comeliness; divine beauty; early joys; happiness; pensiveness; rusticity; winning grace; winsome beauty; 'You are sweeter even than this spring flower'; 'Your grace and beauty have charmed me completely'

Myth and Legend

An old Norse myth dedicates the cowslip to Freya, wife of Odin, goddess of happiness and of sexual delight, as it was said that carrying the plant would allow the bearer to gain admission to her treasure palace. The shape of the pendant flowers suggests a bunch of keys and the tradition remains that carrying a bunch of cowslips will help one find treasure. With the spread of Christianity the plant became associated with St Peter, as the key bearer, and with the Virgin Mary.

Legend tells that one day St Peter, the Rock of the Church, became aware of a rumour that some souls were entering heaven by a back door, and not by the 'Pearly Gates', to which he held the keys. He became so agitated at the irreverence of such actions that he dropped his keys. They fell to Earth and became embedded in the soil, taking root. They grew into the cowslip and hence this little flower is said to be the 'key' that opens rocks and reveals hidden treasure.

Magic and Lore

The common name may derive from the Saxon *cuslippe* as the scent has been likened to a cow's breath. Another option is that it is a corruption of 'cow's leek', leek being derived from the Saxon *leac*, meaning 'a plant', thus 'cow's plant'. Another suggestion is that it is a corruption of 'cow pat'.

Cowslips have long been associated with the world of faeries and the interactions of the 'little people' with humans. Faeries are said to love and protect them and to hide in the flowers when frightened. Carrying the flowers is generally said to bring good luck, and one belief, clearly linked to the association with Freya, claims that it will enable the bearer to find faerie gold. The plant is credited with the ability to split rocks containing faerie treasures.

In Lincolnshire and Northamptonshire, it was traditional on Whitsunday for girls to make balls of cowslip flowers, sometimes called a 'tisty tosty', and throw them to each other. As they did so they would recite:

Cowslip ball tell me true,
Who shall I be married to?

Each girl in turn would then call out the name of one of the young men of the village. When the ball was dropped, the last name called indicated who would be the husband of the girl who threw it. Alternatively the girls might call out the possible careers of their future spouses, much as is often done with cherry stones. Hence,

Rich man, poor man, beggar man, farmer,
Tinker, tailor, plough boy, thief.

A variation of this method of divination requires that the 'tisty tosty' be threaded onto a string, the ends of which were held by two girls several yards apart. The ball would be sent to and fro along the string whilst the boys' names were called out. The name the girl called when the ball came to her would be the person she would wed. This could be done across a street, as referred to in the poem 'Clowslips' by Northamptonshire poet John Clare.

For they want some for tea and some for wine
And some to make a 'cuckaball',
To throw across the garland's silken line
That reaches o'er the street from wall to wall.

One of the oddest superstitions relating to cowslips is that, if they are planted upside down on Good Friday, they will flower red or even become primroses. Another way of ensuring that they flowered red was said to be to feed them with bull's blood. Polyanthus, it is claimed, can be obtained by planting the cowslips upside down in soot. These superstitions possibly originate from the fact that primulas tend to hybridize easily. The results of having a mixture of different species close to each other would be a variety of hybrids growing from seed.

Cowslips were said to be the favourite flower of nightingales and these birds were said only to frequent places where they grew.

They could be used to attract and deter. If a woman washed her face with milk that had been infused with cowslips, her beloved would be drawn closer to her. And a cowslip placed beneath the front doormat would discourage anyone from calling.

Medicinal

Cowslips have been used as a sedative as well as for the treatment of cramps, rheumatic pains and paralysis, hence the common name of palsy wort. Wearing or carrying the flowers is said to restore and preserve youthfulness. An ancient belief also claimed that they had a magical property that enhanced the complexion, and the scent was supposed to have healing properties and be especially good for calming the nerves. Wine in which cowslip flowers have been soaked is supposed to be useful for the treatment of insomnia, amnesia and migraines; cowslip tea, drunk before retiring to bed, was also recommended as a treatment for insomnia. Eating the flowers was said to strengthen the brain.

PRIMROSE

Botanical Name: *Primula vulgaris*
Family: Primulaceae
Common Names: Buckie-faalie; butter rose; darling of April; early rose; Easter rose; first rose; golden rose; golden stars; Lent-rose; May-flooer; May spink; ole voaldyn; paigle; password; pimrose; simmerin
Gender: Feminine
Planet: Venus
Element: Earth
Meaning: 'Believe me'; dawning love; early youth; fears; grace; inconstancy; innocence; lover's doubts; 'My heart is beginning to know you'; sadness; silent love; pensiveness; 'Your inconstancy saddens me'

Myth and Legend

In mythology the primrose, called *Paralisos*, referred to the son of Priapus, the god of reproductive powers, and Flora, the goddess of flowers. The story tells how Paralisos pined away and died after the loss of his lover Mericerta. His parents preserved him by transforming him into a primrose.

Magic and Lore

The primrose is the 'prime rose', the first rose of spring. It has long been seen linked to the faerie folk. Growing both red and blue primroses will attract faeries into the garden and, because they are beloved of faeries they should never be allowed to languish and die, as it will earn the faeries' enmity. It was also supposed to have the unique power of making the invisible visible; In Ireland and Wales it was said to be used by faeries to make them invisible, and so could be used to make them visible again. At one time children would eat the flowers to see the little people. Another way to see faeries was to carry a bunch of primroses and to peer over the top of them.

It was claimed that wearing the right number of primroses (sometimes said to be five) would act as the key to the faerie kingdom. A variant of this says that, to open the way in, you must touch a 'faerie' rock with the right number of primroses, but that to use the wrong number would cause great misfortune. Elsewhere, touching a 'faerie' rock with a single primrose would open up a pathway leading to the faerie world but touching the wrong rock would open a path leading only to doom. The belief that primroses were the keys to access faerieland was also common in Germany. A German folk tale relates how a young girl found a doorway covered with flowers. When she touched the door with a primrose it opened, revealing the way to an enchanted castle. The German name for the primrose is *schusselblume*, 'key flower'.

If a primrose posy was hung above the door or left on the doorstep it was an invitation to the faeries to enter the home. Consequently the house, and those living in it, would be blessed by them. Scattering primroses outside the doors would keep them away. On farms, where a bunch of primroses was left in a cowshed, faeries would not enter and steal the milk.

Primroses were believed to be protective herbs that would keep away witches and protect against their malice. At Beltane and May Day, when witches were thought to be active, bunches of the flowers would be placed in cowsheds across the Isle of Man and primrose balls might be hung on cows' tails. An abundance of primroses was supposed be an excellent guard against evil spirits and protect against all adversity. In Ireland the flowers were used to decorate churches at the beginning of May and placing a primrose on the doorstep on the eve of May Day would prevent witches from entering.

Carrying or wearing the flowers also had different benefits. Carried by women they would attract love, and worn they would cure madness. If they were sown into a child's pillow they would ensure the child's undying devotion and loyalty.

According to Welsh tradition it is an indication of impending misfortune if the primroses flower in June, and in Cheshire it was said that it was an omen of death if a primrose flowered in autumn or winter. It is also said to be unlucky to give a single bloom as a gift as this will bring about the death of a member of the family. Amongst farming communities it was said that it was courting misfortune to pick a single primrose at the beginning of spring. The number of primrose flowers taken into the house would dictate the number of eggs hatched by each chicken or goose during the coming season. Malicious neighbours might give infants a single bloom to take home just out of spite. Elsewhere in Britain it was said to be unlucky to take primroses into the home before April for much the same reason.

In common with cowslips there are various superstitions relating to the variations in the colour of the flowers. It is suggested that planting them upside down will result in them producing red flowers, whilst planting them in cow dung will cause the flowers to be purple.

In 1883, the Primrose League was formed to preserve the memory and support the political beliefs of the former Prime Minister Benjamin Disraeli, Lord Beaconsfield, the primrose having been his favourite flower. The league encouraged the wearing of primroses on Primrose Day, 19 April, the date of Disraeli's death.

The blue primrose is the birthday flower of 7 May and the red the flower of 22 October.

A dream of primroses suggests happiness in a new friendship.

Medicinal

Hundreds of years ago, these plants were grown as much for their medicinal and sweetening qualities as for their decorative effect. In folk medicines they were used as a remedy for a number of ailments. The Romans used them to treat malaria, and early herbalists recommended use of the root to make an expectorant. No doubt it was the yellow flowers that led to their use to treat jaundice. The primrose was also used for the treatments of ailments such as ringworm. It has been used, boiled with lard, for treating cuts and minor wounds. The sap from the stem could be rubbed onto the face to remove spots, freckles and other blemishes. A tincture of primroses could be made to treat restlessness and insomnia. In fact simply placing a primrose onto the pillow was said to be a cure for insomnia. Primroses were also used to treat a similar range of problems in livestock, typically fits, paralysis, rheumatism and worms.

SELF HEAL

Botanical Name: *Prunella vulgaris*
Family: Labiatae
Common Names: Black-man's-flowers
Planet: Venus

In Hampshire children used to be warned not to pick these flowers as they belonged to the Devil. Anyone who ignored this advice would annoy the Devil who, in the dead of night, would come and carry them away.

Even though the name might suggest that this plant has wide usage as a medicinal herb there is little evidence to support this. In folk medicine it has been gathered at Midsummer and dried for use in the treatment of chest complaints.

WILD PLUM

Botanical Name: *Prunus americana*
Family: Rosaceae
Common Names: August plum; goose plum; hog plum
Gender: Feminine
Planet: Venus
Element: Water

In North America the Dakota Indians use the shoots of the wild plum as prayer sticks. They were gathered, the bark peeled off and the sticks painted. An offering for the gods, usually of tobacco, was tied at the top. The sticks were then used in healing rituals and set up about an altar, or simply stuck into the ground, outside the sick person's home.

APRICOT

Botanical Name: *Prunus armeniaca*
Family: Rosaceae
Common Name: Apple of Armenia
Gender: Feminine
Planet: Venus
Element: Water
Meaning: Doubt

Apricots were introduced to Britain in 1524, by one Woolf, a gardener to King Henry VIII. They originate from China, where they have long been used in traditional medicine. One medicine, called 'apricot gold' was supposed to be able to extend a person's lifespan to as long as seven hundred years. It was made of kernels from apricots collected from trees growing in auspicious situations; those fruits that produced double kernels were especially favoured. More commonly Chinese women would use the flowers and fruits in the treatment of 'women's ailments'.

The sweet fruit is supposed to give a sweetness of disposition to anyone who eats it. The leaves and flowers can be added to charms and sachets to attract love, and even the kernels can be carried to bring love.

To dream of apricots is a good omen indeed. It signifies good health, a speedy marriage and a very successful life.

CHERRY

Botanical Names: *Prunus avium*; *Prunus cerasus*
Family: Rosaceae
Common Names: Bird cherry; brandy-mazzard; crab-cherry; gaskin; gean; hawk berry; mazzard; merry; merry-tree; sweet cherry; wild cherry
Gender: Feminine
Planet: Venus
Element: Water
Deity: Vertumnus
Meanings: Good education; increase; spirited beauty (tree); hope; increase; insincerity; 'May our friendship wax firm and true' (blossom)

Myth and Legend

In Christian tradition Mary was asked by Joseph to pick a cherry for her. He replied that she ought to ask the father of her unborn child. Jesus, hearing this from inside the womb, caused the tree to bow down so that she could pick the fruits herself. Seeing this, Joseph was aware of the truth of the child's parentage and was filled with remorse for his harsh words.

Magic and Lore

Most people will have used cherries in a form of divination at some time. To discover how long you can expect to live run around a cherry tree which is in full fruit and then shake it violently. The number of cherries that fall from the tree will show the number of years that you still have to live. A cherry tree in full flower would be shaken, with the words:

Cuckoo, cherry tree,
Good bird tell me,
How many years before I die?

The next burst of calls from a cuckoo would provide the answer.

As with many other flowering fruit trees, if the plant flowers out of season, in the late summer and early autumn, it portends the death of both the tree and its owner.

There are a number of divinatory rhymes that make use of cherry stones to discover what the future holds. For example if a young woman wishes to learn when she will wed she has only to count off the cherry stones remaining after her meal to the words, 'This year, next year, sometime, never.' And if someone wished to discover what profession they would follow they might count the stones off to the rhyme:

Tinker, tailor, soldier, sailor
Rich man, poor man, beggar man, thief.

A woman might use a similar method to learn the profession of her future husband. She would count off the stones saying:

Army, navy, peerage, trade,
Doctor, divinity, law.

Considering the range of potential careers the list is somewhat limited.

The cherry crop is also an indicator of good fortune. In Somerset it was traditional that, on hearing the first cuckoo in spring, people would turn over the money in their pockets saying:

Cuckoo, cherry tree,
Catch a penny and give to me.

In other regions it is said:

A cherry year, a merry year,
A plum year, a dumb year.

Various methods have been recommended to bring a cherry tree into fruit that has, for some little time, failed to bear. One is to hammer an iron nail into the trunk. Another suggests boring a hole into the trunk and inserting a peg of green oak, which is supposed to threaten the tree with Thor and Jupiter, that is thunder and lightning. The oddest way might seem to be to hang bottles on the branches, but in fact, causing the branches to bend over would, indeed, help to bring a tree into fruitfulness. To ensure that a young cherry tree will always be fruitful the first fruits that it produces should be eaten by a woman who has recently given birth to her first child. Similar superstitions can be found linked to other fruit trees.

Cherries may be used in spells and charms to gain the affection of others. A Japanese spell requires that a single strand of hair be tied on a cherry tree. For a more complex charm it is necessary to gather as many cherry stones as your age in years. Beginning at the night of a new moon drill a hole in each, one each night until the moon is full (the most which can be drilled in any one month is therefore fourteen). Once all have been drilled it is necessary to wait until the next full moon before threading them onto a red cord which is tied about the left knee every night for the next fourteen nights and removed in the morning.

A dream of cherries is an indication of disappointments and inconstancy in life, and unhappiness in all love affairs. Dreaming of picking and throwing cherries is a portent of a financial loss.

PLUM

Botanical Name: *Prunus X domestica*
Family: Rosaceae
Gender: Feminine
Planet: Venus
Element: Water
Meanings: Fidelity; independence; promises to keep

Plum trees are said to have some protective properties. Branches can be cut and hung over doors and windows to prevent any evil entering. If a tree flowered in December it was a sign that there would be a death in the household during the winter months.

The fruits of the plum are said by some to have aphrodisiac properties; consuming them not only inspires love but also maintains it. Swallowing the plum stone, according to various warnings, may cause a plum tree to grow in your head or stomach.

To dream of plums is a warning that you should expect ill health, losses, infidelities and marital difficulties. It also suggests that there will be the sickness of a friend or relation, and a death in the family.

In various parts of Britain there is a saying which links the size of the wheat harvest with that of the plum crop:

A good wheat year,
A fine plum year.

Trees may be encouraged to bear fruits in a number of ways. Hanging old buckets on the branches might not look very decorative but, as with cherries, it should work. Considerably less likely to be effective is a suggestion that in spring you should throw salt on the roots.

ALMOND

Botanical Name: *Prunus dulcis* (Synonyms: *Amygladus dulcis*; *Amygladus communis*)
Family: Rosaceae
Common Names: Aaron's rod; wake: wakeful tree
Gender: Masculine
Planet: Mercury (Jupiter, according to Culpeper)
Element: Air
Deities: Artemis Caryatis; Attis; Carmenta; Hermes; Mercury; Metis; Thoth
Meanings: Heedlessness; hope; 'I hope to meet you again'; indiscretion; stupidity; thoughtlessness; 'Your friendship is pleasant'

Myth and Legend

The story of the origin of the almond, in classical mythology, is told in the tale of a young man named Demophon. He had fought at the siege of Troy and was returning home when his ship was wrecked on the coast of Thrace. He was taken to the court of Sithon, the King of Thrace, and whilst there fell in love with the king's daughter, Phyllis. In due time the young couple's wedding was arranged but before it could take place news arrived of the death of Demophon's father, Theseus of Athens. He therefore, had to return to Athens, but before he left he set a date for his return. Unfortunately, the various affairs of state that needed to be dealt with took far longer to complete than he had expected, and he was unable to return to Thrace by the date that he had set with Phyllis. However, she went out to the seashore on the appointed day, and the next nine days, to watch for his return. With the passing of each day her grief became more intense until, finally, convinced that Demophon would not return she hanged herself. The gods intervened at this point and transformed Phyllis into an almond tree. Demophon had not forgotten his love and after three months was able to return to Thrace, whereupon he learned what had happened. To mark his undying love he offered a sacrifice at the almond tree. The spirit of his lost love responded by causing it to burst into blossom spontaneously. The Greek name of the plant is *phylla*, from Phyllis.

Another classical myth claims that Attis (Atys) was born after his mother, Nana, who was a virgin, placed an almond into her bosom (some alternative versions of the story claim that it was a pomegranate seed). At his death violets sprang up from his spilt blood and an almond tree grew out of his body. This tree bore bitter almonds as a symbol of grief.

In the biblical stories we are told that the staff of Aaron, when it rooted, sprouted and brought forth almond blossoms. Moses, in order to establish which of the tribes of Israel would furnish the high priest of the temple, placed twelve rods in the tabernacle, one representing each of the tribes. Next day the rod representing the tribe of Levi was seen to have sprouted and so this tribe was set aside to furnish the priesthood. The rod was formed from almondwood.

Magic and Lore

The Hebrew name for the almond is *shakad*, meaning 'awakening', as it grows quickly and blossoms early in the year in Israel. In scripture it is called one of the best of fruit trees in the land of Canaan. The making of staffs and wands from almondwood is a well-founded tradition, as can be seen from the biblical references. The Pope's staff is likewise made from almondwood. The shape of the nut is symbolic of the intersection of the circles of heaven and earth, of spirit and matter. It therefore encloses God, Christ and the Virgin in medieval art. Hence, the almond is a symbol of the Virgin Mary and of the priesthood.

The Romans called the almond 'Greek nut' because of the Greek custom of eating bitter almonds before meals in order to promote drinking afterwards, the theory being that the drying nature of the almonds would dispel the moisture of the drink. A similar belief can be found expressed in *Gerard's Herbal*: 'Five or six [almonds], being taken fasting do keep a man from being drunke.'

It is said that eating them will grant wisdom. Climbing an almond tree is supposed to be a guarantee of success in business activities, whilst carrying almonds in your pockets is supposed to lead you to treasures.

Almonds have been valuable for weather forecasting since the time of the Romans.

> Mark well the flowering almond in the wood,
> If odorous blooms the heaving branches load,
> The glebe will answer to the sylvan reign,
> Great heat will follow, and large crops of grain:
> But if a wood of leaves o'er shade the tree
> Such and so barren will the harvest be.
>
> – Virgil, *The Georgics*

Dreaming of eating almonds indicates that you will soon be going on a journey. If they are sweet ones the journey will be a profitable one, but if they are bitter then the journey will be unpleasant.

Medicinal

In folk medicine eating almonds has been suggested as a remedy for fevers.

CHERRY LAUREL

Botanical Name: *Prunus laurocerasus*
Family: Rosaceae
Common Name: Laurel

There is a good deal of confusion about this plant and the bay laurel because of their common names. Bay laurel was used in ancient Greece to crown victors and poets. In modern times it is most usually wreaths of the cherry laurel with which victorious sporting heroes are bedecked and it is a wreath of cherry laurel leaves that is traditionally placed on Nelson's flagship, the *Victory*, on the anniversary of the Battle of Trafalgar.

Over the years some of those superstitions that were originally attributed to bay have inevitably been applied to the cherry laurel. One method by which you might discover whether someone loves you may have originally made use of bay but is now most often applied to this laurel. It requires you to write the name of your sweetheart on a laurel leaf with a pin and then place it next to your heart. If the writing turns red then the person named returns your affections. If it turns black they do not.

BIRD CHERRY

Botanical Name: *Prunus padus*
Family: Rosaceae
Common Name: Hack berry
Meaning: Hope

In the north-east of Scotland the bird cherry was often thought of as a tree of witches and witchcraft. The wood was, therefore, rarely used, even for making walking sticks.

PEACH

Botanical Name: *Prunus persica*
Family: Rosaceae
Gender: Feminine
Planet: Venus
Element: Water
Deities: Harpocrates; Horus; Vertumnus
Meaning: 'I am your captive' (blossom)

Myth and Legend

The folklore of the peach shares a good deal with that of the apricot. Like the apricot it is seen as a Chinese symbol of longevity and immortality because the peaches which grow in the celestial gardens are said to flower only once in every three thousand years.

The origin of the Chinese flower calendar comes from a story that Canopus, the god of longevity, fed a peach of immortality to Ho Hsien-Ku, the daughter of a simple shopkeeper. Having gained immortality Ho Hsien-Ku became the deity associated with flowers. A different flower is taken as sacred in each month in deference to her.

Magic and Lore

The tree is supposed to have protective properties. Hanging the kernel as a pendant about the neck of a child will draw out any illnesses and guard them from evil demons. A sprig of peach blossom hung over the doorways and entrances to a property on a silk, cotton or linen thread will protect the house, preventing any evil from entering, even when the blooms have withered away. The twigs have been used to make magic wands and divining rods, and the blossoms used in decorations at the Chinese New Year celebrations. As should be expected for a plant associated with longevity, for the tree to flower out of season is an indication that there will be a death.

In Japan it is said that eating peaches will increase fertility. They have been seen as an aphrodisiac inducing love or lust in those who consume them, and might be given as a gift to gain the recipient's affection. Eating peaches was also said to impart wisdom. A dream of peaches signifies contentment, pleasure, wealth, reciprocated love, good health and many surprises.

There are various pieces of advice relating to the cultivation of peaches. It is claimed that there are male and female trees, which can be identified by the width of the leaves, the male trees having much wider leaves than the female ones. Those that have small flowers will produce big fruits, whilst the trees with big flowers will only produce small fruits. When planting a tree old buckets and shoes must be placed in the planting pit.

The weather will inevitably play its part in how well the tree crops. It will have no fruit if there is thunder on 12 February, for instance, but sleet in December will ensure a good crop. If the wind blows from the south on Christmas Day there will be high quality fruits and you need not worry about the frost damaging the flowers in the period before the full moon as 'peach trees in bloom will never freeze in the light of the moon'.

Medicinal

Carrying a piece of peachwood will ensure a long life and may even bring immortality.

BIRCH BARK CHERRY

Botanical Name: *Prunus serrula*
Family: Rosaceae
Common Names: Japanese cherry; Tibetan cherry

The birch bark cherry is a striking small tree with bright copper-coloured bark. In Japan it has become the symbol of remembrance for the thousands of people who died in the atomic bomb attacks at Hiroshima and Nagasaki.

BLACKTHORN

Botanical Name: *Prunus spinosa*
Family: Rosaceae
Common Names: Mother of the woods; sloe; wishing thorn
Planet: Saturn
Element: Fire
Deity: Eris
Meanings: Difficulties; obstacles; 'Our path is beset with difficulties'

Magic and Lore

The blackthorn represents the power of the invisible and visible worlds. It bestows the strength to resist adversity and overcome defeat, and the energy to control or ward off supernatural powers. It can enable the user to overcome all resistance to their will.

It has been called the 'sister tree' of the hawthorn. Like the hawthorn it was once commonly used in decorations for the May Day celebrations, the pagan Beltane festival. The phallic maypole was topped off with a wreath of intertwined blackthorn and hawthorn, called 'the mother of the woods'. Like hawthorn, it should never be taken into the house as it will result in a death in the family or bring about misfortune for the whole household.

All thorn trees, especially the hawthorn and blackthorn, are revered by faeries and said to be suitable meeting places for them. Using wood from a faerie thorn on top of a faerie mound is supposed to force the little folk to return a stolen child. As with other trees, the blackthorn has its own faerie guardians. The blackthorn guardians are called the 'luantishees'.

The twigs of blackthorn have been used to make divining rods and 'wishing rods', a sort of all-purpose magic wand, which might be used during the ritual of wassailing apple trees but could have a far darker use. They have, over the centuries, gained a grim reputation, which seems to have come from the use of a 'pin of slumber' being used by those wishing to usurp power by removing those in their way. The spines on blackthorn are particularly long and sharp, and so ideal for the purpose. The wound inflicted by the thorn can turn septic but the addition of a little poison on the tip would make sure of the job. In many of our faerie tales there is a wicked witch who causes the beautiful princess to fall into a long sleep by pricking her with just such a thorn or a needle.

In Ireland the blackthorn was the traditional wood for making shillelaghs, walking sticks or wands. In legend these could be used to escape from giants; if a twig from a magical blackthorn was thrown between the hero and pursuing giant, a great thorn forest would instantly grow up, and the giant could not penetrate it. Furthermore a scratch from the spines of the blackthorn was thought to cause blood poisoning.

In the story of Sleeping Beauty these two aspects of the blackthorn come together. The wicked witch uses the 'pin of slumber' to cause the princess and all the others in the castle to fall into a long sleep. The castle is then surrounded and cut off by a deep and impenetrable blackthorn forest through which the hero has to cut his way.

At the time of the great witch trials in Britain and Europe it was claimed that some of those accused of being witches were guilty of using 'black rods', carved blackthorn sticks and blackthorn wands with thorns fixed into the end, to cause miscarriages or other harm to others. If convicted of witchcraft the witch might be burned at the stake and blackthorn was often used as kindling for the fire. This was seen as the final insult to the poor victim.

It seems somewhat paradoxical, given its dark side, that the blackthorn has also been used as a protection against evil. Just as with other species of *Prunus,* sprigs of the wood placed over doorways and windows will prevent any evil from entering the property. It may also be carried as an amulet against demons and evil spirits.

In addition to its ominous reputation it also enjoys a number of holy associations. It has been claimed that the crown of thorns placed on Christ at his crucifixion was formed from blackthorn twigs, and because of this it was never taken into churches. In common with the Glastonbury thorn, it has been said by some to flower at midnight on old Christmas Day.

Its use in fertility rites has already been illustrated by its use at May Day and in apple wassailing. In parts of Worcestershire it was traditional to bake branches in ovens before carrying them out to the fields, where they would be burned and the ashes scattered across the earliest sown cereal crops. In Herefordshire a wreath of blackthorn was made on New Year's Day and then scorched over the fire before being decorated with mistletoe and added to the other decorations about the house.

The flowering time of the blackthorn has been taken as a guide to when barley should be sown:

> When the slae [sloe] tree is as white as a sheet
> Sow your barley, wither dry or wet.

Or:

> When the blackthorn blossom is white
> Sow your barley day or night.

However there is often a period of cold weather following the blackthorn flowering, the so called 'blackthorn winter'. The short spells of warm weather in early spring will bring the blackthorn 'hatch', the flowering of the tree, but it is often recommended that nothing tender should be planted until after the blackthorn winter which follows.

Medicinal

The juice from green sloes was once used as a purgative and recommended for the treatment of colic. The ripe fruits are, of course, used to flavour sloe gin.

FERNS

Botanical Name: *Pteridium* spp.
Family: Polypodiaceae
Common Names: Devil's brushes
Gender: Masculine
Planet: Mercury
Element: Air
Deities: Laka; Puck
Meanings: Sincerity; stormy passions; 'You have deprived me of your heart and left mine an a wilderness'

Myth and Legend

A legend from eastern Europe relates that on St John's Eve one year a farmer found that his cattle had gone. He set out to try and find them and passed through a forest just as the fern undergrowth was coming into flower. A flower fell from the fern onto his shoe and he instantly knew where to find the missing cattle. With the fern in his shoe he was able to see where all kinds of treasures where hidden. He immediately went home to tell his wife so that they could go together and collect up all the wealth he had been able to see. Before they set out he quickly went to change his stockings because they had become wet during his search for the missing livestock. The flower fell from his shoe and he instantly forgot everything.

Magic and Lore

The simple fact that ferns do not flower and produce seeds was not always well understood. In the past this lack led to the belief that the plant, and its seeds, held magical powers. Fern was said to flower only at midnight on summer solstice (or on St John's Eve at the exact time of the saint's birth) when it would produce golden blooms. From these blooms would drop the seeds which, if collected properly, would enable the bearer to become invisible at will.

> But on St John's mysterious night,
> Sacred to many a wizard spell,
> The hour when first to human sight
> Confest, the mystic fern seed fell.

The spirits tried to prevent mere mortals from gathering the seeds, and the wonderful fern flowers could give great wealth. It was said that if a maiden spreads a cloth beneath a fern flower red gold would fall into it. To maintain one's wealth one could mix a little fern seed amongst some money, and the amount of money would never decrease, regardless of how much you spend. Another, and possibly easier, way of attracting wealth was 'fern watching'. Selecting a site that had a good covering of fern one must sit there at midnight and wait. Puck might then appear and give one a bag of gold. A similar superstition was that the seeds of ferns could only be gathered on Christmas night and that gathering the seed then forced the Devil to appear and bring a bag of money.

The ability of fern seeds to bring treasures to light is repeated in the tradition from Brittany that if fern seeds are gathered at midnight on Midsummer's Eve, and kept until Palm Sunday, they can be scattered in the general area of where you believe treasure might be concealed in order to bring it to light.

Great care should be taken in the gathering of fern seeds.They must be collected in total silence at midnight on Midsummer's Eve. The golden fern blossoms, which are only visible at that time, can then be bent over with a hazel twig to allow the seeds to fall into a pewter dish. Some authors recommend a stack of twelve pewter dishes, representing the twelve apostles. The magical fern seeds would pass through some of the dishes to rest, still visible, on others. Some sources assert that the seeds should be gathered on the family Bible, but most agree that it would be dangerous to touch the plant itself. A German tale claims that a hunter obtained his fern seeds by shooting at the sun at noon on Midsummer's Day. Three drops of blood fell from the sun and these were the fern seeds.

It is said that cutting ferns will bring about a storm, and burning them, or their seeds, could cause rain. This belief led the then Lord Chamberlain, Philip Herbert, 3rd Earl of Pembroke, to write the following letter to the High Sheriff of Stafford in 1636:

> Sir:
> His Majesty taking notice of an opinion entertained in Staffordshire that the burning of Ferne doth draw down rain, and being desirous that the county and himself may enjoy fair weather as long as he remains in those parts, His Majesty hath commanded me to write to you, to cause all burning of Ferne to bee forborne, until His Majesty be passed the country. Wherein not doubting but the consideration of their own interest, as well as of His Majesty's will invite the country to a ready observance of this His Majestey's command, I rest
>
> Your very loving friend
> Pembroke and Montgomery
> Belvoir 1st August 1636
>
> To my very loving friend the High Sheriff of the County of Stafford

Ferns were believed to have protective properties and so might be added to vases of cut flowers or planted close to doorways. Being associated with thunderstorms, they were grown on roofs, or dried and hung in the house, to protect properties from being struck by lightning. The smoke caused by burning fern on hot coals was said to drive away evil spirits and all venomous creatures. There is supposed to be a particular link between ferns and snakes, which led to their use in the treatment of adder bites. A Welsh superstition, however, claimed that if you were to carry a piece of fern in your pocket it would cause you to become lost and you would be followed by snakes. Elsewhere it was said that if a traveller trod on a fern they would become disorientated and lose their way.

Another way in which the fern can be used as a protection is in the prevention of infidelity in a loved one. All that is required is that a fern frond be carried in the pocket.

In parts of Cornwall it is claimed that pixies, the faerie folk, were especially fond of ferns and could be found near them. One tale tells of a young woman who accidentally sat upon a fern. Instantly a faerie man appeared before her. He extracted a promise that she would enter faerie land and watch over his son for a year and a day. The promise was bound by her kissing the fern and reciting, 'For a year and a day, I promise to stay.'

The unfurling fern frond resembles a violin neck and so is sometimes called a 'crosier' or 'fiddle neck'. In medieval times it was called 'St John's hands', or 'lucky hands'. These were considered to be lucky amulets and were carried in pockets to ward off witches. The fronds could also be placed over doorways to protect against lightning strikes.

Ferns should never be given as gifts, as if someone accepts the gift of a fern they will never settle down in life. It is most often said that to give someone a fern is to give them sorrows.

Putting a few oats in the bottom of the pot in which ferns are to be planted will ensure that the plants will thrive. Another way to ensure this is to sprinkle them with water in which a woman has washed her clothes after her period. If the plant starts to die it may be revived by putting a spoonful of castor oil onto the roots. Care must be taken when handling the plants. Touching the tip of a frond will cause the plant to die.

Medicinal

The use of fern in folk medicines seems very limited. It has been used in the treatment of snakebites and, when gathered at the time of the waning moon, was used for treating inflammations and sprains. The sap has been claimed to be an elixir of life and so was drunk to ensure eternal youthfulness.

The first fern shoot of the spring has its own special properties. Snapping it will bring good luck, and biting on it will guard against toothache.

In the USA the Cherokee tribe believed that the uncurling shoots of the fern fronds could be used to aid the tightened limbs of arthritis sufferers to uncurl. This is a type of sympathetic magic that can be seen elsewhere in the Doctrine of Signatures.

ADDER'S TONGUE FERN

Botanical Name: *Ophioglossum vulgatum*
Family: Ophioglossaceae

The name *Ophioglossum* comes from the Greek and means 'snake-tongue'. These ferns are so called because the spore-bearing stalk is said to resemble a snake's tongue.

Magic and Lore

One odd belief relating to this fern claims that if a frond is wrapped in new wax, and then placed into a horse's left ear, it will cause the horse to fall down dead. The horse will immediately recover once the fern is removed.

BRACKEN

Botanical Name: *Pteridium aquilium* (Synonym: *Pteris aquilina*)
Family: Polypodiaceae
Common Names: Brake; pasture brake
Gender: Masculine
Planet: Mercury
Element: Air
Meanings: Enchantment; 'You enthral me'

Magic and Lore

There is a degree of confusion with bracken, as it is sometimes simply referred to as 'fern'. Some of the beliefs cited above have been applied to bracken, and much of what is said about bracken is common to other species of fern.

In some areas of England bracken has been known as 'King Charles in the oak tree', from the belief that if the stem is sliced though a clear impression of an oak tree can be seen in the cut stem. In Scotland the stem is said to bear the impression of the Devil's foot. Being considered by many to be a particularly pernicious weed, it seems most appropriate that it is regarded as one of the Devil's plants. On the other hand it was called 'fern of God' in Ireland from the belief that if the stem were cut into three sections the letters 'G', 'O' and 'D' could be found on the cut ends. Witches are supposed to abhor bracken because hidden in the stem are the initials 'J.C.' for Jesus Christ. Another source claims that when severed the letter 'X' can be seen, the initial of the Greek word for Christ. There was a custom to try to discover the initials of a future spouse by cutting the stem just above soil level. In view of all the initials that are supposed to be hidden in the stem the result might prove very interesting.

In common with other ferns, bracken seeds will enable the bearer to become invisible at will.

If you sleep with a bracken root beneath your pillow the answer to a tricky problem will appear in your dreams.

Medicinal

In common with the other ferns, bracken is used for its protective properties as well as to increase fertility, and in medicinal preparations. Decoctions of the sliced root were used in the treatment of worms and constipation. New bracken fronds were used to cause miscarriage in pregnant women and animals. Crushed bracken seeds, taken with water collected from a fern growing on a tree, was supposed to alleviate stomach ache. Given its poisonous nature, this may prove to be permanent.

FLEABANE

Botanical Name: *Pulicaria dysenterica* (Synonym: *Inula dysenterica*)
Family: Compositae
Common Names: Camels; harvest flower; Job's tears; mare's fat; middle fleabane; pig daisy
Gender: Feminine
Planet: Venus
Element: Water

From ancient times fleabane has been used as a protective herb. Together with St John's wort, wheat and capers it was placed over doorways to prevent the entry of evil forces into the home. It was also used during exorcisms to dispel evil spirits.

> Fleabane on the lintel of the door I have hung,
> St John's wort, caper and wheat ears,
> With a halter as a roving ass
> Thy body I restrain.
> O evil spirit, get thee hence!
> Depart, O evil demon.
> – Translation of *Utukke Limnûte*, tablet 'B', in R.C. Thompson,
> *Devils and Evil Spirits of Babylonians*

The name 'Job's tears' is an indication of another use. Job is supposed to have used decoctions of the plant to treat the ulcers and sores that plagued him, and it has been recommended for this purpose. It was also used in the treatment of dysentery.

The seeds have particular properties. A sure way to ensure the chastity of a loved one is to sprinkle a few fleabane seeds onto their bed sheets.

LUNGWORT

Botanical Name: *Pulmonaria saccharata*
Family: Boraginaceae
Common Names: Joseph and Mary; Joseph's coat of many colours; lady's milksile; Mary's tears; soldiers and sailors; spotted Mary; Virgin Mary's milk drops

Myth and Legend

The cream-coloured spots on the plant's leaves are said, in some regions, to have been caused by drops of breast milk from when Jesus was being fed. In other parts of Britain they are said to have been caused by the Virgin's tears as she wept at Christ's crucifixion. The flower colours, red and blue, were said to match her eyes, which were sore because she had wept so much.

Medicinal

Lungwort, as its name suggests, was used in the treatment of lung disorders. The botanical name is also derived from this use.

PASQUE FLOWER

Botanical Name: *Pulsatilla vulgaris*
Family: Ranunculaceae
Common Names: Blue emony; blue money; Coventry bells; Dane's flower; Dane's blood; flaw flower; passe flower; rut; wind flower
Gender: Masculine
Planet: Mars
Element: Fire
Deities: Adonis; Venus
Meanings: Denial; expectation; sickness; 'You have no claims'; 'You are without pretension'

The pasque flower, which is said to flower first at Easter, is often thought of as a protective plant. When it is grown in the garden it acts as a guardian for both the house and the garden, especially true with red-flowered forms. To prepare an amulet the flowers must be gathered in spring when they first appear and carefully wrapped in a red cloth. These amulets can be kept and carried in order to ward off illnesses.

Pasque flowers are said to have sprung up on battlefields where the blood of Danes (some say the Saxons) had been shed. Other myths tell that they grew from the blood of Adonis and are symbolic of his death.

POMEGRANATE

Botanical Name: *Punica granatum*
Family: Punicaceae
Common Names: Carthage apple; grenadier; malicorio; pound garnet; rimmon
Gender: Masculine
Planet: Mercury
Element: Fire

Deities: Adonis; Attis; Ceres; Damuzi; Hera; Judah; Ninib; Persephone; Rimmon; Tammuz
Meanings: Mature elegance (flower); elegance; foolishness

Myth and Legend

In classical mythology the first pomegranate is supposed to have been planted by the goddess Venus on the island of Cyprus. Elsewhere it is said that it grew from the blood of Dionysus (Bacchus) just as anemones had from the blood of Adonis. One of the stories of the birth of Attis (Atys) says that he was conceived when his mother placed a pomegranate seed into her bosom.

Myth tells of a young Scythian maiden who visited the diviners and was told that she would one day wear a crown. She was seduced by Bacchus when he promised that he would give her a crown, but after a while he tired of her and started to ignore her. The girl pined until she eventually died. Touched by guilt Bacchus transformed the maid into a pomegranate tree and placed on the fruits the crown that he had promised her during his seduction of her.

Another story tells of a man who, having lost his wife, became enamoured of Side, his own daughter. In an effort to avoid his advances, she killed herself. The gods, seeing this, transformed her into a pomegranate tree and her father into a sparrowhawk. It is for this reason that the sparrowhawk is said to avoid the pomegranate and never rest in the tree.

The pomegranate plays a very important role in the story of Persephone (Kore) as it appears in Greek mythology. Pluto, god of the Underworld, had chanced one day to see Kore (meaning 'the maiden'), daughter of Demeter, the goddess of agriculture. He was utterly besotted by her beauty and left the Underworld to travel to the throne of Zeus in Olympus to demand that she be given to him as his wife. This placed Zeus in something of a predicament. Pluto was his brother and Kore his daughter, the result of a short dalliance with Demeter. If she were given over to Pluto it would condemn her to a life in the Underworld, which would not please Demeter. Zeus, therefore, did nothing and hoped that the matter would resolve itself. Pluto became tired of waiting and kidnapped Kore one day when she was out collecting flowers. Helios, who followed the sun across the sky, saw what happened and informed Demeter, who then caused a famine, letting nothing grow until Zeus was forced to agree to intervene and persuade Pluto to release Kore.

However, during her time in the Underworld Kore had eaten seven pomegranate seeds, though she had eaten nothing else. No one who eats the fruits of the Gardens of the Dead can be allowed to return to the land of the living and so Kore could not return to the world above. To placate Demeter a compromise was struck. Kore, now Persephone, would be allowed to live on the surface for nine months each year but for the other three months she must return to Pluto and the Underworld. During her daughter's long absence Demeter still blights the world with winter, when nothing will grow.

Magic and Lore

Despite this slightly dark tale pomegranates are usually considered to be lucky and a wish should always be made before eating one. Carrying a branch of pomegranate will help one to locate hidden treasures and will attract money. When a branch is hung over a doorway it will prevent any evil from entering. A Turkish tradition says that when a woman wishes to know how many children she will have she should throw a pomegranate on the floor. The number of seeds that fall out will indicate the number of children she might expect.

Dreaming of pomegranates is an indication of good fortune and success. To a lover it indicates a faithful and accomplished sweetheart. For a married person it is an indicator of wealth, children and success in business.

Pomegranates appear as ornamentation in art, needlework and architecture. The great pillars at the entrance to King Solomon's Temple were said to be adorned with rows of them and they were depicted on the *ephod* of the Jewish high priest. The fruit was also used in some Jewish ceremonies.

Medicinal

Pomegranates have been cultivated since prehistory. The plant and its fruits have been held sacred throughout recorded history and, more often than not, have been seen as emblematic of fertility. In Christian symbolism they were said to be an emblem of hope and, because of the profusion of their seeds, plenty or fertility. The fruit appears as an important medicine in the ancient papyrus *Ebers*, the earliest medicinal papyrus. Eating pomegranates is supposed to increase fertility and the skin of the fruits might be carried for the same reason.

PEAR

Botanical Name: *Pyrus communis*
Family: Rosaceae
Common Names: Pyrrie
Gender: Feminine
Planet: Venus
Element: Water
Deities: Athene; Hera; Priapus
Meanings: Affection; comfort; 'faeries' fire'

The pear is a tree of female birth. There is a custom in Switzerland that an apple tree is planted by the family at the birth of every boy child and a pear tree for every girl. Thereafter the growth, development and health of the child is linked to the well-being of their tree. It might worry the girls to know that pears are usually thought of as slow growing trees. As one country saying puts it, 'Plant your pears, for your heirs.'

Perry, pear cider, was made and consumed by the Romans. They particularly valued it as a beverage to accompany mushrooms, largely because it was believed to be an antidote for accidental poisonings. The fruits are also claimed to be an aphrodisiac.

They were also seen as being even more powerful against the forces of evil than the rowan. In some parts, the tree was thought to be especially potent as a protection for cattle and other livestock. Branches might therefore be cut to take into the home but it was commonly considered to be unlucky to take the flower blossom into the house, as this would cause a death in the family. It seems somewhat ironic that it is also claimed that witches chose them as the centrepiece for some of their ceremonies.

In divination, if a young woman wished to know whom she would marry she had to walk backwards to a pear tree on Christmas Eve. She should then continue to walk backwards around the tree nine times, and an image of her future husband would appear before her.

If pears appear in your dreams it is a sign that you can expect happiness and riches to come to you, and a new friendship. A dream of baked pears indicates that you should expect success in business ventures. When a single woman sees pears in her dreams it is said to mean that she will marry 'above her rank'.

OAK

Botanical Name: *Quercus robur*
Family: Fagaceae
Common Names: Ackeron; aik; ake; akran; cups and ladles; cups and saucers; duir; frying pans; hatch horn; Jove's nuts; juglans; mace; macey tree; mast; ovest; pipes; Tom Paine; wuk; yackrans; yeaker; yik
Gender: Masculine
Planet: Sun (Jupiter, according to Culpeper)
Element: Fire
Deities: Allah; Aria; Artemis; Athena; Baldur; Belenos; Blodeuwedd; Cardea; Cernunnos; Circe; Cybele; The Dagda; Demeter; Dia; Diana; Dianus; Dione; Donar; Egeria; El (Middle Eastern oak god); Erato; Esus; Hecate; Hercules; Herne; Hou (oak god of Guernsey); Janicot; Janus; Jehovah; Jove; Jupiter; Mars; Mary; The Morrigan; Pan; Perkunas; Perun; Picus; Rhea; Taraa; Taranis; Teutates; Thunor; Thor; Viribius; Zeus
Meanings: Courage and humanity; 'Take heart, love will find a way' (leaves); hospitality; 'Your face and person will always be welcome at my door'

Myth and Legend

The oak was held in high esteem by the druids, but it has also been considered sacred for many centuries. It was seen as the most powerful of trees, sacred to all the

European sky gods. In ancient Greece it was dedicated to the god Zeus, 'King of the gods', and one of the most famous oracles of the ancient world, the oracle of Zeus at Dodona (sometimes called the Chaonian forest) in Epirus, was situated in an oak grove at the base of Mount Tomarus, and was one of the most hallowed sanctuaries in ancient Greece. It had originally been dedicated to Dione (Diana) until being seized by Zeus.

Legend tells that two black doves flew out of Thebes in Egypt. One landed at Ammon in Libya and the other at Dodona in Greece. Each landed in an oak tree and both brought about oracular oak cults. The origins of this legend, and of the oracular oak cults, may be found in a very similar story from which it developed. Herodotus tells us that Phoenician merchants carried away two Egyptian princesses of Jupiter. One was taken to Libya where she established the temple of Jupiter Ammon. The other was carried to Greece and took up residence in the forests of Dodona. There, at the base of a great oak tree, she established a temple to Jupiter. This was the Pelasgic oracle, the first prophesies being delivered there by a woman called Pelias, a name that means 'dove'. The Roman god Jupiter can easily be identified with the Greek god Zeus. This temple was enlarged over time and when it was fully established the number of priestesses was increased to three, known as the Peliades, and it is from this that the belief arose that sacred doves inhabited the oak tree at the base of which stood the first temple.

The priestesses would interpret the will of the god from the movement of the leaves and the sounds of a brook that bubbled up from between the roots of the great oak, an evergreen tree described as having sweet, edible acorns. Bronze gongs were hung in the surrounding trees and the sound from these was probably supposed to imitate the thunder that resounded in the surrounding mountains. Thunderstorms were supposed to be more prevalent in the Dodona region than elsewhere in Europe and so this would be a particularly suitable place for a centre for the worship of a thunder god.

The goddess Athena, the daughter of Zeus and Metis and so the embodiment of power and wisdom, placed an oak beam, made from the wood of the oracular tree, in the bow of the *Argo*, the ship used by Jason for his epic voyage. It acted as a talisman and would whisper advice to the Argonauts during their journey, warning them of impending dangers.

Some Greek and Roman writers suggest that oak trees were the first 'mothers' of mankind, a role other mythologies ascribe to the ash.

These woods were first the seat of sylvan Pow'rs
Of Nymphs and Fauns, and savage men, who took
Their birth from trunks of trees and stubborn oak.

– Virgil, *The Aeneid*

The ancient Greeks called the oak the 'mother tree' because, in their mythology, Zeus (Jupiter) had slain the giants and the oak sprang up from the body of one of them, named Rhoecus. The Greeks looked upon it as the first tree which had grown upon the Earth and which provided mankind with nourishment by way of its edible acorns.

Fed with oaken mast
The aged trees themselves in age surpassed.

This would tend to suggest that it was thought that the eating of acorns (oak mast) would ensure a long life.

Another Greek myth tells how Dionysus saved the lives of the Maenads, a group of frenzied priestesses, who were fleeing the wrath of the gods. He was somewhat responsible for their need to evade the gods as he had lured them into destroying Orpheus and his sun cult under the god Apollo. Dionysus saved them by transforming them into oak trees that were rooted in the ground.

The oak, or rather the acorn, was associated with Demeter, goddess of agriculture (in Roman myth called Ceres). The acorn was supposed to have been one of the staple foods of mankind until the goddess replaced it with corn. To mark this oak leaves were worn by participants in the Mysteries at Eleusis at the Temple of Demeter. There are echoes of this in the British tradition of the character of the Green Man in the May Day Mummers plays (see Corn).

In Celtic Myth the oak is associated with the druids. Maximus of Tyre, writing in AD 2, reported that they worshipped Zeus in the form of an oak tree. Having already said that the ancient Greeks and Romans dedicated the oak to the gods Zeus and Jupiter respectively it is no surprise that they would view the practices of the Druids as relating to one of these gods. Even the name 'Druid' may be derived from the name of the tree; it may be from a translation into Greek of a Celtic name, or a name that sounded like 'son of oak'. The ancient Gauls, when adopting Latin, rejected the Latin name *quercus* for oak, calling it *cassanus*, meaning 'sacred oak', from which comes the modern French name of *chêne*.

In Windsor Great Park stood one of the most famous, or infamous, oaks in England. During the reign of King Richard III a forest ranger named Herne was employed there. We are told that he owned two great St Hubert dogs and was held in especially high regard by the King because of his great knowledge of woodcraft and hunting. His fellow rangers were jealous and plotted his downfall. One day, when out hunting, the King was almost gored by a stag. He was saved by Herne, who threw himself between the beast and the king and took the full blow from the great animal. To all appearances he was killed until a man named Simon Urswick came on the scene. He asserted that he could restore Herne to full health in exchange for a reward. With the King's approval he cut the head from the stag and bound it over that of Herne before having the hunter transported to his hut at Bagshot Heath. The King, in gratitude for Herne saving his life, announced that if he survived, he would make him the chief keeper. The other rangers, still jealous, regretted that Herne had not died. Urswick offered them an opportunity for revenge if they would grant the first request he made. When they agreed he promised that although Herne would recover he would lose all his skills and abilities. Herne duly made a full recovery and was appointed as the King's chief keeper. However, King Richard became annoyed, when he saw that his new appointee had none of his former skill, and revoked the appointment. Herne, in his frustration and despair, hanged himself from a great oak tree, but his body mysteriously disappeared.

The next two rangers to succeed him as chief keeper also lost all their skills on taking up the appointment. They appealed to Urswick to remove the curse, but he said that to be freed from it they must go to the oak; there they would learn what they must do. This they did, and saw the spirit of Herne who, leaping astride a mighty horse, ordered them to follow. He led them to a beech tree, where he invoked Urswick, who burst forth, in flames, from the tree. He demanded that they fulfil their promise to grant his request, he made them swear a fearful oath to obey Herne and organize a gang to follow him. This they did and each night this band would go into the forests killing deer. When the King learned of this he, too, went to the oak, where the spirit of Herne appeared to him. Herne demanded vengeance on his enemies, and said that he would then stop causing any more trouble in the forests during the King's reign. Those people identified by Herne were found and hanged.

True to his word nothing more was heard from Herne during the rest of the King's reign. However, following his death he reappeared to ravage the forests of Windsor throughout the reigns of the next eight monarchs.

Other versions of this same tale say, variously, that Herne committed suicide for some heinous crime he had committed or that he was a demon with a stag's head, who haunted the forest trying to persuade the keepers to sell their souls to him. Another variant claims that he was a practitioner of black magic before ultimately hanging himself.

Even the felling of this tree has its tales. It is said to have been felled in a clearance of dead trees by order of George III. This story is unclear as we are told that the King was so distressed to find the tree gone that he chose another tree nearby and asserted that *that* was Herne's tree. Edward VII had another tree planted as a replacement in order to ensure that there was continuity. Other writers assert that the great oak fell from natural decay on the night of 31 August 1863 and on the place where it had stood Queen Victoria planted a replacement young oak.

> There is an old tale goes, that Herne the hunter,
> Sometime a keeper here in Windsor forest,
> Doth all the wintertime at still midnight,
> Walk about an oak, with great ragg'd horn;
> And there he blasts the tree, and takes the cattle;
> And makes milch-kine yield blood, and shakes a chain
> In a most hideous and dreadful manner.
>
> – Shakespeare, *Merry Wives of Windsor*

Christian myth has the tale of an oak at Kenmare, which was dedicated to St Columba. It was blown down in a great storm but nobody dared to touch the wood until a local tanner came along. He stripped some of the bark and used it for tanning leather. With the leather he had cured he made himself a new pair of shoes. However, the first time he wore them he was struck with leprosy and he remained a leper for the rest of his life.

The Virgin Mary was worshipped as 'Our Lady of the Oak' at Anjou in France. When she appeared to shepherd children in Portugal she was referred to as 'Our Lady of Fatima; she was crowned with roses and hovering over an oak tree.

Finally there is a legend relating to the evergreen oak that says that it betrayed Christ. All the trees agreed not to allow their wood to be used to form the cross once the crucifixion had been ordered; when anyone tried to cut the trees the timber simply shattered into fragments. The exception was the evergreen oak. In consequence Greek woodcutters are supposed to have a horror of the tree and are fearful of sullying either their axes or their hearths with the wood.

Magic and Lore

The oak is the king of the forest in much the same way as we consider the lion to be the king of the beasts. It is seen as a tree of great strength and has been thought of as a symbol of England. There are a great many superstitions and tales, from all around Europe, which involve oak trees and worship of the oak tree deity seems to be a common feature of all the European races. In pagan Europe it was believed that oak fire strengthened the sun. Midsummer fires would be made from oakwood. Right through to the Middle Ages sacrifices were made to sacred oaks for various reasons. In Lithuania sacrifices were made to them for plentiful crops. In Siberia groves were swathed in cloth and offerings left, including reindeer hides, kettles, spoons and other valuable household articles. Orthodox Christians continued to revere a holy oak in Russia until the 1870s.

Hills, mountains and other high places were considered to be suitable places on which to worship the gods, presumably because they were closer to heaven or to where the gods resided. In time trees were planted on these sites, and from these we get consecrated groves. In the Old Testament of the Bible we are informed that these high holy places were prohibited (Deuteronomy, 16:21) as the planting of trees near an altar could be seen as idolatry. However, in Genesis we are told that Abraham retired into the oak grove at Mamre to meet God.

Mount Caelius, one of the Seven Hills of Rome, was at first covered by oak woodland, and there was an oak grove dedicated to Venus at the foot of the Palatine Hill. The Romans dedicated the tree to Jupiter and the god was worshipped in the form of a giant oak tree that grew upon the Capitol. The association with Jupiter came from a story that told how an oak tree had shielded the god at his birth. Crowns formed from the leaves were bestowed on Romans who had saved the life of another citizen during battle. Called the Civic Crown, this was one of the highest honours the state could confer on an individual. Those who had won the right to wear it could do so whenever they wished. It meant that they were exempt from all civil burdens and, if they entered a room, all present, regardless of their rank or office, had to stand as a mark of respect.

It is said that the Druids would only meet where there was an oak tree. Their creed held that it was unlawful to construct temples to the gods or to worship anywhere constrained by walls and a ceiling. Consequently their places of worship were usually in the open air, often on high ground, from where they had a good view of the constellations. The planting of woodlands about their meeting sites would give privacy and help prevent them being distracted from their ceremonies.

The Druids revered the oak over all other trees as it contained the power, strength and energy of their god Esus. Their name for it, *Duir*, can be seen as being derived from the Irish *dur* meaning 'hard', 'unyielding' and 'durable'. The English word 'door' is derived from the same source, and the oak represents the doorway into other worlds. Pliny claims that the Druids believed that anything growing on an oak tree was a gift from heaven; mistletoe growing on oaks was therefore considered to be particularly sacred, and the white berries were seen to represent the sperm of the god. Thus the mistletoe contained both its own powers and those of the oak, and the oak was linked to the procreative forces of the universe. The mistletoe used in the Druids' mystical rites was cut from an oak tree in preference to any other source. They cut it each year with a golden sickle, in a ritual emasculation of the tree. Robert Graves, in *The White Goddess*, suggests that the oak and the holly were opposing tree deities, the oak ruling through the summer and the holly through the winter. He further suggests that the winter ruler may have been the holm or holly oak, *Quercus ilex* rather than the holly itself.

The religious use of the oak extends much closer to the present day than we might at first think. In parts of England there are still trees known as 'gospel oak'. This relates to a time when clergymen, accompanied by their parishioners, would walk the parish boundaries during Rogation Week, called 'beating the bounds'. They would halt at selected sites, oaks being particularly auspicious, to read psalms and passages from the gospels. Then they would pray, seeking blessings for the people of the parish. This ceremony may have had pagan origins, deriving from the festivals of the god Terminus.

> Dearest, bury me
> Under that holy oak, or Gospel tree,
> Where, though thou seest not, thou may'st think upon
> Me, when thou yearly go'st procession
>
> – Robert Herrick, *To Anthea (II)*

There are many references to sacred oaths being sworn beneath oaks, suggesting that the presence of the tree rendered the oath binding. In the Old Testament we are informed that Joshua and Israel made their covenant with the Lord beneath an oak tree. Following his defeat at Hastings one legend claims that, far from being shot in the eye, King Harold was forced to take an oath of loyalty to William the Conqueror beneath an oak at Rouen. He broke his oath and at the same moment the tree shed all its leaves. Another famous oath was that of the Spanish kings to respect the rights and beliefs of the Basque peoples. This oath was sworn beneath the sacred oak of the

Basques and was effectively broken by General Franco when he allowed the German Condor Squadron to bomb Guernica.

There are numerous folk tales and legends that relate to the oak. As a sacred tree misfortune will befall whoever cuts one down or otherwise causes it harm. A Shropshire superstition claims that it blooms on Midsummer's Night but that by morning the flowers have disappeared. It is a potent protection against evil. A protective amulet can be formed by binding two twigs together with a red thread to form a cross, the arms of which should be of equal length. This can then be hung in the house to ward off evil and protect the whole household.

Oak Apple Day is 29 May. Following the Battle of Worcester in 1651 Charles II was being pursued and famously hid himself in an oak tree, the Boscobel Oak. In memory of this, oak leaves were worn on that date, his birthday. It was also on 29 May in 1660 that he entered London to resume the throne. Traditionally oak leaves would be worn until noon and then discarded to be replaced by ash leaves, which were worn until sunset. In Hampshire 29 May was known as 'Shick-shack Day'. In the early morning men would gather up oak twigs to hang on doorknockers and wear on their hats. After breakfast they would parade and be welcomed with glasses of beer. If no beer was forthcoming they called out:

Shick, shack, pen a rag,
Bang his head in Cromwell's bag
All up in a bundle.

Anyone not wearing oak leaves, after midday, on that day could reply, 'Shick-shack has gone past'.

Elsewhere whole branches of oak were used to decorate public buildings. Gamekeepers would make no complaint about the collection of the wood and no orders for trespass were made on that day. In some parts of Britain anyone found not wearing the oak on Oak Apple Day would be whipped with nettles. It seems clear that many of the superstitions and celebrations that use the oak pre-date the restoration of the monarchy but later became associated with it.

The celebration of Oak Apple Day is now much less common. However, at Royal Hospital, founded by Charles II in 1682 as a home for old soldiers, on or about 29 May, known as 'Founder's Day', the pensioners and their guests wear the sprigs of oak, and the statue of Charles II is decked with oak branches. The Founder's Day parade, which is presided over by a member of the Royal Family or a high-ranking army officer, has been held annually sine 1692.

Another link between the Royal Family and the oak tree can be found in the legend of the 'topless' oaks of Charnwood Forest in Leicestershire. It is said that these trees lost their tops at the time that Lady Jane Grey was put to death. She resided at Bradygates Hall, which is nearby, and the trees were pollarded as a sign of mourning.

Other famous trees include the Robin Hood Oak in Sherwood Forest, which is probably not as old as the Robin Hood legends, and Merlin's Oak in Carmarthen.

It is said that the great magician, being a necromancer, worked his magic in a graveyard, and for a wand used the topmost branch of an oak tree. Merlin is supposed to have been born in Carmarthen, the name meaning 'Merlin's fortress'. The words of the ancient spell still resound:

> When Merlin's oak shall tumble down,
> Then shall end Carmarthen town.

The tree that was credited with being 'Merlin's oak' stood in Priory Street.

In British folklore old hollow trees that stood in sacred groves are called bull oaks in England, bell oaks in Scotland and Ireland. These were often said to be home to faeries, elves, spirits or demons. When near these trees people were advised to turn their coats inside out to counter the faerie magic:

> Turn your cloaks
> For fairy folks are in old oaks.

The oak apple, a gall caused by a small fly, has superstitions applied to it. When you find an oak apple you should carefully cut it open. If there is the grub of the fly inside you can expect to become wealthy. If you find the actual fly, it is an indication that misfortune will follow, and if you find a spider it is a sign of impending illness. Other traditions suggest that these galls should be cut open at particular times in spring and autumn. What was found there foretold the sequence of the coming season. If an ant was found it showed there would be plenty of grain. A maggot, or white worm, meant murrain in the cattle. If the worm flew then there would be war, if it crept the harvest would be poor and if it turned about there would be an outbreak of plague. These galls might also be used to discover whether a child was bewitched. Three were placed in a basin of water that was then placed beneath the child's cradle. If the galls floated the child was all right. However, if the galls sank it signified that the child was probably afflicted.

The oak apples that we use today are not the same as those used in times past. They would have been a larger, spongy gall that appears on the oaks in May and then matures through late June and July. The gall wasp responsible for the smaller, woodier, oak apples we see more of today was introduced into Devon from the Middle East in about 1830, so that the galls could be used in tanning and ink making, as they are rich in tannic acid.

Acorns may be used in predicting the outcome of romances. The couple involved must each drop an acorn into a bowl of water. If the acorns float together the romance is sure to continue, but if they drift apart then so will the couple. Young women can use acorns to induce prophetic dreams to discover whom they will wed. On the third day of any month between September and March, an odd number of young women, less than nine, must gather, and each must thread nine acorns onto a length of string. This must then be wrapped around a piece of wood and put on a burning fire precisely at midnight. The women should sit absolutely silent, until all the acorns are burned away. They should then take out the ashes before they retire, whilst saying:

May love and marriage be the theme,
To visit me in this night's dream;
Gentle Venus, be my friend,
The image of my lover send;
Let me see his form and face,
And his occupation trace;
By a symbol or a sign
Cupid forward my design.

In the fourteenth century magicians might advise men who wished to ensure their wives' chastity to place two halves of an acorn in the woman's pillow. However, if she was clever she could overcome this spell by finding them and binding them together again. Then, after six days, she and her lover should each eat one of the pieces.

In pagan Europe it was believed that oak fire strengthened the sun. Midsummer fires would be made from oakwood. In pagan Ireland it was a crime to fell an oak tree, and at Kildare, in Christian times the nuns of St Brigit maintained a sacred oak fire. Kildare is derived from the Old Irish *cell dara* meaning 'church of the oak'.

To dream of the oak indicates that a long and happy life can be expected. If, however, the oak tree appears to be withered you should expect poverty in old age. A thriving tree promises male offspring who will win acclaim in their lives. To dream of an oak tree bearing acorns betokens wealth and a blasted oak is an omen of death.

The oak is sacred to thunder gods in various pantheons. The Scandinavian myths associated the oak with Thor, the Norse god of thunder, and Baldur, god of light. In classical mythology Hercules attracted thunderstorms by rattling an oak club in a hollow oak, or by stirring a pool with an oak branch. In British folklore, woodpeckers were thought to be attracting rain when they knocked on oak trunks and in pagan rites black animals were sacrificed to the thunder god for rain.

Oak's connection with Thor and the other thunder gods, has left a legacy of superstitions relating to thunder. It seems that the electrical resistance of oak is particularly low, so it is struck more often than some other species. Whether this is true or not, an oak that has been struck by lightning will usually continue to grow, and the timber of such a blasted oak is especially prized. Even an acorn is said to act as a shield against being struck by lightning. For this reason acorn-shaped blind pulls, preferably made from oakwood, will protect the home.

Beware the oak,
It draws the stroke,
Avoid the ash,
It courts the flash,
Creep under the thorn,
It will save you from harm.

The oak, along with the ash, is used by country folk to forecast the weather. As we saw under 'Ash', depending on which of these two trees comes into leaf first the rhyme tells how wet the summer will prove to be:

Oak before Ash, in for a splash;
Ash before Oak, in for a soak.

Or, more cryptically, from Surrey: 'Oak smoke [mist], Ash squash.' Another rhyme says:

If oak and ash leaves show together,
Us may fear some awful weather.
This be a sight but seldom seen,
That could remind us what has been.

Medicinal

The old wives' remedies included such superstitions as that catching a falling oak leaf in autumn will prevent a cold in winter. Burning oak logs on an open fire in the room where a sick person lies will draw off all their ailments as well as warming the room. A more realistic remedy, perhaps, comes in the recommendation that ground acorns should be used as a treatment for diarrhoea. Carrying an acorn is said to prevent ageing. For women this is most effective when the acorn is put into a pocket or handbag. Acorns, being associated with the goddess of fertility, Ceres or Demeter, would also increase sexual potency and virility if carried. The dew collected from under an oak tree is supposed to be an especially powerful beauty aid.

In the eighteenth century, driving a nail into an oak tree or scratching the gum near the tooth with an oak splinter was believed to cure toothache. A more severe version of this suggests putting the nail into the tooth or gum until it bleeds, and then driving it into an oak tree. Rubbing your hand across a piece of oakwood on Midsummer's Day will, supposedly, ensure that you will remain healthy all year.

In a similar spell to the one involving an ash sapling for the cure of ruptures in children, oak could be used to treat malformations in growth. An oak branch was split and wedged open, and the child passed through three times. If the branch healed then so would the child.

MEADOW CROWFOOT

Botanical Name: *Ranunculus acris*
Family: Ranunculaceae
Common Names: Baffiners; bassinet; blistercup; blister plant; butter and cheese; buttercup; butterflowers; clovewort; crow flower; crowfoot; dew cup; frog's foot; gilt cup; gold cup; gold knobs; gold knops; goldy; gowan; *grenouillette*; kingcup; king's knobs; lady's slipper; locket-gouleons; polts; troil flowers; wart flower; yellow bachelor's buttons

Planet: Mars; Sun
Deity: Hymen
Meaning: Riches

Modern authors are more likely to differentiate between crowfoot (*Ranunculus acris*) and buttercup (*Ranunculus repens*) but the writers of more ancient texts were much less fussy and applied the same common names and folk tales to each. Consequently it is now difficult to separate them. I have tried to do so where it is worthwhile, or where a clear distinction can be made between the two species.

Magic and Lore

The botanical name *Ranunculus* is likely to have been derived from the Greek *rana*, meaning 'a frog', either because both plant and frog share a love of damp meadows or from a supposed similarity between the leaf shape and that of a frog's foot. *Acris* is derived from the Greek *akros*, meaning 'bitter', a reminder that members of the buttercup family tend to be poisonous.

Where the meadow crowfoot is found growing in combination with *Caltha palustris*, the marsh marigold, they are known as 'publicans and sinners' – presumably it is the marsh marigolds, also called 'drunkards', that are the sinners.

This is another plant associated with lovers and marriage. Sometimes called yellow bachelor's buttons, it was once traditionally worn by lovers at their betrothal.

The meadow crowfoot is the birthday flower for 18 September. It is said to symbolize childishness, ingratitude, mockery and spite in addition to riches.

Medicinal

Some writers identify this as a plant of the sun and say that it has the power to calm melancholy or frantic people. In herbal medicine, it has been used mainly as a treatment for stomach disorders. The juice was applied to the nostrils in order to provoke sneezing as this was supposed to help relieve some types of headaches.

BUTTERCUP

Botanical Name: *Ranunculus repens*
Family: Ranunculaceae
Common Names: Bachelor's buttons; bread and cheeses; bur-crowfoot; butterbump; butter cheese; butterchurn; butter daisy; butterflowers; butter rose; caltrops; cat's claws; cowslip; crawfeet; crazies; crazy; crazy Bet; crazy-more; crazy-weed; creeping crazy; crowfoot; crow toe; cuckoo buds; dale cup; dell cup; delty cup; devil's guts; dill cup; faeries' basins; gild cup; gill cup; gilted cup; glennies; gold balls; gold crap; gold cup; golden cup; gold knop; golden knop; gold weed; goldy; golland; guilty cup; kennel; kenning herb; kingcup; king's clover; kings cob; kraa-tee; lantern leaves; lawyers' weed; Mary buds; many feet; Meg-many-feet; Meg-wi'-many-teaz; old man's buttons; soldiers' buttons; tangle grass; tea cups;

tether toads; yellow caul; yellow creams; yellow crees; yellow cups; yellow gollan
Planet: Mars; Sun
Deity: Hymen
Meanings: Cheerfulness; childishness; ingratitude; radiance; riches; 'What golden beauty is yours'; 'You are immature, a heartless and ungrateful flirt'

Myth and Legend

According to legend, Ranunculus was a young poet who had a warm, melodious voice. He charmed all who heard him, including the nymphs. He was given to wearing beautiful clothes of bright yellow and green silks. One day, whilst singing to himself, he became enraptured by the sound of his own voice. He expired in a joyful ecstasy and the gods transformed him into the flower that bears his name. A less generous variation says he was lovely to hear for a short time, but that he never stopped running and singing. This was disturbing the peace of the forest and all the creatures in it so the wood nymphs, to restore peace and harmony, turned him into a buttercup and sent him out into the open meadow to live.

Magic and Lore

Many of the common names given to this plant seem to relate to the golden yellow colour of its flowers. The lesser-known names of 'crazies' or 'crazy Bet' arise from a belief in the Midlands that the scent could cause madness. According to another superstition, holding a tall buttercup flower against the neck, on the night of a full moon, would also cause insanity. The common name 'crowfoot' refers to the resemblance of the leaf to the footprint of a large bird.

The common name 'buttercup' is, of course, derived from the colour of the flower, but it was also thought that the richness of the butter might depend on the number of buttercups in the cows' pasture. In Ireland it was once common for farmers to rub buttercup flowers on the udders of their cows during May to increase the quality and quantity of the milk. Presumably it was believed that it would boost the buttermilk content of the milk. The flowers were also used in May Day celebrations.

Buttercups and daisies are two of the first flowers most children learn to recognize. From the daisies they make daisy chains, while they hold buttercups up to the faces of their friends and family to see if the golden yellow colour is reflected in their skin. When I was a child, this reflected colour was said to be a sure sign that the person liked butter but an older, more traditional meaning is that the person can expect to gain riches. At one time butter was more expensive and would have been more readily accessible to the affluent. Another suggestion is that if the colour of the flower is reflected on the chin of a child it indicates that they are 'as good as gold'.

Buttercup is associated with begging by wounded soldiers. To make their wounds appear worse than they really were, they would scratch them with limpet shells and then rub buttercup sap into the wound. This was painful but inflamed the wound. The sap can cause contact dermatitis and so should be handled with care.

Buttercups appearing in your dreams signify that any business enterprise you are involved in will succeed.

The tern 'sardonic', meaning grimly jocular, bitterly mocking or cynical, is often said to have derived from the name of *Ranunculus sardous*, commonly called sardonion, which is toxic, and when eaten causes the face to contort into an expression resembling a look of scorn.

Medicinal

Names such as 'kennels' and 'kenning herb' have been derived from the use of the plant to treat kennets, a local name for eye ulcers. It has also been used to treat such ailments as burns, sores and rheumatic pains.

RADISH

Botanical Name: *Raphanus sativus*
Family: Cruciferae
Gender: Masculine
Planet: Mars
Element: Fire

It may be difficult to believe but the humble radish was once considered to be a mighty aphrodisiac. This probably has more to do with its colour and shape, arguably resembling a testicle, than with any noted results following its consumption. In sympathetic magic eating radishes is supposed to promote lustfulness.

Carrying radish is a protection against the evil eye. In Germany, there is a tradition that carrying a particular species of wild radish will allow the bearer to determine whether there are sorcerers present.

MIGNONETTE

Botanical Name: *Reseda odorata*
Family: Resedaceae
Common Names: Egyptian bastard rocket; Frenchman's darling; fragrant weed; little darling of Egypt; *mignonette d'Egypte*
Meanings: Dull virtues; 'You are better than handsome'; 'Your qualities surpass your charms'

Magic and Lore

This native of North Africa was a fragrant weed of ancient Egyptian gardens. It was introduced to Britain in the mid eighteenth century. The name is derived from the French meaning 'little darling'. It is claimed that the seeds of the plant were gathered by Napoleon during his Egyptian campaign and sent back for Josephine to grow in the gardens at Malmaison. It was she who popularized its cultivation. In Egypt the flowers were used to decorate couches in tombs. It acquired the name 'yellow flowered'.

Mignonette was widely grown on balconies in London and elsewhere, in an attempt to mask the foul odours of summer in the city. Some people thought that the scent would also help to protect against diseases.

Lovers can ensure that they will always enjoy good luck by rolling three times in a bed planted with mignonette.

Mignonette is the birthday flower of 30 January.

Medicinal

The botanical name of the plant comes from the Latin *resedo*, 'to heal' or 'to calm'. The plant was said to have the power to heal all kinds of bodily disorders. The ancient Romans used the seeds as a sedative.

BUCKTHORN

Botanical Name: *Rhamnus catharticus*
Family: Rhamnaceae
Common Names: Hart's thorn; *herba stellaria*; herb Eve; herb ivy; highway thorn; ramsthorn; sanguinaria; swine-cresses; waythorn; wort cresses
Gender: Feminine
Planet: Saturn
Element: Water

Like many spiky or spiny plants, buckthorn has been identified with the crown of thorns at Christ's crucifixion.

Dioscorides informs us that, where branches of buckthorn are placed about doors and windows, it will drive away all enchantments. More generally, it is seen as simply bringing good luck to the household. It may be carried or worn by those involved in proceedings in court in the hope that it will assist in bringing about the desired result.

Tradition has it that if you form a circle of pieces of buckthorn, an elf will appear. As soon as he does so you must say, 'Halt and grant my boon'. If you say this quickly enough the elf must grant you one wish before he can disappear again.

CASCARA SAGRADA

Botanical Name: *Rhamnus purshianus*
Family: Rhamnaceae

Like its close relative the buckthorn, cascara sagrada can be beneficial for those involved in legal proceedings. Sprinkling an infusion of the plant around the house before going the court will help ensure that the case is won.

Twigs of cascara can be worn as a protective amulet against evil forces. It may also be used in spells cast in order to attract wealth.

RHUBARB

Botanical Name: *Rheum X hybridum*
Family: Polygonaceae
Gender: Feminine
Planet: Venus
Element: Earth
Deity: Gayomart

Myth and Legend

In a Zoroastrian myth, Gayomart shone like the sun but was killed by the force of evil, Ahriman. His 'seed' fell to the earth and from it grew rhubarb. In turn, from the rhubarb came the first human couple.

Magic and Lore

In order to ensure the fidelity of a lover, serve them a piece of rhubarb pie.

Medicinal

Rhubarb is one of the oldest Chinese medicines. A piece of the root about the neck, as a necklace, will protect the wearer from all manner of stomach pains or similar disorders. William Coles says, 'Rhubarb is so effectual for the liver that it is called the Life, Soul, Heart and Treacle of the liver, purging from thence choler, phlegm and watery humours.'

DULSE

Botanical Name: *Rhodymenia palmata*
Family: Rhodimeniaceae
Gender: Feminine
Planet: Moon
Element: Water

The name of this plant is derived from the Irish name *dūileasg*. Traditionally it was used in sea rituals, being thrown onto the waves in heavy seas, to placate sea spirits. It could also be thrown from cliff tops during rites to raise wind spirits.

When small quantities are added to drinks it will form a love potion, inducing lustfulness. Simply sprinkling it around the house will induce a more harmonious atmosphere in the home.

CASTOR OIL PLANT

Botanical Name: *Ricinus communis*
Family: Euphorbiaceae
Common Names: Kiki; *Palma Christi*

This plant, native to Africa, was introduced into Britain in the sixteenth century. As might be expected from a plant called *Palma Christi*, it has been seen as a protective herb against all forms of negativity and evil. The seeds, or beans, are said to be able to absorb evil and all parts of the plant may be used to protect against the evil eye.

ROSE

Botanical Name: *Rosa* spp.
Family: Rosaceae
Gender: Feminine
Planet: Venus
Element: Water
Deities: Adonis; Aphrodite; Aurora; Cupid; Demeter; Eros; Harpocrates; Hathor; Horus; Hulda; Isis; Venus
Meanings: Beauty; 'I love you' (full blown); ambassador of love; 'My Heart is in Flames' (cabbage rose/ Provence rose); 'Beauty always new'; transient brilliance; departure (China, monthly); bashful love (deep red); grace (hundred-leaved); 'Beauty is your only attraction' (Japan); 'If you love me you will find it out' (maiden's blush); precocity (May rose); affection; capricious beauty (musk rose); simplicity (single rose); early attachment (thornless rose); 'I will love you forever' (amethyst rose); 'I love you vigorously' (orange rose); 'My love for you is innocent' (pink rose); 'I love you because you are fair and innocent' (wild rose); decrease in love, infidelity; 'I love another'; jealousy; misplaced affection (yellow rose); war (Roses of York and Lancaster); award of merit (a crown of roses); simplicity and beauty (burgundy); a smile (daily rose); 'Beauty is fleeting' (faded rose); unity (red and white together); pure and lovely (red rosebud); 'Call me not beautiful' (unique rose); pure,

innocent love; 'I confess my love' (rosebud); secrecy (full-blown rose over two rosebuds); happy love (bridal rose); 'Thou art all that is lovely' (Austrian rose); 'Only deserve my love' (campion rose); 'Love is dangerous' (Carolina rose); grace (multiflora rose); 'If you love me, you will find it out' (maiden rose); 'Capricious beauty' (musk rose); charming (cluster of musk roses); genteel; incorrupt; love; silence; 'The loving heart of humanity'

> For it doth deserve the chief and prime place among all floures.
>
> – John Gerard

Mankind has been on the Earth for approximately 250,000 years whereas, according to the fossil evidence from Asia, Europe and North America, roses have an ancestry dating back more than 35 million years. Evidence of rose cultivation can be found in the earliest civilizations. Examples of the flower can be found clearly depicted in the 'Fresco with Blue Bird', unearthed at Knossos and dating from 1900–1700 BC.

It was probable that the Greeks introduced the rose to Egypt as Ptolemy (323–283 BC), the successor to Alexander the Great, insisted on having them grown. Certainly, the remains of wreaths made from roses have been removed from tombs in Lower Egypt, dating from 170 BC These flowers are thought to be *Rosa richardii* (formerly known as *Rosa sancta*, the St John's Rose).

The birthplace of serious rose cultivation is most probably Persia, although there are important species from China, Japan, India and the countries bordering the Mediterranean Sea. Roses were a central feature in many Persian gardens, as it was considered a flower of luxury and sanctity. When the Arabs overran Persia and Syria in the sixth century they were so impressed with the gardens they found there that they set about copying them. Both Persians and Arabs alike celebrated the rose as the 'messenger of gardens of the soul'. It took more wars, the Crusades, for the adoration of the rose to be brought back to Christendom.

Shakespeare might have been correct when he said, 'That which we call a rose by any other name would smell as sweet', but the name itself may have some significance. It is probably derived from the Greek *rhodon*, meaning 'red', as the ancient cultivated rose was a shade of deep red. *Rhodon* relates to *rhein*, meaning to flow, as the scent hangs in the air until it is blown away by the wind. An alternative derivation is that it is taken from *rota*, a wheel, as the regular five petals form the spokes in the circle of the flower.

Myth and Legend

Greek mythology tells that the rose was created by Chloris, Greek goddess of flowers, from the lifeless body of a nymph she loved, whom she found in a woodland clearing. She sought the aid of Aphrodite, the goddess of love, who gave beauty. Dionysus, the god of wine, helped by adding nectar to give a sweet scent. The three Graces gave the flower brightness, charm and joy. Finally, Zephyr, the god of the west wind, blew away the clouds to enable Apollo, the sun god, to shine down and make the flower bloom. The Roman version attributes its creation to Flora, their goddess of spring and flowers. Following the death of one of her nymphs she called upon the gods to transform her into a flower. Apollo gave her life, Bacchus bestowed nectar and Vertumnus a wonderful perfume. Flora gave the flower a 'crown' of petals and Pomona gave it an attractive fruit.

When Cupid shot arrows at the bees that had stung him, thorns grew on the rose stems from where his arrows hit it. At one time all roses were thornless, but one day Cupid, the son of Aphrodite, was stung on the lip by a bee whilst bending to smell the fragrance of a newly opened rose flower. He went weeping to his mother who, to pacify him, strung his bow with bees. She removed their stings and stuck them onto the stem of the rose. Another myth tells us that whilst dancing amongst the gods, Cupid knocked over a cup of nectar which, when it splashed onto the rose, not only turned it red but also gave it the marvellous scent we now associate with it.

Many legends claim that all roses were once white. One explains that Persephone and Aphrodite were rivals for the love of Adonis and shared his favours. Aphrodite decided to prevent her rival from taking Adonis with her into the Underworld. Persephone sought assistance from Ares, the god of war who, in the shape of a wild boar, gored Adonis while he was out hunting. It was said that red roses sprang up from where the blood of Adonis fell or that the white roses turn red in sympathy. An alternative telling claims that the white roses grew from the tears of Venus, or from the foam that fell from her body when she came onto the land from the sea. A variation of this states that when Venus was trying to save Adonis from harm rose thorns pierced her feet.

Her naked foot a rude thorn tore
From sting of briar it bled,
And where the blood ran evermore
It dyed the roses red.

Variations on this myth claim that red and white roses grew where the goddess walked, indicating the light and dark sides of her character. A different Roman legend says that Venus blushed when, one day, Jupiter saw her bathing, and the white roses that grew near were turned to red in her reflection.

Elsewhere it is said that Rhodanthe, Queen of Corinth, was so beautiful that she could have no peace from the number of men who followed her about the palace. She fled into the temple of Diana, where she captivated all who saw her, making even the goddess jealous. Apollo, Diana's brother, in revenge turned her into a rose bush. Three of Rhodanthe's admirers were more persistent than the rest and had followed her into the temple, and Apollo turned them into a worm, a gnat and a butterfly, all of which can still be found near a rose.

A Christian myth suggests that originally all roses were red until, one day, the Virgin Mary placed a veil she had washed on a bush to dry. Thereafter that bush always produced white blooms.

Islamic myth says that the rose grew from beads of sweat that fell from the brow of the prophet Mohammed. Red roses also grew from the sweat that fell from the head of the Archangel Gabriel, who was the prophet's travelling companion, and even the sweat from the prophet's donkey brought forth yellow ones. One variant of this story says that the roses grew from the sweat of a woman, named Joan, who was travelling with the angel and the prophet. In the morning, when her skin was pale, the roses were white but, by midday, her skin was rosy from spending so long in the sun, and the roses that grew were red.

Another Islamic story relates that the Prophet Mohammed suspected that Aisha, a favourite wife, was being unfaithful to him, so he sought the advice of the Archangel Gabriel. When he returned home Aisha greeted him with some red roses. On the instructions of Gabriel, Mohammed told her to throw them into the river. The roses turned from red to yellow, confirming the prophet's suspicions.

According to a Romanian folk tale, the sun was crossing the sky one day and, looking down, observed a beautiful princess bathing in the sea. So captivating was the sight that the sun stopped to watch, and made no further progress for more than seventy hours. God, knowing that the workings of the Universe were at risk, transformed the girl into a rose bush, then caused the sun to go on its way. Some say that as a result all roses will blush and bow their heads as the sun passes over them.

Yet another legend tells of a young Jewish girl named Zillah, from Bethlehem. She was falsely accused of witchcraft and sentenced to be burned at the stake. She prayed fervently that the Almighty might intercede and save her life, but it seemed that her prayers were in vain. Then, as the flames flared up around her feet, the fire was completely extinguished just as it touched her skin. To show that she was completely innocent all the sticks that had been alight were turned to red roses whilst the sticks yet to catch fire became white roses. These were the first such flowers to be seen since man's fall from grace.

There is a story that when the young Abraham was introduced to court by Azar, Nimrod threw him into a blazing fire. Instantly the fire became a soft bed of red roses, on which Abraham went to sleep.

In a Persian legend the flowers complained that the lotus, the 'Queen of the Flowers', slept at night. To make amends Allah named the white rose Queen of the Flowers. The nightingale, a significant bird in Persian myth, was enchanted and flew down to embrace it but pricked itself on a thorn. From the drops of the bird's blood that fell onto the ground grew three red roses.

A tale, similar to the Hans Christian Anderson story 'The Princess and the Pea', tells of Sminivides, an inhabitant of Sybaris in southern Italy. The people of Sybaris were known for their self-indulgent, luxurious lifestyle. Sminivides complained of being unable to sleep because, in his rose-strewn bedroom, he had lain on a petal which was folded in two, and the discomfort had kept him awake.

A Christian myth tells us that all the roses in the Garden of Eden were originally thornless. However, after the fall from grace, when Adam and Eve were expelled, thorns appeared.

Magic and Lore

In ancient Rome, as in Greece and Egypt, the rose enjoyed a special significance. Where other civilizations had been simply enthralled by its beauty the Romans developed an absolute mania for it. In common with the Greeks, Persians and others, they dressed the effigies of their gods with the flowers. Statues of Venus, Cupid and Bacchus were all crowned with wreaths of it. Rose petals were scattered at the festivals of Flora and Hymen. However, the Roman obsession with roses went much further. Roman soldiers might go into battle adorned with rose wreaths and with their chariots and the prows of their ships bedecked with the flowers. On their triumphal return into Rome, victorious soldiers could expect rose petals to be scattered in their path or dropped beneath their chariot wheels. Roman generals were permitted to decorate their shields with rose flowers, a custom that persisted even after the fall of the Roman Empire. In the home, foods, wines and honeys were flavoured with roses. They were used in cosmetics and medicines. Rose water might gush from fountains. Pillows could be stuffed with rose petals. Floors and couches could be covered by a liberal layer of petals, the depth of the layer being a symbol of status. It is reported that when Cleopatra, Queen of Egypt (69–30 BC), went to meet her lover, Mark Anthony, in 42 BC the floor of the banqueting hall was covered with a two foot deep layer of rose petals. The Emperor Nero, it seems, began a fashion for showering guests at banquets with rose petals. This fashion reached a climax during the reign of the emperor Heliogabalus, who commemorated the start of his reign by locking his guests in the banquet hall (to maintain their full attention) before showering them with such a weight of rose petals that several people suffocated.

This tradition of showering rose petals on important people persists today. The petals are now more often made from tissue paper, and we see them as the confetti thrown at weddings. Throwing confetti instead of rice seems to have been popularized following the wedding of Princess May, Duchess of York, in 1928.

In addition to dressing their gods with roses the citizens of Rome would wear the flowers themselves. It is claimed that it was Julius Caesar who popularized the wearing of chaplets of rose blossoms in order to conceal his baldness. It is quite possible, of course, that he chose to wear the flowers because he considered himself a god. At weddings, the bride and groom were garlanded with roses, as the fresh flowers were believed to have aphrodisiac properties, presumably due to their association with Venus and Cupid. Young Roman playboys might refer to their lady loves as *mea rosa* and present them with the first spring rose as a love token. This has obvious parallels with the giving of roses on St Valentine's Day.

Under the Caesars, Romans annually celebrated a festival of the rose, called Rosalia. This was when the greatest excesses of rose adoration could be witnessed. (One of the dates given for this festival is 23 May and it is still said that if you want to grow the finest quality roses you should plant them between 23 and 25 May.)

The Romans associated the rose not only with love, mirth and joy, but also with sorrow, grief and death. The use of roses in medicines may have come from a belief that they could somehow ward off death. It appears that the Romans and the Greeks saw them as a guard against death and a protection for the spirits of the dead. It was common practice to plant a rose when members of the family had gone away or died. One might be planted, for example, when a soldier went away to war, in the hope of his safe return. The flowers would also be put into graves and tombs, and wealthy citizens set aside sizeable sums of money to enable their children to plant roses in commemoration of their death.

The incessant demand for increasing numbers of rose blooms in ancient Rome proved a massive problem. Orchards and olive groves were neglected and cereal crops left untended in favour of spending more time on the cultivation of the flowers. Steam-heated greenhouses were developed to meet the demand through the winter months. Vast quantities were imported from Egypt and other countries. The island of Rhodes, it seems, owes its name to its being a centre of rose cultivation to meet the Roman demand.

At the fall of the Roman Empire the rose fell from grace. In the early Church it was rejected, being seen as emblematic of the decadence of the Roman Empire. Growing and wearing roses was forbidden; Clement of Alexandria expressly forbade the wearing of rose garlands by the early Christians. There is almost no mention of roses in Europe from the time of the fall of the Roman Empire until the rise of the Emperor Charlemagne; even then the rose's use was strictly regulated. Its cultivation was restricted to monks, who grew the plants for medicinal purposes. However, like so many other plants that had a pagan past it was eventually 'converted' to Christianity. St Jerome (342–420 AD) and Paulinus of Nola (345–431 AD) both revoked the ban. Paulinus viewed the five petals of the flowers as symbolic of the wounds of Christ and its red colour as an emblem of the blood of the martyrs.

This flower, once the emblem of Venus and Aurora, and a symbol of youthfulness, beauty and love became associated with the Virgin Mary and martyrdom.

> Mystic Rose! That precious name,
> Mary from the church doth claim.

In Botticelli's *Adoration*, for example, angels are depicted sprinkling rose petals over the holy infant in the midst of a rose garden. At St Albans, in Hertfordshire, roses are placed on the shrine of the first Christian martyr on the Sunday closest to 22 June, his feast day. We are told in the account of his journey from Verulamium, up the hill to the site of his martyrdom, that flowers sprang up as he passed by, but it is not stated that these were roses. Roses are also associated with St Ambrose and St Basil, who declared the plant the most perfect of flowers and said that they were thornless in Paradise until man's fall from grace.

By 1200 roses had become widespread and much appreciated. The returning crusader knights had seen architectural decorations using the rose motif, and brought the style back to Europe. Stained glass rosette windows became fashionable in churches and cathedrals. At about the same time, possibly during the pontificate of Pope Leo IX, a tradition started whereby on Laetare Sunday, the fourth Sunday in Lent, the Pope would award a 'holy rose', made from gold and rubies, or red enamel, to someone who had performed some great service to the church. This holy rose has taken many forms over the centuries, once being a complete golden rose bush set with precious gemstones. It has occasionally been bestowed on deserving rulers, but only two monarchs received such holy roses during their reigns. One was the Emperor Sigismund and the other was Henry VIII of England.

It is quite probable that this tradition was based on a far older custom whereby a simple rose was given, a custom nearly as old as the rosary, a string of beads used by Catholics to represent fifteen paternosters, fifteen doxologies and 150 aves. The origin of the name 'rosary' is uncertain but probably dates to an early collection of prayers to the Virgin Mary, called *Our Lady's Rose Garden*, rosary literally meaning 'a rose garden'. The transfer of the name from garden to beads is understandable, as the early rosaries were formed from tightly pressed rosebuds. Later, paste or roses carved from wood were used. The use of rosaries in churches dates back to at least 1100 and gave the word 'bead' to the English language; a bead originally meant a prayer. The use of prayer beads pre-dates Christianity as they can be found in India associated with the worship of the gods Kali and Shiva.

At one time Christians were called 'golden roses', 'the flower sprung from the root of Jesus'. They were likened to the flowers, considering them to be symbols of Christ in the shining splendour of his majesty. The Rosicrucians ('rose of the cross'), a secret society of Christian magicians founded in the fifteenth century, carried this allusion still further. They were concerned with alchemy and the occult and for them the rose symbolized nature. They adopted the red rose as their symbol.

In Unitarian churches, roses have frequently been used in the naming ceremonies for children. Unitarians do not accept the doctrine of original sin and so the use of water at christenings, to symbolically wash away sin, is unnecessary. The symbolism of the rose's purity is preferable.

Attar of rose, the rose oil used in the perfume industry, was discovered between 1582 and 1612. For the wedding of the Mogul Emperor Djihanguyr to the princess Nour-Djihan a great canal was excavated, encircling the whole of a garden, and filled with rose water. The heat of the sun caused the separation of the water from the essential oil of the rose. When the bridal couple rowed across the waters, the rich fragrance was noticed. The oil was skimmed off and found to have an exquisite perfume. Attar of rose is worth more, weight for weight, than platinum. In Bulgaria, a centre of rose production for attar, the roses are often under-planted with garlic to improve their scent.

The rose has been thought of as emblematic of many different virtues. It was dedicated to the goddess Aurora as a symbol of youthfulness and to Aphrodite (Venus) as a symbol of love and beauty. As an emblem of danger and fugacity it was dedicated to Cupid, and as an emblem of silence to Harpocrates. These last three deities are brought together as Cupid is said to have given the rose to Harpocrates as a bribe for his silence regarding the love affairs of the goddess Venus.

Being a symbol of silence, the Romans would hang a rose from the ceiling, over a table to signify that anything discussed at the table should be kept secret, or *sub rosa*.

The echoes of this practice are still found in modern houses, where plaster ceiling ornaments and pendant light fittings are still known as ceiling roses.

'Rose' was, at one time, a euphemism for the hymen or maidenhead. In Provence, over 100 years ago, prostitutes were referred to as 'roses'. Consequently, several European cities had 'red-light' areas with road names such as Rose Alley or Rose Street. Even in ancient Athens, according to Solon, the prostitutes would wear flowery clothes to distinguish them from the 'pure' ladies who wore white dresses.

In divination, a young woman wishing to discover whom she would wed would cut a rose at midnight on Midsummer's Eve and carefully wrap it up in a clean, white sheet of paper. She would then keep it safely stored away until Christmas Day when, if it was still fit to wear, she would pin it to her dress. The first young man she met who admired the flower, would be her husband. One variation on this says that when she looked at the flower on Christmas Day her future spouse would suddenly appear and snatch the rose away. Needless to say that if the flower had rotted it was an ill omen.

Another way for a young woman to discover whom she would marry was to name a rose for each of her lovers and to toss the roses onto water; the last one to sink would indicate whom she would wed.

To learn whether or not a lover is faithful, you should stick the curled petal of a rose to your forehead. If the petal splits the person is true. Similarly, the petal can be placed onto the palm of the hand and if it splits with an audible 'pop' when struck with the other hand it shows your lover loves only you. Rosebud petals can be added to bath water to conjure up a lover, and red rose petals placed in a red velvet bag and pinned under the clothes will attract love.

Roses, rose oil and rose incense were also used in love potions and philtres. Before preparing the potion the thorns were removed as they were used to symbolize the coming gratification. Washing the hands in rose water before mixing the potions would make them even stronger. One old spell to enable a woman to entrap the man of her choice required her to wear three roses – one pink, one white and one red – next to her heart for three days. She had then to steep them in wine for a further three days in order to produce a love potion. The man who drank the wine would be hers for ever. A string of rosehips might also be worn like beads around the neck or ankle, or a little rosewater sprinkled on the bed sheets or in the bath to inspire love, acting as an aphrodisiac.

Dreaming of roses can be an indication of a happy marriage. If, in your dream, you are pricked by the thorns when gathering roses it shows that you will eventually find your true love. For a farmer to dream of roses portends prosperity and independence. To dream of wither roses is an omen of disappointments and ill fortune.

In Scandinavian mythologies, the rose is said to be under the protection of faeries and dwarfs, and its cultivation is still claimed to be a way of attracting faeries into the garden. To make themselves invisible faeries might eat a rosehip and then turn anticlockwise three times. To become visible again they must eat another rosehip and turn clockwise three times.

Numerous superstitions indicate how a rose may herald misfortune; for example to smell roses in a room where there are none is an omen of a death to follow. It also portends a death if all the petals fall off a rose, leaving only the stem, whilst it is being held. For a rose to be seen blooming in the late autumn, that is to say out of its season, is an omen that misfortunes are to follow early in the coming year. Although scattering petals on the ground is an ancient tradition, this too has been considered as tempting misfortune.

If a rose was pruned on May Eve it was said that it would bloom again in the autumn. The most worrying superstition for garden centre owners, however, is the one that says that roses will thrive best if they are stolen.

There is a superstition that claims that wearing a rose as a buttonhole will ensure that the man wearing it has good luck with the women that he meets. The buttonhole may owe its origin to the rose. The gift of a flower between lovers has its own magic, especially on St Valentine's Day, and often the flower given is a red rose. The modern buttonhole flowers are usually wired and bound around with tape or, occasionally with foil, but in the early twentieth century small metal holders were designed, which could hold a little water, to keep real flowers fresh. In the eighteenth and nineteenth centuries it was common practice to wear flowers, particularly roses, in the buttonhole designed for closing a coat at the neck. Consequently, by the 1870s, it was the flower that had become known as a 'buttonhole'.

WHITE ROSE

Botanical Name: *Rosa X alba*
Family: Rosaceae
Common Name: Rose of York
Planet: Moon
Meanings: Candour; death preferable to loss of innocence (dried white rose); 'I am not worthy of you'; refusal; secrecy (white rose over two buds); ever beautiful; girlhood; heart ignorant of love; too young to love; (white rose bud); transient impressions (withered rose); charm and innocence; 'Her heart knows naught of love'; 'I love you not'

Myth and Legend

The white rose is known as a symbol of the House of York. However it did not come about as depicted by Shakespeare, by the plucking of red and white roses from the Temple gardens in London, at the start of the Wars of the Roses:

> And here I prophesy; this brawl today,
> Shall send between the red rose and the white
> A thousand souls to death and deadly night.
> – Shakespeare, *Henry IV Part I*

It was the symbol of the House of York since its foundation by Edmund Langley, the fourth surviving son of King Edward III and the first Duke of York. It is, arguably somewhat younger that the House of Lancaster as Edmund Plantagenet, known as Crouchback, the younger brother of Edward I, was made the first Earl of Lancaster and Leicester. That said, the first Duke of Lancaster was John of Gaunt, the third surviving son of King Edward III and therefore Edmund Langley's elder brother.

The white rose also became a symbol of the Jacobites. One reason suggested for this is that the supporters of the rebellion could not declare themselves openly but had to act *sub rosa*. Another possible explanation is that James II was Duke of York prior to ascending the throne in 1685. The white rose is the flower of the day for 10 June, thus:

The tenth of June I hold most dear,
Then sweet white roses do appear
For the sake of James the rover.

Often the form of white rose associated with the Jacobites is *Rosa X alba* 'Maxima', also known as the Cheshire rose.

From 1975 1 August has been designated as Yorkshire Day to commemorate the 1750 battle of Minden, in Germany, when soldiers of the Yorkshire regiments picked white roses from local fields to wear in tribute to their fallen companions.

In North America, the Cherokee tribe of Native Americans have their own legend about the origins of the white rose. They tell of a brave and handsome warrior, named Tuswenahi, who was the tribal leader. He returned from a hunting trip to find his people's settlement destroyed and his sweetheart, Dowansa, missing. His first fear was that she had been kidnapped but he later learned that she had been transformed into a white rose bush by the Nannshi, the 'little people'. The following year, when the rose bush put out its pure white blossoms, Dowansa had to ask the Nannshi to give her thorns so that she could protect herself as everyone, Tuswenahi included, kept trampling on the flowers. She was given such sharp thorns that even the wild animals were afraid to touch her.

Magic and Lore

White roses are not the luckiest of plants. It is said that if you are out walking in woodlands before Mother's Day and come across a white rose, it indicates that you will die before Mother's Day arrives. Another superstition says that if a white rose blooms out of season in your garden, it augurs badly for everyone in the family

In Scotland, it is said that to find a white rose blooming in autumn is a sign that there will be an early wedding.

DOG ROSE

Botanical Name: *Rosa canina*
Family: Rosaceae
Common Names: Briar; briar rose; brimble; buck-breer; canker; canker-rose; cat rose; cat whin; choop-rose; cock bramble; common briar; common brier; dike

rose; dog breer; dog briar; hip briar; hipseyhaws; hip tree; horse bramble; humack; klonger; klunger; lawyers; neddy-grinnel; pig-rose; rose-briar; wild briar; wild rose; yoe-briar
Planet: Jupiter; Moon
Meanings: Maidenly beauty; pleasure mixed with pain; simplicity; 'You are as fair and innocent as this pure bloom'; 'You have enchanted me'

Myth and Legend

Dog roses can survive to a great age. One plant at Hildersheim, Germany, was said to be over a thousand years old. The flowers were said to be sacred to the White Goddess, Mother Earth, because of having five petals – in magic five is a number of deep significance.

Some early traditions identified the plant as that used to form the crown of thorns placed on the head of Christ before his crucifixion. However, many other plants are more likely candidates.

Medicinal

It is difficult to ascertain whether the dog rose was so named because it was supposed to be a powerful treatment for the bites of mad dogs, or whether it was used as such a treatment because of its name.

The seeds of the dog rose have been used to expel intestinal parasites and, in North America, the Apaches used infusions made from the buds to treat gonorrhoea.

Periodically the plant will suffer from galls, caused by the action of insects. These 'moss-galls', 'pin-cushion galls' or 'green tossels' were carefully collected in Wales and used as a treatment for insomnia. Placed under the sufferer's pillow the galls would ensure that they would fall asleep. However, it is important to remove the galls afterwards, to ensure that they wake up again.

To cure an infant who suffered from whooping cough these galls were hung in the house. They were also carried as charms against ailments as diverse as rheumatism, toothache and shingles. In some parts of Northamptonshire schoolboys once carried or wore them as a protection against beatings.

MOSS ROSE

Botanical Name: *Rosa X centifolia muscosa*
Family: Rosaceae
Common Names: Cabbage rose; Provence rose; Holland rose
Meanings: Confession of love; 'I love you' (bud); ecstasy of enjoyment; 'I admire you from afar'; pleasure without pain; shy love; voluptuous love

Myth and Legend

Christian legend tells us that moss roses were created when the blood of the crucified Christ dripped onto the moss at the foot of the cross.

Magic and Lore

The moss rose used to be used in love divination to try to predict how long a relationship would last. The flower was picked on Midsummer's Eve and then kept in a safe place. Every so often it was inspected. If it had faded, then so would the relationship, but if the colour remained strong then so would the love.

> When faded in its hue
> She reads – the rustic is untrue,
> But if its leaves [petals] the crimson paint
> Her sick'ning hope no longer faint,
> The rose upon her bosom worn
> She meets him at the peep of morn.
>
> – *The Cottage Girl*

DAMASK ROSE

Botanical Name: *Rosa X damascena*
Family: Rosaceae
Common Names: Holy rose; rose of Abyssinia; summer damask rose
Planet: Venus
Meanings: Beauty ever new; brilliant complexion; maidenly blushes

Myth and Legend

According to Herodotus, the damask rose was said to have a scent surpassing that of any other, and it was probably the sweet scent that led to its introduction into Britain. Its origin seems to have been from a chance hybridization. Tradition has it that it was collected by a crusader knight, Robert De Brie, while fighting in the Holy Land, and was brought back by him on his return to his chateau at Champagne in the thirteenth century, from where it was dispersed across the breadth of Europe.

Magic and Lore

In Christian symbolism the damask rose is emblematic of the love of God for the world. It is often to be seen in pictures of the appearance of the Virgin Mary to St Bernadette at Lourdes.

The autumn damask rose is *Rosa X damascena semperflorens*, sometimes referred to as the four seasons rose, or monthly rose.

EGLANTINE

Botanical Name: *Rosa eglantineria* (Synonym: *Rosa rubiginosa*)
Family: Rosaceae
Common Name: Sweet briar
Planet: Jupiter

Meanings: Simplicity (full blown flower); fragrance; 'I wound to heal'; Poetry; 'The perfume of this flower brings sweet memories of you'

Myth and Legend

The eglantine is usually taken as a symbol of true love that will survive and overcome adversities. It is as such that it is referred to in the romantic story of Tristran and Isolde. From the grave of Tristran there grew an eglantine, which twined around the image of Isolde. No matter how heavily the plant was pruned back it would always quickly reshoot, and grow back over the figure.

> From this bleeding hand of mine,
> Take this sprig of Eglantine
> Which, though sweet unto your smell,
> Yet the fretful briar will tell,
> He who plucks the sweets, shall prove
> Many thorns to be in love.
>
> – Robert Herrick, *The Bleeding Hand*

In ancient Greece a wreath of eglantine was presented as a prize to the victorious poets at the Floral Games.

Medicinal

Eglantine has been used in herbal medications. In Iran it has been used to treat diarrhoea and stomach pains.

RED ROSE

Botanical Name: *Rosa gallica*
Family: Rosaceae
Common Names: French rose; rose of Provence; rose of Provins
Planet: Jupiter
Element: Water
Meanings: Love; pure and lovely (bud); bashful shame (deep red); 'Our love is over' (withered); 'I love you'

Myth and Legend

Although the red rose is now a floral symbol of England, it is not native to Britain, but was introduced by the Romans, via Gaul. It first appeared as an emblem on the shields of Persian warriors 3,000 years ago.

It was also the symbol of the House of Lancaster, but the story of how it came to be is a little confused. Edmund Plantagenet, called Crouchback, First Earl of Lancaster and Leicester, was the younger brother of King Edward I, and the son of Henry III and Eleanor of Provence. Provence has long been famed for its roses, and Queen Eleanor used a rose, probably red, as her family symbol. In due time, both her sons adopted roses as their emblems. The elder, Edward, adopted a golden rose, and Edmund the red rose. The first Duke of Lancaster was John of Gaunt, the third surviving son of King Edward III and therefore the elder brother of Edmund Langley, Duke of York.

There is another tale that claims that Edmund, the Earl of Lancaster, was also Count of Provence, and took the red rose as his family symbol as it had been that of his wife's family. At the time of Elizabeth I it was said that the rose had been derived from the crest of the Italian noble family of Orsiao.

According to many legends all roses were once white. One myth explains that red roses came about when these white roses were stained by the blood of Venus.

Magic and Lore

In Somerset there is a traditional spell that states that if a young woman seeks a husband she should scatter rose petals in a churchyard on Midsummer's Eve. As she scatters the petals she must chant the words;

Rose petal, Rose petal,
Rose petals I strew
He that will love me
Come to me soon.

An apparition of her future husband will then appear immediately behind her.

Another method of divination required the maid to gather her red rose in June, presumably at Midsummer, no later than 7.00 a.m. The bloom was then placed in a plain white envelope and the flap sealed with wax. Finally, she would mark the wax with the nail of the third finger of her left hand. If she placed this envelope beneath her pillow she was guaranteed to have a dream that would predict her future marriage prospects. If she dreamed of water, fields, mountains, glass, silver, the moon, children, parents, or organ music it indicated that she would be married within the year. Dreams of giants, animals, birds, fish, mirrors, papers or the sun showed that she would not wed for about five years. Gold, bells, soldiers, reptiles, or storms meant that she would never be married. A dream of a red rose would signify that she would soon be married to a handsome man.

A rose charm recommends that the young woman gathers up the brightest and best red rose she can find, between 3.00 and 4.00 a.m., ensuring that no sees her. The flower must be taken back to the safety of her bedroom and held over a chafing dish, or other similar bowl, in which a mixture of charcoal and brimstone is smouldering. The flower must be held in the smoke for about four minutes then, before it can cool, it must be folded up into a sheet of white paper on which is written the young woman's own name and that of the man she loves. Also on the paper should be the date and the name of the morning star currently in the ascendant. The paper must be sealed with three separate seals and then buried beneath the tree closest to the bush from which the rose was originally taken and left untouched until 6 July. It must then be placed beneath her pillow to ensure that she will enjoy a prophetic dream. The rose and the paper may be retained for three days without spoiling the charm, after which time both rose and paper must be burned.

To choose between three suitors a girl should select three rose petals and mark the initials of each suitor on one of the petals. These she must then store in a safe place but inspect them at regular intervals. Whichever petal stays freshest longest indicates which of the suitors she should select.

The wearing of red roses on St George's Day has never really been common amongst the English, probably because it is not naturally in flower in English gardens at that time of the year. The wearing of forced or imported roses is now largely limited to newsreaders and politicians.

Red roses were sometimes planted or placed onto graves to indicate continuing devotion.

The ancient link to the Roman Festival of Rosalia continues. The red rosebud is the birthday flower of 7 July (the 6th being one of the dates suggested for the ancient Roman feast day). The red rose in full bloom is the birthday flower for 13 October.

ROSA MUNDI

Botanical Name: *Rosa gallica* 'Versicolor'
Family: Rosaceae
Common Name: Rose of the world
Meaning: Variety

Myth and Legend

The Roman feast of the Rosalia was held to honour Rosa Mundi, the goddess, as rose of the world, the heart of creation, the consuming fire, the maker and shaper of souls. The rites included decorating the altar with roses and candles, and meditations on the inner meaning of the manifest power.

There are some wonderful twists in the story of the rosa mundi. The basic tale is that Rosamund Clifford, daughter of Walter de Clifford, was the mistress of King Henry II. He would meet her in secret in a hunting lodge, in the middle of a labyrinth which he had constructed near his palace at Woodstock, in Oxfordshire. Within the gardens of the palace there grew a free-flowering variety of the red rose, and this was known as 'Rosamund's rose', the rosa mundi.

The Queen found her, waiting for her lover in the rose bower at the centre of the maze one day, and offered her a choice, poison or a dagger. However, although the story that she was poisoned by a jealous Eleanor gained popularity in Elizabethan England it is certainly untrue, as is the tale of the labyrinth. Rosamund Clifford was buried at Godstow Priory, and her epitaph reads:

Hic jacet in tumba
Rosa mundi, non rosa munda
Non redolet, sed olet,
Quae redolere solet.

Peter Coates translates this as:

Here rose the graced, not rose the chaste, reposes;
The scent that rises is not scent of roses.

A rather more free translation puts it as:

A rose lies here within this tomb,
More chased than chaste, methinks;
She once exhaled a sweet perfume,
But now, alas, she stinks.

Unfortunately the lady's name was not Rosamund, but Jane. The only rose in the original story was that mentioned in her epitaph, engraved on her tomb. 'Rosamund' is derived from Teutonic roots and means a 'horse guard', or 'protector of horses'.

BURNET ROSE

Botanical Name: *Rosa pimpinellifolia*
Family: Rosaceae

On the small islands in the Bristol Channel, off the coast of southern England, the blooming of the burnet rose is observed with a certain amount of trepidation. When it flowers out of season it is always considered to be an ill omen, indicating that there will be shipwrecks and other disasters.

ROSEMARY

Botanical Name: *Rosmarinus officinalis*
Family: Labiatae
Common Names: Bride's herb; compass weed; dew of the sea; elf leaf; friendship bush; guard robe; *hasah leban* (Arabic); incenser; libanotis; polar plant; *romarin* (French)
Gender: Masculine
Planet: Sun (in Aries)
Element: Fire
Deities: Mary; Venus
Meanings: Fidelity in love; gladness of the spirit; remembrance; 'Your cherished memory will never fade from my heart'; 'Your presence revives me'

Rosemary has been used for thousands of years. Ancient Arabs and Egyptians grew it as a border for rose gardens. In folklore it is a holy and magical plant finding favour as a medicinal and culinary herb, a decoration and a talisman. It was introduced to Britain in the fourteenth century and was reputedly first grown in England by Philippa of Hainault, the wife of Edward III.

Myth and Legend

In Italy there is an unusual legend relating to this plant. There was once a queen who, whilst walking in her gardens one day and bemoaning her inability to conceive children, stopped in front of a rosemary bush covered in lush fresh shoots. The queen wished that she too might be as fruitful as the plant. Some time later she gave birth – to a small rosemary plant. She cherished it and lavished on it the best of care.

Her nephew, the King of Spain, however, played a trick on her and stole the plant. He made sure that it was still well cared for, watering it with goat's milk. One day, whilst he sat in his garden playing tunes on his flute, a beautiful girl sprang forth from the plant. He was captivated by her beauty and fell hopelessly in love with her. A short time later, he was called away to war and lamented that he must leave his secret beauty of the rosemary plant. Wanting no harm to come to the plant, and its secret occupant, he instructed his gardener to give it all possible care, at the same time instructing him that no one was to play any music nearby. The gardener did care for the plant but forgot the second part of his master's instructions. One evening, when he was playing his flute in the garden, the young woman again appeared from the rosemary bush. The king's sisters, observing the maiden's beauty, hit her and abused her until she disappeared. The plant then began to wither and wilt. The gardener was fearful of the king's wrath and ran away to hide in the forest. There he came across a dragon and learned that the only way he could revive the plant was to water it with dragon's blood. Using the poor weapons he had, he killed the dragon and took its precious blood to water the rosemary bush. The bush revived and the king returned to find his mysterious ladylove as healthy and as lovely as she had been when he left her. The king praised his gardener for his care and, in due time, married the maiden, the fair Rosa Marina.

Ros-marinus is 'the dew of the sea'; Pliny said, stating that the plant grew on dewy sites. In classical mythology it is linked to the goddess Venus, as both were children of the sea; it was therefore linked to love and lovemaking.

> The sea his mother Venus came on;
> And hence some reverend men approve
> Of Rosemary in making love.
>
> – Samuel Butler, *Hudibras*

In Christian mythology it is sacred to the Virgin Mary. *Ros marinus* became *rose maris*, the flower of Mary. According to Christian legend, she spread the wet clothes of the baby Jesus to dry on a rosemary bush and it burst into bloom. Another legend has it that the bush became aromatic when she spread the clothes over it. During the flight into Egypt, she is said to have laid her robe over a rosemary bush, whereupon its flowers turned from white to the blue of her robe. The plant was particularly revered in Spain for having given shelter to the Virgin and child during their time in Egypt.

Another old Christian belief is that a rosemary bush cannot grow taller than a man or exceed the lifespan of Christ, thirty-three years.

Magic and Lore

Rosemary has accompanied people at the best and worst of times. Its power as an aid to memory or remembrance had led to its use at weddings and at funerals. Brides, for example, might wear or carry wreaths of rosemary, dipped in scented water, just as Anne of Cleeves did at the wedding to King Henry VIII. This was symbolic of love and fidelity. Sprigs of rosemary, bound with ribbons, were sometimes presented to bridesmaids and guests by the groom, as a souvenir of the wedding day and a token of friendship. Where the rosemary was included in the bride's wedding bouquet it should be planted in the garden following the wedding, like myrtle. Pieces from the same plant could then be used in the wedding flowers of daughters and granddaughters. It might also be dipped into the wine before the bride and groom drank of it as a token of their faithfulness and lasting devotion to one another.

> Rosemary is for remembrance
> Between us daie and night;
> Wishing that I might always have
> You present in my sight
>
> – Clement Robinson, *A Nosegaie*

The classical associations with Venus have led to it being used in potions to induce love and lust A young man might, for example, press a sprig of the plant into the palm of his true love as a token of his undying love.

> Who passeth by the Rosemarie
> And careth not to take a spray,
> For woman's love no care hath he,
> Nor shall he though he live for aye.
>
> – Spanish Proverb

Perhaps the most commonly known use of rosemary is as a symbol of the remembrance of the dead. Robert Herrick puts it succinctly when he says:

> Grow for two ends, it matters not at all,
> Be't for my bridal or my burial.
>
> – *The Rosemary Branch*

Sprig of rosemary were left in the room where a corpse was laid out and these were taken up by the mourners at the funeral to be dropped into the grave, on top of the coffin, as symbols of remembrance. In times when corpses were not as well prepared as today the strong scent of the herb must have helped to conceal the odour of decaying flesh.

Rosemary sprigs were worn by the returning Australian and New Zealand soldiers on Anzac Day as a mark of remembrance of their fallen comrades. A more recent tradition has been for it to be worn on 23 April in Stratford-on-Avon, to mark the birthday of William Shakespeare. The herb, and its significance as a symbol of remembrance, is referred to in several of his plays, including by Ophelia in *Hamlet*:

> There's rosemary,
> That's for remembrance

It may be because of its associations with the holy family that rosemary continued to be used widely as a protective herb, although its use in this way pre-dates the spread of Christianity. It was burned as incense in churches. It is one of our oldest incenses and has been used as a substitute for frankincense to purify areas of negativity and evil before magical rites (hence its French name, *incenser*). It has been said to be powerful enough to protect the church, the dead and the living from all evils. It was grown in churchyards and used to decorate churches at Christmas.

> Down with the rosemary, and so
> Down with bays and mistletoe;
> Down with the holly, ivy, all,
> Wherewith ye dress'd the Christmas Hall.
> – Robert Herrick, *Ceremony on Candlemas Eve*

Sprigs of rosemary and oranges stuck with cloves were sent to special friends as gifts on New Year's Day. Just as the herb could be dipped into the wine at a wedding it might also be added to the wassail cup, emblematic of the loyalty of all who partook of the wine. People believed that the Christmas rosemary from churches had special powers of protection against evil spirits.

It might be worn as an amulet to protect against illness and injury as well as to ensure success in all enterprises. It was also used as a protection against storms and nightmares. Placing a small sprig of it under the bed or under the pillow was said to prevent bad dreams and ensure the sleeper's safety throughout the night. A mixture of rosemary, garlic and sea salt meant 'Do not cross my path', and would stop a ghost. Burning a mixture of rosemary, juniper and thyme will clear witches and clean the air in a room where someone has been ill.

Growing it by the entrance to the home protects the house. Hanging a sprig above doors and windows, or placing it onto doorposts, will protect a home from burglary, as it prevents thieves from entering. It also serves as a guard against evil spirits, witches and witchcraft. It was hung in clusters over the cradles of babies to protect them from the evil eye and gained the name 'bride's herb' as it was included in the bride's bouquet at weddings, also to ward off evil and negativity. Growing rosemary, and placing sprigs of it about the home, will also keep malicious faeries away, but burning the dried herb will *attract* them. Sicilian tradition tells that young faeries would lie amongst the branches of a rosemary bush disguised as snakes.

It was believed that rosemary would strengthen the brain and aid the memory. Students used to sniff sprigs of it, or wear small pieces in their hats, while they studied in order to better remember their lessons. Roger Hackett, writing in 1607, says, 'It helpeth the brain, strengthenth the memorie, and is very medicinal for the head.'

Rosemary has sometimes been called the friendship bush. It was long believed that if you grew the plant in the garden you would never be short of friends. Sir Thomas More is said to have allowed it to spread right across his garden because it symbolized friendship, and because he liked it. 'As for Rosmarine I lett it runne all over my garden walls, not onlie because my bees love it, but because it is the herb sacred to remembrance, and, therefore, to friendship; whence a sprig of it hath a dumb language that maketh it the chosen emblem of our funeral wakes and in our burial grounds.'

In love divination a young woman might discover who her future husband was to be by dipping a sprig of rosemary into a ground glass vessel containing a mixture of wine, gin, vinegar, rum and water on the eve of St Mary Magdalene's Day (22 July), in an upper room of the house. Two other young women must accompany her, and all three of them should be under the age of twenty-one. All three should pin a piece of the plant to their bosoms and take a sip of the mixture before retiring to bed. If the entire ritual is conducted correctly, and in total silence, then they will all be assured of having prophetic dreams.

Alternatively the girl might place a sprig of rosemary and a crooked sixpence beneath her pillow on St Agnes Eve (some writers suggest Halloween). She must dip sprigs of thyme and rosemary into water and then place one herb in each of her shoes and put them each side of the bed with the words:

St Agnes that is to lovers kind,
Come ease the troubles of my mind.

That night she is sure to have visions of her future husband in her dreams.

Yet another way for her to discover whom she will marry is to place a bowl of flour beneath a rosemary bush at midnight on Midsummer's Eve. In the morning she would find the initials of her future husband written in the flour.

There is a belief that rosemary grows well in gardens where the wife dominates and poorly where the husband is dominant.

Where rosemary flourishes,
Misses will be master.

If the plant roots readily it is an indication of the degree of power of the matriarch of the family. Another proverb proclaims that the plant will only grow well for a good woman. Rosemary is said not to grow in the gardens of those who are not just and righteous: 'Lavender and rosemary is as a woman to a man, and white rose is to red. It is a holy tree and with folke that hath been just and rightfulle it groweth and thryveth.' Other traditions contradict this, however, claiming that the plant will always thrive best in the worst people's gardens.

The most obvious use of this plant for most of us is as a culinary herb, but eating food from a spoon made from rosemary wood is also said to prevent poisoning, and any food prepared using such a spoon will always taste good, regardless of how badly it started out. A small amount of rosemary added to a barrel of beer is said to prevent the drinker becoming drunk.

A dream in which you see rosemary means that everlasting love will be yours.

Rosemary is the birthday flower of 17 January.

Medicinal

Whenever *officinalis* is appended to a plant's botanical name, it indicates that it has been an important herbal remedy. It has been claimed that rosemary was introduced to Britain in 1548 but, as it was important to the Romans, it is more likely that it was first introduced much earlier.

It has a reputation of being something of a panacea. It was used in tisanes and treatments to aid the mind, body and the spirit. Physicians of old gave it as a stimulant against depression, an aid to digestion and a cure for chest complaints, gout and diseases of the brain. It was even claimed that a comb made from the wood would be a treatment for baldness and prevent giddiness, and Arab physicians prescribed it to bring back speech following strokes.

One the best-known uses of the herb was as a treatment for the plague and for 'jail fever'. People carried sprigs during plagues, smelling them to ward off infection, and sprinkled their bodies with rosemary water. Writing in 1603 Decker, in *The Wonderful Year*, reports that the price of rosemary rose from 1 shilling an armful to 6 shillings a handful because of the spread of the Black Death. An old tradition claims that in the fourteenth century Queen Phillippa's mother, the Countess Hainault, sent her supplies of the plant as a protection against the plague. The Queen's posy, carried on Maundy Thursday, contained rosemary and thyme, a standard plague remedy.

Rosemary is believed to preserve youth. 'Spirits of Hungary', or Hungary water, was made from rosemary in spirits of wine. It was formulated for the invalid Queen Isabella of Hungary in 1235, and she claimed it restored her youth and cured her paralysed limbs. Bathing in rosemary water is supposed to keep one lusty, lively, joyful and young. Regularly sniffing at a box made from the wood might help you to discover the secret of eternal youth.

Culpepper recommended that a decoction of rosemary in wine should be drunk or rubbed on the temples as a treatment for lethargy, epilepsy and weak memory, and to quicken the spirits, ease cold diseases of the head and brain, combat dullness of mind and senses like stupidness, giddiness or swimming of the head.

BLACKBERRY

Botanical Name: *Rubus fruticosa*
Family: Rosaceae
Common Names: Bly; bramble; brambleberry; bramble-kite; brameberry; brummel; bumble-kite; cloud berry; dewberry; goutberry; scald head; thimbleberry
Gender: Feminine
Planet: Venus
Element: Water
Deities: Blodeuwedd; Brigit
Meanings: Envy; death; grief; injustice; 'I was too hasty, please forgive me'; lowliness; pain; remorse; weariness; wickedness

Myth and Legend

In Greek mythology Bellerophon attempted to fly across the heavens on the winged horse Pegasus. Zeus brought him down by sending a gadfly to sting the horse so that it threw him. Bellerophon landed in a blackberry bramble, which blinded and maimed him, thereafter he became an outcast, shunned for the remainder of his life.

One tradition claims that the burning bush from which God spoke to Moses was a bramble. In Christian mythology it was used to form the crown of thorns at Christ's crucifixion, the dark berries symbolizing the saviour's blood. In Christian art the blackberry symbolizes spiritual neglect or ignorance. St Matthew, in his gospel, uses the image of the blackberry in a warning about false prophets.

Magic and Lore

The name bramble derives from the Old English *braembel*, or *brymbyl*, meaning prickly. In some pagan European religions it was considered to be sacred and was used in rituals and worship. Blackberry pies are still made by some modern Wiccan groups at Lughasadh (2 August) to celebrate the harvest. The Celts considered them to be taboo. When the plant was 'Christianized' this taboo was maintained, at least in part.

In a tale similar to the story of Bellerophon, it was said that when Lucifer was expelled from Heaven following his struggle with St Michael, he fell to earth in a patch of blackberries. He spat on them and cursed them for pricking him. Traditionally, therefore, to pick blackberries after 10 October (old St Michael's Day) was to court misfortune. In some parts of France country folk would not eat the fruit as it was said that the dark colour was a result of the Devil spitting on them. In Brittany they were considered to be untouchable because they were faerie fruit.

Different places set different dates on which the plant becomes unsafe to eat.

Some identify Halloween as the cut-off date and others state that they should not be eaten after 1 November so that the Devil may have his share. In the Lake District it was simply said that blackberries should not be eaten after they have been frosted, as they then become the Devil's fruits. Disfiguring marks on the foliage, caused by leaf miners, are known as 'Devil's marks'. In Scottish tradition blackberries were poisoned after Old Holy Rood Day, 26 September. The warning states:

Oh weans! Oh weans the morn's the fair
Ye may na eat the berries mair.
This nicht the Deil gangs ower them a'
To tough them with his poisonous paw.

Older traditions sometimes link blackberries to witches or faerie folk. They would not be used after Halloween as it was said that witches urinated on them or that Pooka, a mischievous imp, would inhabit them.

There is a tradition that before Lucifer cursed the bramble it was a plant of great beauty. It is suggested that, if gathered at the correct phase of the moon, the fruit may still prove to be a great protection against evil. It was used as a protection against such earth-bound spirits as vampires, a use that pre-dates garlic. The method depends on a supposed fascination among physical demons with counting things. Placing blackberry, or an elder, over a doorway or on a windowsill will keep the demon at bay, as it will count the berries and the thorns until the morning.

The bramble's link to the Devil has yet another twist. Passing through a hoop or arch formed by the bramble rooting along its length is a folk cure. It has been suggested that people would pass through such a bramble hoop in order to 'give themselves to the Devil'. Thereafter they could expect to enjoy good health, great strength and good luck at cards.

Blackberries were planted on or near graves in order to prevent the dead from walking. They are described as 'that which holds the rose and beauty of the soul from answering the call of the deity.'

There are a number of links between animals and blackberries. Kittens born on St Michael's Day are known as 'blackberry kittens' and are supposedly small, weak and difficult to rear as they are especially mischievous. It is especially lucky to own a tortoiseshell blackberry kitten. Generally, it is claimed that cats, horses and chickens are never well at blackberry time. However, as with blackberry kittens, chickens born at blackberry time are always supposed to be the best.

If blackberries ripen late, or slowly, then the winter will be particularly long and cold. The blackberry crop will be poor if there is rainfall on 1 May or 2 June. Cold weather during May is known as a 'blackberry winter', as it is sure to bring a bumper blackberry crop. As a general guide, frosts when the blackberries first come into bloom are rarely severe enough to kill plants. A spell of fine weather at the end of September and early October is sometimes called a 'blackberry summer'.

To dream of passing through a place covered in brambles means troubles ahead. If they prick you then secret enemies will do you harm and if they draw blood it indicates a loss in trade. To dream of passing through unharmed shows that you will be triumphant over all your enemies.

Blackberry is the birthday flower of 19 July.

Medicinal

As we have seen, passing through an arch formed by a bramble rooting itself at either end brings good health and it is most effective when the two ends are rooted in ground belonging to two different people. It has been claimed as an effective cure for ailments such as whooping cough, rheumatism, hernia, and rickets. It has also been said to be a treatment for blackheads and even slowness in learning to walk.

There are a number of variations on the ritual of passing a child through the arch. The most straightforward requires that the child be passed east to west, as the sun passes east to west across the sky. Another variation requires the child to be passed through three times on each of nine mornings (some say nine times on each of three mornings) before sunrise. If this is done to treat whooping cough the following should be repeated as the child is passed through.

Under the briar and over the briar,
I wish to leave this chin-cough here.

Or:

In bramble, out cough;
Here I leave this whooping cough.

On the Welsh borders it was traditional that an offering of fresh bread and butter was left beneath the bramble arch after the child had been passed through. In Herefordshire the patient was require to eat some of the bread and butter whilst the adults accompanying them recited the Lord's Prayer. The remaining food would be given to an animal or bird on the way home. This creature would then die and the illness would die with it.

In East Anglia forms of bramble that have especially large fruits are known as 'mulberries', and it has been claimed that passing children through a bramble arch was the origin of the song, 'Here we go round the mulberry bush', but this is very unlikely. It was said that an amulet to prevent whooping cough could be formed by cutting the briar into the shape of a cross and putting it over the breast of the child.

The Greeks and Romans considered ripe blackberries to be a cure for gout, and eating the young shoots was supposed to fasten loose teeth. To treat burns take nine blackberry leaves dipped in water drawn from a holy well and apply them to the afflicted areas, repeating three times on each leaf:

There came three angels out of the East
One brought fire and two brought frost
Out fire, in frost;
In the name of Father, Son and Holy Ghost

RASPBERRY

Botanical Name: *Rubus idaeus*
Family: Rosaceae
Meanings: Envy; remorse

Raspberries appearing in your dreams indicate success and happiness in marriage, fidelity in a sweetheart and good news from abroad. They might also suggest that you might suffer great disappointment but consolation from an unexpected source.

Raspberries have been used in folk medicine to treat a number of disorders. Consuming the leaves, or a tea made from the leaves, was supposed to ease labour and period pains, prevent miscarriage and increase lactation. Raspberry leaf tea has also been recommended for those who have been caught out in the rain and soaked to the skin. Drinking raspberry leaf tea before going to bed was supposed to prevent colds.

An eyewash may be made from raspberries to relieve tired or sore eyes.

SORREL

Botanical Name: *Rumex acetosa*
Family: Polygonaceae
Common Names: Green sauce; soorik; sooricks; sorrel grass; sour-dock; sour-dockling; sour grass; sour leeks; sour sabs
Planet: Venus
Element: Earth
Meaning: Ill-timed; wit (wild); paternal affection; with affection

Sorrel is one of the oldest of culinary herbs. John Evelyn says of it:

> Sorrel sharpens the appetite, assuages heat, cools the liver and strengthens the heart; is an antiscorbutic, resisting putrefaction and in the making of sallets imparts a grateful quickness to the rest as supplying the want of oranges and lemons. Together with salt, it gives both the name and relish to sallets from the sapidity, which renders not plants and herbs only, but men themselves pleasant and agreeable.

Medicinal

> No plant better cleanse the body of feculent humours, if the plant be eaten green or its juice drunk; it helps an offensive breath, fastens loose teeth, cures putrefaction of the gums, and is extremely beneficial in all cases where the blood is too fluid and the vessels lax.
>
> – Herman Boerhaave

The juice of sorrel has been used to treat spots, and the Romans chewed the leaves to allay thirst.

DOCK

Botanical Name: *Rumex obtusifolius*
Family: Rosaceae
Common Names: Batter dock; bulmint; bulwand; butter docken; celery seeds; common wayside dock; cushy cows; docken; doctor's medicine; donkey's oats; kettle dock; land robber; ranty tanty; redshank; smart dock; sour dock
Gender: Masculine
Planet: Jupiter
Element: Air
Meaning: Patience

Magic and Lore

The seeds of the dock have been used in spells and incenses aimed at aiding trade. Sprinkled around a place of business it is claimed that they will attract customers.

Medicinal

Dock is best known as a folk remedy for treating nettle stings, but part of the charm has been lost. In different parts of Britain there were various incantations to be said when using dock, amongst them:

Nettle in dock,
Dock in, nettle out,
Dock rub nettle out.

Nettle out, in dock,
Dock shall have a new smock,
Nettle out, dock in,
Dock remove the nettle sting.

In dock, out nettle,
Don't let the blood settle.

Dock, dock shall have a smock,
Nettle shall ne'er have a one.

It has also been used as a cure for boils, cuts, and sunburn. The leaves were cut and dried before being bound to joints as a remedy for rheumatism. And William Coles, in the *Art of Simpling*, says, 'The seeds of docke tyed to the left arme of a woman doe help barrennesse.'

RUE

Botanical Name: *Ruta graveolens*
Family: Rutaceae
Common Names: Bashoush; garden rue; German rue; herb of grace; herb of grace o' Sundays; herbygrass; hreow; mother of the herbs; rewe; witchbane
Gender: Masculine
Planet: Sun
Element: Fire
Deities: Aradia; Diana
Meanings: Distain; docility; domestic happiness; 'Do not annoy me with your unwelcome attentions'; grace; purification; repentance

Myth and Legend

Since the late seventeenth century rue has been used as a heraldic device on the collars of those holding the Order of the Thistle. In 1811, Fredrick Barbarossa permitted the first Duke of Saxony to adopt, as his heraldic symbol, a chaplet of rue across bars. This, in time, passed on to become the emblem of the suit of clubs on a pack of cards. The Order of the Rautenkrone, or rue crown, was established by King Frederick Augustus I of Saxony in 1807.

Magic and Lore

Rue is symbolic of sorrow, remorse and repentance. It has been used variously to help, bless and protect as well as to cause harm and to curse. The ancient Greeks and Romans viewed it as a protective herb. The Romans believed that it could be used to avert the evil eye, preserve sight and to give second sight. It was the Romans who introduced it to Britain. Pliny tells us that it was eaten in great quantities by artists to preserve their sight, sharpen their vision and ease strained eyes.

The scent was said to sharpen the mental processes, clearing the head, and so be especially beneficial when considering matters of the heart. It has been suggested that it encourages chastity, although perhaps not in all who use it: 'This noble hearbe [while] making men chaste, women fils with luste.'

Another writer sums up the benefits of the plant as:

Rue maketh chaste and eke
Preserveth sight,
Unfuseth wit, and fleas doth
put to flight.

It was, indeed, used at one time to kill fleas and the sap was rubbed onto areas to prevent them entering.

In magical practices rue was used as a protection. Practitioners would sprinkle the fresh juice of rue, mixed with the morning dew, in a circle around them to prevent demons and spirits from harming them. The juice of the herb, added to bath water, would break curses sent against the bather and rubbing the fresh leaves on the floorboards in the home would send curses back on the person who cast them. It was also used in the rites of exorcism. Holy water was sprinkled with a sprig of rue and the practice of doing this before the celebration of high mass on a Sunday led to the plant being called 'herb of grace o' Sundays'. Growing it in the garden is supposed to be sufficient to protect the whole household, while carrying a sprig protects the bearer against poisons, werewolves and many other ills. Hanging it above doorways and windows would ward off evil spirits and prevent them from entering the house. Bunches of the herb might also be worn at the waist to repel witches, earning it the name 'witchbane'.

In Herefordshire, if a woman dropped a nosegay of rue wrapped in a half eaten piece of bread and butter at the porch of a church (or at the gate), the mid point between the hallowed and unhallowed ground, it denoted that she had been deserted or was in an unhappy marriage.

Like roses, rue is said to always grow best when it has been stolen.

Shot that has been washed in the juice of rue, and arrowheads that are dipped in the sap, are said always to find their target. Eating rue is said to prevent one talking in one's sleep.

Medicinal

Like many other strongly scented herbs, rue was used as a precaution against 'jail fever' and plague. In July 1760 a rumour spread that plague had broken out at St Thomas's Hospital in London, and by the following morning the price of wormwood and rue at Covent Garden Market had risen by 40 per cent. Judges attending the assizes were given rue to wear or place on the bench according to records from the Central Criminal Court in the mid-seventeenth century. 'Jail fever' was most likely a euphemism for typhus spread by lice.

John Gerard recommends the use of rue to prevent harm from almost any type of bite or sting. If a man were anointed with the juice of rue, the poison of wolf's bane, mushrooms or toadstools, the biting of serpents, stinging of scorpions, spider, bees, hornets and wasps would not hurt him. He also claims that weasels will eat rue before they attack snakes to avoid being harmed by their venom.

Rue tea was given to treat coughs and colds and to improve the appetite, and the fresh leaves were placed against the forehead to ease headaches. Wearing of necklaces of the herb following illness, was supposed to aid recuperation and prevent future illnesses.

Great care must be taken in gathering rue, as it causes a photo-allergic reaction on the skin, resulting in severe blistering. It used to be said that it was safe to collect early in the morning but became poisonous later, but it would be unwise to handle the plant at all in bright sunlight.

WILLOW

Botanical Name: *Salix alba*; *Salix* spp.
Family: Salicaceae
Common Names: English palm; pussy willow; saille; salicyn willow; saugh tree; tree of enchantment; with; withy; witches' aspirin
Gender: Feminine
Planet: Moon
Element: Water
Deities: Anatha; Arawn; Artemis; Athena; Bel; Belenos; Belili; Belin; Cerridwen; Circe; Diana; Europe; Geshtinanna; Gwydion; Hecate; Helice; Hera; Ishtar; Jehovah; Luna; Mercury; Minerva; Orpheus; Osiris; Persephone; Zeus
Meanings: freedom (water); melancholy; mourning (weeping); bravery and humanity (French); 'Be mine again'; forsaken

Myth and Legend

A great many deities are associated with the willow in classical mythologies. According to Pausanias, a historian from AD 2, Persephone, the goddess of the Underworld and daughter of Demeter and Zeus, had a sacred grove in 'far western Tartaras' (a synonym for Hades). Robert Graves, in *The White Goddess*, tells us that this grove was remarkable for its 'black poplars and aged willows'.

Helice, a goddess particularly associated with water magic, and therefore with willows, was beloved by Zeus. Hera, his consort, was known for her extreme jealousy, and transformed Helice into a large bear. Zeus placed her in the night sky as the constellation Ursa Major. Her priestesses were believed to use willows in every aspect of water magic. An ancient willow muse (a willow faerie) was called Heliconian, after Helice. The willow muse is supposed to be sacred to poets as the sound of the wind in the willows inspires the human mind.

Orpheus, considered by the ancients to have been the most celebrated of poets, gained his gifts of eloquence and communication by carrying willow boughs during his journey into the Underworld searching for Euridice, his love. He also carried a lyre which Apollo had presented to him, and the muses had schooled him in its use. Playing this lyre enchanted wild beasts, plants and even the stones on Mount Olympus.

The association with elements of communication and eloquence take on another twist, as the tree is linked to St Brigit and said to aid the development of visions and ease communications. This association extends further with the tree also linked with Mercury, the messenger of the gods.

Hera herself was born beneath a willow tree on the island of Samos and this tree was long preserved with her temple there, which was alleged to contain a wooden image of the goddess that would disappear each year. When it was searched for it would always be found on the seashore bound to a willow, the branches of which would partially conceal the image. The priestesses would unbind the image, wash it and convey it back to the temple.

The enchantress Circe, we are told, lived in a thick willow grove on the island of Aeaea. There, by her magic arts, any people who set foot on the island were transformed into swine. We are also told that Circe had a riverside cemetery dedicated to Hecate and her moon magic, planted with willows. It is in the cemetery of Circe that the darker aspects of the tree can be seen. Corpses wrapped in ox-hide were placed into the treetops for the elements to claim and the birds to devour.

Hecate, the witch goddess, was the most powerful of the willow-moon goddesses. She was descended from the race of Titans and was the only one of them to retain her power under the rule of Zeus. She taught sorcery and witchcraft, and haunted tombs and crossroads. It is said that her presence is announced when dogs howl at the moon.

In the myths of Osiris, one of the trees identified as having caught up his coffin as it floated down the Nile is the willow. The spirit of the god flew up into the tree's branches in the form of a phoenix. The ancient fertility rite of raising the willow is connected with Osiris and there are similarities with the pagan rites of 'Green George' (see page 448).

In the rites of the goddess Artemis there are obvious echoes of the pagan rituals of Osiris and 'Green George'. In the Spartan fertility rites a male celebrant was bound with strips of willow to a sacred image or a tree trunk. He was then flogged until the lashing produced an erotic reaction and he ejaculated, thereby 'fertilizing' the ground with his blood and semen.

Belili was a Sumerian goddess of trees, especially willows, and presided over springs and wells. At some point she was dramatically superseded by her consort, the willow god Bel. He went from being a local willow deity to being a sun god worshiped in countries across the globe. He became the supreme god of the universe, the patriarchal deity replacing the matriarchal one, the sun replacing the moon. In Europe, the Celts acknowledged Bel as Belin, the sun god, lord of life and death. His worship replaced lunar worship, and the feast day was 30 April – 1 May, Beltane, when fires were lit in his honour. In many traditional tales, young sun gods were set adrift upon the waters in baskets woven from willows. Those best able to help them plucked them from the waters. The biblical story of the infant Moses clearly has a very similar basis.

Anatha (a derivative of Athena or Anat) was the centre of a willow cult based at Jerusalem. This goddess was ultimately ousted by the priests of Jehovah who claimed the 'rain-making' willows as 'Jehovah's trees' at the Feast of the Tabernacles. The time of the fire and water ceremony is still referred to as the Day of the Willows.

In the Muslim telling of the story of the David, the Israelite king, and his marriage to Bathsheba, we are informed that, some time after the wedding David was playing his harp in a private chamber and looked up to find three strangers. He was initially surprised as he had given strict instructions that he was not to be disturbed, but learned that his uninvited visitors were angels. They had come to convince him of the seriousness of his crime, as he had sent Bathsheba's husband into battle, so that he might be killed and she would be free to marry him. Realizing the heinousness of what he had done, David prostrated himself on the floor and wept tears of repentance.

There he stayed for forty days, weeping and trembling before the judgement of the Lord and it is said that in those forty days he wept as many tears as the whole of the human race have shed, and will shed, on account of their sins from his time until the Day of Judgement. Throughout the whole time that David wept he recited psalms of penitence. His tears formed two streams that ran from his chamber through the anteroom and into the garden. Where they drained down into the soil, up sprang two trees, one a weeping willow and the other a tree of frankincense. The willow weeps and mourns whilst the frankincense tree drops big tears in memory of David's sincere repentance.

A Christian folk carol, 'The Bitter Withy', links the willow with the story of Christ and explains why the tree's centre rots comparatively quickly. In the tale the infant Jesus asks his mother's permission to go and play, and she agrees but instructs him to stay out of trouble. The young Jesus meets up with three highborn children and invites them to play with him. Their response it that the sons of 'lords and ladies' cannot play with 'a poor maid's child, born in an ox's stall'. The child Christ then uses a bridge formed from sunbeams to cross a river, but when the other children try to follow him, the bridge does not support them and they are drowned in the waters below. Their mothers complain to Mary, who punishes Jesus by beating him with willow twigs.

> So Mary mild fetched home her child,
> And laid him across her knee,
> And with a handful of willow twigs
> She gave him lashes three.

The child then cursed the willow saying:

> Ah bitter withy, ah bitter withy,
> You have caused me to smart,
> The willow must be the very first tree
> To perish at the heart.

A Czech folk story tells of a nymph who appeared amongst men during the day but each night returned to her willow tree. Eventually she met and married a man, and had a child by him. All went well until one day, unaware of its significance, the man cut down the willow tree. At that point his wife died. The willowwood was used to make a cradle for the child and when the infant was placed in it, it would instantly fall asleep. The baby was still able to communicate with its mother by means of a pipe made from the sucker twigs taken from the stump.

In Irish mythology, the willow is one of the seven sacred trees.

Magic and Lore

This is a tree of enchantments and magic. Musicians, poets, artists, priests and others who sit beneath it will, like Orpheus, gain eloquence and inspiration.

Its pagan associations have always been strong. It was revered, as it was sacred to the various manifestations of the moon goddess, she who reflected her lunar magic on the waters of the Earth. Across north-western Europe there are stories of the trees being sought out by witches, or village wise women, as a herbal remedy, or for flexible shoots to bind their brooms.

Willow being a tree of the moon is also a tree of dreams and dreamers. Its powers fluctuate with the phases of the moon, ebbing and flowing like the tides. It has always been considered feminine, with influence over the vision-generating subconscious. Its visionary aspects are strongest at two points in the year, Beltane and Samhain. Used in divination it will be more powerful by night than by day, except when the moon is visible in the day.

The story of Orpheus in the Underworld clearly links the willow with the realm of the dead, as do the associations with Persephone, Circe and Hecate. Ancient burial mounds, sited by lakes and wetlands, were lined with willows to protect the spirits of the departed. Flints shaped as willow leaves have been found in megalithic burial mounds. In country lore it was said:

> Plant as willow and allow it to grow,
> To ease the passage of your soul at death.

Branches of willow were placed on coffins and willow saplings planted on graves. This echoes a Celtic tradition whereby the soul of the corpse in the ground rises into the sapling planted above, which then retains the essence of the departed person. Scandinavian folklore says that no soul may depart the Earth in peace and no child be safely born unless a willow wand is suspended somewhere nearby.

> By the rivers of Babylon we sat down and wept when we remembered Zion.
> There on the willow tree we hung our harps
> For there, those who carried us off demanded music and singing
> and our captors called on us to be merry. 'Sing us one of those songs of Zion'.
> How could we sing the Lord's song in a foreign land?
>
> – Psalm 137

Thus the willows wept at the captivity of the Jewish people. Although more recent translations of the Bible identify the willows upon which the exiled Jews hung their harps as poplars or Euphrates aspen, it is the image of the willow as a tree of mourning that persists. Even Linnaeus, the father of plant nomenclature, may have been deceived as he named the plant *Salix babylonica*, even though it is not native to Babylon.

'Wearing the willow' once meant grieving openly, and garlands of mourning were traditionally woven from willow branches. It also became a symbol of someone who had been jilted by his or her lover: 'In love, the sad forsaken might the willow garland wearth.' Swan's *Speculum Mundi* of 1635 also says, 'It is yet a custom that he which is deprived of his love must wear a willow garland.' This custom persisted for several centuries There have been a number of suggestions as to the thinking behind it, although there is an obvious link with the willow as a tree of mourning. One suggestion is that wearing or carrying willow will help the bearer to attract a new lover, as willow was used in spells to attract love and lovers.

This tradition may have given rise to the wearing of the willow by people who were parted from their loved ones for a considerable period. In this case, we are told that it would prevent any jealous thoughts.

It is often considered unlucky to have the tree growing in the garden or to take the foliage or catkins into the house, the only possible exception to this being on May Day (Beltane), when it was used in decorations. It was also at Beltane that 'willie wains', willow wands, were cut in the north of Britain. These were supposed to contain the power of water, never truly still, and were used as a protection against evil.

It is perhaps then that a young woman might discover the identity of her future husband. She must run, three times, around the house whilst carrying a willow wand, and an image of her future husband will then be seen holding the end of the wand. Willow wands were given as tokens of love and friendship on May morning and the wood was considered to be lucky.

In Shropshire the pussy willow was also known as 'goosy goslins' and it was said that if the flowering shoots were taken into the house there would be no goslings that year.

The Chinese considered the willow a lucky plant, magical and capable of warding off all illness and harm. They identified it as an emblem of immortality, as even the smallest piece of the wood pushed into the ground will take root and grow. This makes it appropriate foliage to be carried at Easter, in the English Christian celebrations, as 'palms'. Its use as a church decoration on Palm Sunday may be based on an earlier pagan symbolism, possibly from its use in May Day decorations. Easter itself was originally a pagan festival date and can be compared to festivals marking the 'resurrection' of Persephone at the beginning of spring after her six-month exile in the Underworld. In northern Russia the week immediately prior to Palm Sunday was known as Willow Week, the 'palms' in this case being decorated with ribbons and sold for use on Palm Sunday.

On 23 April (St George's Day), Romanian gypsies celebrate the festival of 'Green George'. This character is very similar to the British Jack-in-the-Green or the Green Man, and is represented by a man wearing a wicker frame covered by greenery. He epitomizes the spirit of vegetation that brings fruitfulness to the cornfields. Unlike

Jack-in-the-Green, Green George propitiates the water spirits through the willow trees. To enable this to happen great preparations are made. A young willow tree is cut down and set up in the centre of the festivities. It is bedecked with garlands in a party atmosphere. On the same night, all the pregnant women gather around the tree and each lays one item of clothing beneath it. If, overnight, a leaf has fallen onto one of the garments it is taken as an omen that the willow goddess will grant them an easy delivery for their child.

At dawn, Green George appears and approaches the willow, and knocks three nails into its trunk. He then removes them, takes them to a nearby stream and throws them into the waters to propitiate the water spirits. He returns to collect the willow tree, taking this down to the stream and dipping its branches in the current until they are heavy with the water. Livestock is then brought to him so that he can shake the water onto the animals as a blessing from the willow and the waters.

When all this has been done the willow tree is once more erected in the centre of the festivities so that it may be approached by those seeking healing. It is considered especially favourable for the elderly and for infants, and gives relief from ailments such as rheumatism and the pains of childbirth. When the festivities draw to a close the elderly can once again approach the willow to seek a further blessing. In a bid for another safe year they must spit on the tree three times saying, 'Willow tree, willow tree, you will soon die. Let us live.'

The malevolence of the willow is seen in a belief that the trees would uproot themselves at night and, muttering to themselves, stalk any unwary travellers.

Ellum do grieve,
Oak he do hate,
Willow do walk
If yew travels late.

As we have seen, willow was once considered a protective plant. The 'good' willow tree kept witches away and pieces of the wood hung over a doorway would ward off witches and protect against the evil eye. Touching or knocking on wood for luck is a common superstition, and willow is the preferred wood; knocking on it will avert the evil eye and bring good luck.

To make a wish you must formulate it in your mind and explain it to the tree. Then select a pliable shoot and tie a knot in it whilst reciting the wish. Remember to thank the tree for its help before departing, and go back and untie the knot in the shoot when the wish has come true.

To discover whether you will marry within the year, throw one of your shoes up into a willow tree on New Year's Eve. If the shoe lodges in the branches in the first nine attempts you will be married within twelve months, if not, you will not.

When witches were making conjurations in the open air, in order to call up spirits, they burned a mixture of sandalwood and crushed willow bark during the waning of the moon. The wood could also be used in dowsing, being one of the woods used for making divining rods.

The willow is a sacred tree and, like many others, it must never be burned as to do so will result in grief. To dream of mourning over some great calamity beneath a willow tree is, somewhat paradoxically, an indicator of good news to come.

If a child was struck with a withy stick it could cease to grow afterwards.

Medicinal

Willow epitomizes the moon goddess, with powers over lunar magic and the earth's waters. Being a tree of wet sites it offers protection against the 'damp' diseases and was sought by healers to treat such ailments as rheumatism. Its medicinal and religious uses come together in its name of 'witches' tree'. There are tales of old crones visiting the tree at dead of night, when they are watched by the moon.

The tree was originally the sole source of salicylic acid, the base of aspirin (now synthesized), which is used in the treatment of rheumatic and other pains. This relates back to another once widely held belief that any endemic illness must have its remedy close by, rheumatism being associated with the damp and the willow a tree of damp places.

The seeds can be used as an aphrodisiac, or to boost the male libido. They must be steeped in spring water, and the water drunk. Thereafter, although the man's sex drive will improve, he will father no sons, only barren daughters.

It is sometimes said that the priests of Ascelepius, god of healing, used a variety of willow, *agnus castus*, to cure barrenness. It is more likely that this relates to the chaste tree, a totally unrelated species, *Vitex agnus-castus*.

SAGE

Botanical Name: *Salvia officinalis*
Family: Labiatae
Common Names: Common sage; garden sage; swage; the saving herb
Gender: Masculine
Planet: Jupiter
Element: Air
Meanings: 'I think of you' (blue); 'Thine forever' (red Salvia), domestic virtue; esteem; good health; household; long life; 'We have a wonderful family'; wisdom

> Sage strengthens the sinews, feavers heat doth swage
> The palsie helps and rids of mickle woe
> In Latin, Salvia, takes the name of safety,
> In English, sage, is rather wise than craftie;
> Sith then the name betokens wise and saving,
> We count it Nature's friend and worth the having.
>
> – *English Doctor*, 1607

Myth and Legend

The Romans held sage sacred as a healing plant, believing that it could create life, and that eating sage would make one immortal. '*Cur moriatur homo, cui salvia crescit in horto?* (Why should a man die who has sage in his garden?)' *Regimen Sanitatis Salernitanum.*

A young nymph lived as a sage plant and inhabited a hollow oak tree. In the shade of her tree garish jonquils grew and detracted from her beauty. The nymph was not jealous and lived shyly in the woods until, one day, she espied the king out hunting with his men and dogs. She was immediately smitten by him, and he by her, but she knew it would mean her death to love a mortal man. She told him that her best days were over and suggested that they should remain in the beauty of the woods, telling him, 'You ask my love and I give you my life.' The king had no idea what she meant by this but as he held her passionately in his arms, she paled and her head dropped. Despite his efforts to revive her with water from a nearby pool, the warmth of their love was more than the fragile sage could bear and she died. The king left her there and went on his way confused and mourning the loss of his love.

In Christian myth, sage is one of the plants said to have sheltered the holy family from Herod's soldiers during their flight into Egypt. For its aid the Virgin blessed it and thereafter it became a healing herb.

Magic and Lore

Sage is a plant of wisdom, fertility, longevity, immortality and protection. It is used in magic for spells to ensure success in business and the granting of wishes. In psychic work it is supposed to aid relaxation, concentration and memory.

To make a wish come true you must write it on a sage leaf, then place this beneath your pillow and sleep on it for three nights. If, during that time, you have a dream then your dream will come true. If you do not dream of the things you wish for you must bury the leaf in the garden so that you will come to no harm.

In the Middle Ages it was believed that sage augured prosperity, and that as it flourished and declined so would the business prospects of the master of the house. The vigour of the plant was also linked to the health and wellbeing of members of the household, so where the plant grew well the family would also flourish. If a member of the household was going away on a long journey a sprig of sage would be hung up in the house: so long as it remained healthy, showing no signs of wilting or rotting away, the traveller would also remain well.

It is said to be courting bad luck to plant sage in your garden yourself. Ideally, you should ask a stranger to do so. To grow nothing but sage in a garden bed is also unlucky; it should always be grown with another plant (perhaps rosemary or rue, whose growth it is said to stimulate). Growing sage with rue has the added benefit of guarding against 'noxious toads'. In some countries it is thought to be unlucky to let sage run to flower in the garden, as this will bring misfortune to the whole family. You must never give someone sage plants as it means that you will quarrel with them.

Putting sage seed in the garden is supposed to bring death to the family. The country saying warns, 'Plant sage seed and plant your sorrow'. This saying could, of course, be interpreted to mean the exact opposite, that having sage in the garden will dispel sorrows. And there is an old belief, especially in France, that sage could alleviate grief. Presumably this and the belief that, like rosemary, it aids memory, led to it being strewn on graves as a mark of remembrance. In addition, as a symbol of immortality, it was often planted on graves as it was said to live forever and, therefore, in a form of sympathetic magic, could ensure the dead gained 'eternal life'. Pepys wrote of it, 'Between Gosport and Southampton we observed a little churchyard where it was customary to sow all the graves with sage.'

If a young woman wishes to discover whom she will marry she must collect twelve sage leaves at midnight on Christmas Eve (some sources say Midsummer's Eve) – great care must be taken that the plant is not damaged. If the leaves are then scattered to the four winds an image of her future husband will appear before her. In various parts of Britain alternative dates for this ritual are given. In Lincolnshire it is St Mark's Eve and red sage is the preferred plant. In Staffordshire it is All Hallow's Eve (Halloween). There is a further refinement of the ritual requiring that the woman must collect one leaf at each stroke of the clock.

Sage is said to grow best for the wisest people. Other traditional superstitions say that it grows most vigorously for the dominant person in a marriage or, more directly, in gardens where a woman ruled the household.

> If the sage tree thrives and grows,
> The master's not master, and he knows.

Men, we are told, would cut down the sage in their garden lest their neighbours mocked them. One suggestion was that a bride and groom should both plant sage in their garden to see who would run the home. Perhaps as a result of this superstition it was said that sage gave unnatural strength to the wilfulness of women. The plant can also be used as a protective amulet. Carrying or wearing a small horn filled with it will avert the evil eye.

In weather lore, burying sage in the garden until it rots may end long spells of storms and rain.

To dream of sage indicates that you will soon be married.

Medicinal

The common name is derived from *salvus*, meaning 'safe' or 'heal', or 'in good health'. There is a tradition that the health of the plant is linked to the health of the head of the household. If the head of the house becomes ill, the sage in the garden will wither.

> He who would live for aye
> Must eat sage in May.

It is supposed to be an aid to feminine fertility. Ancient Egyptian women drank sage tea to increase their ability to conceive. Likewise Roman women would drink sage juice for four days whilst abstaining from sexual intercourse, and then have intercourse on the fourth day. Hippocrates recommended that, following battles or plagues, women should take sage tea to make them more fertile in order to repopulate the country.

It was considered to be a panacea and has been used to treat most ailments at some time. Hippocrates also recommended that sage tea should be taken as a general tonic and to treat indigestion. It was eaten by elderly Romans to improve memory, and both the Greeks and the Romans used it to treat snakebites. During the Middle Ages, it was used as a cure for colds, fever, epilepsy, cholera and even constipation. In periods when epidemics raged, essence of sage, possibly with other herbs, was sprinkled over the body to ward off the infection.

Culpepper recommends a conserve of the flowers to help the memory and quicken the senses. He also claims that sage is good for the liver. Elsewhere it has been suggested for treating flatulence, migraines and rheumatic pain. Arthritis was treated by drinking water in which sage leaves had been boiled. Sage mouthwashes were said to treat mouth ulcers, bleeding gums and sore throats. Toothpaste could be made by mixing fresh sage leaves and salt in equal parts. Dioscorides claims that sage should be used to cure kidney problems and says that it will also fade black hair. In a remedy for St Anthony's fire and fevers; seven leaves must be eaten, on seven consecutive mornings, before breakfast.

In a rather strange way the herb could be used to numb parts of the body.

> This herb, being purified under dung of cattle in a glass vessel, bringeth forth a certain worm, or bird having a tail after the fashion of that bird called a Black Mack or Ousel, with whose blood, if any man be touched on the breast, he shall lose his sense or feeling the space of fifteen days and more. And if the aforesaid Serpent be burned, and the ashes put in a fire, anon shall be there a rainbow, with an horrible thunder, and if the aforesaid ashes be put in a lamp, and be kindled, it shall appear that the house is full of Serpents, and this hath been proved of men of late time.
>
> – *The Book of Secrets of Albertus Magnus*

It is symbolic of immortality and is reputed to contain oils to enable those that use it to reach a ripe old age. John Evelyn says, 'It is a plant indeed with so many and wonderful properties that the assiduous use is said to render men immortal.'

DWARF ELDER

Botanical Name: *Sambucus ebulus*
Family: Caprifoliaceae
Common Names: Blood hilder; Dane ball; Dane's blood; Dane weed; Dane wort; deadwort; dwarf elden; ground elder; plant of the blood of man; she elder; stink plant; walewort; wallwort

No doubt, over the years, there has been some confusion between the dwarf elder and the common elder, in consequence of which a number of the folk names of the plant, and some of the folklore, will have been mixed up.

Magic and Lore

Dwarf elder differs distinctly from common elder by virtue of being herbaceous rather than woody. It is said to have sprung spontaneously from the blood of Danes killed in battle. However, some writers claim that it was introduced to Britain by the Danes, who brought it to plant on the graves of their dead countrymen.

Medicinal

Perhaps the link to the Danes comes from a slightly different, and far less attractive connection: it has been used as a strong purge, inducing 'the danes', i.e. diarrhoea. 'It is supposed it tooke the name Danewort from the strong purging quality it hath, many times bringing them that use it into a fluxe, which then we say they are troubled with the Danes.' (John Parkinson, *Theatrum Botanicum*)

ELDER

Botanical Name: *Sambucus nigra*
Family: Caprifoliaceae
Common Names: Aeld; alhuren; battree; black elder; bole tree; boon tree; boor; bore tree; borral; boun tree; bourtree; bull tree; common elder; devil's wood; dog wood; elderberry; eldern; eldrum; ellaern; eller; ellet; ellhorn; European elder; Frau Holle; hollunder; hylan tree; hylder; God's stinking tree; Judas tree; lady ellhorn; old gal; old lady; old sal; pipe tree; rob elder; scaw; scaw tree; scawen; stink tree; sureau; tea tree; tramman; trammon; tree of doom; whit-alder; *yakori-berigeskro* ('Devil's eye')
Gender: Feminine
Planet: Venus
Element: Water
Deities: Hulda; Venus
Meanings: Compassion; 'My efforts will remain unremitting'; sorrow; zealousness

Myth and Legend

In the legends of some tribes of Native Americans the elder is identified as the mother of the human race.

It is commonly associated with witchcraft, and witches were even said to be able to transform themselves into its shape. One such 'tree witch' is accredited with the creation of the Rollright stone circle in Oxfordshire. According to legend, an invading Danish army was marching to do battle with the English when they met a witch. The Danish king asked the witch to tell his fortune and her response was:

Seven long strides thou shalt take,
and if Long Crompton thou canst see
King of England thou shalt be.

Being close to the top of the hill already he strode forward with great confidence but, on his seventh stride, a long mound rose up from the ground before him blocking the view. The witch continued:

As Long Crompton thou canst not see
King of England thou shalt not be.
Rise up stick and stand still stone
For king of England thou shalt be none.
Thou and thy men hoar stones shall be,
And I, myself, an elder tree.

In an instant the king and all his soldiers were changed to stone. Those warriors most loyal to him became the King's Men, the main stone circle. Those soldiers who, in a huddle slightly apart from the rest, had questioned the king's wisdom, became that group of stones called The Whispering Knights. A little way from the rest, standing alone, is the King Stone, and close by grows an elder tree.

From this folk tale there developed a number of customs. It was said that, on Midsummer's Eve, people sought out the elder witch and danced about with elder garlands on their heads. At midnight, the King Stone would acknowledge the festivities, turning his head to watch the dancers. With the spread of Christianity and the outlawing of pagan rites, many such customs were lost or changed to reflect a new 'Christian' structure. The festivities continued around the King Stone and, at midnight, people would cut the elder to bleed the witch. The King Stone would then nod its approval.

Local custom said that if you discovered the elder witch and broke off a branch, you could discover her disguise. An added incentive was the promise that if you

broke off a branch and it turned red, you could claim a boon from the witch (presumably in return for not disclosing her secret).

Christian tradition has it that the wood used for the cross on which Christ was crucified was elder and, thereafter, the tree was for ever doomed to grow small, crooked and weak. Even the tree's fruits degenerated from their original size and flavour to become small and 'worthless'.

Bour tree – bour tree: crook it rung
Never straight and never strong;
Ever bush and never tree
Since our Lord was nailed to thee.
– Robert Chambres, *Popular Rhymes of Scotland*

Another tradition claims that, realizing the enormity of his betrayal of Christ, Judas hanged himself from an elder, and the tree has become symbolic of sorrow and remorse. Sir John Maundville claims to have seen the precise tree during his travels, 'And faste by is zit the Tree of Eldre that Judas henge him self upon, for despeyr that he hadde, whan he solde and betrayed oure Lord.' It would seem unlikely that elder could ever be used to hang oneself, however, as it is not strong and tall enough. Nevertheless, it is probably due to the association with the death of Judas that it gained a reputation as having been used for hangings as well as earning it the name of 'God's stinking tree'.

White elder flowers are sacred to the White Goddess at Midsummer so it is unlucky to cut them and bring them into the house for use in decorations. The plant is representative of the dark aspect of the White Goddess, the crone rather than the maiden, and is also associated with the Fates and the Norms, signifying the three aspects of time. According to some calendars it presides over the thirteenth month, described as a three-day period ending at Samhain (31 October). It is linked with the turning of the wheel of life and death. It is a tree of birth and death, beginning and end, and as such was revered by the Druids. It is sometimes called the 'Samhain tree', and the moon at this time is the 'elder tree moon'.

It is the tree of Hylde-Moer or Hylde-Vinde, the elder mother or elder queen in Scandinavian and Teutonic myths. Hylde-Moer is described as a dryad and a relative of Hulda, the goddess of marriage. The tree gains its unique personality from the elder mother dwelling within it. She was strong in earth magic and would cause harm to anyone who harmed her tree.

Elder is the Lady's tree –
Burn it not or cursed you'll be.

In an example of the confusion of folklore between different plants, on the Scottish borders elder is said to grow where blood has been shed. There are clear similarities between this and the folklore of the dwarf elder.

Magic and Lore

In folklore, elder is both feared and revered. It seems to be thought of as a tree of magic and power in every country in which it grows. As we have seen, it was held sacred by the Druids, who viewed it as a tree of the beginning and the end; the flowering of the plant denoted the arrival of summer and the ripening of the fruit its end.

The darker side of the White Goddess's character ties in well with the links to witchcraft. As with the story of the Rollright stones, there are many tales of witches adopting the form of the tree; if the witch tree was cut it would bleed and the woman would still bear the marks when she returned to human form, and so be recognized. An old Northamptonshire tale tells of a man who cut an elder stick to make a toy for his child, and the tree bled. Later the man met a local woman with her arm bound up: such was the witch hysteria of the period that she was ducked as a witch for the alleged offence.

No forester would consider cutting the elder without first asking permission of Hulda three times, and even then he would fear that he might suffer her wrath. Bareheaded, on bended knee and with his arms folded, the forester would implore:

> Lady Ellhorn [or Old Gal] give me of thy wood
> And I will give thee of mine,
> When I become a tree.

Some sources add that he must then spit three times before daring to cut wood from the tree. Even then he must be free of any selfish intent. In some parts of Britain and Europe it was traditional for people to show the tree due respect by touching their hats as they passed by. Scandinavian folklore tells that Hulda lives amongst the roots with her elfin entourage. She is called the 'good mother' of the elves. In many places, especially the Isle of Man, the elder is thought of as a tree of the faeries, another reason for not cutting or burning the wood, even from a fallen tree.

Burning the tree might cause one to become bewitched, especially if one were foolish enough to burn it in a hearth in the house. It was even said that if someone put it on a house fire they would see the Devil sitting on the chimney pot, and he might come down the chimney into the house. Quite possibly, in pre-Christian days, they would have seen the tree's spirit.

The Danes believed that elder dryads would haunt anyone who lived with furniture made of elderwood, and a similar belief can be seen in many other superstitions. It was said that it should never be used in the making of houses or of furniture. If it were used in the construction of the house, perhaps in the floorboards, the occupants would feel invisible hands pulling at their arms and legs. If it was used in making furniture it would warp, twist and break. There are other, more mysterious, problems associated with making furniture of elderwood, especially for children. It was believed that a child in an elderwood cradle would be unable to sleep and would fall out or be thrown out by Hulda. A British belief was that a child in such a cradle would become sickly, pine away, be pinched black and blue, or even stolen away by faeries.

In Scandinavian folklore there is a tale of an elder tree growing in a farmyard, which had a habit of walking at twilight to peep in through the windows at the children when they were alone. Perhaps this was a faerie tree looking for an opportunity to steal the children away.

The link with dryads goes further than the haunting described above. If you want to see dryads, you should spend the night in an elder grove when the moon is full. You must have no evil in your heart and be properly prepared for the experience, presumably with some form of protection against enchantments.

Sleeping in the shade of an elder seems to be hazardous; the tree is said to have a narcotic scent and resting beneath it will result in all types of ills. According to Withering, 'The whole plant hath a narcotic smell; it is not well to sleep under its shade.' These ills include sickness, dreams of death and near death itself. The scent of the flowers would poison anyone who slept in its shade. It was also claimed that simply being under an elder tree would be make one delirious.

This delirium might account for the belief that if you stand beneath an elder on Midsummer night and breathe its fragrance, it opens the portals into the Land of Faerie; there you will see the king and queen of the faeries and their entourage. Be careful, however, for unless you have suitable protection against enchantment with you, such as an implement of iron, you may be swept away into the faerie realm, never to be seen again.

Dreaming of elder is also said to result in sickness.

Archbishop Jacobus de Voragine, in the medieval *Golden Legend*, calls the elder a tree suitable for a graveyard, saying that it is a tree of shame and darkness, without beauty. But its association with death goes back much further. Flints shaped like elder leaves have been found in Megalithic barrows. Windows shaped like elder leaves were sometimes made between two contiguous slabs of a burial chamber. In both the Old and the New Stone Ages arrowheads were made in the shapes of elder leaves. The wood has been used for funeral pyres at cremations. It might also be placed in coffins, or the tree planted onto graves. In parts of the Austrian Tyrol an elder trimmed into the shape of a cross was planted on the grave; if the plant blossomed it indicated that the spirit of the grave's occupant had found heaven. Its use in coffins and on graves might be to protect the spirit of the deceased from witches and evil spirits or, perhaps, derive from an ancient belief that wherever the tree grew was sacred. It could, however, be from an attempt to use the scent of the wood to deter worms and delay decomposition. This scent is supposed to have been due to it growing amongst stinking corpses. When hearses were predominantly drawn by horses elderwood was used for the whip handles by hearse drivers to protect them from the spirits of the recently departed still floating close by, especially where their passing had been particularly tragic or violent.

Elder was supposed to be a powerful protection against earthbound spirits such as vampires, and its use as such pre-dates the use of garlic. In Russia it was said to drive away all evil out of compassion for mankind. One of the odd ideas regarding its effectiveness comes from a belief that demons are fanatical about counting.

Placing the flowers, leaves and berries of elder on window frames and over doors would protect the home as the evil spirit will stop to count the number of flowers or berries on the shoots. When it is planted for its protective properties it should be placed as close as possible to the most often used door of the house. Likewise, it can be planted close to the door of a cowshed to protect livestock. In Ireland it is suggested that the tree will be much more effective when it is self-set rather than planted. It was said that a farm with elder growing on it was blessed as the tree protected the livestock against witches, sorcery, evil spirits and even lightning strikes, as well as promoting their fertility. Elder twigs can be used to form a protective amulet by forming them into a cross and tying them together with a red thread (see also Rowan).

The ultimate protection might be a hedge of elder around a property, although this has a number of drawbacks. There is a tale of a number of houses hemmed in by overgrown elder trees, whose occupants died in quick succession. The situation was only alleviated when the trees were thinned. Tradition also has it that repairing a wattle fence with elder would give witches power over whatever is inside.

Elder could be gathered on May Eve (Beltane) and used to guard against witchcraft. Other sources, possibly from regions where the trees fruit later, state that elderberries must be gathered on St John's Eve (Midsummer). The berries could be fed to those thought to be bewitched or possessed in order to drive the demons to flight or remove the enchantment. Elderwood gathered for use as protective amulets should be gathered on 1 April and is most effective if taken from an area of the garden where the sun has never shone. The wood could be tied on a cord, between two knots, and worn about the neck. Other sources recommend that elder be gathered at Midsummer (21 June or St John's Eve) to use as a protective talisman on Twelfth Night, commemorating the massacre of the innocents by Herod. Another twist this theme reminds us that Twelfth Night, 6 January, is Old Christmas Day and was changed when the calendar was adjusted in 1752 (it is sometimes also known as Bertha Night). On this night the Devil is especially virulent and to be safe you must stand in the centre of a magic circle and have elderberries gathered on St John's Night; if the Devil does appear you can demand from him the magical fern seed that, amongst other attributes, will give you the strength of thirty or forty men. Some writers add that Hulda will ensure that the seed is delivered to you by an unseen hand, wrapped up in a chalice cloth.

Elder could be used to guard against mischievous faeries as well as malicious witches. A Cornish folktale of the early 1800s tells of a farmer from Lostwithiel who, every morning, found one of his ponies tired and travel-stained. He suspected that the poor animal was being hag-ridden at night so set out to keep a vigil. Late in the night he saw five small men having a heated argument, which became a brawl, in a nearby field. The eventual victor then jumped on the farmer's pony and rode it until it was exhausted. The farmer then stabled his pony and pegged the door closed with an elder twig, and his pony was never troubled again.

Having already said that elder could be a witch in disguise, the apparent paradox of the tree protecting against witchcraft was explained by saying that the witch inhabiting the tree was territorial and would ward off other witches. But, although it is a plant of protection against witches and their work it was also woven into the chaplets worn by witches at Samhain, when they made contact with the spirits of the dead. In Ireland it was the wood used for the handles of besoms, witches' brooms.

The protective power of the elder, or the spirit dwelling within it, also extends to more tangible things. In Sicily, it is said to be the foremost plant for killing snakes and driving away thieves. Wearing it will ward off all attackers and shield the bearer from lightning (it was once believed that elder would never be struck by lightning). Carrying elderberries was also supposed to be a protection from poisoning, as well as the inevitable negativity and evil. The flowers were used to discourage flies, and were put into the bridles of carriage horses; the trees were also planted near to outside toilets. When an elder sets root in the garden it is claimed that the Mother Elder protects the garden, and someone who has a tree growing in the garden will die in his own bed. Be warned, however, that having elder growing in your garden is also said to keep friends away!

It has been said that the only 'proper' use of elder is to seek out witches. At Christmas, an elder twig, dipped in oil and floated on water, would act as a compass, pointing towards any witches nearby. Another tradition suggests that thin branches could be sliced into flat discs and these dipped in oil then floated on water in a dish. If the pith was then lit, forming a floating candle, reflections on the water's surface would expose witches. A much simpler method requires a baptized person to dab the juice from the green inner bark of an elder twig on their eyelids. This would enable them to see witches and all of their activities within the community.

As we have seen, Hulda is the goddess of marriage. If a couple drink from a cup of ale to which an elder flower has been added they will, with Hulda's aid, be married within the year. Elder sticks and elder blossoms might also be used in decorations at the wedding to bring luck to the happy couple. Pregnant women were encouraged to kiss an elder tree to gain Hulda's favour in bringing good fortune for their unborn child. Looking at this from a different perspective, those tempted into adultery might carry a piece of twig, or even just a leaf, to alleviate the temptation.

People and places may be blessed; the leaves and berries were scattered to the four winds from the top of a hill, whilst the name of the person (or place) was visualized. Another method suggested, more simply, that the leaves and berries were sprinkled directly onto the person.

It was said to be unforgivable to beat an animal or child with an elder stick, as they would never grow properly afterwards. Perhaps this is why some farmers were reluctant to use elder twigs when driving their cattle.

The soft pith of elder shoots can easily be removed and, consequently, the twigs have been used for making flutes or pan pipes, which could be used to call up spirits. This must be done at midnight in a deserted location, far away from any habitation. Pliny records that the shrillest notes were to be made from the stems of elders that grew out of the 'sound of cock crow'.

Medicinal

Elder has been used medicinally since the times of the ancient Egyptians. It was used by the Romans, the Britoins and the Celts, and has been called a complete medicine chest. It is valued as a diuretic and antispasmodic, and used to reduce fever, stimulate sweat glands and to treat chronic respiratory inflammation. John Evelyn says: 'If the medicinal properties of the leaves, bark, berries etc. were thoroughly known, I cannot tell what our country-men could ail for which he might not fetch a remedy from every hedge.'

In folk medicine, simply poking an elder stick into the ground whilst remaining silent could treat fever, in the belief that the fever would adhere to the twig and would be transferred to whoever pulled it up. Rheumatism was treated, or avoided, by carrying an elder twig that had been tied into three or four knots. Insomnia could be overcome by placing a few elderberries beneath the pillow.

Toothache, however, requires a slightly more complicated cure. The sufferer must chew on an elder twig then place it on a wall saying, 'Depart thou evil spirit'. This works on the assumption, once widely held, that toothaches were caused by evil spirits. A different suggestion states that clay taken from beneath an elder tree can be used to allay the pains of toothache. In Ireland, to ease teething, infants might be given necklaces of nine strigs of green elderberries, or of elder twigs cut into nine pieces. This same treatment was used to treat epilepsy at various times.

Various wart charms can be found that use elderwood. Rubbed with a green elder stick that was then buried in the garden to decay, they would disappear as the wood rotted. (Some sources suggest that the elder twig be buried in horse manure, presumably so that it will rot more quickly). A tale from Waddesdon, in Buckinghamshire, from the early nineteenth century, tells of a young girl who suffered badly from warts. A neighbour secretly counted them and cut that many notches into an elder stick, which she then buried in her garden. As the twig rotted away, so the warts disappeared. Another cure for warts is to bury the same number of elder buds as there are warts. Once again, the warts will vanish as the buds rot. If that should fail three drops of blood from each wart could be put on an elder leaf, and the leaf then buried in the garden to decay. 'Wartes – wash them with the juice of the berries when the berries are black and doe so every night and so binde them to the night.' (W. Langham, *The Garden of Health*)

Elder flowers beaten with lard were used as anointment for scalds and burns. The powdered blossom could be sniffed up the nose as a snuff to clear the head, or mixed with peppermint and honey to produce a sweet tea to ease colds.

The juice and flowers were used to create a skin lotion that would remove wrinkles and freckles. In addition the plant was also seen as a treatment for gout, dropsy, piles and, if collected on May Eve, toothache, melancholia and bites from snakes and mad dogs. Having elder twigs in your pockets was even supposed to prevent saddle sores.

It is not only people who were treated with elder to cure their ills. Lame pigs could be treated by having an elder twig put through their ears.

Links between people and plants have been identified elsewhere (see Ash and Oak). In the Middle Ages people setting out on a journey might place their welfare into the keeping of a tree or, more precisely, the tree spirit. These trees were known as guardian trees and in Copenhagen it appears that each sailor would have an elder guardian tree.

BLOODROOT

Botanical Name: *Sanguinaria canadensis*
Family: Papaveraceae
Common Names: Coon root; Indian paint; king root; paucon; red poucon; red root; snakebite; sweet slumber; tetterwort
Gender: Masculine
Planet: Mars
Element: Fire

The root might be carried or worn to attract love and avert evil spells. When placed near to doors and windows it would protect the home. It has also been used as an ingredient in Native American war paint and in all instances it is the deepest red roots that are considered to be the most effective.

SANDALWOOD

Botanical Name: *Santalum album*
Family: Santalaceae
Common Names: Sandel; santal; white sandalwood; white saunders
Gender: Feminine
Planet: Moon
Element: Water

Sandalwood is used in incense burned during exorcism rites and spells for healing or protection. Mixed with lavender it forms an incense for conjuring spirits, and with frankincense an incense used at moon rituals and séances. Wearing beads of sandalwood will protect the wearer and heighten their spiritual awareness.

Sinistrari, in *De Daemonialitate et Incubis et Succubis*, 1875, lists many ingredients to be used for warding off demons. He includes the following odd concoction to be burned to drive away an incubus, a male demon given to raping women whilst they sleep. It is made up from a mixture of medicinal herbs and, rather inappropriately it would seem, stimulants and aphrodisiacs: 'Sweet flag, cubeb seed, roots of aristolochia, great and small cardamom, ginger, long pepper, clove-pink, cinnamon, cloves, mace, nutmegs, resin, benzoin, aloewood and root, and fragrant sandle. These ingredients were to be brewed in three and a half quarts of brandy and water.'

To have a wish come true, write it onto a chip of sandalwood and then burn the wood in the censer whilst visualizing it.

PEPPER TREE

Botanical Name: *Schinus molle*
Family: Anacardiaceae
Common Names: Californian pepper tree; Peruvian mastic tree; piru
Gender: Masculine
Planet: Mars
Element: Fire

In Mexico, the branches of the pepper tree were used in healing rituals, the essence of which was similar to that described for many other 'healing trees'. The sick person was brushed over with the branch so that the wood might absorb the ailment. The branch was then taken out and burned to destroy the disease.

KNOTTED FIGWORT

Botanical Name: *Scrophularia nodosa*
Family: Scrophulariaceae
Common Names: Carpenter's square; kernel wort; rose noble; throatwort

Myth and Legend

In 1628, during the thirteen-month siege of La Rochelle by the army of Richelieu the garrison survived for a considerable time by eating the tuberose roots of figworts. From this the plant gets its French common name *herb du siege*. Considering that the scent and flavour of the plant are somewhat unpleasant it was only likely to be used in times of siege or famines.

Magic and Lore

In folklore the figwort is used as a protective amulet. Hung about the neck it guards against illnesses and the evil eye. If it is smoked over a summer fire, presumably a ritual Midsummer fire, it may be hung up in the home to protect the whole household.

Medicinal

As I have said, wearing figwort about the neck will ward off illness and promote good health. The plant has also been used as a painkiller.

COMMON STONECROP

Botanical Name: *Sedum acre*
Family: Crassulaceae
Common Names: Bird's bread; biting stonecrop; candles; creeping Charlie; creeping Jack; creeping Jenny; creeping sailor; creeping Tom; crowdy-kit-o'-the-wall; English mouse-tail; French moss; ginger; gold chain; golden carpet; golden dust; golden moss; golden stonecrop; hundreds and thousands; Jack-in-the-buttery; little houseleek; London pride; love entangled; love-in-a-tangle; mousetail; pick pocket; pig's ears; plenty; poor-man's-pepper; prick madam; queen's cushion; rock crop; rock plant; star; wall moss; wall pepper; wall wort; welcome-home-husband; welcome-home-husband-though-never-so-drunk
Planet: Mars
Meaning: Tranquillity

Magic and Lore

Stonecrop was planted around farms and cottages, and often on the roofs, as a protection against lightning, fire and witches, as well as to generally promote good luck. This may be due to confusion with the houseleek, *Sempervivum*, with which it shares many common names.

The name 'crowdy kit'may arise from children in rural regions at one time using the stems as rough fiddles. The stems were stripped of leaves and scraped over each other in the manner of a bow across a violin to produce a squealing sound.

Medicinal

Pliny recommends the use of stonecrop as a treatment for insomnia. It must be wrapped in a black cloth and placed beneath the patient's pillow. This should be done without the patient's knowledge or it will be rendered ineffective.

It the past the plant was used in the treatment of bot worm and the name 'Jack-in-the-buttery' is thought to have been derived from this, 'buttery' being a corruption of *bot-therigue* or 'bot-treacle'.

HOUSELEEK

Botanical Name: *Sempervivum tectorum*
Family: Crassulaceae
Common Names: Ayegreen; ayron; *bauchaill a' tighe*; bullock's eye; foose; fuets; fullen; healing leaf; healing plant; hen and chickens; hockerie-topner; hollick; ice plant; Jan's leaf; Jove's beard; Jupiter's beard; Jupiter's eye; mallow rock; roof leek; sengreen; silgreen; singreen; sungreen; syphelt; Thor's beard; thunder plant; wax plant; welcome-home-husband-however-drunk-you-be; welcome-home-husband-though-never-so-drunk; welcome-home-husband-though-never-so-late

Gender: Masculine
Planet: Jupiter
Element: Air
Deities: Jupiter; Thor
Meanings: Domestic industry; vivacity

Magic and Lore

The use of houseleek tends to have been as a protective plant. It was a popular belief amongst the ancient Greeks and Romans that it was an effective guard against witchcraft. This belief persisted in parts of Britain, where it was thought that the removal of any part of the plant would bring misfortune on the household. Uprooting the plant caused domestic strife and the removal of the flower led to a death in the household. It has sometimes been said to be unlucky to grow houseleeks unless you also have vegetable leeks growing in your garden.

The Emperor Charlemagne ordered that houseleeks should be grown on house roofs as a protection against lightning strikes, fire and sorcery.

'Old wryters do call it Iovis barba, Iupiter's Bearde, & holde an Opynion supersticiously that in what house so ever it groweth, no Lyghtning or Tempest can take place to doe any harm there.' William Bullein

In love magic the herb was used to attract lovers. For this purpose it must be worn fresh and renewed every few days.

Medicinal

Houseleek has been recommended as a treatment for burns and scalds. This may be due to its association with fire, or from its fleshy leaves, a characteristic it shares with many other 'burn' plants, such as aloe.

Juice from the leaves was not only advised for the treatment of burns but might also be rubbed onto warts and corns in order to remove them. It has also been recommended as an abortifacient.

RAGWORT

Botanical Name: *Senecio jacobaea*
Family: Compositae
Common Names: Agreen; balcairean; beaweed; bennel; benweed; bindweed; boholawn; bouin; bowen; bowlocks; bunweed; cankerweed; cankerwort; cheedle-dock; cradle-dock; crowfoot; cushag; Devil dums; dog stalk; dog standard; dog standers; ell-shinders; faeries' horses; fizz-gigg; flea-nit; flea-nut; fly dod; gipsy; grundswathe; grunsel; jacoby-fleawort; James wort; kedlock; keedle-dock; mare-fart; muggert; ragged-Jack; ragged Robin; ragweed; St James' wort; scattle-dock; scrape-clean; seggrums; seggy; sleepy dose; staggerwort; stammerwort; stinking alisander; stinking Billy; stinking Davies; stinking nanny; stinking weed; stinking

Willie; summer farewell; weebo; yackrod; yackyar; yallers; yellow boy; yellow daisy; yellow weed
Gender: Feminine
Planet: Venus
Element: Water

Myth and Legend

Following the defeat of the Scots at Culloden in 1746, the English are said to have named *Dianthus barbatus* 'sweet William' in honour of William, Duke of Cumberland, their victorious commander. The defeated Scots named this pernicious weed 'stinking Billy', or 'stinking Willie', for exactly the same reason. There are even those who maintained that the spread of ragwort across Scotland was due to it being included in the fodder used for the horses of the English troops.

Magic and Lore

The ancient Greeks used ragwort as a protective herb that could be used in charms and spells to form an amulet. In the British Isles, however, it had a vastly different significance. In Scotland and Ireland it was known as faeries' horses as faeries were said to gallop on the golden-flowered stalks at night, enabling them to fly between the islands around the coast.

> Don't call it a weed though a weed it may be,
> 'Tis the horse of the Faeries, the *boholàun buidhe*.

When belief in faeries waned and Christian orthodoxy spread, the same belief was applied to witches. During the persecution of witches throughout the sixteenth and seventeenth centuries it was claimed that they could fly on the stems at midnight. According to John Aubrey, 'Horse and hattock' were the magical words required to make the stems fly.

> Let warlocks grim, an' wither'd Hags
> Tell how wi' you on ragweed nags
> They skim the muirs an' dizzy crags
>
> – Robert Burns, *Address to the De'il*

In Ireland it is said to be unlucky to take a stem of ragwort in your hand. Likewise, it should never be used to strike an animal, as it will bring misfortune on it.

For some the herb is sanctified by being in flower on 26 July, St James's Day, and so is called *herba sancta Jacobi* or 'St James wort.'

GROUNDSEL

Botanical Name: *Senecio vulgaris*
Family: Compositae
Common Names: Ascension; bird seed; canary food; canary seed; chicken

weed; grindsel; ground glutton; groundswelge ('ground swallower'); groundswell; groundwill; grundy-swallow; senlion; sention; simson; sinsion; swallow-grundy; water drums; yellow heads
Gender: Feminine
Planet: Venus
Element: Water

Magic and Lore

One suggestion for the origin of the generic name *Senecio* is that it comes from the Latin *senex*, meaning 'old man', relating to the grey-haired seed heads that are eventually blown away to leave the 'old man' bald.

Medicinal

Groundsel can be used as an amulet which protects against toothache, and it was even said to cure the pains once they had started. Some writers went so far as to claim that carrying the herb would help to keep teeth in good order.

SESAME

Botanical Name: *Sesamum indicum*
Family: Pedaliaceae
Gender: Masculine
Planet: Sun
Element: Fire
Deity: Ganesha

In legend sesame could be used to find hidden treasures, locate secret entrances and to open locks, which may have been the origin of the command in *Aladdin*, 'Open Sesame!'

Having sesame seed in an open jar in the house is said to attract money into the home. The seed must be fresh, and should be changed at regular intervals. Eating sesame seeds is supposed to induce lustfulness.

RED CAMPION

Botanical Name: *Silene dioica*
Family: Caryophyllaceae
Common Names: Adder's flower; bachelors' buttons; Billy buttons; bird's eye; Bob Robin; brid een; bull rattle; bull's eye; cancer; cock Robin; cuckoo; cuckoo flower; cuckoopint; Devil's flower; dolly winter; drunkards; fleabites; gipsy flower; gramfer-greygles; granfer-giggles; granfer Jan; Jack-by-the-hedge; Jack-in-the-hedge; Jack-in-the-lantern; Jan granfer; Johnny woods; kettle smocks; lousy beds; lousy soldiers' buttons; Mary Janes; mintdrop; red gramfer greygles; red mintdrop;

mother Dee; mother-die; plum pudding; poor Jane; puddings; red bird's eye; red butcher; red Jack; Red Riding Hood; red Robin; red Robin Hood; red soldiers; robin flower; Robin Hood; Robin-i'-the-hedge; Robin-run-in-the-hedge; robin redbreast; robin's eye; robin's flower; Rob Roys; round robin; rose campion; Sarah Janes; scalded apples; soldiers; soldier's buttons; sweet Willie; wake robin; water poppies
Meaning: Encouragement

This plant was associated with Robin Goodfellow, a mischievous faerie character, sometimes described as being the son of the faeries. He was portrayed as a Pan-like character, having horns and goat's hooves, and so the plant was soon associated with the Devil. Picking red campion, with the appropriate incantation, was a way to curse one's father, who would ultimately die. White campion could be used in the same way to curse one's mother.

COMPASS PLANT

Botanical Name: *Silphium laciniatum*
Family: Compositae
Common Names: Pilot weed; rosin wood

The leaves of the compass plant are supposed to point due north and south, an attribute that is said to have been noticed by the frontiersmen in North America, who used it for navigation. The dried roots might be burned during thunderstorms in order to protect those nearby from lightning.

MILK THISTLE

Botanical Name: *Silybum marianum*
Family: Compositae
Common Names: Blessed thistle; holy thistle; Marian thistle; Mary thistle; Mediterranean milk thistle; our lady's milk; our lady's thistle; St Mary's thistle; variegated thistle
Gender: Masculine
Planet: Mars
Element: Fire

Dioscorides named the plant *Silybum* in AD 100. Legend has it that, whilst the Virgin was breast-feeding the infant Christ, some milk fell upon the leaves of thistles growing at her feet, which retained the milky white marks on their foliage.

There is a traditional belief that if milk thistle is hung about a person's neck it will induce all snakes nearby to begin fighting. In folk medicines it has been used for many hundreds of years in preparations to treat liver ailments.

COMMON HEDGE MUSTARD

Botanical Name: *Sisymbrium officinale* (Synonym: *Erysimum officinale*)
Family: Cruciferae
Common Names: St Barbara's hedge mustard; singers' plant

In France this common roadside plant is known as the 'singers' plant'. During the reign of Louis XIV it was recommended as an infallible cure for loss of voice.

BITTERSWEET

Botanical Name: *Solanum dulcamara*
Family: Solanacaeae
Common Name: Dulcamara; felonwood; felonwort; fool's cap; granny's nightcap; halfwood; mad dog's berries; mortal; poison berry; poison flower; poisoning berries; scarlet berry; scaw-coo; shady night; snake berry; snake flower; snake's food; snake's meat; snake's poison food; terry-divil; tether Devil; tether-diddle; violet bloom; witch flower; woody nightshade
Gender: Masculine
Planet: Mercury (Saturn, according to Culpeper)
Element: Air
Meaning: Truth

Magic and Lore

Bittersweet has long been considered a protective herb against the activities of witches. Culpeper says that it is 'excellent good to remove witchcraft both in men and beasts; as also all sudden diseases whatsoever.' Shepherds would tie it about the necks of their sheep to ward off the evil eye; a piece of the plant kept somewhere on the body was sufficient to protect man or livestock.

The red berries were strung on strips of palm leaf, alongside blue sequin beads (in groups of twenty-four beads to four berries), to form part of the floral collaret in the third coffin of Tutankhamun. They represented the drops of blood of the goddess Isis and were used to invoke her protection.

Bittersweet is the birthday flower of 19 July.

Medicinal

'Felonwood' or 'felonwort' are names commonly given to plants used in folk medicines to treat felons or whitlows – inflammations near to fingers or toenails. The berries were also gathered in autumn and stored in jars for winter use in the treatment of chilblains; they were rubbed well into the skin.

The dried berries, strung onto threads, were made into necklaces for children to ease teething and prevent convulsions.

TOMATO

Botanical Name: *Solanum lycopersicum*
(Synonym: *Lycopersicon esculentum*)
Family: Solanaceae
Common Name: Love apple
Gender: Feminine
Planet: Venus
Element: Water

At the time of its introduction to Britain the tomato was believed to be an aphrodisiac, hence the name 'love apple'. However, this is not the only power attributed to it. Like many other red-fruited plants it was also considered to be a good protection against the forces of evil. When grown in the garden the red fruits and yellow flowers were supposed to frighten away evil beings. A red tomato placed on a windowsill, or at any entrance into the house, was considered sufficient to prevent evil from entering the home. Furthermore, placing a red tomato on the mantelpiece was said to be a good way of attracting money.

POTATO

Botanical Name: *Solanum tuberosum*
Family: Solanacaeae
Common Names: Flukes; lapstones; leatherjackets; no eyes; red eyes; rocks; spuds; taters; tatties; totties
Gender: Feminine
Planet: Moon
Element: Earth
Meaning: Benevolence

Myth and Legend

Potatoes originate from South America, where they were associated with the creator god. Puritans resisted their use, as they were not mentioned in the Bible.

Magic and Lore

There are numerous superstitions relating to the planting, harvesting and even serving of potatoes. Many dates have been suggested as right for planting. Although, in some parts of Europe, it has been said that if you plant potatoes on Good Friday they are bound to fail, this is the date most commonly identified as ideal for planting. Good Friday falls between 20 March and 23 April and, being a time of resurrection, it will clearly be a blessing for the plants, as they too will be resurrected when they are harvested. Another suggestion is that they ought to be planted 100 days after 1 January, i.e. 10 April.

Another good day identified for potato planting is St Patrick's Day, although sometimes it is said that they should be planted before then. Some old sources state that Irish potatoes should always be planted on 17 and 18 March; others that they should be planted on the first two days of the month, regardless of the weather.

The general considerations are that, as an underground crop, it should be planted during the dark of the moon, i.e. when it is waning. Likewise, it must be planted when whatever astrological sign it is planted under is waning. Planting during the time of Gemini, the twins, will ensure that you get double the crop. In coastal regions, it is usually recommended that it be planted at the time of a rising tide so that it will grow with the tide.A morning planting is supposed to be more auspicious that an afternoon planting. In Looe, in Cornwall, it was standard practice to plant early potatoes on Boxing Day. However, one piece of rural advice says:

Plant your tatters when you will,
They won't come up before April.

There have been certain treatments recommended to ensure a good harvest. In Ireland's County Mayo seed potatoes would be sown with a dressing of human excrement and a pinch of salt. In nearby County Kerry sprigs of cypress were pushed into the tops of the ridges in the potato fields, and another piece of the wood burned at the time of harvesting. Although in some regions it would be heresy to plant onions alongside potatoes, as it would make the potatoes cry, in others it is recommended as the tears from the eyes of the potatoes ensure that there will be no shortage of moisture for the ripening crop.

To ruin a neighbour's crop the Irish suggest placing a pot filled with boiled potatoes in their field, as this will prevent the crop from growing. To be most effective it should be done on Good Friday during the dark of the moon.

The traditional harvest time for the 'first digging' of potatoes was at the end of July or beginning of August, in time with the ancient Celtic festival of Lughnasa (1 August). The dates in the Christian calendar for harvesting have been given as 'Garland Sunday', the last in July or 'Garlick Sunday', the first in August.

When eating new potatoes, or the first potatoes of a new season, wishes could be made. In Tipperary there was a tradition to say, 'May we all be alive and happy this time twelve months' at the first serving of new potatoes each year.

Dreaming of potatoes indicates that you have secret enemies of whom you are unaware. If you dream that you are peeling potatoes, and you also see some money in your dream, it indicates trouble in two days' time. To dream of a row of potatoes protruding from the soil is a sure sign of death and to dream of digging for potatoes denotes gain. Dreaming of eating them is an omen of heavy losses and dishonesty in business.

Medicinal

Potatoes have been used to treat ailments as diverse as rheumatism, sciatica, lumbago, gout, toothache, warts and the common cold. They have even been recommended for use as an aphrodisiac.

Small potatoes were carried as protective amulets and to bring good luck. This was also supposed to be useful as a treatment for rheumatism, sciatica, lumbago and colds. To be most effective they should be new potatoes that have been stolen from a neighbour's garden and allowed to go hard and black with age (in Yorkshire, it was advised that the potato be dried in the sun). At one time women might even sow a small pocket into their skirts to contain this potato amulet. The potato was supposed to able to draw the iron out of the blood, as it was clearly the excessive amount of iron in the blood that caused stiffness in the joints.

To cure sore throats, roast potatoes were wrapped in stockings and tied about the neck. Chilblains could be treated by rubbing them with a slice of potato each night for three successive nights. One treatment for leg ulcers was to grate raw potato on to the wound. Warts were treated by rubbing each wart nine times with a potato cut into two pieces. The two halves of each potato were then put together again and the whole thing buried in the garden. As the potato rotted away, so the wart would also disappear.

GOLDEN ROD

Botanical Names: *Solidago odora*; *Solidago virgaurea*
Family: Compositae
Common Names: Aaron's rod; blue mountain tea; cast-the-spear; golden wings; goldruthe; gonea tea; *verge d'or*; wound weed; woundwort; yellow rod
Gender: Feminine
Planet: Venus (in Virgo)
Element: Air
Meanings: Encouragement; indecision; 'Let us take care that our love remains undiscovered'; precaution

Magic and Lore

Golden rod was used in money spells, and it was once believed that it would point the way to hidden springs of water and treasures of gold and silver. When carried in the hand it is supposed to bend, like a water diviner's rod, towards the water, hidden treasures or lost objects. If it grows close to the main door to the home it is an omen of good fortune for the entire household. Wearing golden rod will result in you seeing your future lover on the coming day.

Golden rod is the birthday flower of 30 November.

Medicinal

The generic name *Solidago* is derived from the Latin *solido*, meaning 'to make whole' or 'to strengthen' and come from the plant's ability to heal wounds.

ROWAN

Botanical Name: *Sorbus aucuparia*
Family: Rosaceae
Common Names: Bour tree; care; care tree; *cas gangythraul*; cayer; chit-chat; delight of the eye; Devil's hate; mountain ash; picken tree; quickbeam tree; quicken; quicken tree; quicken wood; ran tree; ranty; roan; roddan; roddin; roddin tree; roden-quicken; roden-quicken-royen; royen tree; sap tree; sip-sap; shepherd's friend; sorb apple; Thor's helper; twickband; twickbine; whistlewood; whitten; whitty; whitty-tree; wice; wicken tree; wicky; wiggen; wiggy; wiky-tree; wild ash; wild sorb; witchbane; witchbeam; witchen; witchen tree; witch hazel; witch-wicken; witchwood; withen; withy; witty tree
Gender: Masculine
Planet: Sun; Saturn
Element: Fire
Deities: Aphrodite; Akka/Mader; Akka/Rauni; Alys (Elis; Halys); Brigantia; Brigid; Oeagrus; Ran; Thor
Meanings: 'Love should bring us wisdom though our hearts are wild with passion'; prudence

Myth and Legend

In the Celtic legend of Finn MacColl, Finn was seeking a wife for his old age, and found Grainne, the daughter of Cormac, the High King of Ireland. Initially both the King and his daughter agreed to the match and when his ambassadors returned with the news Finn set out to claim his bride. He was received in state, with a feast in the great banqueting hall of Tara. But looking around at the assembled company Grainne saw Diarmait O'Duibhne, Finn's nephew, and she fell in love. She drugged the wine that all but Diarmait drank then, as the others slept, Grainne told him of her love for him. Despite his refusal to return her affections, she put him under a *geasa*, an obligation that no hero could refuse. That night, after the feast, the two fled from Tara and what followed was one of the greatest pursuits described in mythology. Toward the end Diarmait and Grainne took refuge in the woods of Dubhous. They sought shelter beneath a sacred rowan tree, the Quichen Tree of Dubhous, which had grown from seed dropped by the Tuatha de Danann, the ancient gods of Ireland. Eating three of the fruits of this tree could magically return anyone, regardless of how old they really were, to thirty years of age. Having realized that they had accidentally set celestial and immortal food amongst humans the Tuatha de Danann put a faerie guardian to watch over it. He was a Fomor, called Sharvan the Surly, thick boned with a large nose and crooked teeth. His one deep red eye was set in the middle of his coal-black face. By day he sat at the bottom of the tree and at night he slept amongst its branches. With his permission Diarmait built a hut, close by the tree, where he and Grainne might hope to evade Finn, as none would normally approach the Fomor.

Grainne wanted to eat some of the berries from the tree so she persuaded Diarmait to get some for her. He first approached the mighty Fomor and, explaining their situation, asked if he might take some. Sharvan the Surly refused so Diarmait killed him. With the Fomor dead, however, there was nothing to prevent people approaching the tree so ultimately the pursuing Finn and his retinue encircled the tree and trapped Diarmait and Grainne.

Irish legends abound with accounts of rowan trees being guarded by great serpents and dragons. On the other hand, it is also said that rowans guard the earth dragons, which express the life forces of the earth. Christian tradition gives us the saints who kill the dragons and claim these sacred placed as their own.

The rowan is sacred to the White Goddess in her various forms. She can be seen as a goddess of the harvest and of female sexuality, a river goddess and the Queen of the Eleusine Islands. It is also a tree of the sun and associated with the ancient sun goddesses such as Brigid (Ireland) and Brigantia (England), both of whom are linked with the spring, the season of birth, and rule over such arts as spinning and weaving. Traditionally the wood was used for making spindles and spinning wheels. In myth the spindle, because of its circular spinning motion, was likened to the passage of the sun through the year, or to the spinner of the 'thread of life'. This is the wheel of fortune, of nemesis, of life and death. These goddesses also had arrows made from rowanwood that could blaze with fire.

There are similarities between the Western goddess Brigid and the Egyptian goddess Neith, who is said to have ruled Lower Egypt whilst Osiris ruled over Upper Egypt. Both of these goddesses represented the healing greenness of nature; Neith was coloured green and Brigid pictured dressed in green. Like Brigid, Neith also presided over spinning and weaving and her emblem was crossed arrows. Both suckled baby sun gods; for Brigid it was Mabon and for Neith, Horus. With the spread of Christianity Brigid became St Bridget, midwife to the Virgin Mary and wet nurse to the infant Jesus.

Parallels with the tree in the Celtic legend of Finn MacColl may also be found in classical mythology. In Greek legend Hebe, the daughter of Zeus and Hera, the goddess of youth, had the power to make the old young again. She was a cupbearer to the gods and filled their cups with sweet nectar. She was, however, lax in her duties and lost the cup of Zeus, which was stolen by demons. There was great concern over this loss and the gods sent out an eagle to recover the cup. A fierce battle ensued and wherever an eagle feather, or a drop on blood, fell to the ground up grew a rowan tree. This is why these trees have feathery foliage and fruits like drops of blood.

Just as in some Celtic myths a sacred rowan grew from seed accidentally dropped by the Tuatha de Danann, so in other myths the trees that grow in Ireland came from seed given by the Sidhe, the faerie folk, and had been brought directly from faerieland.

In some Scandinavian myths the rowan is seen as the feminine equivalent of the masculine ash tree. Creation stories have the first man formed from the ash and the first woman from the rowan. The tree is also called 'Thor's helper', as it is said to have saved the thunder god by bending low over a rapidly flowing river that was sweeping him away and enabled him to get back onto the riverbank.

In Icelandic myths rowan is especially potent at the time of the winter solstice, that is the beginning of the new solar year. At this time the tree is frosted, as though covered by stars, expressing the outpouring of the light of the spirit during the darkest part of the year. This also shows the reflection of the moon's light and the myths of star-dressed rowan may have developed from a tradition of erecting 'moon trees' in an early representation of the energy of the sky goddess. 'Moon trees' were trees, plants and wooden poles or even truncated pillars (probably evolved from cone-shaped stones) representing the moon deity. They were covered with fruits and lights, and were crowned by a crescent moon. In later years they were dressed with glittering objects to portray and reflect the light of the moon. It is not a giant leap to see how these might have been the original Christmas trees. There was also a Yule-time legend that a special star glowed atop a mythical rowan tree indicating the return of the world from the period of darkness.

In some Christian myths rowan is identified as the tree from whose wood the cross at the crucifixion was made, the red berries no doubt relating to the drops of the saviour's blood. In legends from the ancient world, as can be seen in many of the stories contained in this book, red foods, be they fruit, berries, fungi, nuts or any other kinds of food, were considered to belong to the gods and thus taboo to the common man. Red is the colour of blood, of death and so best avoided.

Just as the red berries ensured that the tree was held in high esteem by the pagans so its veneration by the Druids ensured that its use would be spread throughout Western Europe, and the movement of Scots, Irish and other northern Europeans to North America spread the myths and folklore of this plant to the New World. In the legend of Fraoth a rowan, possibly the same one as in the story of Dairmait and Grainne, was the mythological tree of life. It produced different fruits each month and at each quarter of the year, attuned to the lunar and solar cycles. The fruits staved off hunger with the sustaining power of nine meals, healed wounds and added a year to the life of those who ate from them. A mighty dragon guarded this tree; rowans have often been associated with serpents and dragons.

Magic and Lore

Rowan is the tree of quickening, the 'tree of life', 'quick' being used here in the sense of living. Although we usually associate the oak with the Druids it seems that they also held the rowan as a sacred tree. The Celts associated it with mystery and magic, considering the berries divine fruit, whose wine granted longevity. The Druids of King Cormac of Ireland used rowanwood to make magic fire. It is often found growing, and is indeed said to thrive best, close to stone circles and other ancient sacred sites. The common name 'wicken' may be a corruption of 'quick' or derived from 'witch tree'. In ancient times it was a euphemism for yew, which no doubt means that there are may overlaps between the folklore associated with both these trees. In ancient Rome lectors carried rowanwood wands as badges of their authority.

> In the yard there grows a Rowan
> Thou with reverend care should tend it.
> Holy is the tree there growing,
> Holy likewise are its branches,
> On its boughs the leaves are holy
> And its berries yet more holy.
>
> – *The Kalevala*

Rowan Tree Day, or Rowan Tree Witch Day, was observed on 1 or 2 of May, the time of the old Beltane feast, or on 3 May, Holy Rood Day, in different regions of Britain. On that day branches were cut to be used for the following twelve months as a protection against evil. Those cutting them should never before have seen the tree, or even have had any knowledge of it, and must take a different route home from the outgoing path.

The wood cut on Rowan Tree Day was taken into the home, set over doors and windows or placed over the hearth to protect the house against the forces of evil. It could be put on wells to conserve the quality of the water. On farms small pieces might be added to milk churns, or churn staffs made from the wood, to prevent the milk and butter being stolen. On the Isle of Man and in the Scottish Highlands, special cakes were made over rowanwood fires.

With the spread of Christianity it was said that rowan which is to be used for protective amulets should be gathered on Good Friday. If it was to be carried or worn as an amulet it required the spell:

> From witches and wizards,
> And long tailed buzzards,
> And things that run in the hedge bottoms
> Good Lord deliver us.

One Rowan Tree Day charm for gardeners requires them to gather rowanberries on that day into old leather boots or shoes. (The difficulty was that there are unlikely to be many berries on rowan trees at the beginning of May.) They should then be buried in beds where bedding plants are to be planted whilst reciting:

Rowan fruit, boot and shoe,
Bless my flowers the summer through
Faeries of the quicken tree
Work this growing charm for me.

Especially if done during a waxing moon on a Wednesday or Friday, would ensure bountiful blooms.

Other dates in the calendar are considered auspicious for gathering rowanwood for protective purposes. At each equinox and solstice, or the quarter days, rowan sticks might be collected and laid on lintels. Elsewhere, where Christian tradition has influenced the folklore, Ascension Day or Good Friday were specified as the best days for gathering it.

Rowan has long been thought to be one of the most powerful protections against witches and the forces of evil. In the Scottish Highlands, it was said to be the 'total cure' for all witchcraft, and was used in much the same way as hawthorn might be used in the lowlands and in England. A cross formed from rowan twigs tied together with a red thread could be stitched into the clothing as a protective amulet. This use of rowan amulets is mentioned in a number of Scottish traditional rhymes:

Roan ash and red thread
Haud the witches a' in dread.

Black luggie [a milking vessel], Lammer [amber] bead
Rowan tree and red thread,
Put witches to their speed.
– Robert Chambers, *Popular Rhymes of Scotland*

Rowan tree and red threid,
Gor the witches tyne their speed.

Some other sources recommend making a protective amulet from rowan leaves with rue and basil, all tied together in a cloth of white or gold. This may be carried to ward off evil or negative forces and protect against any form of psychic attack.

In Jutland, 'flying rowans', trees found growing on top of other trees (neither on earth, nor in heaven), were believed to possess the greatest magical power. They were considered potent charms against witchcraft as, like mistletoe, they had never touched the ground and so were not under the power of witches and earthborne spirits. In some regions of Scandinavia flying rowan was taken from high cliffs where the seed had set onto rocks. It was recommended that all who ventured out at night should carry some flying rowan to chew on as it would prevent them from being bewitched and so rendered unable to move.

Rowanwood might be included in the construction of houses in order to gain the protection of the wood, or its spirit. It was frequently said that it was better than iron as a protection. 'Witch posts' made from the wood might be added to the thatch of a cottage to protect against lightning and fires, or used for lintels and over hearths as a guard against witches, who were supposed to enter the house by the door, cross the room and be able to exit via the chimney. Placing rowan above the main doorway of the house was also a guard against the evil eye.

The protective powers of the rowan are illustrated in a folk tale from Herefordshire. Two hogsheads of money were hidden in a subterranean vault beneath Penyard Castle, and supernatural forces guarded the hiding place. A local farmer undertook to find the money and used twenty of his finest oxen, urged on by a rowan goad, to open the great iron doors. Looking inside the vault the farmer saw two massive chests, upon one of which sat a large jackdaw. The delighted farmer exclaimed, 'I believe I shall have it!' but immediately the door slammed shut in front of him again and a ghostly voice from inside said, 'Had it not been for your quickenwood goad, and your yew-tree pin, you and your cattle had all been drawn in!'

The use of rowanwood for the goads used in driving livestock had a benefit beyond that of protecting the drover and his animals from enchantment; it was also said that bewitched horses could be controlled with rowan whips. In Yorkshire it was said:

> Woe to the lad
> Without a rowan tree gad.

On the Isle of Man rowanwood crosses were formed without the aid of a knife, and tied into the tails of cattle to protect them from evil. In spring, probably as a part of the Beltane activities, goats would be driven through hoops made of rowanwood to prevent them from being bewitched. Plough pins and tether pegs used for cattle would be made from rowan, and cows believed to have been bewitched would have garlands of rowan leaves put over their horns and in their stalls. Pigs with rowan leaf garlands placed about their necks were supposed to fatten more quickly than others. Traditionally, of course, materials to be used for magical purposes should never be cut using tools made of base metal such as iron and steel. Farmers and carters might also take to wearing rowan twigs or berries in their hats, or as necklaces as a protection.

In the south-west of England it was claimed that if a witch was touched by a rowan stick the Devil would appear and spirit them away, although some authors suggest that he would only take them every seventh year.

It has been said, 'Woe to those with no rowan tree near', because of the protective and magical properties of the wood. Its presence was supposed to ensure good fortune for the house and the family living there. The tree might be planted in a garden following a house move, especially by newlyweds, to bring honour to the home and help in the starting of a family. It has also been said that if a rowan takes root in your garden your house is blessed as the garden is under the protection of the faeries, and like other sacred trees it would be unlucky to cut down a rowan tree.

As I have said rowan is a protection against lightning and fires as well as the mischief of witches and faeries; clearly this is an association with the Scandinavian thunder god, Thor. The thunder god of the Finns was Ukko, the oak, and the rowan was Rauni, his wife.

Another Scandinavian goddess to whom the rowan is linked is Ran, wife of Aegir, the sea god. She was seen as an enemy of seamen and so rowan was included in Viking longboats as a guard against her malice. There remains a superstition that carrying rowanwood when at sea will prevent storms and prevent the boat from going off course.

Carrying the wood or berries is supposed to heighten psychic powers and aid concentration, helping one develop powers of prediction and improve perception. It enables those who bear it to keep their wits about them, better control their senses and enable them to easily distinguish good from evil. All who set out on journeys ought to carry some with them. Walking sticks of rowanwood will not only protect those that use them but also prevent them getting lost, guiding them safely home again.

A V-shaped piece of the wood should be carried or worn to prevent the bearer from being transported off to faerieland. Those out riding at that time might also place a similar piece of wood onto the bridle of their mount to ensure that the horse is not troubled by malicious faeries.

It has been suggested that rowan was planted in cemeteries because of its supposed connection with the holy cross, as a protection and a warning to negative forces and evil spirits. However, the plant's association with religious and sacred sites long pre-dated the spread of Christianity. In the Highlands and Islands Druids planted rowan near stone circles, possibly for oracular use, although this may even pre-date druidic usage. The later plantings in churchyards were supposed to maintain the peace of the dead. The presence of rowan would ward off evil spirits and prevent the dead from emerging prematurely. Rowan stakes, preferably with berries on them, might be driven through corpses to immobilize the ghosts; and also to ensure that the spirit of the departed passed peacefully from this world into the next. In the Scottish Highlands the wood was used for coffins and biers. When a funeral procession had to stop, the coffin was rested under a rowan tree rather than in the open so that it would not be vulnerable to evil forces. It is said that if you come across a rowan tree with a bountiful crop of berries you may be sure that some saintly person is buried close by.

Rowanwood has been used to make wands and dowsing rods in preference to hazel, particularly when they are to be used for dowsing for metals. These wands were also favoured by some for drawing magic circles.

Some authors have said that the Druids lit rowan fires, and spoke incantations over them, to summon demons to aid them in battle, whilst others state that it was the aid of the Sidhe that was being sought. Burning rowan, with the correct invocation, was also supposed to be used to invoke spirits, familiars, spirit guides or even the White Goddess herself.

Irish Druids used rowan in divination, spreading the newly flayed hide of a bull, bloody side up, on rowan wattles to attract spirits and demons and to compel them to answer difficult questions. Elsewhere groves of rowans were preserved as oracular shrines, where the priestess would imbibe the sacred berries, or a wine made from them, to induce a trance by which she might divine the future.

Rowan rod, forefinger,
By power of divination.
Unriddle me a riddle,
The key's cast away.
– Robert Graves, *The White Goddess*

The Irish expression 'to go to the wattles of knowledge' meant to do one's utmost to obtain information. Rune sticks were generally cut from rowanwood and tradition has it that knowledge and science of all kinds was engraved onto three rowan rods, 'except the name of the god which has originated the bardic secret'.

In Wales it was said that if you step into a faerie ring your only hope of rescue is for a rowan staff to be placed across the circle; otherwise you are liable to remain within the faerie ring for a year and a day.

One old country rhyme relating to the meanings of plants says:

Rowan – Affection;
Holly – Folly;
Briar – Liar;
Plum in bloom – marry soon.

A good crop of rowanberries is an indication that there will be a poor harvest.

Many rains, many rowans,
Many rowans, many yawns [a light grain of wheat].

Medicinal

Rowan has been little used in folk medicine, although the berries have been used to ease the pains of childbirth. To treat other ailments, such as whooping cough, a lock of the sufferer's hair might be pushed into a cut in the bark of the tree. Inhaling the smoke from the burning leaves was a treatment for asthma, and drinking an infusion made from the leaves was recommended for easing rheumatic pains. It has been claimed that carrying the leaves would be sufficient to prevent rheumatism but one has to suspect that this idea might stem from its use as a treatment.

In parts of Russia the berries were soaked in vodka and the resultant liquor drunk as a treatment for stomach aches. The fresh juice of the berries was used as a laxative and an infusion of the berries for treating haemorrhoids and stangury.

Perhaps the best recommendation of this plant is still that each berry eaten will add a year to life.

SERVICE TREE

Botanical Name: *Sorbus domestica*
Family: Rosaceae
Common Name: Sorb-apple
Deity: Oeagrus
Meanings: 'From corruption, sweetness'; prudence

Myth and Legend

It was the fruit of the service tree, not the apple, that was the 'apple of immortality' in pagan France, Spain and Scandinavia. Elsewhere in Europe it was sacred to Oeagrus, the father of Orpheus, who belonged to a 'sorb-apple' cult. The apple of the Thracian Orphic cult was also the sorb-apple.

Magic and Lore

The fruit of the service tree is emblematic of the idea of 'from corruption, sweetness' because, like its relative the medlar, it is not edible until it rots.

> So, in the strange retorts of medlars and sorb-apples
> The distilled essence of hell.
> The exquisite odour of leave-taking.
> *Junque vale*!
> Orpheus, and the winding, leaf clogged, silent lanes of hell . . .
> Medlars and sorb-apples,
> More than sweet
> Flux of autumn
> Sucked out of your empty bladders
> And sipped down, perhaps with a cup of Marsala
> So that the rambling, sky-dropped grape can add its savour to yours,
> Orphic farewell, and farewell and farewell
> And the *ego sum* of Dionysus
> The *sono io* of perfect drunkenness
> Intoxication of final drunkenness.
>
> – D.H. Lawrence, *Medlars and Sorb-Apples*

BETONY

Botanical Name: *Stachys officinale* (Synonym: *Betonica officinalis*)
Family: Labiatae
Common Names: Bidney; bitney; Devil's plaything; wild hop; wood betony
Gender: Masculine
Planet: Jupiter
Element: Fire
Meaning: Surprise

Magic and Lore

In common with other plants gathered for magical purposes betony must be collected without the use of iron tools. It could be carried or worn about the neck to ward off evil spirits, demons and despair. A more extreme method to remove all ills and evil spirits requiried the herb to be burned on Midsummer fires and those needing purification to jump into the ashes of the fire.

According to the writer Apelius, to be most effective, betony must be gathered in August, shaken until clean, then dried thoroughly in the shade before being reduced to a dust. This dust can be scattered near doors and windows to form a protective barrier preventing any evil from entering the house. Wearing betony, or placing it beneath the pillow, would prevent nightmares and 'fearful visions' and protect the sleeper throughout the night.

The plant is said to have the power to reunite quarrelling couples if it is added to their food.

An ancient tradition tells that if two snakes are placed in a circle made of betony they will fight and kill each other.

Medicinal

Betony has been held in high repute for centuries as a panacea for all ailments. The Greeks extolled its benefits and Antonius Musa, physician to the Emperor Augustus, wrote a treatise showing it as a cure for some forty-seven diseases. The physician Robert Turner, writing in the seventeenth century, details thirty ailments for which betony was an effective remedy. Betony was also supposed to strengthen the body and treat a mysterious disease called 'elf-sickness'. Some ancient authors claimed that animals could also recognize the healing powers of the plant; injured stags would seek it out and eat it to heal their wounds. An Italian saying encourages that you should 'sell your coat and buy betony'.

It was thought of as being an enemy of the vine, the tendrils of which would turn away from it. The plant was, therefore, seen as a favourite cure for drunkenness.

GREATER STITCHWORT

Botanical Name: *Stellaria holostea*
Family: Caryophyllaceae
Common Names: Adder's meat; adder's spit; all bones; arva; arvie; Baalam's smite; bachelor's buttons; Billy buttons; Billy White's buttons; bird's eye; brandy snap; break Jack; cuckoo flower; cuckoo meat; cuckoo's victuals; dead man's bones; Devil's corn; Devil's eyes; Devil's flower; Devil's nightcap; Devil's skirt buttons; Easter bell; Easter flower; eyebright; granny's nightcap; hagworm flower; headache; Jack-in-the-box; Jack-in-the-lantern; Jack snaps; lady's buttons; lady's chemise; lady's lint; lady's needlework; lady's smock; lady's white petticoat; lieenferish; little John; May flower; May grass; milk cans; milk maidens; milk maids;

milk pans; miller's star; mischievous Jack; moon flower; moonwort; morning star; mother's shimbles; mother's thimble; murren; Nancy; nightingale; old lad's corn; old man's shirt; one o'clock; pick pocket; pisgie; pisgie flower; piskie; pixie; pixy; pop-guns; pop Jack; poppers; poppy; pretty Nancy; sailors' buttons; satin flower; scurvy grass; shepherds' weather grass; shimmies; shimmies and shirts; shirt buttons; skirt buttons; smock frocks; smocks; snake grass; snake flower; snake weed; snapcrackers; snap Jacks; snapper-flower; snappers; snaps; snapstalks; snapwort; snick-needles; snow; snowflake; snow-on-the-mountain; star flower; star grass; star-of-Bethlehem; star-of-the-wood; starwort; stepmothers; stitchwort; sweethearts; sweet Nance; thunderbolts; thunder flower; twinkle star; watches; wedding flowers; white Bobby's eye; white flower; white flowered grass; white Sunday; wild pink
Meaning: Hermitage

This is a plant under the power and the protection of the Devil, jack-o'-lantern, pixies or even snakes. Cornish children were warned that they must avoid picking it, as they would be in danger of being bitten by adders or led away by pixies. In later traditions, the plant was dedicated to the Virgin and associated with Whitsunday. The pink flowers were supposed to provoke thunderstorms.

It is an indication of good soils.

CHICKWEED

Botanical Name: *Stellaria media*
Family: Caryophyllaceae
Common Names: Adder's mouth; arva; arvie; bird's eye; chicken meat; chicken weed; chicken wort; chickny weed; chick wittles; clucken weed; clucken wort; mischievous Jack; mouse ear; murren; passerina; satin flower; *schickenwir*; skirt buttons; star chickweed; star weed; star wort; stitchwort; tongue grass; white bird's eye; winter weed
Gender: Feminine
Planet: Moon
Element: Water
Meanings: Ingenuousness; rendezvous

The common name 'chickweed' is derived from the practice of feeding this herb to canaries and other caged birds; it supplemented their diets and was supposed to quieten them.

It is used in spells and mixtures to attract love and to maintain loving relationships. Water in which the plant had been steeped was used in the treatment of obesity.

BENZOIN

Botanical Name: *Styrax benzoin*
Family: Styraceae
Common Names: Ben; benjamen; gum benzoin; Siam benzoin
Gender: Masculine
Planet: Sun
Element: Air

Benzoin is a balsamic resin produced following damage to the tree. It is a 'clearing herb' and used either as a purification incense or as a base for other incenses. It may be burned with cinnamon and basil, at places of business, to attract customers.

DEVIL'S BIT

Botanical Name: *Succisa pratense* (Synonym: *Scabiosa succisa*)
Family: Dipsacaceae
Common Names: Angel's pincushion; bachelor' s buttons; bee flower; bitin' Billy; bitten-off root; blue ball; blue bobs; blue bonnets; blue buttons; blue cap; blue heads; blue kiss; blue tops; bund weed; curl-doddy; Devil's buttons; fire leaves; forebitten-more; gentleman's buttons; gypsy rose; hardheads; hoary plantain; hog-a-back; lamb's ears; lough-shule; of bit; pin cushion flower; premorse scabious; stinking Nancy; wooly hardheads
Gender: Masculine
Meaning: Unfortunate love

Myth and Legend

The common name is derived from a legend that the Devil found this plant growing in the Garden of Eden and, envying the good that it might do for mankind as a healing plant, he tried to destroy it by biting away a part of its root – only the Devil would be able to bite away pieces from below ground. An alternative tale tells that the Devil was using the herb for malicious purposes so the 'Mother of God' took it away from him. The Devil, vexed at having lost its power, bit its roots. Care must be taken not to touch the end, as no amount of washing will ever get rid of the Devil's breath.

Magic and Lore

In Fife, Scotland, it was believed that the plant could be used to summon up brownies (faerie folk) to assist with the domestic duties of cleaning the house and clearing out cowsheds. The correct incantation was:

> Curl-Doddy, do my biddin',
> Soop my house and hool my midden.

It is supposed to bring good fortune, especially for women. A piece worn about the neck will ward away evil spirits but, according to a Cornish tradition, great care must be taken in gathering it as picking it will cause the Devil to appear at your bedside.

A rather different warning on the gathering of the plant comes from the farming community. As the leaves are somewhat thicker than those of grasses in a hayfield they take longer to dry out. If the hay is collected too soon after cutting, before it is completely dry, these damper leaves can cause spontaneous combustion in the haystacks.

Devil's bit scabious is the birthday flower of 8 August and garden scabious of 26 June.

Medicinal

The 'good' that the plant would do, that so much annoyed the Devil, was the treatment of numerous ailments.

COMFREY

Botanical Name: *Symphytum officinale*
Family: Boraginaceae
Common Names: Abraham, Isaac and Jacob; ass ear; blackwort; boneset; bruise wort; church bells; coffee flowers; consolida; consormol; consound; gooseberry pie; gum plant; healing herb; Isaac and Jacob; knitback; knitbone; miracle herb; nip bone; pig weed; salsify; slippery root; snake; suckers; sweet suckers; wallwort; yalluc
Gender: Feminine
Planet: Saturn
Element: Water

Magic and Lore

Wearing or carrying comfrey ensured safety when travelling. This protection extended to possessions as well as people; putting a small piece of the plant into luggage prevents it from being lost or stolen.

Medicinal

It is said that knights returning from the Crusades introduced the plant into Britain; it was then grown and used by monks to treat their brethren and travellers. The common names allude to its use in the treatment of broken bones. 'Comfrey' is a corruption of *con-firma*, referring to the uniting of the broken bone. The botanical name, *Symphytum*, is likewise derived from the Greek *symphyo*, meaning to unite or to make things grow together. In herbal remedies the beaten root of the plant was spread onto leather and laid over the affected areas. Gerard says that comfrey should be 'given to drinke against the paine of the backe, gotten by violent motion as wrestling or overmuch use of women'.

LILAC

Botanical Name: *Syringa vulgaris*
Family: Oleaceae
Common Names: Lily-oak; May flower; pipe tree
Gender: Feminine
Planet: Venus
Element: Water
Meanings: Humility (field); first emotions of love; 'You are my first love' (purple); 'A tribute to your beauty and spirituality'; innocence; modesty; purity; virginity; youth (white); brotherly love; forsaken; memory

Myth and Legend

The name *Syringa* is derived from the Greek for a pipe as, when the soft pith is removed from the stem, it forms a pipe. Some claim that the pipes of Pan were formed from lilac and made in honour of the nymph Syrinx, who had evaded his advances by turning herself into the fragrant lilac bush.

Another story of the flower's origin tells that Spring drove away snow from fields and raised the sun higher above the Earth with the rainbow. Then she gathered sunbeams merged with some rainbow beams and threw them down to the Earth. When she reached the north the only colours she had left were white and purple. As she passed over the Scandinavian lands she threw down the purple to colour the flower-covered bushes below. Then Spring scattered the white colour and where it landed white lilac bushes sprang up.

Magic and Lore

Lilac was introduced into Britain during the reign of Henry VIII. In its native Persian the name, *lilaq* simply means 'a flower'.

The white-flowered form of the plant, like so many other white-flowered plants, is associated with the White Goddess. In consequence it is usually considered to be unlucky to cut the blossoms and bring them into the home. Indeed the whole plant is, more often than not, thought to be one of ill fortune. Purple-flowered forms are said to be less hazardous. Some writers state that the plant became thought of as unlucky because of the 'drowsy' scent of the flowers, which could transport those who were overcome by it to faerieland, but it is unclear whether this pre-dates other associations. The scent of lilacs was said to attract faeries into the garden.

Again, in common with other white-flowered, heavily scented plants lilac became associated with death. Giving the flowers to someone convalescing after illness was said to be a sure way to cause a relapse. It is likely that the association of lilac with misfortune and death comes from it having been used to decorate a house following a death, as the strong scent of the flowers would mask any smell from the body laid out in the home.

The only exception to the rule of never taking the flowers indoors appears to be when they are taken into a haunted house, as the presence of the blossoms will drive away any ghosts. This superstition may have led to the plant being considered, in the New England region of the USA, to be a protection against evil. There it was planted close to houses as a guard against dark forces.

In the Welsh border regions it was often claimed that lilac would mourn the death of other plants; if one lilac were cut down others near to it would not flower in the following year.

CLOVE

Botanical Name: *Syzygium aromaticum*
Family: Myrtaceae
Common Name: Zanzibar redwood
Gender: Masculine
Planet: Jupiter
Element: Fire
Meaning: Dignity

Cloves may be burned as an incense to purify a property or area, clearing away all negativity. It has the added advantages of attracting wealth and preventing others from gossiping about you.

Carrying or wearing cloves was said to make the bearer more attractive to the opposite sex. This may date from a time when regular washing was less common and the scent of the cloves masked other odours.

AFRICAN MARIGOLD

Botanical Name: *Tagetes erecta*
Family: Compositae
Common Names: Aztec marigold; French marigold; *la flor de muerto* ('flower of death')
Deity: Tages

Myth and Legend

This plant is named in honour of Tages, grandson of Jupiter, an Etruscan deity who sprang from the newly ploughed earth and taught the Etruscans the nature of omens regarding winds, eclipses and lightning.

There is a Mexican tradition that the plants commemorate the Indians killed by Cortes, and grew from where their blood was spilt.

Magic and Lore

The Spanish, having found these plants in Mexico, took them to Spain and North Africa, where they flourished. In England they were called 'marigolds' because of their similarity to the *Calendula* species. In their native Mexico they are known as the 'flower of death' and put on graves at All Hallows (1 November), the day of the dead.

See also Marigold (*Calendula*).

TAMARISK

Botanical Name: *Tamarix gallica*
Family: Tamariscinaceae
Common Names: Brummel; cypress; French tree; manna tree
Gender: Feminine
Planet: Saturn
Element: Water
Deities: Anu; Isis; Osiris
Meaning: Crime

Myth and Legend

In the Egyptian creation myth Ra discovered that his wife, the sky goddess Nut, had been unfaithful and been made pregnant by the Earth god Ged, so he decreed that she should be delivered of her child in no month and no year. Nut, however, had another lover, Thoth who played draughts with the moon and won from her a seventy-second part of every day. He compounded these parts until he had an additional five days to add to the Egyptian year of 360 days and so bring harmony between solar and lunar time. In these extra five days it was possible for Nut to give birth. On the first day Osiris was born, on the second the elder Horus, on the third Set, on the fourth Isis and on the fifth Nephthys.

Osiris and Isis ruled the pantheon of Egyptian gods but their brother Set was jealous of the power and position of Osiris and resolved to kill him. He had a great sarcophagus made, inlaid with gold and precious stones, and suggested that his brother should try it for size. As soon as Osiris was inside Set closed and locked the lid. He tossed the sarcophagus into the Nile and it was swept away downstream. It floated into the Mediterranean Sea and eventually came to rest in roots of a tamarisk tree growing on the coast of Byblos. The roots and branches of the tree grew out and around the sarcophagus, absorbing it into its wood, where it remained hidden for many years. One day, a local potentate, having ordered extensions be made to his palace, had the tree felled to form a pillar to support the palace roof. It may be that this tree was selected because of its magnificent scent, or perhaps that was only noticed once the pillar was in place. However, news of the miraculously fragrant timber that could be smelt for miles around eventually reached Isis. Because of her knowledge of magic, she realized the secret the pillar must contain. She hurried to the place and ordered the pillar to be cut open, revealing the body of her husband. Just as there are several trees suggested as having given their timber for the cross of Christ's crucifixion so there are a number of trees identified as having enclosed the sarcophagus of Osiris.

Some scholars suggest that the 'manna' received from God by the Israelites during their time in the wilderness came from the tamarisk. *Tamari gallica var. mannifera* exudes droplets of sugary sap during June and July, which solidify, forming 'manna'.

Magic and Lore

Like other woods used in magic tamarisk must be cut with a golden axe or silver knife, never with tools of iron or steel. The wood has been used in exorcisms and other rites for more than 4,000 years. The branches would be held and leaves scattered to ward off demons. Smoke from the burned wood was supposed to drive away snakes and the twigs were used in divining.

COSTMARY

Botanical Name: *Tanacetum balsamita* (Synonym: *Chrysanthemum balsamita*)
Family: Compositae
Common Names: Alecost; balsam herb; balsamita; balsam tree; costmarie; goose tongue; *herbe Sainte-Marie*; mace; maudlin; mint geranium; our lady's mint; sage o' Bedlam (or Bethlehem); tanzy; yellow buttons
Planet: Jupiter
Meaning: Impatience

Costmary used to be popular as a bookmark, and was also used as a strewing herb. The derivation of the name comes from a strange combination of *Costus*, meaning an oriental plant, and 'Mary', which is a reference to the plant being dedicated to the Virgin Mary (some authorities suggest that it is dedicated to St Mary Magdalene). The common name 'alecost' was derived from the use of the plant to flavour ales.

Medicinal

Culpeper said of costmary, 'It is an especial friend and help to evil, weak and cold livers.'

FEVERFEW

Botanical Name: *Tanacetum parthenium* (Synonym: *Chrysanthemum parthenium*)
Family: Compositae
Common Names: Altamisa; bachelor's buttons; featherfew; featherfoil; febrifuge plant; flirtwort; Midsummer daisy; nosebleed
Gender: Masculine
Planet: Venus
Element: Water
Meanings: 'Let me shield you'; protection

The name of this plant speaks for itself. It is a febrifuge, which means that it dispels fevers. The herbalist Nicholas Culpeper recommends it as a general tonic or 'pick-me-up', to combat weariness and melancholy. Other herbalists have, at various times, recommended it for the treatment of vertigo, depression and, of course, fevers. In modern usage it is often recommended as a treatment for migraines. Gerard informs us that it is efficacious whether taken internally or externally. It may be added to drinks or bound to the wrists. This latter application can have the added advantage that it should protect the person from harm as the bearer of a feverfew plant is said to be safe from accidents, fevers and colds.

TANSY

Botanical Name: *Tanacetum vulgare*
Family: Compositae
Common Names: Bachelor's buttons; bitter buttons; buttons; ginger; ginger plant; golden buttons; hind heel; parsley fern; scented daisies; scented fern; stinking elshander; stinking Willie; traveller's rest; weebo; yellow buttons
Gender: Feminine
Planet: Venus
Element: Water
Deity: Ganymede
Meanings: 'I declare war against you'; reflected addresses; refusal; resistance; 'Your feelings are not reciprocated'

Myth and Legend

The botanical name is derived from the Greek *athanaton*, meaning 'immortal'. In classical mythology it was given to Ganymede to render him an immortal and in Christian mythology it is dedicated to the Virgin Mary.

Magic and Lore

Tansy is linked with Easter festivities. It was once a custom, in some parts of England, for clergymen to play a form of handball with members of their congregation at that time. The victors were rewarded with tansy cakes, made from the young leaves of the plant and eggs. The bitter juice was also used to flavour other Easter foods. These were thought to purify the body following the strictures of Lent, or at least to make body odours more pleasant.

> On Easter Sunday is the pudding seen
> To which the tansy lends her sober green.

These tansy cakes, in time, became symbolic of the bitter herbs eaten by the Jews at Passover.

Tansy is the birthday flower of 23 February.

Medicinal

Tansy was strewn to keep ants and flies out of buildings. It was also strewn in public buildings, such as churches and courts, to prevent the spread of the plague and other contagious diseases. In Europe and New England it was once included in the winding sheet when preparing a corpse for burial, apparently to deter worms, but more likely in an effort to control the spread of the disease from which the departed died or to mask the odour of decay.

Boiled tansy is supposed to be good for the heart and was recommended as a compress for bruises and strains. William Coles in *The Art of Simpling* recommends that the leaves be laid in buttermilk for nine days to produce a skin cream 'making the complexion very fair'.

In Sussex, tansy leaves were put into shoes to prevent fevers. Carrying tansy was said to lengthen the lifespan generally.

In parts of Cambridgeshire couples that were anxious to start a family would eat salads containing tansy. The logic of this was that where tansy grew naturally there were frequently large groups of rabbits, and as eating tansy had rendered them very fertile, it must do the same for people. Elsewhere the same plant was recommended as an abortifacient to be used by unwed women who became pregnant.

DANDELION

Botanical Name: *Taraxacum officinale*
Family: Compositae
Common Names: Bitter aks; blow ball; bum pipe; burning fire; canker; cankerwort; clock; clock flower; clocks and watches; combs and hairpins; conquer more; Devil's milk pail; dindle; dog-posy; doon head clock; eksis-grise; faerie clocks; farmer's clocks; four o'clock; golden suns; heart fever grass; horse gowan; Irish daisy; lay-a-bed; lion's teeth; male; mess-a-bed; milk gowan; monk's head; old man's clock; one o'clock; one, two three; pee-the–bed; pishamoolag; piss-a-beds; piss-i-beds; pissimire; piss-in-the-beds; pister beds; pitty beds; priest's crown; puff ball; schoolboy's clock; shepherd's clock; shit-a-bed; stink Davie; swine snout; tell time; time flower; time teller; twelve o'clock; wet-a-bed; wet weed; what o'clock; white endive; wild endive; wishes; witch gowan; yellow gowan
Gender: Masculine
Planet: Jupiter
Element: Air
Deity: Hecate
Meanings: Absurdity; bitterness; coquetry; depart; grief; rustic oracles

Myth and Legend

The Choctaw Indians of North America have a legend about the love of Shawondasse, the south wind, for a beautiful flower maiden. At that time the summers were unmercifully hot, the sun glared down without any hint of relief. Shawondasse

would be found resting himself beneath the shade of a tree, becoming fatter, heavier and still more languid as the summer wore on. Consequently, even if he did move he still gave little relief from the heat.

One day, as he lay beneath his tree, he was startled by a bee that had mistaken his face for a flower. He jumped up from his shady resting place and, in doing so, noticed a beautiful, golden-haired flower maiden dressed in fringed buckskins. It was love at first sight. Thereafter, each day, from his place in the shade, Shawondasse watched over the flower maiden. And each day he promised himself that he would go across and speak to her, but he never did. In this way the summer slipped past. Then Shawondasse noticed that the flower maiden had changed; she was no longer young and no longer beautiful. Her golden hair had turned grey with age and Shawondasse realized that his brother, the north wind, had touched the flower maiden with his icy fingers turning her hair frosty white. Shawondasse, realizing his loss, sighed and his breath scattered the seeds of the dandelion far and wide. In the villages the people were surprised to feel a breath of wind from the south. Each summer afterwards there could be felt a gentle southerly breeze as Shawondasse sighed for the love of the pretty flower maiden he never spoke to.

Magic and Lore

The botanical name *Taraxacum* is of great antiquity. It is traceable back through the Latin and Arabic to the Persian *talkh chakok*, meaning 'bitter pot herbs'. The common name 'dandelion' is derived from the Middle English 'dent de lyoun', a corruption of the French *dent de lion*, the lion's tooth, an allusion to the ragged margin of the leaves.

As a bitter herb, dandelion is a symbol of the passion of Christ. It is a herb of great antiquity and is mentioned in the Bible as well as in the writings of Dioscorides, Theophrastus and Hippocrates.

Dried dandelion roots have been ground and used as a coffee substitute; an infusion of dandelion 'coffee' is supposed to increase psychic powers. And if the infusion, still steaming hot, is placed beside the bed it will call up spirits.

A modern superstition that appears to owe something to the legend of Shawondasse says that to ensure a favourable change in the wind you must bury a dandelion at the north-east corner of the house.

There are several suggested uses for the seed heads of dandelions, the best known of which is probably the dandelion clock. The number of blows needed to remove all the seeds on the head indicates the hour of the day.

It is one of several plants known as rustic oracles, as its flowers always open about 5 a.m. and close at 8 p.m. Another tradition says that if a young girl blows away the top of a dandelion clock the number of breaths required to blow away the seed head indicates the number of years before she will wed, or if she blows away all the seeds in one breath then she is sure to marry. An alternative has a chant to be said with each breath: 'This year, next year, some time, never.' To test the depth of a lover's feeling one can try to blow away all the seeds with one breath. If all the seeds are removed, their love is passionate. A few seeds remaining can indicate unfaithfulness. Many seeds suggests indifference.

Variations on these themes say you must blow three times and then count the number of seeds left on the seed head to discover the time, or that if you blow the dandelion clock three times then count the number of the remaining seeds it will indicate the number of lovers you will have. When you blow the seeds off a dandelion head the direction in which the seeds fly shows the way you must travel in order to find your fortune. When a child blows away all the seeds from a dandelion head, it indicates that their mother will not want them. If a few seeds remain, the child ought to run home quickly. In a final thought on this, if you try to blow away all the seeds in one breath, counting the ones that remain will show you how many years you have left to live. The flower head, with all the seed plumes removed, was known as a 'priest's crown'.

If you want to send a message to a loved one you must visualize the message in your mind and blow the seeds from the head of a dandelion clock in the direction of the person you want to receive it. The seeds will carry your message to them. If you want to discover whether they are thinking of you, blow the seed head a second time. If even a single seed is left on the head it shows you are not forgotten.

The individual seeds were called faeries and considered to be very lucky if caught as they flew past. Having caught the 'faerie', people would make a wish then release it.

Like buttercups, it is claimed that if you tickle someone under the chin with a dandelion flower and they laugh, it shows that they like butter. It is sometimes said that dandelions show the weather by closing before rain.

Where dandelions appear in a dream it is an ill omen, indicating misfortunes, enemies and deceits on the parts of loved ones.

Dandelion is the birthday flower of 27 September.

Medicinal

The juice of dandelion was seen as a cure for every disease. An early interpretation of the name suggested that it came from two Greek words, *taraxis* meaning 'an inflammation of the eye' and *akeomia*, meaning 'to cure', the milky sap of the plant being used to treat eye complaints.

Dandelion was used as an important spring tonic, and the juice was used to treat warts. The names of 'piss-a-beds' and 'wet weed' relate to its diuretic properties. So effective was the herb that picking, or even smelling, the flowers was supposed to be sufficient to cause bed-wetting.

Drinking a cup of dandelion tea each morning has been suggested as a tonic to purify the blood and promote regularity of the bowels. And a daily cup will also ensure that you never have rheumatism.

There is a recipe for a broth made from sliced and stewed dandelion roots, with some sorrel and an egg yolk, which was prescribed for chronic liver problems; this would seem to owe much to the Doctrine of Signatures.

YEW

Botanical Name: *Taxus baccata*
Family: Taxaceae
Common Names: Bow wood; ewe; ewgh; ife; palm; palm tree; tree of the dead; u; ugh; uhe; vew; yeugh (The berries have been called: red snot; snoder galls; snot berries; snot gobbles; snotter galls; snottle berries; snotty gogs; suss)
Gender: Feminine
Planet: Saturn
Element: Water
Deities: Banba; Hecate
Meanings: 'My heart feels with yours'; death; sorrow

Myth and Legend

When all the world was young, and trees were first opening their foliage, the yew tree was unhappy. Looking around it saw the other trees and thought that they, with their lighter, brighter leaves, were better 'dressed'. Faeries, hearing the yew's complaint, tried to please it and changed all its needles to gold. The tree was happy until robbers came by night and stole them. The faeries then gave it leaves of crystal that sparkled in the sun, but a violent storm shattered them all. The faeries gave it broad leaves and again it was content, until a flock of goats came by and ate them all. On reflection, it decided that perhaps its original dark green needles were best after all.

Celtic myth tells the story of Mark, King of Cornwall, who married Iseult, a lady of Ireland, although she did not love him. Following the wedding they set sail and returned to Cornwall. Unbeknown to anyone Iseult's mother had prepared a love potion in a draught of wine for the couple, in the hope that under its spell her daughter would fall in love with her new husband. Unfortunately, in a twist of fate common in these tales, the potion was not consumed by the newly weds but by Iseult and Tristan, the nephew of the king, and these two fell passionately in love. The spell lasted for three years, during which time the couple took whatever opportunity arose to spend nights together. Many times they were discovered and reported to the king, whose love for them both tore him apart. As he was affected, so was the kingdom because of the situation and the gossip it aroused. After many partings and tricks of fate the couple died in each other's arms. The king gave them a funeral with full ceremony. It is said that Tristan and Iseult were buried either side of the nave of Tintagel Castle in Cornwall, above Merlin's Cave. Within a year yew trees had grown from each grave, but the king ordered them cut down. Again they grew and were again cut down. A third time the trees sprouted and were felled. Eventually, King Mark, moved by his deep affection for his wife and his nephew, allowed the trees to grow. At their full height the branches of the yew trees reached over the nave and embraced, never more to be parted.

A similar tale occurs in other legend cycles. Irish myth has the story of Deidre and Naoise, whose graves were staked with yew by the High King of Conchobar in order that they would always be separated. The yew stakes took root and the branches of the resulting trees reached over Armagh Cathedral and interwove in an embrace, linking the lovers beyond the grave. In the *Historical Cycle* the tale of Baile and Ailinn runs a similar course. Baile's grief for the beautiful Ailinn brings about his death. From his grave springs a yew tree, carrying a likeness of his face in its branches. After seven years the tree was cut down and poets made writing tablets from its wood.

In the Irish myths of the Tuathe de Danaan, the earliest magical race, and gods of ancient Ireland, there is the story of Banbha, the last great warrior queen, the sister of Fodhla and Eire. After she was killed she became semi-deified as the death aspect of the triple goddess of Ireland. The yew tree was considered sacred to her and, became known as the 'renown of Banbha'. The 'tree of Ross' was one such 'renown of Banbha' yew and was said to represent the death and destruction that was to come again in Ireland (as Christianity was imposed over Druidic paganism and a pantheistic culture).

Other ancient names for the yew were the 'spell of knowledge' and 'the king's wheel', referring to a brooch worn by the king and passed on to his successors on this death. The brooch represented the cycle of existence and was to remind the king that death could come at any time and it was his duty to maintain a viable realm that he would leave to his heirs.

There stands a yew tree at Congresbury, Avon, which is associated with the story of St Conger, who is believed to have been a sixth-century missionary. Legend claims that the saint wished for a yew tree to give him shade, then thrust his staff into the ground. The following day it had sprouted leaves, forming a spreading yew tree.

At Fortingall in Tayside, Scotland, there stands an ancient yew tree that is said to be over 9,000 years old. It is also claimed that beneath this tree, as an infant, Pontius Pilate was suckled by his mother during his father's tour of duty as a legionary in Britain.

Eighteenth-century charcoal burners at Ambergate in Derbyshire took their young children to work with them, and used to form crude cradles hollowed out of the limbs of yew trees. This tree was also the inspiration for the children's nursery rhyme 'Rock-a-bye-baby'.

In the legend of the thirteenth-century Thomas the Rhymer, Thomas of Ercledoune, it is said that he met with the queen of Elfland. He received from her the gift of prophecy and gifts of clothes in elfin green. It says that he awaits rebirth in a yew grove, somewhere near Inverness, guarded by faeries.

Magic and Lore

The yew is a tree of death and resurrection, the unity of life and death, representing great age, rebirth and reincarnation. It is associated with the wisdom that was, is and always will be. It is regarded as such in many countries across Europe, and as one of only three British native conifers it would have stood out in early Britain. It is one of the five magical trees of Ireland and stands as a 'firm straight deity'. This alludes to the fact that the Irish yew is a fastigate form with an upright habit and vertical branching, unlike the more spreading habit of other European species.

In common with all sacred trees it is considered unlucky to cut down or damage a yew tree. In Ireland only wood from fallen trees would be used when making shrines or bishops' crosiers. Even branches broken by the wind were thought to be unlucky and dangerous to use; they were said to be 'slivered', as referred to by Shakespeare in *Macbeth* (see below).

In ancient Greece and Rome, yew was viewed as sacred to the witch goddess Hecate. The Romans sacrificed black bulls, wreathed in yew, to Hecate at Saturnalia, so that ghosts might feed on the blood, in the hope of a less severe winter and the preservation of the remainder of their herds. Yew guards against all evil, being said by some to have been planted in churchyards to protect the spirits of the dead and bring about rebirth.

The cult of Hecate spread to Scotland and, although she may once have been considered a fair goddess, she became best known for her darker, malevolent moods. In Shakespeare's *Macbeth* the witches' cauldron contained 'slips of yew slivered in the moon's eclipse' and in *Hamlet* 'Heberon' is the 'double fatal yew' Hamlet's uncle pours into the ear of the King. The Scottish island of Iona was thought to have been originally named 'Ioha', the Gaelic word for yew, its traditions, linked to death and rebirth, long pre-dating the coming of St Columba.

One traditional belief was that yew trees were planted in graveyards, because they were consecrated ground. Livestock let loose to graze on the common would not be able to stray into the churchyard and so would be in no danger of poisoning from the yew's highly toxic foliage. Or to look at it another way, its presence prevented livestock from straying into the churchyard, another facet of its protective qualities. The reason for planting yews was to supply the raw material for bow-making in a time when the longbow was the ultimate weapon. This would have been convenient for armies, who met at churches to receive blessing. Unfortunately for supporters of this idea, yews were to be found near to churches and other sacred sites for centuries before their use in weaponry. Druids considered the tree sacred and associated it with places of worship, a tradition continued by Christians when they built their churches on established sites of pagan worship.

Bows made from the wood of the yew are sacred to the Norse bow god, Ullr. Some writers state that archers were 'men of the yew', that is to say, 'yew-men', and hence 'yeomen'. Having already said that as this is a sacred tree, and so considered unlucky to cut down, it has also been at various times an offence to cut down the tree, as this prevented dissenters from making bows to arm themselves.

As an evergreen tree yew represents everlasting life. The traditional belief was that it warded off all evil and protected the spirits of the dead. Paradoxically it is also said to be a plant used when 'raising the dead' and able to prevent the dead 'rising untimely from the grave'. In Brittany traditional belief goes one step further, stating that the tree sends its roots into the mouths of the deceased, pinning them in their graves and so preventing them from haunting the living. An alternative suggestion was that their roots would grow through the eyes of the dead to stop them seeing into this world, and prevent them from wanting to return as spirits. Some claim that the indistinct faces that may be perceived on the trunks of graveside yews as the bark peels are a sign that the dead have been freed from earthly restraints and are able to begin their journey of rebirth.

The tree's link to death, and its use as a symbol of life, extends to its use to scatter in graves, like rosemary. Sprigs were sometimes placed in the shrouds of the dead to protect and restrain the spirit. It was believed that it would prevent any evil forces from influencing the ghosts of the dead. Its evergreen habit and its extremely long life made it the expression of the belief that death was not the end but only a transition into a new life. In pagan graveyards in Ireland measuring rods of yew were kept for measuring corpses and graves. Whether the wood had a great significance in this case, or was used because it is particularly hard and long lasting, is uncertain.

In Ireland, sprigs of yew, like willow, have been carried as 'palms' in churches on Palm Sunday, possibly due to a Christian, or pre-Christian, tradition that its evergreen habit symbolized resurrection. After the church service, the yew would be distributed amongst the congregation, to be worn or put into houses and cattle sheds as good luck charms. This has led to Palm Sunday sometimes being called by Irish speakers *Domhnach an luir*, 'Yew Sunday'. It was claimed that, if a maiden placed churchyard yew from a tree she had never visited under her pillow, then she was sure to dream of her future husband. There is another superstition that it is always unlucky to take yew into a house, although it may be planted outside, usually on the windward side. This was considered to be doubly protective, sheltering the building and warding off evil or mischievous elves.

Just as the planting of yew trees on sacred sites would appear to be a pagan custom which was later adopted by Christians (possibly because early Christian missionaries stood beneath these trees when preaching before churches were built on the ancient sacred sites), so the handing out of sprigs of yew may have been a druidic practice. Pieces of yew were given at Samhain (31 October) to assist people in contacting the spirits of their departed friends and relatives.

Yew-wood was used in rites at other times of the year. It was sought for its protective qualities at Lammas (Lughnasadh) as harvest began. Lughnasadh was the feast of Lugh, an ancient sun deity, and his feast marked the time when the sun neared the end of its summer reign and began to face long periods below the horizon, in the Underworld. It was considered to be at its most potent in midwinter, representing the passage of the sun through the darkest time of the year. As we know, there has long been a tradition of dressing evergreen trees with shiny objects at Yuletide. At the midwinter festivals sprigs of yew were burned on the Yule fire as people cast off outworn things, freeing the spirit. This in itself may be a transfer for the druidic tradition, as the druidic year turned at Samhain much as ours does in midwinter.

There are two colours in the wood of the yew when it is first cut. The sapwood is white, whilst the heartwood is very red. With the spread of Christianity this has been taken to symbolize the body and blood of Christ.

There was a traditional belief in Scotland that if a chieftain held a sprig of yew in his left hand when he stood to denounce his adversaries to their faces, his enemies would hear nothing whilst all others around would hear every word spoken.

Yew has been used to make relic boxes and divining wands for finding lost goods. If held in front of the diviner the wand will jump when close to the missing object. In other forms of divination sticks of yew were cast on the ground and the future discovered by reading the signs they formed.

It is perhaps from the mixture of life and death overtones that are associated with the yew that it is usually considered to be courting misfortune to seek shelter in its shade. Indeed some say that to sleep beneath it will prove fatal.

Dreams of sitting beneath a yew tree mean the loss of a friend through illness. To see a yew tree in your dreams generally indicates a lucky escape from a serious accident.

Medicinal

Yew contains one of the most virulent of vegetable poisons. In ancient Ireland a poison from yew, hellebore and Devil's bit is said to have been used on weapons. Recently taxol, a chemical derived from the yew, has been found to be effective in treating certain types of cancers. The trimmings of ancient yew hedges are now being used in cancer research and treatments.

WILD THYME

Botanical Name: *Thymus serpyllium*
Family: Labiatae
Common Name: Mother of thyme
Planet: Venus (in Aries)

Magic and Lore

Wild thyme is said to aid relaxation during all types of psychic work. It was gathered on Midsummer Day in Bohemia and was then used to fumigate trees at the solstice as a fertility charm, making them grow well.

A brew prepared to enable the user to see faeries requires the following: A pint of sallet oyle … put into a vial glasse … first wash it with rose-water and marygolde water the flowers to be gathered towards the east. Warm it till the oyle becomes white, then put it into the glasse, and then put thereto the budds of hollyhocke, the flowers of marygolde, the flowers or toppes of wild thyme, the buds of young hazel, and the thyme must be gathered neare the side of a hill where fairies used to be; and take the grasse of a fairy throne; then all these put into the oyle in the glasse and sette it to dissolve three dayes in the sunne and then keep it for the use.' Eleanair Sinclair Rohde, *A Garden of Herbs*. It may well be from this charm that there came a superstition that if you wear a sprig of wild thyme you will see faeries. Placing a sprig on your closed eyes when you sleep on a faerie mound guarantees that you will see them. Sprinkling powdered, dried wild thyme on to the doorsteps and windowsills of the house is an invitation to the faerie folk to visit the house.

On a more general basis it was claimed that carrying or wearing thyme would aid the development of psychic powers for the bearer. It dispels negative energies from an area. Sleeping on pillows stuffed with it will ward off all nightmares.

Medicinal

Wild thyme is somewhat stronger than the cultivated forms. The oil has been used as a general tonic, a stimulant, to promote appetite, and as an aid to digestion. Just as common thyme has been used in treating women in labour, the wild form has been used to help regulate menstruation. It is used for treating coughs, colds and even the flu. It has even been recommended as a treatment for headaches, including those caused by hangovers, and to prevent nightmares. Infusions of wild thyme have been used as a scalp rub to preserve the natural colour of hair.

THYME

Botanical Name: *Thymus vulgaris*
Family: Labiatae
Common Names: Bank thyme; common thyme; French thyme; garden thyme; horse thyme; shepherd's thyme; tae grise; *thym* (French); *za'atar* (Arabic)

Gender: Feminine
Planet: Venus
Element: Water
Meanings: Activity; affection; courage; domestic virtue; energy; happiness; strength; 'I need a wife as capable as you'

Myth and Legend

Amongst Zoroastrians it was believed that thyme sprang from the heart of the Sole-Created Ox, the primal beast of Ahura Mazda.

Magic and Lore

> Thyme and Fennel, a pair great in power,
> The Wise Lord, holy in heaven,
> Wrought these herbs while He hung on the cross;
> He placed and put them in the seven worlds
> To aid all, poor and rich.
>
> – *The Nine Herbs Charm*

Thyme is associated with courage and strength; as such medieval ladies would embroider the plant, with a bee hovering above it, as a motif on the scarves of their knights. It is, of course, loved by bees. Over the centuries it has become linked with death, not unsurprisingly, as it was customary to use strongly scented plants to mask the smell of death when corpses were laid out in the home. It was burned as incense in ancient times, and the Egyptians used it in embalming ointments. In Ancient Greece it was used in the purification of temples and also in purification baths to wash away sorrows and ills. Greek soldiers put it in their baths or were massaged with its oil as a charm for bravery. Roman soldiers put it in the bath for energy, and it was planted on the graves of heroes.

In Wales, the tradition of planting it on graves persisted for a much longer period. In Britain it was said that the souls of the dead dwelt in its flowers and in England it was especially associated with murdered men. Some writers claim that the ghosts of the dead inhibit the plant's flowering. In the Order of Oddfellows, a quasi-masonic society, it was carried at funerals and thrown into the grave. This tradition clearly mirrors the use of rosemary and other herbs.

It is quite possible that its use to mask the scent of decay of corpses led to the superstition that it is unlucky to bring it into the house, as it will cause a severe illness or even a death to a member of the family. Its association with the souls of the dead ties it to the festivities at Halloween or more correctly the druidic festival of Samhain. At Samhain the veil between this world and the next is at its thinnest and it is easiest to contact the spirits of the departed.

Thyme is a favourite plant of the faeries and so it is unlucky to bring it in to the house as this may incur their wrath.

On St Agnus Eve, if a young woman places a sprig of thyme in one shoe and a sprig of rosemary in the other, she is sure to see a vision of her future husband. Elsewhere it is stated that if a woman wears a sprig in her hair it will render her irresistible to the opposite sex. The wearing or burning of thyme was also supposed to attract good luck and good fortune.

Burning thyme is supposed to put all venomous creatures to flight, an idea that may have originated from its use in incense during exorcism rites. The Devil is still linked to the image of the venomous serpent.

Medicinal

Thyme has been used as a medicinal herb, an antiseptic and a disinfectant since ancient times. It was once believed that it could cure all ailments that were caused by the malice of the Devil. Throughout the Middle Ages it was used in posies of aromatic herbs to ward off disease, including plague, and was to be found amongst the herbs placed on a judge's bench at the assizes to ward off jail fever. In folk medicine, it is used for fevers, digestive problems and respiratory ailments. Culpeper informs us that it strengthens the lungs and is an excellent remedy for shortness of breath. He adds that it helps women in labour, granting a speedy delivery and expelling the afterbirth. It might be used to 'comfort the stomach' by expelling wind. Culpeper goes on to recommend it for various other ailments including warts and sciatica.

MEADOW RUE

Botanical Name: *Thalictrum* spp.
Family: Ranunculaceae
Common Name: Flute

Amongst the Native Americans, meadow rue was formed into protective amulets that were carried or worn about the neck. Carrying meadow rue was also supposed to attract the love of others.

CHOCOLATE

Botanical Name: *Theobroma cacao*
Family: Sterculiaceae
Common Names: Cacao; cocoa; food of the gods

The botanical name, given by Linnaeus, translates as 'the food of the gods'. It was brought to Europe in the sixteenth century, by the Spanish, as an aphrodisiac and stimulant. It was even claimed that eating chocolate would increase fertility. The Aztecs used chocolate beans as a form of currency for small transactions.

BE-STILL

Botanical Name: *Thevetia peruvianum*
Family: Apocynaceae
Common Names: Trumpet flower; yellow oleander

The beans of this attractive member of the periwinkle family are strung on necklaces and bracelets in Sri Lanka. There they are called 'lucky beans' and are worn as charms.

ARBOR VITAE

Botanical Name: *Thuja occidentalis*
Family: Cupressaceae
Common Names: False white cedar; tree of life; western white cedar; yellow cedar
Planet: Saturn
Meanings: 'Live for me'; old age; unchanging friendship

Arbor vitae was introduced into Britain in the middle of the sixteenth century. The name *Thuja* is derived from the Greek for a fumigant because that is how the plant was used. It was also one of the plants strewn on to the floors of churches and old halls so that it scented the air when it was walked on.

Medicinal

In folk medicines the arbor vitae has been recommended as a cure for warts.

LINDEN TREE

Botanical Name: *Tilia vulgaris*
Family: Tiliaceae
Common Names: European lime; *flores tiliae*; lime tree; lin; lind; line tree; pry; tilleul; whitewood (fruits called hens' apples)
Gender: Masculine
Planet: Jupiter
Element: Air
Deities: Lada; Venus
Meaning: Conjugal love

Myth and Legend

The gods Zeus and Hermes took on human form and went down from Mount Olympus into Phrygia. There they sought the hospitality of the people, but were turned away from every door at which they called. Finally they came to the home of a devoted elderly couple, Philemon and his wife Baucis. Here they were welcomed and given the best of hospitality. The grateful Zeus took them to the top of a hill, thereby saving them from a flood that devastated the low-lying ground for miles around. Everything was destroyed except their home, which was transformed into a glorious temple. Zeus made them the guardians of the temple and agreed to grant their only wish, that they should die at the same time. As they died, Zeus transformed them into two trees; Philemon was changed into an oak, a tree sacred to Zeus, and Baucis became a linden tree, the emblem of conjugal love.

Magic and Lore

Its image as a tree of love is an ancient one, and the leaves and flowers have been used in love spells. Its character as a symbol of love can be seen intimated in the following medieval poem:

> Under the lime-tree, on the daisied ground,
> Two that I know of made their bed;
> There you may see, heaped and scattered round,
> Grass and blossoms, broken and shed,
> All in a thicket down in the dale;
> Tandaradei – Sweetly sang the nightingale.
>
> – Walther von der Vogelweide, *Under der Linden*,
> translated by Thomas Lovell Beddoes

It has commonly been seen as feminine in character. In Lithuania, during religious rites, women would make sacrifices to the tree deity, just as their men folk would to the oak, which was identified as its masculine counterpart.

It is supposed to have protective properties, and planting it in the garden will afford protection to the whole household. For those without sufficient space to plant it, branches may be cut and hung over doorways to prevent any evil from entering.

In common with many other species, individual trees may be seen to be linked to particular people, the character and wellbeing of the tree being an index of the life of the person with whom it is associated. It is also occasionally said to symbolize immortality; the leaves were used in spells. The wood was also used to make good luck charms.

Its inner bark was once used as a writing material. From this use came the Latin word, *liber*, meaning 'book', and hence a collection of books is a library.

The father of botanical nomenclature, Carl Linné, (Linnaeus), owed his family name to the tall linden trees that grew close by his family home.

Medicinal

Carrying the bark was supposed to be a sure way to prevent intoxication. The flowers were used in folk medicine as a treatment for epilepsy; just sitting beneath the tree was supposed to improve the condition. In the treatment of insomnia equal quantities of linden and lavender were mixed together and put into the patient's pillow.

RED CLOVER

Botanical Name: *Trifolium pratense*
Family: Leguminosae
Common Names: Bee bread; broad clover; broad grass; claver; clover rose; cow-cloos; cow grass; honey; honey stalks; honeysuckle; honey sucks; king's crown; knap; lady's posies; marl-grass; pinkies; plyvens; red cushions; sleeping Maggie; sugar bosses; sugar plums; trefoil; trifoil
Planet: Mercury
Element: Air
Deities: Olwen; Rowen
Meanings: 'Do not trifle with my affections'; injured dignity (pink); provident (purple); entreaty; industry; petition; 'Will you be faithful to me though oceans part us?' (red)

Myth and Legend

On the eve of her wedding a young woman was stricken by a sudden illness and died. She was entombed in the cemetery of St Roch, New Orleans, USA, next to her infant sister. The bridegroom, overcome with grief, sat for hours weeping at the graveside. One day, whilst sitting beside young bride's grave, he shot himself. The clover growing there was stained red by his blood. In an amalgamation of legends, on Easter Sunday children there sell to any passer-by the 'blood of Christ' and anyone who pays them is given a red clover.

Magic and Lore

The very nature of folklore, an oral tradition, results in many of the aspects of the red clover being shared with the white variety.

In cleansing rites the sprinkling of an infusion of red clover around a property will remove all negativity. It may be added to bath water to aid the bather in financial dealings. Being a trefoil, it is associated with the Holy Trinity. In consequence, it brings good luck to those who keep it in the house. It also forms a protection against witchcraft, protecting men and livestock from all spells.

WHITE CLOVER

Botanical Name: *Trifolium repens*
Family: Leguminosae
Common Names: Baa lambs; bee bread; bobby roses; broad clover; broad grass; claver; curl doddy; Dutch clover; Dutch honey; honey stalks; honeysuckle; honey sucks; Kentish clover; lambs suckling; milkies; mull; mutton rose; pussy foot; quillet; shamrock; sheep's gowan; smara; smoora; sucklings; white sookies; three-leaved-grass trefoil; trifoil; wild white
Planet: Mercury

Element: Air
Deities: Olwen; Rowen
Meanings: 'Be mine'; petition (four-leaved); happiness; 'I'll be true to you'; promise; 'Think of me'

Myth and Legend

There is no certainty as to which trifoliate plant should be identified as the shamrock; whilst it is usually associated with *Trifolium minus* or *Trifolium repens*, other three-leaved plants, including wood sorrel and bird's foot trefoil, have been suggested. An Irish tradition claims that it is not a clover as, being a mystical plant, it never flowers and will refuse to grow on alien soils. The shamrock was a natural contender to be the floral symbol of Ireland having been used by St Patrick to explain the concept of the Trinity. As a boy, he had been captured in Scotland and taken to Ireland as a slave. He grew up hoping for the opportunity to convert the Irish King, Loaghaire, to Christianity. In defiance of a royal decree he lit a fire in celebration of the resurrection of Christ at Easter. This act infuriated the local Druids, who brought him before the king for judgement. Before the passing of sentence, St Patrick was allowed to speak in his own defence. He spoke of his beliefs, and of the Trinity. He made use of the shamrock leaf to explain the possibility that three could be one. In consequence of this act of witness the king converted to Christianity and the shamrock became a national symbol. The word 'shamrock' is derived from the Arabic word *shamrakh*, the symbol of Persian trinity. The wearing of shamrock on St Patrick's Day can be traced back to 1651.

Magic and Lore

White clover, as an emblem of the Trinity, can be worn, carried or scattered about the house to protect against witchcraft, spells and curses. It will bring good fortune to all those in a house where one is kept. A grain of wheat wrapped in a clover leaf gives second sight, but it is not stated whether it should be eaten or carried. An expansion of this says that laying seven wheat grains on a four-leaved clover will enable the owner to see faeries, as will wearing a four-leaved clover in your hat. A field of clover was said to attract faeries, and a four-leaved clover is supposed to give protection against the faeries and be able to break faerie spells.

Most people will be aware of the superstition that a four-leaved clover brings good luck. One reason given for this is that Eve took one with her when she was expelled from the Garden of Eden. A four-leaved clover is never found, but rather makes itself known to the finder, so generally it is suggested that to be lucky it must have been found by accident. A traditional rhyme says:

> One leaf for fame,
> One leaf for wealth,
> One leaf for a faithful lover
> And one leaf for to bring glorious health,
> All in a four-leaved clover.

There is a sting in the tail, however:

Blessed is the eye that seeth a four-leaved clover,
And cursed is the hand that plucketh it.

Furthermore, it is sometimes said that it is unlucky to collect a four-leaved clover at any time in May, although there is another superstition that if you pick it on 1 May and keep it safe, you will gain whatever your heart desires. The good luck will be enhanced if it is passed on to someone else, or when it is placed inside a Bible. One superstition says that when you find a four-leaved clover you must place it inside your left shoe; thereafter you will enjoy good luck so long as it remains there. With the clover in your left shoe, if you shake hands with the first person you meet it may show something of your future. If the person is of the opposite sex then you will soon be married, but if of the same gender it shows that you will never wed.

There are other superstitions relating to four-leaved clovers. If, as soon as you have found one, you rush home and hang it above the doorway into your home then, if the first person through the door is unmarried it indicates that your wedding will take place within the following twelve months. In some parts of Britain, it is said that finding a four-leaved clover is a sign that you will meet your future spouse before the end of the day.

If you find a four-leaved clover,
Your true love will you see ere the day is over.

When a young woman finds a four-leaved clover she should place it into her right shoe, then the next bachelor she meets will be the man that she will marry. Even if she is already involved with a man, the clover might prove invaluable, as wearing it over her breast will ensure that the affair runs smoothly. If her sweetheart is leaving on a journey, the clover in her shoe will ensure that he will remain faithful whilst they are apart.

In southern Europe, if a man placed a four-leaved clover beneath his pillow he was sure to dream of his true love. For those disappointed in love the wearing of a four-leaved clover in a piece of blue silk next to the heart will ease all heartaches. Wearing a four-leaved clover is also said to help men avoid military service. When two people eat a four-leaved clover together, it means that they will fall in love. If only one person swallows one then they will wed whoever they desire.

A four-leaved clover will protect against insanity and strengthen psychic powers. It enables the wearer to detect the presence of witches and spirits, and if placed in a shoe and forgotten about it will protect against all evils. When it is concealed in a cowshed it protects the cattle from evil, ensures that their milk will come easily and gives rich butter. The four-leaved clover is said to give one the power to see faeries and to break faerie spells. The faeries of the clover are those that will help in finding love and fidelity.

When it is worn on the breast, it will give success in all undertakings and assist in the detection of money or treasures. Sometimes it is suggested that finding a four-leaved clover indicates that one will soon inherit wealth. Perhaps because of the association of St Patrick with clovers, it is claimed that a four-leaved clover will keep snakes away from any property. Tradition has it that where one four-leaved clover is picked another will soon grow in its place.

It is unlucky to find a five-leaved clover, as it betokens bad luck or illness. Alternatively, a five-leaved clover may be passed on to someone else as, for the recipient, it will attract wealth. Finding a six-leaved clover always means misfortune.

Finding a two-leaved clover indicates that you will be kissed by your sweetheart or that a new lover will be found. To be most effective the following rhyme should be said when it is found:

A clover, a clover of two,
Put it in your right shoe;
When the first young man you meet,
In field, or lane or street,
You'll get him – or one of his name.

Clover seed must be sown in the dark of the moon if it is to grow, as when it is sown in its light it will not sink into the soil. Other authors recommend that it should be sown on a 'no moon', that is to say in the twenty-four fours between the waning of one moon and the waxing of the next. If it is sown during the sign of Cancer (mid-June to mid-July) it will neither freeze nor die in the winter.

To dream of clover (some say of a four-leaved clover) indicates health, happiness, prosperity and a happy marriage.

Medicinal

Clover has been used in the treatment of whooping cough and is supposed to be an antidote for all poisoning. It was said to 'cleanse the blood' and was used in preparations to aid fertility.

FENUGREEK

Botanical Name: *Trigonella foenum-graecum*
Family: Leguminosae
Common Names: Bird's foot; Greek clover; Greek hayes; Greek hayseed;
Gender: Masculine
Planet: Mercury
Element: Air
Deity: Apollo

An old magical spell used fenugreek to attract wealth. A small jar was partially filled with fenugreek seeds and left open – the spell was bound to fail if the top of the jar was closed. Each day a few more seeds were added until the jar was completely full, at which point it was emptied out and the whole thing started again with fresh seeds.

An alternative, possibly easier charm, required a few fenugreek seed to be added to the water used to mop floors.

BETHROOT

Botanical Name: *Trillium erectum*
Family: Liliaceae
Common Names: Beth; birthroot; ground lily; Indian balm; Indian root; Indian shamrock; lamb's quarters; true love; wake Robin.

The names 'bethroot' and 'beth' are, most probably, corruptions of another common name, 'birthroot'. The plant earned this name from having been used by the American pioneer settlers as a treatment for haemorrhaging in women following childbirth.

In magic the plant is used to attract love and wealth. In order to win the love of another you must rub the root over you body. Carrying the root will ensure that you enjoy good luck and attract money.

WHEAT

Botanical Name: *Triticum* spp.
Family: Gramineae
Common Name: Corn
Gender: Feminine
Planet: Venus
Element: Earth
Deities: Adonis; Ceres; Demeter; Ishtar; Isis; Min; Osiris; Zeus Polleus
Meanings: Fruitfulness; 'I offer you all I have'; riches

Magic and Lore

Many of the traditions and superstitions that are associated with corn actually refer to any grain crop, and so may be applicable to wheat. It is taken as a symbol of fertility and fruitfulness. Eating or carrying the grain is supposed to improve fertility. If it is eaten by a woman it is claimed that it will ensure conception. Carrying the grain or placing sheaves about the house will attract money, which means that all arable farmers ought to be millionaires!

A simple spell for good luck involves walking backwards and throwing your handkerchief over your shoulder into the wheat field.

An old sowing rhyme applied to wheat, and many other crops, suggests that you must sow four seeds for every two that you would have grow.

One for cutworm, one for crow,
Two to plant and two to grow.

In Suffolk there is a saying, 'Sow in slops, heavy in tops.' This suggests that the wetter the soil is at sowing time the heavier the crop will be at the harvest

In common with many grain crops there are several omens of good or poor yields. Heavy snows in the winter indicate that there will be a good harvest in the following year, as does a good nut crop. Likewise, if the nut crop is poor so will the wheat crop be. A good year for wheat is also a good year for plums and in Dorset it is said that the price of wheat can be estimated by counting the number of flowers on Madonna lilies.

To ensure a good harvest, boys would be sent through the growing crop carrying flaming torches to 'blaze' it, thereby driving out any evil spirits that might lurk there. At harvest there was a custom for four to six young unmarried men to walk between the ricks, criss-crossing their paths, carrying whips of straw which they swung alternately on their left then right. The end of the harvest was filled with ritual significance. When harvesting was carried out by hand, the reapers would work systematically, in ever-decreasing circles, until only a small swathe remained. They would be reluctant to cut this final swathe as it was said that the spirit of the wheat would have taken refuge there. Sometimes it would be wrapped in ribbons and the sickles thrown into it; elsewhere it was cut, decorated with ribbons, and carried back triumphantly atop the last load being taken into the rickyard. In Kent the last swathe was cut and shaped into a rough human form which was dressed in silk and lace. The figure was known as the 'ivy girl' and appeared in the harvest home supper later in the day. In Northumberland, the last rick was set up on a pole and carried back by the harvesters to be erected in the barn.

Perhaps the best-known form of the final swathe of wheat is the 'corn dolly' or 'kern baby'. In different counties, and in different villages, there were different patterns. The dolly was hung up in the farmhouse kitchen and kept until the following year to bring fertility to the house and the farm.

NASTURTIUM

Botanical Name: *Tropaeolum major*
Family: Tropaeolaceae
Common Names: Cress of Peru; gold nugget; Indian cress; lark's heel; yellow larkspur
Meanings: Splendour (scarlet); artifice; 'Beauty unadorned I seek'; optimism; patriotism; warlike trophy

Pliny named the plant 'nasturtium' from the Latin *nasus*, meaning 'nose' and *tortus*, meaning 'twisted' because of the plant's pungent scent.

Myth and Legend

The Incas of Peru called the plant 'gold nugget' from an Andean legend. In the days of the Spanish conquistadors an Inca convert with the Christian name of Juan decided he would replace the gold looted from the Inca temple. He went to a hidden spring to look for any gold that might have been washed out in recent floods and was rewarded with a heavy bag. On his way to the temple, however, he was attacked by Spanish riders. He prayed to the god of the mountain for protection and tossed the bag of gold into the woods. One of the Spaniards went to find it but was bitten by a venomous snake sent by the god and, instead of gold, all he found were golden-flowered nasturtiums. Later Juan recovered his gold, restored by the mountain god, and took it back to his village. There the village blacksmith beat out and moulded the metal to remake the idol for the temple.

Magic and Lore

An old legend says planting three red nasturtiums will protect the garden and keep all unwanted visitors off your land. They can be used in feng shui to bring balance. They are planted to help bring the energies between the buildings and the land into harmony following alterations. During the Second World War the seeds were dried and ground as a substitute for black pepper.

Medicinal

In early English herbals it was referred to as Indian cress. A lotion made from the leaves, flowers and seeds, mixed with nettle leaves and three oak leaves, could be applied to prevent baldness. Wearing a nasturtium in your lapel will boost your energy levels if you are feeling sluggish and run down. The Peruvian Indians used the leaves in teas as a treatment for coughs, colds and the flu, as well as respiratory and menstrual problems.

TULIP

Botanical Name: *Tulipa gesnerana*; *Tulipa* spp.
Family: Liliaceae
Gender: Feminine
Planet: Venus
Element: Earth
Meanings: Declaration of love (red); undying love (purple); beautiful eyes (variegated); hopeless love (yellow); avowal; consuming love; eloquence; fame

Tulips originate from Turkey. The name tulip is derived from the Turkish word for gauze, *tuliband*, from which turbans were made. It is said that the black centre of the tulip represents a lover's heart, darkened by the heat of passion.

Myth and Legend

A Persian folk story relates how a young couple, Ferhad and Shirin, discovered that the path of true love never runs smoothly. A jealous king sent Ferhad to a far-flung corner of his kingdom on the pretext of having him oversee the completion of a piece of elaborate statuary. The statue was so vast and intricate that it was believed that it was impossible to complete it in one man's lifetime. Ferhad set to work with a will and after some years spies reported to the king that it was a wonderful piece of work and was nearing its completion. The king panicked and sent word to Ferhad that Shirin had died. Ferhad, demented with grief, mounted his horse and rode it straight over the edge of a high, rocky precipice. Where his blood was splashed onto the earth a red flower, resembling his turban, sprouted from the ground. The saddest part of this tale is that the message telling of Shirin's death was a hoax. She was actually still alive.

An ancient Iranian belief says that if a young soldier dies patriotically, a red tulip will grow on his grave.

Magic and Lore

The story of the rise to renown of tulips has become almost a legend. They have been grown in the gardens of Persia for centuries, and were introduced to the West from Turkey. Ogier Ghislain de Busbecq, the ambassador of the Holy Roman Empire to the court of Suleiman the Magnificent, encountered the flowers and was the first Westerner to write of them. He sent seed of the tulip to his friend Clusius (sometimes called Charles de L'Ecluse), a Flemish doctor and renowned botanist of his time. Carolus Clusius was Prefect of the Royal Medical Garden in Prague and at some point also worked in Vienna. He wasted no time in sowing the seed sent to him. He enjoyed great success with the new plants, and also received bulbs from Matteo Caccini, a Florentine botanist and dealer in rare plants. By 1592, when he moved to Leiden to become Hortulanus, or Prefect, of the botanic gardens he had a collection of over 600 bulbs along with many other plants. He planted them at the rear of the university buildings, in a garden of little more than 1,200 square metres. His reason for accepting the post at Leiden, Western Europe's first botanic garden, appears to have been because, as a Protestant, he was finding it increasingly difficult to practice his faith and work in Prague or Vienna.

Clusius appears to have been of the opinion that his plants were solely for scientific study. Although it is said by some sources that he did offer some tulip bulbs for sale, but at great price, most writers say that he would neither sell nor give away any from his collection. However, one winter's day, a great part of his collection was stolen. So the Dutch bulb industry was born on the basis of a theft and Clusius, disheartened by the whole affair, is reported never to have grown the plants again.

Tulipa gesnerana was a collective name given to those tulips introduced to Europe from Turkey from 1554 onwards, which are of great significance as the ancestors of many of our garden tulips today.

In 1562 a cargo of tulip bulbs from Constantinople was landed at Antwerp. The somewhat bemused burghers of the city, according to local legend, tried to eat them but, finding them unpalatable, threw the remaining bulbs into the local midden. There they later flourished and flowered. In both England and Germany there were fleeting attempts to use the bulbs, prepared in various ways, as a delicacy but these had little success. Precisely when the tulip was introduced into Britain is uncertain but it must have been before the end of the 1570s as a practical guide to their cultivation is to be found in the *Gardener's Labyrinth* by Thomas Hill, published in 1577.

In Holland the cultivation of the tulip began slowly, mainly in the land between the North Sea and Amsterdam, an area still known as Bollenstreek or 'the bulb growing region'. At first the flowers were so rare that they were the preserve of the aristocracy or the wealthy merchant class. They were expensive, and so became a status symbol. 'Tulipomania' had begun. It seems to have started in France in the 1620s and spread to Holland and then to Turkey as the demand for tulip bulbs outstripped the supply. In 1611 the Earl of Salisbury sent John Tradescant to buy plants for the gardens at the stately home, Hatfield House, he was building. He purchased 800 tulip bulbs at 10 shillings (50 pence) per hundred, but he soon returned to Holland to buy them by the ton. By comparison thirteen years later, in 1624, the renowned white and maroon Rembrandt type 'Semper Augustus' commanded a price of 3,000 guilders (equivalent to approximately £1,000 today). At the height of the tulip craze in Holland, from 1634 to the market crash in 1637, the price of a bulb of this type had risen to 4,500 guilders plus a horse and cart. A bulb of another cultivar, 'The Viceroy', sold for 'two lasts of wheat, four tuns of beer, four lasts of rye, two tons of butter, four fat oxen, one thousand pounds of cheese, three fat pigs, twelve fat sheep, two hogsheads of wine, a complete bed, a suit of new clothes, a silver tankard and the wagon to haul it all away'.

Bulb selling became an activity for all kinds of people. Vast fortunes were made and lost. In the 1620s tulips were priced by the bulb, but in 1634 sales were made by weight, the same 'grain' weights as used by goldsmiths. Dutchmen from all walks of life were buying and selling bulb promissory notes, which promised the supply of a stated number of bulbs, and increased in value as they were sold on. The sale of bulbs no longer had much to do with growing flowers – it now had a great deal more to do with making money. In Holland transactions were carried out in taverns, which became the basis of the Dutch stock exchange, just as the coffee houses had in Britain. In Bruges the buyers and sellers of promissory notes met at the home of the van der Beurse family and from this the French word *bourse*, for the stock exchange, derived.

In 1637 the Dutch Government stepped in and decreed that the promissory notes had to be honoured and all bulbs supplied. The market crashed, with millions of pounds being lost. In other countries, which had not been caught up in the foolishness of 'tulipomania', the prices remained high. They remained relatively expensive until the nineteenth century when floods of cheap bulbs from America drove prices down and made it a flower of the masses. Holland, however, remains the centre of tulip production and the flower is the Dutch national symbol.

In Turkey, where the story started, homage was paid to the flower at the annual 'feast of the tulips'. Sultan Selim III, the son of Suleiman, may have been partially responsible for the first round of 'tulipomania' by sending orders to the remotest areas of Turkey for as many as 50,000 bulbs for the royal gardens. This may have been the cause of the extinction of the parent species in the wild. Turkish love for the flower peaked in the reign of Sultan Ahmed III in the eighteenth century, in a period known as Lalé Devri, the 'tulip period'. The Sultan paid the greatest price of all for his love of the tulip flowers; he was brought to trial on charges including having spent too much money on the annual feast of the tulip and was beheaded.

The wearing of the flowers will prevent poverty and give protection from misfortune. In Turkey, lovers declared their love by the gift of a red tulip, much as we would do with a single red rose.

In Devon folklore, it was said that faeries used the silky tulip flowers as cradles for their children. Local legend has it that, one night, an elderly woman found the faerie babes and so planted even more tulips in her garden. The flowers flourished ensuring that there were enough cradles for all the faerie infants. In gratitude the faeries caused them to take on even brighter colours than normal and to have the scent of roses. When the woman died the faeries removed the scent in the tulip flowers; a local farmer took over the land, dug up the tulips and planted parsley. The faeries pinched off the roots, ripped up the plants and danced on the leaves, which is why parsley leaves are so ragged. Nothing thereafter would thrive in the garden but tulips flourished on the old woman's grave.

The purple tulip is the birthday flower of 21 March, the yellow of 17 May, the red of 7 June and the variegated of the 8 June.

Medicinal

Dioscorides attributed aphrodisiac properties to tulip seeds. This may be due to 'sympathetic magic', given the shape of the bulb.

COLTSFOOT

Botanical Name: *Tussilago farfara*
Family: Compositae
Common Names: Ass foot; baccy plant; British tobacco; bull's foot; butterburr; calves' foot; clatterclogs; cleat; coughwort; dishilago; donnhove; dove-dock; dummy leaf; dummy weed; English tobacco; farfara; field hove; *filis ante patrem*; foal's foot; fohanan; hall foot; hogweed; hooves; horse hoof; *pas d'ane*; poor-man's baccy; son-afore-the-father; sow foot; sponne; sweep's brushes; tushalagies; tushalan; tushy lucky gowan; wild rhubarb; yellow stars; yellow trumpets
Gender: Feminine
Planet: Venus
Element: Water
Meanings: 'Justice shall be done'; maternal care

This little flower was once the symbol of the apothecaries in Paris and would be painted on the outside of their shops. In magic it is used in spells to create an atmosphere of peace and tranquillity.

The botanical name is derived from the Greek *tussis*, meaning 'a cough' as it was widely used as a cough treatment. Common names such as 'British tobacco' and 'English tobacco' suggest that the leaves have been dried for smoking. In herbal tobaccos, it is added as a treatment for asthma and in magical use smoking coltsfoot tobacco is said to induce prophetic visions. 'Bechion', the plant mentioned by Dioscorides and taken to be coltsfoot, was smoked to aid dry, hacking coughs. The leaves, boiled in water and sweetened with honey, were prescribed to be drunk twice daily to treat common cold.

REED MACE

Botanical Name: *Typha angustifolia*; *Typha latifolia*
Family: Gramineae
Common Names: Bull rush; cat's tail; nail rod
Gender: Masculine
Planet: Mars
Element: Fire
Meanings: Docility; indiscretions; rashness

This plant is now often popularly referred to as a 'bull rush' although, more correctly, the rushes are a different group of plants altogether. It is also not related to the bull rushes referred to in the biblical story of Moses. The plant in that case was most likely to have been *Cyperus papyrus*.

It is said to be unlucky to take reed mace into the home. One rather unusual superstition relates to its use as a form of aphrodisiac. It suggests that where a woman has ceased to enjoy sexual intercourse, she should carry a reed mace with her at all times.

GORSE

Botanical Name: *Ulex europaeus*
Family: Leguminosae
Common Names: Broom; fingers and thumbs; French-fuzz; frey; furra; furze; fyrs; gorst; goss; hawth; honey-bottle; hoth; ling; pins and needles; prickly broom; ruffet; thumbs and fingers; whin
Gender: Masculine
Planet: Mars
Element: Fire
Deities: Jupiter; Thor
Meanings: Anger; enduring affections; love for all seasons

Gorse is either very good or very bad, depending on the advice you receive. To some it is a terribly unlucky plant, a harbinger of death. Merely to hang a sprig in the house is to invite misfortune to befall some member of the household.

To others it is considered to be a powerful protection against evil. In Wales hedges of gorse might be planted around properties to keep faeries out. The flowers were included in 'summer', an amulet that the Irish placed above doorways in order to ward off faeries and witches. Another Irish tradition is that if you wear the 'blessed furze' you will not stumble.

In the druidic calendar the plant typifies the young sun at the spring equinox. At Midsummer, fires of gorse were lit and the blazing branches used in the rites and celebrations, being carried around the cattle to ensure good health throughout the coming year.

Gorse is said to be the first plant each year to be visited by bees, as a result of which it was thought to be enchanted and so was used in charms against witches and witchcraft. It is in flower almost every month of the year and so became a symbol of continuous fertility. Its possession helps one to carry out one's work, and to stand out from the background. No doubt the symbolism of fertility, together with the golden yellow flowers of the gorse, have led to its use in spells to attract money. The main use of this plant, however, comes from its association with love. It is just as well that the gorse has a long flowering period considering the old country saying:

> When Gorse is out of bloom
> Kissing is out of season!

Or, put another way:

> Love is never out of season,
> Except when Furze is out of bloom.

The evidence that it was still 'kissing season' came in the use of gorse amongst the flowers in a bride's bouquet.

Gorse was used to commemorate St Stephen's Day (26 December), sometimes called 'Wrenning Day'. On this day, wrens were stoned to death in commemoration of the saint's martyrdom. The dead birds were then carried around on the prickly gorse bush by boys who would beg for money.

Gorse is the birthday flower of 28 November.

ELM

Botanical Name: *Ulmus glabra*
Family: Ulmaceae
Common Names: Alm; bough elm; chewbark; drunken elm; elem; elm-wych; elven; emmal; halse; helm; holme; hornbeam; horn-birch; ime; olm; quicken; switch-elm; Warwickshire weed; witan-elm; wych elm; wych-halse; wych-hazel; wych tree; wych wood

Gender: Feminine
Planet: Saturn
Element: Water
Deities: Dionysus; Hoenin; Lodr; Odin
Meanings: Dignity; 'Your queenly bearing and elegance delight me'

Myth and Legend

The creation story in Teutonic mythology tells us that three gods came across an ash and an elm when they were out walking. From the ash they formed Aska, the first man, and from the elm they formed Embla, the first woman. The gods, Odin, Hoenir, and Lodr (Loki) gave breath, warmth and soul respectively.

Orpheus, on his return from Hades, lamented the loss of his beloved Eurydice and played his lyre in his grief. As soon as the first notes sounded a forest of elms sprang up and under one of these trees Orpheus rested.

The Thracian bard a pleasing elm-tree chose,
Nor thought it was beneath him to repose
Beneath its shade
When he from hell returned.

– Anon.

Magic and Lore

In folklore, the elm is often associated with images of sleep, death, the grave and theories of rebirth. Its wood was used for coffins. It was once also considered to be particularly favoured by elves and was sometimes called 'elven'. In some parts of Britain, elves were supposed to give protection against lightning strikes and to assist in attracting the love and affection of others. Presumably because of this association, a Devonshire saying advises that lightning never strikes an elm. There are, of course good and bad in everything, and faeries and elves are no different. In Scotland the elves may be described as the 'seelie' or 'unseelie', that is the blessed and unblessed or the good and the bad.

It may be the thought of mischievous elves living in or near the elm or the belief that the tree itself is somehow malevolent in nature that led to the French saying, 'He will wait for me beneath the elm.' This is taken to mean that the person will not be there, as only a very foolish person would chose to stand beneath an elm. Rudyard Kipling referred to this in 'Tree Song' when he wrote:

Ellum she hateth mankind and waiteth
Till every gust be laid,
To drop a limb on the head of him
That anyway trusts her shade.

It is, of course, not all bad; for the farmer and gardener it can be a useful guide of sowing times.

When the elmen leaf is as big as a mouse's ear,
Then sow barley and never fear;
When the elmen leaf is a big as an ox's eye,
Then I say 'Hie, boys, Hie!'

When the elm leaves are as big as a shilling,
Plant kidney beans, if to plant 'em you're willing;
When elm leaves are as big as a penny,
You must plant kidney beans if you're to have any.

It also acts as a warning for the farmer, as if the leaves fall prematurely, before the autumn, it is an omen that some disaster will befall his cattle.

Theophrastus, writing in the third century AD said that elm was the best wood for the bent timbers used in shipbuilding, and it was common for sapling elms to be bent over for future use. One of the benefits of using it in shipbuilding is undoubtedly because it rots very slowly when immersed in water. These bent trees would also provide wood for making ploughs and similar implements.

Vineyard owners also benefit from growing elms. Since Roman times it has been suggested that there are benefits to vines from growing them near to or over pollarded elms, as the trees give support and shade to the vines. This mixture of vines and elms appears in some legends, particularly those relating to Bacchus, the god of wine.

The botanical name, *Ulmus*, has been used to refer to an instrument of punishment; it comes from the use of elm rods to chastise slaves.

Medicinal

In folk medicine the boiled bark of elm was placed on burns; it was also boiled in milk, which was then left for two hours before being used in the treatment of jaundice. The inner bark of the elm could be chewed to ease colds and sore throats.

SLIPPERY ELM

Botanical Name: *Ulmus rubra*
Family: Ulmaceae
Common Names: Indian elm; moose elm; red elm
Gender: Feminine
Planet: Saturn
Element: Air

Elm is not usually noted as being good to burn on open fires. Tradition has it that:

Elmwood burns like churchyard mould
E'en the very flames are cold.

It does, however, have its uses. In order to prevent malicious gossip being spread about you, simply throw a knotted yellow cord onto a fire burning slippery elm.

Another attribute of the tree is its supposed ability to bestow the gift of eloquence. If a necklace is made of the bark, and hung about the neck of a child, it ensures that they will have a persuasive tongue in adulthood.

PENNYWORT

Botanical Name: *Umbilicus rupestris*
Family: Crassulaceae
Common Name: Navelwort

Magic and Lore

Pennywort has been used in rustic weather forecasting. Select two large leaves, spit on them liberally, them stick them together and toss them into the air. If they stay together when they hit the ground rain is on the way, but if they part then dry weather may be expected.

Medicinal

Pennywort, mixed with unsalted butter, has been used in folk medicine as a treatment for burns and scalds. It has also been used as a remedy for corns, saddle sores, spots and pimples, and to draw out thorns and splinters.

SQUILL

Botanical Name: *Urginea maritima*
Family: Liliaceae
Common Names: Red squill; sea onion
Gender: Masculine
Planet: Mars
Element: Fire

The squill has long been used in magical rites, usually as a protective amulet. Carrying it will break curses or spells cast against you. Hanging the plant over a window will protect the whole house from evil. Silver coins placed with a squill, in a box, will attract money.

NETTLE

Botanical Names: *Urtica dioica*; *Urtica urens*
Family: Urticaceae
Common Names: Cool faugh; Devil's claw; Devil's leaf; Devil's plaything; heg-beg; hidgy-pidgy; hoky-poky; Jenny nettle; naughty man's plaything; stinging nettle; tangling nettle; white archangel
Gender: Masculine
Planet: Mars

Element: Fire
Deity: Thor
Meanings: Coolness; cruelty; pain; slander; 'You are cruel'; 'You are spiteful'

Myth and Legend

Nettle has nutritional value, being eaten after blanching to remove the sting. It used to be considered as a 'poverty food'. In an Irish legend St Columcille was out walking one day when he met an old woman who was gathering nettles. He asked her what had reduced her to eating them, and she replied that she was waiting for her cow to calve so that she should have milk. St Columcille resolved to survive on a diet of nettles while he waited for the Kingdom of God. It is said that he became so thin that 'the tracks of his ribs could be seen on the Strand when he used to lie out there through the night'.

In the Highlands of Scotland it was said that nettles grew from the bodies of the dead; consequently they would continue to survive long after the people had gone. The traditional belief in Denmark was that it grew from the blood of those who had died innocent victims.

It is a plant of healing, and the white nettle is sometimes depicted as having been brought to Earth by the angel of mercy.

Magic and Lore

There can be few gardeners who will find anything good to say about this pernicious weed, but some people believe it to be a 'plant doctor' aiding the growth of anything planted near to it. The more cynical gardeners would suggest that the only plant that tends to grow next to a nettle, is another nettle.

Its most notable feature, as far as many people are concerned, is its sting. We learn early in our lives that the ideal treatment for a nettle sting is to rub the area with a dock leaf. 'Grasp the nettle' is a common encouragement to those who keep putting off an unpleasant task, but perhaps they should take the advice of Aaron Hill, who said, in 'Verses Written on a Window in Scotland':

> Tender handed stroke the nettle
> And it will sting you for your pains,
> Grasp it like a man of mettle
> and it soft as silk remains.

It is worth remembering that the sting stays active long after the plant has dried out. When the Linnaean herbarium was being photographed in 1941, as a record in case of damage during the war, one of the photographers was stung by a nettle specimen which had been pressed and dried out some 200 years before. A superstition says that if you see a piece of nettle pointing towards you, and you pick it, you can expect to enjoy good luck.

Considering how prevalent this plant is in Britain it is strange to think that the Romans brought their own with them when they invaded. They brought a species called *Urtica pulilifera*, native to southern Europe, which they used to 'chafe their limbs' apparently to prevent numbness in the cold weather. This Roman nettle is all but extinct in Britain now. It may, however, have been the source of the superstition that nettles give courage in times of danger.

It has protective powers, and can be used to dispel curses and spells cast against you, returning them onto the person who sent them. In a Yorkshire tradition, it was used to exorcise the Devil but quite how is another question. Sprinkling a little chopped nettle around the house will ward off evil, and if it is thrown onto an open fire it is said to avert danger. Holding or carrying a nettle will keep ghosts away and prevent the bearer from being struck by lightning. If a small sprig of yarrow is carried along with the nettle it will allay all fears and keep all negativity at bay.

It is said that nettles mark the dwelling places of elves, and that nettle stings are a protection against witchcraft and the mischief caused by them. Indeed, nettles can be used in the dairy to prevent the milk being affected by witches or house trolls. In a slightly different form of protection, throwing nettles into the face of an oncoming storm was supposed to protect property from being struck by lightning.

In Cheshire, nettles have been used, rather than rennet, to curdle milk in cheese-making. Elsewhere they were hung in larders to deter flies and grown close to beehives to drive away frogs.

As we have seen, nettles were at various times a 'poverty food', but in more recent years they have become a health food as they have a particularly high iron content. They are best eaten young, and in the New Forest it was said that on May Day the Devil collected them to make his shirts, so after that date they were unfit to eat (nettle fibre was commonly used to make clothes). May Day was sometimes called Nettle Day. The white nettle tends to come into flower around 8 May (St Michael's Day) and has been called 'white archangel'.

Dreaming of being stung by a nettle is a portent that there are disappointments ahead, and gathering nettles in a dream indicates that someone has formed a favourable opinion of you. If the dreamer is married, a dream of gathering nettles is an omen that their family life will be a happy one.

Medicinal

Although we normally think of the nettle as more likely to cause a problem than cure one, it has been widely used in folk remedies. It is variously recommended for treating dog bites, and bee stings, and to reduce rashes. They have even been suggested as a remedy to stop nosebleeds, and combing nettle juice through the hair was said to be a cure for baldness. In a throwback to the Roman use of the nettle to prevent numbness, it is claimed that nettle stings will 'warm' cold legs and soothe the pains of rheumatism. In general terms, when it is to be used in folk remedies it will be at its most effective if gathered in silence at midnight. To treat arthritis the joints are whipped with the stems, on the basis that it will stimulate the adrenal glands and so reduce swelling and pain. The juice of boiled nettles is supposed to be good for the blood and the pulp can be used in a poultice to aid sciatica.

It is used in folk medicine to cure fevers. If a member of the sufferer's family pulls up a nettle by the roots whilst repeating the name of the invalid they will be cured (nettles growing in the shade are supposed to particularly effective). Alternatively, placing a pot of fresh nettles under the patient's bed is said to be a sure aid to their recovery.

If they would drink nettles in March
And eat mugwort in May,
So many fine maidens
Wouldn't go to clay.

It is said to aid male fertility, but conversely perhaps the most outlandish claim is that it could be used as a contraceptive; a man had to place a thick 'sole' of nettles inside his socks and wear them for the twenty-four hours before indulging in intercourse.

BILBERRY

Botanical Name: *Vaccinium myrtillus*
Family: Ericaceae
Common Names: Airelle; arts; blackberry; blackhearts; black wortles; blaeberry; bleaberry; blueberry; brylocks; coraseena; cowberry; crowberry; fraughan; frougs; hartberry; huckleberry; hurtleberry; hurts; moss-berry; whinberry; whortleberry; wimberry; worts
Planet: Jupiter

In Ireland the gathering of bilberries was associated with the festival of Lughnasa, celebrated at harvest time. To keep unwelcome callers from coming to your door simply hide a few beneath the doormat. This also afforded some limited protection against evil forces that might seek to enter there.

In folk medicine the berries were used to treat dysentery and diarrhoea.

VALERIAN

Botanical Name: *Valeriana officinalis*; *Valerian phu*
Family: Valerianaceae
Common Names: All-heal; amantilla; Belgian valerian; black elder; bloody

butcher; bovis and soldiers; capon's tail; capon's trailer; cat's love; cat's trail; cat's valerian; cut finger; cut finger leaf; cut leaf; drunken slots; English valerian; filaera; fragrant valerian; garden heliotrope; German valerian; God's hand leaf; heal-all; herb Bennett; phu; red valerian; St George's herb; sete wale; set well; valara; vandal root: wild valerian
Gender: Feminine
Planet: Venus (Mercury, according to Culpeper)
Element: Water
Meanings: Accommodating disposition; concealed merit; 'Conscious of my lowliness, I aspire none the less to wed you'; merit in disguise

Sachets of powdered valerian root, hung about the house, are said to afford protection against lightning. If they are placed under pillows they aid restful sleep. The root was claimed to be an aphrodisiac, provoking love. It may be powdered and added to sachets and philtres, and if a woman wears a sprig on her clothing men will feel impelled to follow her.

It has been suggested that the Pied Piper of Hamlin used valerian, carrying a little of the root in his pockets, in order to induce the rats to follow him. Cats and rats are supposed to love the plant, particularly the root, and will dig it up.

Another attribute is that of 'peace maker'. To prevent people quarrelling it should be brought near them. For magical purposes the powdered root has been used as 'graveyard dust'.

It is the birthday flower of 16 March.

Medicinal

The name 'valerian' is derived from the Roman Emperor Valerius who is said to have been the first to use it medicinally. As a herbal remedy the powdered root can be applied to cuts and scratches to aid healing.

VANILLA

Botanical Name: *Vanilla inodora*
Family: Orchidaceae
Gender: Feminine
Planet: Venus
Element: Water

Vanilla is regarded as an aphrodisiac, the scent and flavour inducing lustfulness. One can use it covertly by placing a vanilla bean into sugar to flavour it, and then using this sugar to sweeten love potions. More generally the carrying of beans is claimed to restore vitality and boost energy.

MULLEIN

Botanical Name: *Verbascum thapsus*
Family: Scrophulariaceae
Common Names: Aaron's flannel; Aaron's rod; Adam's flannel; beggar's blanket; beggar's stalk; blanket leaf; blanket mullein; bullock's lungwort; bunny's ears; candlewick plant; clot; clote; clown's lungwort; cuddy lugs; cuddy's lungs; Devil's blanket; doffle; duffle; faerie's wand; feltwort; figwort; flannel jacket; flannel plant; fluff weed; golden rod; graveyard dust; hag's tapers; hare's beard; hedge tapers; high taper; Jacob's staff; Jupiter's staff; lady's foxglove; *lus-mor* ('great herb'); mullein dock; old man's fennel; Peter's staff; rag paper; shepherd's club; torches; torch plant; velvetback; velvet plant; wild ice leaf; woollen
Gender: Feminine
Planet: Saturn
Element: Fire
Deities: Jupiter
Meanings: Good nature; friendship (white); comfort; take courage

Magic and Lore

The soft, downy stems and leaves of the mullein plant burn relatively easily, a characteristic which led to it being used for candle wicks before the use of cotton. The Romans, who knew the plant as *candelaria*, dipped the shoots in tallow and used them as lanterns. Traditionally, the candles and lanterns used by witches to light their meetings should all have mullein wicks.

Like ragwort, mullein could also provide a steed for witches to ride. The following rhyme is often applied to the plant:

> The Hag is astride
> This night for to ride,
> The Devil and she together.

So it is ironic, that in parts of Asia and Europe it is considered to be a protective plant that will drive away all evil. In India it is given much the same prominence as St John's wort was in Europe, being hung over doorways to prevent evil from entering, or carried as a protective amulet. Carrying it was also supposed to ward off wild beasts. In classical Greek mythology Ulysses carried mullein with him to shield him from the power of Circe. In folk magic it was claimed that it could be used to gain back children who had been abducted (presumably by faeries).

In the Ozark Mountains of the USA it has been used in love divination. In order for a mountain man to discover whether his love for a woman is returned he should seek out mullein growing in a woodland clearing and bend the stem over so that it points toward the woman's home. If, when he returns to it, the stem has straightened again then his love will be returned, but if the plant has died the woman does not care about him.

It is from the soft downy leaves that the plant gets its botanical name, as *Verbascum* is derived from the Latin, meaning 'bearded'. The leaves were wrapped around figs that were put into storage to keep them soft and moist, hence the name 'figwort'. They were also once used by the poor to line their shoes in winter in order to keep their feet warm.

Medicinal

Mullein is used in herbal medicines (and some proprietary brands), as an expectorant for coughs and bronchitis. This is, perhaps, the grain of truth in the old wives' tale that suggests placing a few leaves in your shoe will prevent you from catching colds. An old remedy for earache required mullein oil made by steeping the flowers in olive oil and exposing them to sunlight.

The regular consumption of a small piece of mullein is also supposed to ensure a long life.

VERVAIN

Botanical Name: *Verbena officinalis*
Family: Verbenaceae
Common Names: Britannica; cerealis; *crubh-an leoghain* ('dragon's claw'); enchanter's plant; *herba sacra*; *herba veneris*; herb of enchantment; herb of grace; herb of the cross; holy herb; holy plant; Juno's tears; Mercury's blood; persephonion; pigeon's grass; pigeonweed; pigeonwood; simpler's joy; tears of Isis: van-van: verbens; vervan
Gender: Feminine
Planet: Venus
Element: Earth
Deities: Aradia; Cerridwen; Isis; Juno; Jupiter; Mars; Mercury; Persephone; Thor; Venus
Meanings: Enchantment; purity; reconciliation; superstition: witchcraft; 'You have stolen away my soul'

Myth and Legend

Vervain was a sacred herb in many different cultures, including ancient Greece. Pliny called it the *herba sacra*, the 'sacred herb'. It had many other names, however, including 'cerealis', 'persephonion', and 'tears of Isis', all of which suggest links to fertility rites and festivals of renewal. It was dedicated to Isis as goddess of birth, and the statues of Venus in Rome were crowned with myrtle interwoven with vervain.

In mythology, the goddess Hera was jealous of the amount of attention Zeus, her consort, was paying to Callisto, a beautiful Arcadian nymph; she therefore turned her into a large she-bear. Whilst wandering in the forest, Callisto's son, Arcas, saw the bear and killed it with a spear, not knowing that it was really his own mother.

Zeus placed both mother and son in the heavens as the constellations Ursa Major and Ursa Minor, the Great Bear and the Little Bear.

This only served to inflame Hera's anger still further, and she hastened to seek the advice of Oceanus. On her journey she shed tears of pain and anger, and wherever her tears fell there grew a vervain plant. Oceanus had no interest in Hera's marital problems and so Callisto and Arcas remain in the night sky.

In Christian myth, vervain first grew on Calvary and it was used to staunch the blood of Christ as he hung on the cross.

Magic and Lore

In ancient Rome vervain was a symbol of good faith and was worn as a badge by heralds, messengers and ambassadors. The heralds, called *verbenarius*, carried a bough when attending peace negotiations. The Romans went so far as to have an annual feast, Verbenalia, in honour of the plant and its virtues. The ancient Persians also valued it, carrying branches in their rites of sun worship.

Following the Middle Ages, the religious significance of the plant diminished in comparison to its use in alchemy, sorcery and magic. It is said that it will protect against the evil eye, hinder the activities of witches and be a powerful weapon against the Devil.

> Vervain and dill
> Hinder witches from their will.

Paradoxically medieval sorcerers used it when casting spells, possibly as a protection when conjuring up spirits and demons. The juices, smeared over the body, enabled them to gain the affection and confidence of everyone, including their enemies. It also ensured that their every wish would be fulfilled and that they would be able to see into the future. It cured any ailment that they might be suffering from at the time and prevented any future illness.

The protecting and cleansing attributed to the herb are reflected in its use in exorcism incenses and purification baths. It has been used as a cleansing herb since ancient times, and small bundles were being used by the priests of Jupiter to brush and cleanse altars.

> Lift your boughs of vervain blue
> Dipt in cold September dew;
> And dash the moisture, chaste and clear,
> O'er the ground, and through the air.
> Now the place is purged and pure.
>
> – W. Mason, *Caractacus*

This use as a purging and purifying herb in religious rituals led to it being made into amulets and charms to keep evil away from houses or to protect individuals. Vervain water could be sprinkled about properties in exorcism to cast out evil and to prevent malevolent forces from entering. It was used in purification, healing, peace, money spells and for relaxation in psychic work.

To the Druids it was second only to mistletoe as a magical herb. They used it for divination and all manner of other magical rites. Sacrifice was made and honey poured onto the soil in exchange for the plant. Of course, to have such powers the plant had to be collected with due ritual, and at the most propitious time. It had to be cut with the left hand at Midsummer (possibly St John's Eve, 23 June), or at the rising of Sirius, the Dog Star, when neither the sun nor the moon are visible (some state that it should be done when the moon is full) although it might have been collected at different times for different purposes. One writer suggests that the following should be said as the plant is gathered:

> Hallowed be thou, Vervain, as thou growest on the ground,
> For in the Mount of Calvary where thou was first found
> Thou healedst our saviour Jesus Christ, and staunchest his bleeding wounds,
> In the name of the Father, Son and Holy Ghost, I take thee from the ground.

Long winded though this seems, it may be but the edited highlights, as elsewhere the charm runs:

> All hele, thou holy herb vervain,
> Growing on the ground;
> In the Mount of Calvary
> There was thou found;
> Thou helpest many a griefe,
> And stenchest many a wound.
> In the name of sweet Jesus,
> I take thee from the ground.
> Oh Lord, do effect the same
> That I do go about.
> In the name of God, on Mount Olivet
> First I found thee;
> In the name of Jesus
> I pull thee from the ground.

This may be the original spell, or else an over-embellished version of the original. In Lancashire, a further embellishment was that, before gathering it, one had to make the sign of the cross over it.

It has also been used in the preparation of love potions. In some parts of Europe, at one time, it was considered to be lucky for brides, and a hat of woven vervain was given to them to put them under the protection of Venus, or the Earth Mother. This would also ensure fertility, and might also be a throwback to the crowns worn by the Druids' daughters, which were woven to mark their rank. In northern France, shepherds would collect vervain and use it to charm the sheep, and the country maids. The *Operation des sept esprits des Planetes* says that, to win the love of another, one should rub vervain sap on the hand, and then touch the desired person. A sixteenth-century love charm says: 'Rubbe vervin in the bale of thy hande and rubbe thy mouth with it and immediately kysse her and it is done.' Albertus Magnus tells, 'It is also of greatest strength in venerest pastimes, that is, the act of generation.'

Ironically the plant can also be used to ensure chastity. If the sap is collected before sunrise on the first day of a new moon and is drunk it will, according to ancient instruction, result in the loss of all sexual urges for seven years. Equally odd was the belief that when the powdered root was placed between a loving couple it would cause malice and strife between them.

To bring peace and tranquillity scatter the dried herb about the house. This will calm even the wildest of emotions. Burying it in the garden will attract wealth, while placing a sprig in the cradle of a baby will cause it to grow up with a happy nature and a love of learning. To regain stolen items, wear it when you confront a thief, this will lead to all the items being returned. It was also said to protect against lightning strike and so was collected and placed in the house. To drink a little of the juice, and hang the plant about the neck, before retiring to bed will ensure a sleep free from dreams.

Vervain was regarded as a necessary ingredient in the preparations of alchemists and sorcerers. It was also reported to be used to open locks. A small piece could be placed in a cut on the hand, it was only necessary to then touch a lock in order to open it. The rites observed in preparing the 'hand of glory', a severed human hand used for magical purposes, required vervain and other herbs. The hand of glory was also reputed to be used to open locks and discover hidden treasure. Considering vervain's application in such magical rites, it seems strange that, in combination with dill and rue, it came to be used as a protection against the evil affects of witchcraft.

A number of odd superstitions were also attached to this herb. Put into dovecotes it was said to attract doves. Powdered and put into the sunlight it was supposed to make the sun appear blue. Put in the house, grounds or vineyards it would attract wealth. The strangest belief is probably that, when planted in fertile soils, it would bring worms within eight weeks. However, if anyone touched these worms, they would die.

It is suggested that the plant that we now commonly refer to as 'vervain' is not the same plant known by that name in antiquity. The change may to have occurred at some point in the Middle Ages, the identity of the 'true' vervain being lost. All the attributes identified with that plant were then applied to *Verbena officinalis*, our vervain.

Medicinal

The common name of 'vervain' is derived from the Celtic *ferfaen*, *fer* meant 'to drive away' and *faen*, 'a stone', implying that it was used to treat stones in the bladder and kidneys. An alternative derivation suggests that it is taken from *herba vener*, as it was considered to be an aphrodisiac in ancient times.

It is, perhaps, because illnesses were at one time thought to be caused by evil forces that vervain has been claimed to be a cure for ailments varying from cancer to the plague. It has even been suggested as a cure for the bites of snakes and rabid animals. Tumours and ulcers could be dealt with by cutting a root into two. One part was smoked over an open fire and the other hung on a thread about the sufferer's neck. As the root dried out the tumour would disappear. If, however, someone kept the smoked part of the root and dropped it into water the tumour would return as the root swelled. When used in the treatment of epilepsy the vervain must be gathered whilst the sun is in Aries and then combined with a grain of corn, or a one year old penny.

For most other ailments, such as fevers, toothaches, headaches, jaundice, heart diseases, snakebites and scrofula (the king's evil), it was merely necessary to hang the bruised root about the sufferer's neck on a white cord. It was also recommended for the treatment of poor eyesight.

Carrying vervain could also grant eternal youth. To discover the extent of an illness a piece could be pressed into the invalid's hand with sufficient pressure so that it was undetected, and then the patient asked how they felt. If they were hopeful, they would soon be restored to full health; if not, they might not survive.

Culpeper recommended the use of the herb in the treatment of gout, wheezing or shortness of breath, ailments of the stomach, spleen and liver, problems of the womb and menstrual cramps.

SPEEDWELL

Botanical Name: *Veronica officinalis*
Family: Scrophulariaceae
Common Names: Angel's eyes; bird's eyes; bright eyes; cat's eyes; milk maid's eye; Paul's betony; thunderbolts
Planet: Mercury
Meanings: Female fidelity; 'I remain faithful'; resemblance; true love

Myth and Legend

The botanical name *Veronica* is said to derive from the saint who wiped the blood from Christ's face at the crucifixion. The image of his face remained ever after in her handkerchief. A result of this is that the plant is associated with miracles. The handkerchief became a holy relic called the veronica, or *vern icle* (*vera* and *icon*), meaning 'the true image'. The flower, therefore, may have been named for a resemblance to the image of Christ.

Magic and Lore

The common name of 'speedwell' might have been derived from the plant's rate of spread as a weed, but is often said to be from an obsolete way of saying 'goodbye', as the flowers quickly drop when they are picked.

In various parts of England it was claimed that misfortune followed the picking of speedwell, and it should never be taken into the home. In Cheshire the superstition was that it would bring on thunderstorms and elsewhere children were warned that if they picked it birds would swoop down and peck out their eyes, or that their mothers' eyes would drop out.

VERTIVERT

Botanical Name: *Vetiveria zizaniodes*
Family: Gramineae
Common Names: Khas-khas: khus-khus grass; vertiver
Gender: Feminine
Planet: Venus
Element: Earth

Vertivert is alleged to have aphrodisiac properties and was added as an ingredient in love potions. It could also be added to bath water in order to make one more attractive to members of the opposite sex.

Carrying it should bring good luck, and it might be placed in a cash register so as to attract money and make a business more successful.

DEVIL'S SHOESTRING

Botanical Name: *Viburnum lantaniodes*
Family: Caprifoliaceae
Common Name: Hobble bush

The root of the Devil's shoestring is used as a lucky charm and is 'activated' by placing small pieces into a jar filled with a mixture of whisky and spirit of camphor. Where money is needed the activated root should be put beside some coins, or into a purse. This amulet is also reputed to be able to ease work-related problems.

BEAN

Botanical Name: *Vicia faba*; *Phaseolus vulgaris*
Family: Leguminosae
Common Name: Poor man's meat
Gender: Masculine
Planet: Mercury (Venus, according to Culpeper)
Element: Air
Deities: Cardea; Ceres; Demeter

Magic and Lore

Beans were revered as sacred in various parts of the ancient world. In ancient Egypt, they were objects of veneration and so were a forbidden foodstuff. At the festival of

Cardea, 1 June, beans and pork were included amongst the offerings made. Throughout Britain and northern Europe beans were associated with the White Goddess. Amongst the ancient Scots, only the highest-ranking priests were allowed to plant and cook beans. Jewish high priests are forbidden to eat beans on the Day of Atonement. Apparently Pythagoras founded a religion that had, as one of its tenets, that the eating of beans was sinful.

For centuries, beans have been associated with death and so considered to be a plant of ill omen; its shape was supposed to reinforce this association. It has been said that the spirit of the plant knows the secrets of life, death and rebirth. The spirit of the plant is supposed to sing out to spirits in order to guide them on their way to the next world. The Romans believed that the ghosts of the dead, called lemures, would throw beans at houses during the night and so bring misfortune on the people who lived there. In an effort to placate these troubled spirits black beans were thrown onto graves or burned, so that the smoke would banish the ghosts. They were distributed and eaten at funerals and it is still traditional in parts of Italy to distribute them on the anniversary of a death. A similar tradition could be found in the north of England as late as 1890; children would distribute beans at the funeral and recite the rhyme:

God save your soul,
Beans and all.

Elsewhere it was said that the souls of the dead inhabited bean fields or dwelt within the beans themselves. Consequently, sleeping overnight in a bean field would result in insanity, or at the very least horrendous nightmares. If, when beans germinate, one should come up white instead of green it is an indication that there will be a death in the family within the coming year.

In mining towns colliers would say that mining accidents were more likely to occur when the beans came into flower; their fragrance is said to cause light-headedness and dizzy spells. Sleeping where bean blossom can be smelt will result in hideous nightmares.

To protect yourself from witches, take a bean in your mouth and spit it at the first witch you see. Carrying beans will protect against evil magic and can be used to ward off those evil spirits that cause ill health. Prepare a rattle by putting three beans in a bladder. To scare away the demons rattle the bladder three times with the words:

Three blue beans in a blue bladder,
Rattle, bladder, rattle.

To clear a house of ghosts for twelve months, scatter beans all about the property on the last day of the year, whilst saying, 'With these beans I redeem me and mine.' Any ghosts will collect up the beans and trouble you no more.

A good number of superstitions apply to the cultivation of beans. Broad beans would be sown on St Thomas's Day (21 December). In south-west England, it was traditional that beans should not be sown before 3 May and sowing should be completed before the end of the month.

Be it weal or be it woe,
Beans should blow before May go.

Another source gives the following guidance for sowing kidney beans:

When Elum leaves are as big as a farden [farthing]
It's time to plant kidney beans in the garden.

Elsewhere, it is suggested that you wait a little longer before sowing the seeds.

Plant kidney beans if you so be willing,
When elm leaves are as big as a shilling.
When elm leaves are as big as a penny,
You must plant beans if you mean to have any.

Broad bean, on heavy soils may need to be sown a little earlier.

On St Valentine's Day
Beans should be in clay.

However, heavy and wet soils have not always been thought ideal for the cultivation of beans and you will be well advised to wait until the ground is fit to work on.

Sow beans in mud,
And they'll grow like wood.
– William Ellis, *The Modern Husbandman*

To tie down the sowing time a little better you should follow the advice that has been given to gardeners and gravers for hundreds of years.

Sow peasen and beans in the wane of the moon
Who sows them sooner he soweth too soone,
That they with the planet may rest and arise,
And flowereth with bearing most plentiful wise.

Even the time of day is important. The advice is to sow seed in the morning or at noon, in order to get a bigger yield. Perhaps the silliest recommendation is that they should be planted upside down in a leap year.

In order to discover the future, prepare three beans at Midsummer. The first should be unaltered, the second be half peeled and the third completely peeled. They should then be hidden and the enquirer left to find them. Whichever is found first is a guide to their future prospects. If the unaltered one is found first it indicates wealth, the half peeled one indicates a comfortable life and the fully peeled one indicates poverty.

Beans have been used as an aphrodisiac because of their supposed resemblance to testicles. They were recommended as a cure for impotence, either eaten or carried.

A woman could use them to entrap the man she loved. All she needed to do was place seven beans, of any type, in a circle on the ground. If she could persuade the man of her choice to step into or over the circle he would immediately be attracted to her. The scent of broad bean flowers is said to stimulate the passions of men and willingness of women. To find out who in a group will marry first, hide a small bean in a dish of peas that is served with the main meal. Whoever receives the bean will be the first to marry.

Seeing beans in your dreams is an indication that there will be troubles ahead. Where a couple are frequently quarrelling the woman can cause their difficulties to be resolved by carrying three lima beans strung on a silken thread for two days.

Medicinal

In folk remedies, beans can be used to cure warts. Each wart must be rubbed with a dried bean during the waning of the moon whilst saying:

> As the bean decays
> So wart, fall away.

Alternatively, the white fluff inside the pods of broad beans can be rubbed onto warts to remove them. The pod used must then be buried beneath an ash tree whilst saying the rhyme:

> As this bean shell rots away
> So my wart shall soon decay.

Carrying beans is supposed to be a cure for thirst and carrying a baby between rows of bean plants so that they can breathe the scent is recommended as a treatment for whooping cough.

GIANT WATERLILY

Botanical Name: *Victoria regia* (Synonym: *Victoria amazonica*)
Family: Nymphaeaceae
Common Name: Star of the waters

Myth and Legend

A princess of the Tupis-Guaranis people, of northern Brazil, was told an ancient legend of the moon by her father. He said that a handsome and powerful warrior-god lived in the moon, and each night when the moon dipped behind the mountains far off on the horizion, he was going to live with his favourite young women. If he particularly liked a girl he would carry her away, transforming her into a star in the sky.

The princess believed the romantic tale and fell in love with the warrior-in-the-moon. Thereafter she spurned the advances of any man from her own tribe. All the efforts of her family to persuade her to marry were in vain. She waited for night to fall, and the moon to appear in the sky. She would then spend hours looking at it trying to see the face of her imagined love and running in the jungle in an attempt to catch the embrace of a moon ray. Despite her best efforts, however, the moon never seemed to notice her.

One night, when she had gone deeper into the jungle, she came across a great smooth lake in which she could see the moon's reflection. Without hesitation, believing that the warrior in the moon had come down to bathe, she dived into the lake to meet him and was drowned.

Taking pity on the girl, but without the power to return her to life, the moon transformed her into a star on the Earth. He made her the water lily, the 'star of the waters', whose flowers open fully at night.

Magic and Lore

Following its introduction, *Victoria regia*, named in honour of Queen Victoria, became the subject of rivalry between gardeners eager to be the first to bring it into flower in England. The first to succeed was Joseph Paxton, head gardener to the Duke of Devonshire at Chatsworth House. The Duke was pleased to be able to present Queen Victoria with one of the flowers.

Paxton had an interest in the construction of glasshouses and conservatories. Having designed the building at Chatsworth in which the giant waterlily was grown he was later to take inspiration from the ribbed undersurface of the leaves when he designed the structure of the Crystal Palace for the Great Exhibition of 1851.

LESSER PERIWINKLE

Botanical Name: *Vinca minor*
Family: Apocynaceae
Common Names: Bachelor's buttons; bluebell; blue Betsy; blue buttons; blue Jack; blue smock; *centocchio*; cockles; cockle shells; cut finger; Devil's eye; dicky dilver; flower of death; joy of the ground; old woman's eye; parvenke; penny winkle; pin patch; St Candida's eyes; sengreen; sorcerer's violet; tutsan; virgin flower
Gender: Feminine
Planet: Venus
Element: Water
Meanings: Early friendship; sweet remembrances; tender recollections (blue); pleasures of memory (white); first love; 'My heart was mine until we met'

Magic and Lore

Several authors write of the power of the lesser periwinkle against evil, and of its use in love charms. In *Herbarium* (1480), Apuleius tells of the efficacy of the plant in treating 'Devil's sickness', evil spirits and demonic possession, the bites of snakes and wild beasts, and all forms of poisoning. He gives strict guidance for its collection. The collector must be 'clean of every uncleanness'. The plant is to be picked 'when the moon is nine nights old and eleven nights old and thirteen nights old and thirty nights old and when it is one night old'. As the plant is picked the following incantation must be said:

> I pray thee vinca pervinca, that thee art to be had for thy
> many useful qualities, that thou come to me glad blossoming with thy
> mainfulness, that thou outfit me so that I be shielded and
> prosperous and undamaged by poisons and water.

In Italy the plant is called *centocchio*, 'hundred eyes'. It is also called the 'flower of death', from the practice of placing garlands of the flower on the funeral biers of children, or about the necks of the dying. In Germany it is the 'flower of immortality', and in France it is symbolic of friendship, but called 'sorcerer's violet'. There was a common belief that the spirits of the dead inhabited the flowers. In Britain, it was said that if you uprooted, or picked the flowers of a plant, growing on a grave you would be haunted by the grave's occupant.

Culpeper tells us that if a man and woman eat periwinkle together it will bring love between them. Albertus Magnus in his *Book of Secrets* adds that where it is placed between where two lovers lie it will ensure that they never feel any malice toward each other. 'Perywynkle when it is beate unto powder with worms of ye earth wrapped around it and with an erbe called houselyke, it induceth love between man and wyfe if it bee used in their meals.'

It has long been used in love charms. A bride, to ensure fertility, and happy marriage, might wear the flower on a garter, hence:

> Something old, something new,
> Something borrowed, something blue.

It might be planted in the gardens of newly weds for the same purpose and it is a common plant in old cottage gardens. One warning, however: if an adulterous wife or promiscuous woman wears the flower it will shrivel up and die.

The name 'St Candida's eyes' comes from Dorset, where the plant is associated with the saint, sometimes called St Wite. There is a long obsolete meaning of the word 'periwinkle' as 'the fairest'. In the fourteen century it was applied to someone who excelled.

To dream of periwinkle indicates that a spirit watches over you. It is the birthday flower of 31 January.

Medicinal

Binding periwinkle leaves about the legs was supposed to prevent cramps and the plant has generally been used in remedies for boils, toothaches and nosebleeds.

VIOLET

Botanical Name: *Viola odorata*
Family: Violaceae
Common Names: Blue violet; sweet violet
Gender: Feminine
Planet: Venus
Element: Water
Deities: Aphrodite; Attis; Io; Priapus; Venus; Zeus
Meanings: Candour; innocence; purity of sentiment (white); faithfulness; love (blue); 'You occupy my thoughts' (purple); constancy; faithfulness; modesty (sweet); rare worth; rural happiness (yellow) innocence; 'Pure and sweet art thou'

Myth and Legend

There is some disagreement about the origin of the name *Viola*. Some authorities claim that it is derived from the Greek Ione. In classical mythology the god Zeus, who was known to be rather promiscuous, had a mistress called Io, the beautiful daughter of Inachus, king of Argos. In order to hide her from his wife, Hera, who was extremely jealous, Zeus transformed her into a white heifer and caused the violets to grow for her as a fitting food for her to eat. Another myth says that Cupid greatly admired the white violet because of its purity and sweetness. Venus, in a fit of petty meanness, turned it blue so that it would not be as attractive. Her action proved unsuccessful as the pretty blue flower is still amongst the best loved of our garden flowers.

> Io, the mild shape,
> Hidden by Jove's [Zeus's] fears,
> Found us first i' the sward, when she
> For hunger stooped in tears;
> Wheresoe'r her lips she sets,
> Said Jove, be breaths called violets

Elsewhere in mythology it is claimed that violets grew from the blood of Aias (Ajax), a warrior of great stature, second only to the mighty Achilles as a hero of the Trojan Wars, who committed suicide after having, in a fit of pique, killed the sheep of the Greeks when the armour of the slain Achilles was bestowed on Odysseus rather than him. In versions of other myths they are said to have grown from the spilt blood of Attis, the Phrygian deity of vegetation, who was transformed into a pine tree after breaking his vow of chastity to Cybele and marrying the daughter of Sangarius, the river god. Venus, we are told, stamped her rivals into blue violets after Cupid declared that they were even more beautiful than her.

There is the tragic tale of Orpheus. Eurydice, his wife, had been carried off into the Underworld, and his attempt to bring her back had proved unsuccessful. While he grieved for her loss he was completely immune to the overtures of the Thracian women. In jealous rage they tore him to pieces, throwing his head into the River Hebros. It was carried out to sea and was eventually washed up on the shore of Lesbos. From where the famous lyre of Orpheus fell to the ground in this attack, there grew up violets as the embodiment of pure music. It is, perhaps, this story of the love of Orpheus for Eurydice, faithful unto death, which made the plant a symbol of faithfulness and fidelity.

> Violet is for faithfulness,
> Which in me shall abide
> Hoping likewise that from your heart
> You will not let it slide;
> And will continue in the same
> As you have now begun,
> And then forever to abide,
> Then you my heart have won.
> – William Hennis, *A nosegay always sweet, for lovers to send for tokens of love at New Year's Tide, or for fairings*

Magic and Lore

Although an attractive flower, the violet is considered to be a plant of misfortune, probably because, like the primrose, it is associated with death, especially the death of children. Traditionally, if the flowers are taken into a house it brings bad luck for the whole household. If they are taken into a farmhouse it will cause all the poultry to die. It is also unlucky for landowners if violets flower in autumn anywhere on their land. The scent is supposed to attract fleas, and as bearers of the plague there could easily be a link to the superstition that for violets to flower in the autumn or winter is a warning of an epidemic of plague and pestilence, or a sign that there will be death, possibly that of the householder. An old Greek prayer was that violets, marjoram and roses would flourish on a new grave, as it indicated that the occupant had settled in Paradise.

Over 2,000 years ago the Greeks chose the purple violet as the symbol of Athens, and the flowers were worn in chaplets, like roses. This had an additional benefit, or drawback depending on your point of view, as the scent of the flowers is supposed to be suggestive of sex. Violets were a plant of the garden of Aphrodite and her son, Priapus, god of gardens and reproduction.

The sweet violet is traditionally a protection against evil spirits. It is associated with the spring equinox and the first violet of spring should be picked and kept as a lucky amulet. Another tradition, however, says that it is unlucky to pick the first violet of the spring. It is only safe to do so when there are sufficient for a small bunch, at least a dozen blossoms. Wearing a wreath of violets will prevent deception.

The violet was a favourite plant of the Empress Josephine, and following her death Napoleon had them planted on her grave. The flower was adopted as the symbol of the Imperial Napoleonic Party, and when Napoleon was sent into exile he said that he would return with the violets in spring. His supporters continued to toast him, in secret, as 'Caporal Violette'. On his death a locket was found containing violets and a lock of Josephine's hair.

In common with many other spring flowers, it is often stated that bringing a few flowers into the home will cause an adverse effect on the laying habits of chickens. In Gloucestershire it was said that the flowers harboured fleas and so should never be taken indoors. All of this is very unfortunate, as violets have long been a popular gift on Mothering Sunday, when children would go-a-mothering, gathering wild violets in the country lanes.

A dream of violets is a good omen as it portends an improvement in fortunes, advancement in life and marriage to someone younger than you.

The blue violet is the birthday flower of 11 March, the white violet of 14 March and the yellow violet of 28 April.

Medicinal

The scent of violets is said to soothe the temples. The ancient Greeks wore wreathes of the flowers to induce peace and sound sleep. They may be worn, or a medicine based on them taken, to comfort the heart, calm anger, and cure headaches and dizzy spells. Wearing violets in a sachet was also said to aid the healing of wounds and prevent evil spirits from making them worse.

Necklaces of violets may could worn to prevent inebriation. The flowers have been recommended for the treatment of diverse ailments including headaches, melancholy and insomnia. To aid the sleep of children they should be laid on their pillows at night. The scent of the flowers, however, could bring about swelling of the vocal cords.

PANSY

Botanical Name: *Viola tricolor*
Family: Violaceae
Common Names: Banewort; banwort; biddy's eyes; bird's eye; bonewort; bouncing Bet; bullweed; butterfly flower; call-me-to-you; cat's face; coach horse; cuddle me; cull me; Cupid's flower; eye bright; flamey; flower o' luce; gentleman-tailor; godfathers and godmothers; heart pansy; heartsease; herb constancy; herb trinitatis; herb trinity; horse violet; Jack-behind-the-garden-gate; Jack-jump-up-and-kiss-me-over-the-garden-gate; Johnny jumper; Johnny-jump-up; Johnny-jump-up-and-kiss-me; kiss-behind-the-garden-gate; kiss-her-in-the-buttery; kiss me; kiss-me-behind-the-garden-gate; kiss-me-quick; Kit-run-about;

Kit-run-in-the-fields; Kitty-run-the-street; lark's eye; leap-up-and-kiss-me; live-in-idleness; love-a-li-do; love-in-idleness; love-in-vain; love-lies-bleeding; loving idol; love idol; lover's thoughts; love true; meet-her-in-the-entry-kiss-her-in-the-buttery; meet-me-in-the-entry; monkey face; pensee; pink-eyed John; pink-of-my-John; pink-o'-the-eye; pussy face; *shasagh-na-criodh*; stepfathers and stepmothers; stepmother; three-faces-under-a-hood; tittle-my-fancy; trinity flower
Gender: Feminine
Planet: Saturn (in Cancer)
Element: Water
Meanings: Souvenirs; memories; 'The thought of happy days spent with you are my greatest treasure' (purple); thoughts of love; 'I cherish loving thoughts of you' (white); remembrance; 'Oceans part us but my heart stays with you' (yellow); 'Kind thoughts of you'; meditations; tender thoughts; 'Think of me'; 'You occupy my thoughts'

Myth and Legend

Originally the pansy had white flowers until Cupid wounded it with one of his love-inducing darts.

> This flower, as nature's poet sweetly sings,
> Was once milk-white, and Heartsease was its name
> Till wanton Cupid poised its roseate wings,
> A vestal's sacred bosom to inflame.
>
> – Mrs Sheridan, *Heart's Ease*

Magic and Lore

Across Europe, according to tradition, the pansy was once greatly valued for its fragrance, which was sweeter and stronger than that of the violet, as well as for its medicinal properties. The pretty little flower grew freely in cultivated land and, consequently, valuable crops were damaged by people walking through them in search of it. The flower then appealed to the Holy Trinity to remove its scent and make it less desirable. Its plea was answered and, thereafter, it was called *dreifaltig-keitsblume* – 'The flower of the trinity'. In some parts of the Rhineland it is known by a different, and equally odd, name, *je-länger je-leiber* – 'the longer the dearer'. It is said that the flower has a 'painted maiden's face' and is therefore emblematic of disappointment in love.

> Of all the flowers that come and go
> The whole twelve months together,
> This little purple pansy brings
> Thoughts of the sweetest, saddest things.
>
> – Mary Bradley, *Heart's Ease*

The name 'pansy' is derived from the French *pensee*, which means 'idle thoughts'. The name 'love-in-idleness' has the same root, meaning to love in vain. This is the plant to which Shakespeare refers in *A Midsummer Night's Dream*. When Oberon, king of the faeries, sends Puck to find it he tells him:

> … And maidens call it love-in idleness …
> The juice of it on sleeping eyelids laid, will make [a]
> man or woman madly dote upon the next living creature that it sees.

The essence of this claim can be found in the superstition that carrying or wearing a pansy will attract love. It can also be used to ensure that the relationship continues to develop and grow. If a heart-shaped plot is planted with pansies the condition of the plants will reflect that of the relationship; if the plants flourish so will the love between those involved.

To ensure that whilst sailors are away they think of home, a handful of sand, taken from the seashore, should be buried in the pansy bed. The plants must then be watered daily, before sunrise. It is pure folly to pick pansy flowers on a dry day, or when they are still damp with the morning dew, as this will cause it to rain later in the day. It was also said that to pick a pansy still wet with dew was to invite the death of a loved one.

The name 'stepmothers' is illustrated in the flower itself. The single, large, lower petal of the flower is the stepmother. The two petals above it, one to either side, are her own daughters. The upper two half-hidden petals are her stepdaughters, whom she neglects.

To discover the future, the number of lines on a petal could be counted (it must be on a flower that has been picked or given but not one bought):

Four lines – your wish will come true.
Five lines – there is trouble ahead, that you will overcome.
Six lines – a surprise is coming to you.
Seven lines – you have a faithful sweetheart.
Eight lines – you have a fickle sweetheart.
Nine lines – you will travel over water to marry.

To cause pansies to grow tall, and always look good, the seed must be sown at 6.00 a.m. and the plant always watered at the same time. They must be sown on the north side of the house and the seed placed exactly 2 inches (5.08 cm) below the soil surface.

Medicinal

Pansies were used in a folk remedy for treating the skin disorders of babies. This is continued in modern herbals, as the wild flower is recommended for the treatment of such complaints as eczema, acne and pruritus. As the name 'heartsease' suggests, an infusion of the herb has been recommended as a treatment to heal a broken heart.

MISTLETOE

Botanical Names: *Viscum album (Phoradendron flavescens* or *serotinum* in North America)
Family: Viscaceae
Common Names: All-heal; all healer; bird lime; churchman's greeting; Devil's fugue; *donnerbesen*; *Druad-lus* (Druid's plant); Druid's weed; golden bough; heal-all; *herb de la croix*; holy wood; kiss-and-go; kiss-and-tell; *lignum sanctae crucis*; masslin; misle; mislin-bush; misseltoe; mystyldene; *onnia sanantem*; thunderbesem; witches' broom; wood of the holy cross
Gender: Masculine
Planet: Sun (Jupiter, according to Culpeper)
Element: Air
Deities: Apollo; Asclepius; Baldur; Freya; Friggaa; Ischys; Ixion; Manannan Mac Lir; Odin; Venus
Meanings: Kisses; 'I am determined to overcome difficulties in this life, if it cannot be as we wish, we will meet in the next world'; 'I overcome obstacles'; 'I send you my kisses, as many as stars'; overcome

Botanically mistletoe is a semi-parasite, absorbing water and minerals from its host but making its own food through photosynthesis. The common name is derived from the belief that the plant spontaneously grew from bird droppings. – 'mistel' is from the Anglo-Saxon *mistel* meaning 'dung', and *tan* meaning 'a twig', so could be translated as 'dung-on-a-twig'.

Myth and Legend

The golden bough was a branch taken from a tree in the sanctuary of Nemi in a woodland grove sacred to the goddess Diana. No one could take wood from this tree except a runaway slave, and in breaking the branch the slave earned the right to challenge the priest-king of Nemi, in a fight to the death, for the throne. If he won he would reign until a new challenger came forward. Aeneas took this golden bough, as instructed by the Sibyl, to light his way as he journeyed into the Underworld. It may have been more than a mere lantern; it might have been the key, as elsewhere in myth we are informed that Persephone, Queen of the Underworld, opened the gates of Hades with mistletoe berries. It was called 'thunder-besem' by some as it was supposed that it was caused to grow in the treetops following lightning strike; it therefore carried the seeds of fire and would have been a logical choice for Aeneas to use as a torch.

Roman and Norse mythologies both tell similar tales about the plant. The Scandinavian myth links it with the story of Baldur the beautiful, god of light and peace. A son of Odin, he was terrified by a dream in which he saw his own death, and he feared that it might prove to be prophetic. To calm him his mother, Friggaa sought out all things – stones, plants, fire, water and animals, and exacted a promise from them not to harm him. This led to the gods playing with Baldur, firing weapons at him knowing that they would not harm him. However, Loki, the god of fire, discovered that the mistletoe had not been asked, being considered too young to take an oath, and so had made no such promise. He therefore took some mistletoe wood and from it made a spear that he tricked the blind god, Höldur, into throwing at Baldur. The wound proved fatal, and Baldur died, the tears of Friggaa at her son's death forming the berries on the mistletoe.

There are two basic endings to this tale. In one, the gods grieve at the death of Baldur and set his body on to the burning pyre of his ship. On seeing this his wife, Nanna, falls dead from grief, and her body is added to the pyre. Thereafter the Scandinavians lit bonfires, called Baldur's fires, at Midsummer's Eve and no doubt, human sacrifices or effigies of Baldur were burned on these in ancient times.

The alternative ending tells us that after various interventions Baldur was resurrected, and the mistletoe given over to Friggaa, the goddess of love, to be her symbol. She decreed that it should be hung from the ceiling and that all who passed beneath it should kiss, to show that it was no longer an instrument of hate but now one of love. Some say that this was also a part of the plant's punishment, to look on as beautiful women are kissed. It would never again be a cause for evil unless it touched the earth, Loki's domain. The story in Roman mythology is basically the same except that, rather than Friggaa and Baldur, it involved Venus, the goddess of love and Apollo, the god of music, poetry, prophecy and medicine.

Mistletoe's parasitic nature has been said to be a punishment for being the wood from which the cross used at Christ's crucifixion was made. Although this seems a little improbable, there are a number of similarities between the Norse story of Baldur and the Christian story of the resurrection.

Magic and Lore

It was at Midsummer that mistletoe to be used in magical rites was collected. The Baldur's fires are probably a more civilized variation of the older practice of burning a human representative of the god. It was one of the Midsummer rites to ensure that the sun went on shining, crops flourished, trees grew and that the people were protected from the influences of witches and warlocks, faeries and trolls. Baldur was, in this sense, a tree deity, the personification of the oak. Mistletoe was symbolic of the oak's life force and, as long as it was not removed, the tree was invulnerable. In winter, when the oak lost its foliage, its life force was thought to move into the mistletoe.

Mistletoe has been widely considered to be a sacred plant throughout history. It was *druad-lus*, the Druid's plant, which fell from heaven into the top of the oak. The Druids revered the mistletoe that grew on oak trees, the oak being their main tree deity. The viscous white berries resembled semen and embodied the power of both the mistletoe and the oak on which it grew. The oak, in this case, was twice as sacred as it symbolized that the immortal was a small step closer to mankind. All sorts of misfortunes would befall anyone foolish enough to cut down or damage a tree on which mistletoe was growing. For sorcerers, the best magic wands were those made from mistletoe that had been grown on oak.

It was associated with the festivals of the solstices. It was used in the midwinter festivals and collected by the Druids at Midsummer or when the moon was six days old (in ancient Italy it was gathered on the fifteenth day of a new moon). It had to be cut with a single stroke from a golden knife and, as in the Norse myth, it was not allowed to touch the ground, but was caught in a white cloth. In common with other magical plants, tools made of base metal could not be used in the gathering of mistletoe (it is quite possible that the Druid's 'golden' sickle was actually made from brass or bronze). The sickle is in the shape of a crescent moon, the emblem of the lunar goddess Diana, and the Druids followed a lunar calendar. Mistletoe collection is often said to have occurred in the seventh month of a thirteen-month year, that is to say a lunar year. It should not be forgotten that the mistletoe is dedicated to the moon, whereas the oak is dedicated to the sun; the combination of the two would imply balance. Likewise, growing as it does on another plant, it is of neither heaven nor earth and so falls somewhere between.

It was only collected when the Druids felt themselves directed to do so. If a long period elapsed between collections, or if the plant was allowed to touch the ground, it was believed that great disasters would befall the nation. It was also believed that it would bring misfortune if all the mistletoe was removed from any one tree. The material gathered would be divided up and might be hung over doorways as an emblem of peace and hospitality.

Pliny the Elder described the druidic ritual:

> The Druids, for so they [the Gauls] call their wizards, esteem nothing more sacred than the mistletoe and the tree on which it grows, providing only that the tree is an oak ... For they believe that whatever grows on these trees is sent from heaven, and is a sign that the tree has been chosen by the god himself. The mistletoe is very rarely to be met with; but when it is found they gather it with solemn ceremony. This they do above all on the sixth day of the moon, from whence they date the

> beginning of their months, of their years, and of their thirty years' cycle, because by the sixth day the moon has plenty of vigour and has not run half its course. After due preparations have been made for a sacrifice and a feast under the tree, they hail it as the universal healer and bring to the spot two white bulls, whose horns have never been bound before. A priest clad in white robes climbs the tree and with a golden sickle cuts the mistletoe, which is caught in a white cloth. Then they sacrifice the victims, praying that god may make his won gift to prosper with those upon whom he has bestowed it. They believe that a potion prepared from the mistletoe will make barren animals to bring forth, and that the plant is a remedy against all poison.
>
> – Quoted in *The Golden Bough*

In Aargau, Switzerland, mistletoe was gathered when the sun was in the constellation of Sagittarius the archer and when the moon was waning, and was obtained by shooting it from the tree with an arrow, then catching it in the left hand. In other parts of Europe it had to be knocked from the tree with stones.

It is probable that its use as a protective amulet dates back at least to the Norman conquests. It was strategically placed throughout the home or worn to ward of all manner of ills, including witches and other evil forces, lightning strikes, diseases and fires. One spell to form a talisman to protect against witchcraft required that the mistletoe be cut with a new knife after the collector had walked thrice around the oak tree at Halloween. This is possibly a compromise with the rituals of the Druids. Pieces of the plant might also be placed into the cradles of young babies to prevent the faeries stealing the child and leaving a changeling in its place. Soldiers might carry the leaves as a protection against wounding. In Brittany it was supposed to impart strength and courage to wrestlers and athletes, a use that might stem from its use by soldiers. In Austria, a sprig of mistletoe laid in the doorway would prevent nightmares, presumably by stopping the evil that caused them. In a Welsh tradition, placing mistletoe beneath a pillow would induce prophetic dreams, good and bad.

In the classic text, *The Golden Bough*, Sir James Frazer suggests that mistletoe was called 'golden bough' because when old the entire plant takes on a golden hue. However, as in the case of the 'golden bough' of apple, where there is fruit and blossom on the same branch, the mistletoe was, in mythology, a key to the Underworld.

Mistletoe is now a central feature in our Christmas celebrations. It was claimed that bringing it into the home before Christmas would result in a death in the household. In Worcestershire, it was traditional that the mistletoe should only be gathered at midnight on Christmas Eve. It was also traditional for the last male servant to have joined the family to gather it. It was then dressed with ribbons, nuts and apples before being hung from the ceiling. In Herefordshire mistletoe might be put up at New Year and retained for twelve months, the old piece being removed and burned. In Staffordshire a bunch of mistletoe was retained after Christmas as a witch repellent and burned beneath the Christmas pudding the following year. When the Christmas greenery was taken down, farmers in England and Wales would give the Christmas mistletoe to the first cow that calved in the New Year, as this bestowed good luck to the whole herd.

In Victorian times, having mistletoe in the house at Christmas gave people a rare opportunity to kiss in public. Kissing beneath the mistletoe is usually thought to have been a peculiarly British tradition, or at least to have originated in Britain. It may, however, date back to the Roman festival of Saturnalia, or to the Scandinavian belief that it was a plant of peace and harmony. It was considered lucky to be kissed under the mistletoe, and very unlucky to try to avoid it. With each kiss, a berry should be pulled off the plant and tossed over the left shoulder. When there were no more berries, there could be no more kisses. One superstition claims that if you kiss beneath the mistletoe and pull off a berry, you will have a child within the following year. A kiss beneath the mistletoe was a pledge of love and could even be a promise of marriage. One superstition says that if a girl is not kissed beneath the mistletoe at least once before she marries she will remain barren her whole life and another says that she will die a spinster. The head of the household should kiss every woman in the house to ensure fertility. All the mistletoe used at Christmas should be burned before Twelfth Night, or all those who kissed beneath it would quarrel before the New Year, and would definitely not marry.

Being considered a plant of peace and harmony, not all kisses beneath the mistletoe were those of love and affection. A kiss in greeting might be given by enemies, meeting in woodland, to seal a truce. They would lay down their arms and the truce would stay in force until the following day. Warring spouses might choose to kiss and make up under a branch of mistletoe.

In Sweden divining rods were made of mistletoe and used to discover gold and treasure. It is also said that mistletoe can be used in spells to open all locks, possibly in the form of the 'hand of glory' (see Vervain).

Being a fertility symbol it was traditionally believed that if the plant was intentionally cultivated, the seed being placed in the bark of the tree, the daughters of the person responsible would remain barren. It was also said that its growth in an apple orchard would improve the crop. It would be included in the garlands of Jack-in-the-Green, the 'green man' character in the fertility pageants in May. The main foliage used in the garlands was, of course, oak.

Mistletoe is the birthday flower of 6 February.

Medicinal

Mistletoe, like many another plant, has been viewed as something of a panacea, curing all ailments. This may be largely due to it having been viewed as a sacred plant able to ward off evil and, therefore, the illnesses caused by the forces of evil. Amongst Native Americans mistletoe teas are given to aid in the cure of measles, toothaches and dog bites.

The greatest use of the plant is most probably as an aphrodisiac, the arrangement of the berries implying testicles, and their colour semen. Various applications were used by women to ensure conception. Kissing beneath the mistletoe might be sufficient, although some writers recommended that it might be worn about the arms or neck. Other authors required that the berries be eaten and that draughts of the juice be given to animals to cure infertility. Great care must be observed before following such advice as it has narcotic properties.

Mistletoe was used in the treatment of epilepsy and, it was claimed, it could overcome the coma caused by an epileptic fit. In Sweden, sufferers of epilepsy would carry knives with handles of mistletoe wood to prevent them having fits and, elsewhere, children might have pieces of the wood strung on threads about their necks to ward off epilepsy. This use may result from sympathetic magic, as the plant cannot be allowed to fall to earth.

CHASTE TREE

Botanical Name: *Vitex agnus-castus*
Family: Verbenaceae
Common Names: Abraham's balm; hemp tree; monk's pepper
Deity: Asclepius
Meanings: Apathy; chastity; coldness; indifference; 'I prefer to remain a virgin than let you come near me'; to live without love

With a name like 'chaste tree' it is not surprising to discover that this plant is supposed to suppress sexual desires. In ancient times the priestesses of the goddess Ceres, Mother Earth, slept on couches which were covered with the leaves of this tree in order to 'subdue sensual feelings'. Some members of the early Christian church adopted the same practice in order to calm their 'unhallowed thoughts'.

According to Pliny, at the Thesmophoros, the Greek festival of Demeter Thesmophoros (Demeter the law-giver), Athenian wives sat on the ground on branches of *agnos*, or strewed *agnos* leaves in their beds to increase their fertility. They would also refrain from 'sexual congress' with their husbands at this time.

VINE

Botanical Name: *Vitis vinifera*
Family: Vitaceae
Common Name: Grape vine
Gender: Feminine
Planet: Moon (Sun, according to Culpeper)
Element: Water
Deities: Bacchus; Dionysus; Hathor; Jesus; Osiris
Meanings: 'Christ our life'; drunkenness; intoxication; mirth; prosperity; 'This is my blood'

Myth and Legend

The origin of the vine is unknown and consequently various stories have arisen to account for its creation. One tells how, when Adam and Eve were driven out of the Garden of Eden, the angel sent to carry out this task grieved for mankind and, on his return to Paradise, rested on his staff and wept. The staff took root and grew as a new plant. The angel collected the seeds from this plant and took them to Adam, instructing him to sow them, and from these seeds grew a vine.

In classical mythology the vine is a symbol of Bacchus (Dionysus), as god of wine, but despite this it was Saturn who is credited with having introduced it to Crete. In Egyptian mythology, viniculture was taught to mankind by the god Osiris, who is also identified with the pine, and pine resin could be used in drinks when the grape harvest failed.

One story tells how Astyages, King of Medes, had a dream in which his daughter 'brought forth a vine'. It proved to be prophetic of the wealth, grandeur and prosperity of Cyrus, born to his daughter shortly after the dream.

In North America the Mandan tribes of Sioux Indians believed that their nation had once lived in a subterranean village beside a great lake. The roots of a giant vine penetrated into their village from the world above, and one day some of them climbed up the roots to discover what was in the world above. They returned about one week later and told such wonderful stories of the land above ground that the whole tribe set out on the same journey but only about half had reached the surface before the roots broke with the weight of a particularly corpulent woman. To the Mandan, the worthy will, in death, return to their subterranean village by way of the lake but the bad must abandon all hope of returning because of the weight of their sins.

Magic and Lore

Greece appears to be the birthplace of vine cultivation and some sixty different varieties were known to the ancient Greeks and Romans. Much of their instruction relating to the cultivation of vines is still available to this day, and can be found in the writings of such authors as Virgil. For example, in the *Georgics*, Virgil instructs that vines grow well where they are trained over elms as a support. The elm and the vine were described as 'wedded', as the branching habit of the tree was ideal as a support for the vine as well as shading the plants.

The vine was emblematic of fertility, prosperity and abundance. With the olive and the fig it symbolizes the prosperity that can accompany peace. It might be painted as a mural on garden walls to ensure the fertility of the land, and of those who dwelt within the household.

The relaxation of inhibitions resulting from drinking wine allowed more subtle intuition to surface and so the vine became linked to the use of psychic powers and prophetic dreams. It is also associated with divinity, and the 'energy of life'. Jesus called himself the 'true vine' and, in the Old Testament, the Israelites are described as the vine that was brought out of Egypt.

Odd superstitions surround the cultivation of grapes. Vines should be planted in the light of the moon, that is the period between a new moon and a full moon. They should be trimmed following the first frost in the autumn. If they are trimmed just before the full moon, in the sign of Cancer, they will be unaffected by attacks from caterpillars and birds. Leaving two notches on each vine when trimming it will ensure better quality grapes. Black grapes grow better with black tom cats buried beneath them.

If the white underside of the leaves were visible, it was a prediction of rain, and rain at Easter, or on 4 July, meant that there will be no grapes on the vine that year.

YUCCA

Botanical Name: *Yucca* spp.
Family: Agavaceae
Common Name: Adam's needle;
Gender: Masculine
Planet: Mars
Element: Fire

There was a belief, among some of the tribes of Native Americans, that yucca could be used to transform people into animals; wearing of a wreath made from the plant would allow the wearer to adopt the shape of any animal they wished. Passing someone through a hoop made from the twisted fibres of the yucca would also cause them to be transformed into an animal.

Yucca was also used in protective magic. A cross made from the fibres, placed against the hearth in the house, would protect the home from all evil. Lather from the juices could be used as a purifying wash before, and after, magical rites to remove illnesses or curses.

ARUM LILY

Botanical Name: *Zantedeschia aethiopica*
Family: Araceae
Common Names: Altar lilies; Easter lilies; lily of the Nile; trumpet lily; white arum

This is a flower associated with mourning, and it was used to decorate graves. In consequence it was often thought of as an inappropriate choice to decorate houses or to take into hospitals as it was said it would result in a death.

In Ireland it is associated with the 1916 Easter Rising, although there is a certain amount of confusion as to which plant is actually meant.

CORN

Botanical Name: *Zea mays*
Family: Gramineae
Common Names: Dolly; giver of life; kern; kern baby; mare; maiden; maize; mealie; old wife; pig; sacred mother; seed of seeds; strawcock; wolf
Gender: Feminine
Planet: Venus
Element: Earth
Deities: Abuk; Acca Laurentia; Adonis; Anath; Ashnan; Attis; Bhim Deo; Bhimsen; Ceres; Cerridwen; Chitariah Tubueriki; Cronus; Dagon; Damuzi; Danae; Demeter; Dionysus; Enlil; Ezinu; Fornax; Gauri; Heqet; Ino; Isis; Jehovah; Jesus; Kore; Llew; Lugh; Maneros; Metsik; Nebri; Neper; Ninlil; Ninhursag; Nisaba; Osiris; Persephone; Robigus; Sita; Sud; Tailltiu; Tammuz; Tubueriki; Uma; Viribius; Volos
Meanings: Agreement (straw); quarrel (broken) plenty; riches; holy communion

Myth and Legend

In mythology, as in popular speech, 'corn' is a name given to almost any cereal crop, causing confusion between maize, wheat, barley, rye and even oats. Many of the traditions and superstitions described below could refer to wheat and barley rather than corn. The use of the name 'corn' to mean exclusively *Zea mays* dates from the eighteenth century.

In myth corn is one of the principals signifying the return of summer and the defeat of winter, and so the victory of life over death. This is established within its role in the story of Demeter and Persephone from Greek mythology. Hades, the god of the Underworld, had abducted Persephone, the daughter of Demeter, the goddess of agriculture. In her grief Demeter renounced all her divine duties and travelled to search for her missing child. Eventually she arrived at the city of Eleusis and found her way amongst the women at the court of King Celeus. One of these women, Iambe, made Demeter smile in spite of her grief, so she decided to break her fast and made a drink from barley water and mint, called *kykeon*. Soon after, she was sought out by the king to be a nurse and bring up his son, Demophoon. She gave him the task of spreading the knowledge of corn cultivation to mankind. Therefore, the barrenness of the world caused by the grief of Demeter was ended with the coming of the corn.

Corn also signified the successful production of food and was used in the rites of Adonis and Osiris. It is usually seen as symbolic of plenty or fertility, and brides in the Middle Ages would carry ears of corn to ensure their own fertility. There are obvious connections with the showering of newly weds with rice, and thus with confetti.

One way of seeing the three aspects of the triple goddess is with Persephone as the maiden, Demeter as the mother and Hecate as the crone. Persephone is the green corn, Demeter the ripened corn and Hecate the harvested corn. The descent of Persephone into the Underworld is represented by the sowing of the corn, and the growth of the plants represents her resurrection.

The corn goddess was widely worshipped, in the East, in Europe and in North America. A Pawnee Indian legend tells that it was the 'Corn Mother' who led the first men up to the surface of the Earth from their home in the Underworld.

> Before the World was we were all within the Earth.
> Corn Mother caused movement. She gave life.
> Life being given we moved towards the surface:
> We shall stand erect as men!
> The being is become human! He is a person!
> – Dr M. Gilmore, quoting Pawnee Indian, Fair Rings, in *The World's Rim*

In parts of central Europe the spirit of the Corn Mother was said to wander through the fields of ripening grain at night. In Germany, children were told that the movement of the corn in the breeze was caused by the Corn Mother passing by.

A North American myth tells of a young man who, in a vision, saw a beautiful fair-haired maiden. She instructed him to set the prairie on fire and then to drag her by her hair across the scorched ground. He did so, and corn sprouted out of the earth, and in each ear of corn was a tuft of the maiden's hair.

The Corn Mother is the fertile, food-giving aspect of the White Goddess. She is the Good Mother but she also reaps as well as sows, and may grant or withhold the harvest. In various parts of the world different goddesses have been seen as this aspect of the deity. For the Celts it was Cerridwen, the Old White One, whilst the Greeks called her Demeter, the Barley Mother, Athene Alta (Patroness of the Mill) or Alphito (Hag of the Mill, goddess of barley flour). To the Romans she was Ceres, who might be invoked at harvest time, and who presided at the grain markets. In Sumerian myth, the Corn Mother was Ninlil or Ashnan, the guardian of the grain. The Minoans knew her as Ariadne, the high fruitful mother, the daughter of King Minos, who had helped to save Theseus from the Labyrinth. For the Hindus she is Sita, the Furrow or Uma, from the body of whom grain emerges after the rain. In the home of maize the Corn Mother is Onatha amongst the Iroquois and in Peru she is Zara-Mama, the Mother of the Maize.

Magic and Lore

In ancient Mexico, a festival was held to celebrate the corn goddess as the 'long-haired mother'. It began at the time when the maize had achieved its full height of growth and the 'hair' shooting from the head indicated that the ear was fully formed. At the festival the women wore their long hair unbound and tossed it in the dances that were the main feature of the festivities in the hope that the tassles of the corn might grow in equal profusion. In Europe there were similar festivals, with many jumping and leaping dances to encourage the grain to grow tall.

Often in European tradition the Corn Mother, or spirit of the grain, was embodied in the last sheaf of grain left standing in the fields and with the cutting of this last handful she was either caught, driven away, or slain. This last sheaf played an important part in harvest customs and the spirit of the Corn Mother embodied in it was given various names.

In Wales the corn dollies were traditionally made from the last sheaf of corn to be cut in the harvest. This sheaf might be thought to symbolize a human or animal spirit and was said to contain the 'spirit of the corn' which had fled before the harvesters, to end up caught in the final sheaf. The dolly made from it would be kept until the following year to ensure another good harvest. In Cornwall, the last sheaf is called the 'neck' and was traditionally cut by the oldest reaper. On cutting it, he would call out, 'I have it' three times. When asked what, he would reply 'The neck', again three times. This sheaf would then be carried back to the farm to be bedecked with ribbons and kept until the following year.

In parts of France and amongst the Slavic peoples, the Corn Mother was called the mother of rye or the mother of wheat. The last sheaf of the harvest was made into a poppet and dressed in farmer's clothes. It was given a crown and bedecked with a blue or white scarf. In a clear reference to the Roman goddess it was called Ceres and a tree branch was stuck in its breast. At the harvest celebration the farmer who had reaped fastest danced round this figure with the girl of his choice. Finally, the figure was burned on a pyre and prayers were said to ask it for a bountiful year.

The last sheaf being used as a emblem of fertility is more clearly shown in Brittany where a small corn figure inside a larger one, called the mother sheaf, was presented to the farmer's wife. In Prussia, the Corn Mother, Zitniamatka, 'gave birth' to the corn baby as a fertility charm. Somewhat less charmingly the figure formed from the last sheaf of the harvest was called 'the bastard' in West Prussia. In East Prussia, the corn mother was called the grandmother and was decked in ribbons and flowers and dressed with an apron before being given to the woman who bound this last sheaf. In Magdeburg it was believed that the recipient of the grandmother would marry someone old within the year.

In Belfast she was affectionately known as the Granny. Here the last sheaf was cut by having sickles thrown at it and it was braided and kept until the next harvest. In Poland a giant sheaf formed from twelve sheaves tied together was called Baba or old woman. It was said that in the last sheaf resided the Baba. The Corn Mother was also called Baba in Bohemia, where the last sheaf was shaped to look like a woman with a big straw hat. It would seem that the age of the person who cut the last sheaf might be the determining factor over the name given to this embodiment of the corn spirit. The corn mother or grandmother was cut by the oldest woman, whereas the maiden was cut by a young girl. In parts of northern England and Scotland the final sheaf was called either the 'Maiden' or the kern (or kirn). The harvester was said to have 'won the kern', and once again this sheaf was dressed up as a 'kern-dolly'. Sometimes in Scotland the Corn Mother is referred to as 'Carline' or 'Carlin'. She was the old woman when harvested after Halloween and the maid if harvested before. The worst luck was to be expected if the harvesters had to work into the dusk and the last sheaf was cut after sunset; then she was simply called 'the witch'. Other names for the final sheaf of the harvest include the bitch (Bikko in Orkney), corn baby, end of harvest mare, hag, harvest child, harvest mare, kern baby, and ivy girl. In Holstein the doll was soaked with water as a rain charm, but elsewhere it might just as easily be the person who cut the last sheaf that was drenched or thrown into water to bring the rains.

Various animals have been associated with the spirit of the corn. In Germany and France it is the wolf, and when pictured running through the cornfield with its tail held high it is a fertility symbol. The harvester of the last sheaf was said to have 'caught the wolf'. Another name was the barley horse and in the area around the French town of Lille it was known as the cross of the horse. It was sometimes said that the last sheaf harvested was the hiding place of an evil witch who was eventually caught in the last sheaf on the last farm to complete that year's harvest.

The traditions of the corn dollies persist amongst the practitioners of Wicca, who still make or buy them for use at their sabbats and in rituals. A corn dolly is seen as an appropriate altar decoration at Lughnasadh, Mabon or Samhain, and one made from oat straw is perfect for Imbolc. In the Christian Church corn dollies might also be found in the church decorations for the harvest festival.

As I have said the Good Mother can grant or withhold the harvest. At one time the failure of a farmer's crop might have been considered a punishment by the Corn

Mother for his misdeeds. In her darker personification as the Old corn woman, she was said to strangle those children who trampled the grain, and in Germany children were warned against pulling up blue cornflowers or red poppies lest the Corn Mother catch them and punish them.

Corn can be used to assess lifespan. An ear of corn should be selected from the store and the number of grains counted. Allow twelve grains to a year to gain an indication of your life expectancy. To ensure good luck, hang a bunch of corn stalks over a mirror. Grains of red corn, strung and worn as a necklace, will prevent nosebleeds. The ancient Meso-American peoples used corn pollen in ceremonies to cause rain and, paradoxically, to prevent poor weather.

Superstitions relating to corn include the following. If you should find a white silk on the first ear of corn you see in the season it warns that someone in your household will become ill and may die. Likewise missing a row when planting corn also foretells a death in the family before the next sowing. If the first ear of corn you see in the season is red, it indicates that you will hear of a wedding before you hear of a death. Finding a red ear of corn is usually taken as an omen that you will be married within the year. If you should come across a 'Sally corn', that is a blue-spotted ear, when husking the corn, it is a sign that you will enjoy exceptionally good luck. If, amidst a growing field of corn, you find some white stalks, pulling them up will cause the death of as many cows.

Guidance on the cultivation of corn or the size of the yield can be found in many sayings.

A wet and windy May
Fills the barn with corn and hay.

Rain in April; Rain in May
Or Mainsworth Fare well corn and hay.
[Mainsworth, in County Durham stands on gravelly soil].

Thunder in November is said to be sure sign of a bumper crop, as is dry weather in June. And when sowing seed the advice varies from 'Two for crow, two to rot and two to grow' to 'Two for company, one for himself, and three for the rats and mice', or 'One for blackbird, one for crow, one for mole and two to grow.' A more thrifty farmer or gardener might prefer to follow the advice that if there are more than three grains in any sowing they will not grow.

The seed ought to be planted during the light of the moon for a good yield. This will also cause the ears to be carried high up on the plant, whereas sowing in the moon's dark will result in the ears being borne low down on the stalks. It has been claimed that to gain maximum yield it is essential that one male seed is planted to every female seed, although no guidance is given as to how to tell the difference.

Advice on the timing of planting varies from the precise 'plant on 26 March' to the more general, 'Sow when the dogwoods are in flower.' It is also said that the weather is not suitable for growing corn until after Whitsunday, that is seven weeks after

Easter. You could plant when oak leaves are the size of a squirrel's ear or when the woodpecker first appears but you should never plant during the first three days of May.

A thin texture to the cornhusks and few silks, indicates a mild winter, whereas if the husks are thick and the silks are abundant the winter will be severe.

Dreaming of corn, or of a cornfield, foretells an addition to the family. Any dream of corn is a sign of wealth and a dream of gathering corn is very good luck in everything. As in the biblical dreams of Pharaoh, interpreted by Joseph, a dream of full ears of corn indicates that you can expect as many good years as there are ears of corn. If you dream of thin ears, expect as many bad years. When you dream of blighted or mildewed corn you will be a considerable loser and if the corn in your dream suddenly becomes ripe you will gain an unlooked-for inheritance. If you dream of a burning stack of corn it indicates famine and death.

Medicinal

To cure warts you must collect a corn stalk with a joint on the stem to represent each wart. Cut the joints from the stem and parcel them up. Throw the parcel over your head and bless yourself, and the warts will soon disappear.

GINGER

Botanical Name: *Zingiber officinale*
Family: Zingiberaceae
Common Names: African ginger; Jamacian ginger
Gender: Masculine
Planet: Mars
Element: Fire

Ginger is used in magical rites to 'warm up' spells, increasing their power. It may be used to attract money, either by growing the plant or by sprinkling the dust into purses and pockets. This should also bring success in all undertakings.

The Dobu islanders, in the Pacific, make great use of ginger in their magical rituals. To cure ailments, for example, they chew the spice and spit at the 'seat' of the illness. They will also spit ginger into the wind to stop oncoming storms at sea.

ZINNIA

Botanical Name: *Zinnia* spp.
Family: Asteraceae
Common Names: Brazilian marigold; ever-changing woman; youth and old age plant
Meanings: 'I mourn your absence'; thoughts of absent friends; 'Where there is love there can be no separation'

Amongst the Native Americans, notably the Apache and Navajo, Zinnia is known as the 'ever-changing woman', the personification of nature. Mother Nature, being the goddess, would grow old and then become young again. Zinnia ray flowers age, to be followed by the appearance of young ray florets.

A Navajo legend describes a tribe whose crops were destroyed by insects and droughts. The tribe sought advice from the Ever-changing Woman, who instructed them to seek guidance from the Spider Woman. The task of seeking out the Spider Woman fell to Straight Arrow, the twelve-year-old son of the tribe's chief. On his journey to find her, sulphur yellow zinnias sprang up along his trail. When eventually he located her, she told him that all the tribe's difficulties were the work of the Chindi, the Devil, who must be exorcised from the fields if the tribe ever wished to harvest a crop. She told him that they must plant the sulphur yellow zinnias in their fields at night. Ever after, the Navajo have planted the yellow zinnia in their cornfields and never harm the spider or destroy its webs stretched across the stands of zinnias.

LISTS AND APPENDICES

The Language of the Flowers

> Teach thee their language ? Sweet I know no tongue,
> No mystic art those gentle things declare;
> I ne'er could trace the schoolmaster's trick among
> Created things, so delicate and rare:
> Their language ? Prythee; why they are themselves
> But bright thoughts syllabled to shape and hue;
> The tongue that erst was spoken by the elves,
> When tenderness as yet within the world was new.
>
> – *The Language of the Flowers*

The use of plants and flowers as symbols of different sentiments dates back into the far corners of time. The Egyptians, Greeks and Romans gave plants a symbolic as well as a physical significance. The language of flowers, as we would recognize it, probably has its origin in the Ottoman Empire. A culture where members of the opposite sex were unable to mix, and speak, freely was fertile ground for such a code. It was, perhaps, inevitable that such a code would also take root in the strict households of the Victorians. The idea of this 'language' caught on to such an extent that in 1879 a book, solely given over to the subject, was published (written by a certain Miss Corruthers of Inverness). This was followed in 1885 by *The Language of Flowers*, by the illustrator Kate Greenaway, which proved to be one of the most popular texts on the topic. Over the years there have been additions and adaptations of the original sentiments and, as with any good game of 'Chinese whispers' a number of errors appear to have crept in. All this has now come down to what we now call the language of the flowers. Many of the significances given to the various plants and flowers have long since been lost from our collective memories, but others remain. It is, however, not a simple language. Not all sources agree on the meaning to be attributed to any individual flower or plant. But a quick glance at the following lists will show that many of the meanings attributed to the plants are classical and still in use even today: the olive branch of peace, the red rose to mean love, and the ear of corn to show the plentiful harvest.

Every time I have come across additional material for this list I have included it. From this, two points have become significant. In one old text I found the directions given below. These make the whole thing take on a new power, as each bunch of flowers given now has an added significance depending on how it is given and received. This could make the lives of florists difficult, if they were required to report back to the sender on how the recipient accepted the flowers. In collecting from various old sources I have come across some plants by their common name, such as abercedary and abatina, which I cannot cross reference to a botanical name. To put it another way, I have no idea what plant they are, nevertheless I have left them in the list in the hope that the names will mean something in a some corner of Britain.

'Twas the clime of the east, 'twas the land of blue hours,
That taught its fair maidens the language of Flowers;
Basking sweet in the gleam of the sun's setting rays,
And love's exchange making in flower-writ lays.

Directions for use

1. When a flower is given, inclining it to the (giver's) left implies 'I' and inclining it to the right 'thou'.
2. A flower presented with leaves still on the stem positively expresses the sentiment of which it is emblematic. Stripped of leaves it carries the negative meaning. To express a negative meaning using a non-flowering plant the tips of the leaves are removed.
3. If the gift of flowers suggests an answer to a question then presenting it in the left hand would be a negative response and in the right hand a positive one.
4. The manner in which a flower is worn may alter its meaning. Worn on the head it conveys caution and over the breast, it would convey friendship or remembrance. When worn over the heart it means love of course.
5. When sending flowers, the knot of the ribbon should be on the left (when the blooms are looked at from the front) to imply 'I' or 'me' and on the front to imply 'thee' or 'thou'.

Abatina		Fickleness
Abecedary		Volubility
Acacia		Beauty in retirement; chaste love; concealed love; friendship; platonic love
Acacia, pink		Elegance; 'You possess a queen's majesty'
Acacia, yellow		'Let us disclose our hearts to no one'; secret love
Acalia		Temperance
Acanthus		Arts
Achimenes		'Such worth is rare'
Acorn		Nordic symbol of life and immortality
Adonis	(pheasant's eye)	Painful recollections; sorrowful remembrance
Agapanthus	(lily of the Nile, African lily)	Flower of love
Agrimony		Gratitude; 'Please accept my thanks for your token'
Alathea frutex		Persuasion
Alfalfa	(lucerne)	Life
Alkanet	(bugloss)	Falsehood; mendacity
Allium		Unity; humility; patience
Allspice	(calycanthus)	Benevolence; 'My heart feels compassion'
Allysum, sweet		'I admire your nobility of character'; virtue and worthiness; worth beyond beauty
Almond		Hope; heedlessness
Almond blossom		Discretion; hope; 'I hope to meet you again'; indiscretion; stupidity; thoughtlessness; 'Your friendship is pleasant'
Aloe		Acute sorrow; affliction; bitterness
Alstroemeria	(Peruvian lily, lily of the Incas)	Devotion; good fortune; prosperity; wealth
Amaranth		Desertion; hopeless and heartless; immortality; unfading love
Amaranth, crested	(cock's comb)	Foppery; singularity; with affection

Amaranth, globe		Immortality; 'My love for you is undying'; constancy
Amaryllis		Beautiful but timid; determination; pride; sparkling; splendid beauty; timidity
Ambrosia		Love returned
Amethyst		Admiration; 'My love for you is undying'
Andromeda		Self-sacrifice
Anemone, garden		Abandonment; daintiness; estrangement; forsaken; love; lust; 'Your charms no longer touch my heart'
Anemone, wild	(wind flower)	Anticipation; expectation; sickness
Anemone, wood		Forlornness
Angelica		Inspiration; 'You are my perfect inspiration'; 'Your love is my guiding star'
Angrec		Royalty; rudeness
Apocynum	(dog's bane)	Deceit; falsehood
Apple, blossom		Beauty and goodness; 'Fame speaks him great and good'; preference; 'You are as great as you are lovely'
Apple, fruit		Temptation
Apricot		Distrust; doubt
Arbor vitae		'Live for me'; old age; unchanging friendship
Arbutus		'Be mine I beg you'; love; 'Thee only do I love'
Arum lily	(calla ethiopica)	Feminine modesty; magnificent beauty
Asclepias		Cure for the heartache
Ash		Grandeur; 'My love is iron-hearted, high as the mountains and deep as the ocean'
Aspen		Fear; 'How could you leave me thus?'; lamentation; sighing
Asphodel		Mourning; 'My regrets follow you to the grave'; 'Our love shall endure after death'
Aster, China	(Chilean starwort)	Afterthought; variety
Aster, double		'I partake of your sentiments'
Aster, single		'I will think of it'

Anthurium	(flamingo flower, painted tongue)	Hard working; hospitality; think of me.
Auricula		Painting; pride; 'Wealth is not always happiness'
Auricula, green edge		'Importune me not'
Auricula, scarlet		Avarice
Austurtium		Splendour
Azalea		Moderation; temperance
Azalea, Indian		True to the end
Babies' breath posies		Everlasting love
Bachelor's buttons		A snare; betrayal; hope in love; single blessedness; youthful love
Balm		Fun; 'I was but jesting'; pleasantry; social intercourse; sympathy
Balm of Gilead		A cure; healing; relief
Balsam		Ardent love; 'I can hardly live 'til I see you again'; impatience
Balsam, red		Impatient resolve; touch me not
Bamboo		Happiness; longevity; wealth
Barberry		Hot temper; petulance; sharpness; sourness of temper; tartness
Basil		Animosity; hatred; 'I cannot like you'; poverty; sympathy
Basil, sweet		Good wishes; love
Bay, berry		Instruction
Bay, branch		Glory; 'Your elegance and majesty dazzle me'
Bay, leaf		'I admire you but cannot love you'; 'I change but in death'
Bay, wreath		Award of merit; 'Your determined suit has won my heart'
Bear's breeches	(acanthus)	Artifice; love of fine arts
Beech		Grandeur; prosperity; 'The halcyon days of our love are at hand'
Begonia		Warning; 'We are being watched'
Bell flower		Constancy; morning

Bell flower, small white		Gratitude
Bells of Ireland	(shellflower)	Good luck; whimsy
Belvedere		'I declare against you'
Bergamot		'Your whims are quite unbearable'
Betony		Surprise
Bilberry		Treachery
Bindweed, blue	(blue convolvulus)	Repose
Bindweed, field	(convolvulus)	Bonds; uncertainty
Bindweed, great	(*convolvulus major*)	Dangerous insinuator; dangerous situations; extinguish hopes; insinuation; persistence
Bindweed, pink	(pink convolvulus)	Worth sustained by judicious and tender affection
Bindweed, small	(*convolvulus minor*)	Humility; obstinacy; night
Birch		Gracefulness; meekness
Bird of paradise	(crane flower)	Joyfulness; paradise
Bird's foot trefoil		Retribution; revenge
Bittersweet		Truth
Blackberry	(bramble)	Death; envy; grief; injustice; 'I was too hasty, please forgive me'; lowliness; pain; remorse; weariness; wickedness
Blackthorn	(sloe)	Difficulty; obstacles; 'Our path is beset with difficulties'; keep promises
Bladder senna		Frivolous amusements
Bluebell		Constancy; everlasting love; humility; 'I am faithful'; kindness; solitude and regret
Bluebottle centaury		Delicacy
Borage		Bluntness; brusqueness; courage; energy; roughness of manners; talent; 'You have the courage of your own convictions'
Bouquet of flowers		Gallantry
Bouvardia		Enthusiasm
Box		Delicacy; firmness; stoicism
Bracken		Enchantment; 'You enthral me'
Brooklime		Fidelity

Broom		Ardour; devotion; humility; neatness; severity
Bryony		Prosperity
Bryony, black		'Be my support'
Buckbean		Calm repose; 'May sweet sleep attend you'
Bugloss		Falsehood
Burdock		Importunity; persistence; touch me not; 'Your suit is rejected'
Burning bush		Birth
Burr-marigold		Predictions
Buttercup		Charming; cheerfulness: childishness; 'I am dazzled by your charms'; ingratitude; radiance; riches; 'What golden beauty is yours'; 'You are immature, a heartless and ungrateful flirt'
Butterfly weed		'Let me go'
Cabbage		Gain; profit
Cacalia		Adulation
Cactus		Ardour; endurance; 'I Burn'; 'Our love shall endure'; warmth
Cactus, serpentine		Horror
Caladium	(angel wings, elephant ear)	Great joy and delight
Calamus	(sweet flag)	Fitness; resignation
Calceolaria		'Keep this for my sake'; money
Calycanthus	(allspice)	Benevolence; compassion
Camellia		'Beauty is your only attraction'; gratitude; 'How radiant is your beauty'; loveliness; 'My future lies in your hands'; perfection
Camellia, blue		'You are the flame in my heart'
Camellia, pink		'Longing for you'
Camellia, red		Unpretending excellence; 'You are a flame in my heart'
Camellia, white		Perfect loveliness; purity; unpretending excellence; 'You are adorable'
Camphor		Fragrance

Campion, meadow	Poverty; 'Though of humble station I dare to admire you from afar'
Campion, red	Encouragement
Campion, white	Evening
Canariensis	Self-esteem
Canary grass	Perseverance
Candytuft, perennial	Diffidence; 'I am indifferent to your wiles'; indifference
Canterbury Bell, blue	Aspiring; constancy; faithfulness; gratitude
Canterbury Bell, white	Acknowledgement
Cardinal flower	Distinction
Carnation	A mother's love; devoted love; fascination; pride and beauty; pure love; womanly love
Carnation, pink	Encouragement; 'I will never forget you'; 'Your charming token pleased me well'
Carnation, red	Admiration; 'I must see you soon'; 'My heart aches for you'; passionate love
Carnation, deep red	'Alas for my poor heart'; heartache; heart break
Carnation, purple	Capriciousness
Carnation, striped	Refusal
Carnation, white	Devotion; lovely; pure affection; pure and ardent love; sweet; woman's love
Carnation, yellow	Contempt; disappointment; disdain; 'You disappoint me'; rejection
Cashew nut	Perfume
Catalpa tree	Beware of the coquette
Catchfly	Snare
Catchfly, red	Youthful love
Catchfly, white	Betrayed
Cedar leaf	'I live for thee'
Cedar of Lebanon	Incorruptibility; faithfulness; strength
Celandine	Joys to come; 'Let this harbinger of spring speak to you of my love'; reawakening
Centaury	Delicacy; felicity

Cereus, creeping		Horror; modest genius
Cereus, night blooming		Transient beauty; true affection
Champignon		Suspicion
Chamomile		Energy in adversity; fortitude; 'I admire your courage, do not despair'; initiative; love in austerity
Chaste tree	(*Agnus castus*)	Apathy; chastity; coldness; indifference; to live without love; 'I prefer to remain a virgin than let you come near me'
Cherry, bird		Hope
Cherry, blossom		Increase; insincerity; 'May our friendship wax firm and true'
Cherry, tree		Good education; increase; spiritual beauty
Cherry, winter		Deception
Chervil		Sincerity
Chestnuts		Luxury
Chestnut leaves		'Please be just and fair in your dealings'
Chestnut tree		'Render me justice'
Chickweed		Ingenuousness; rendezvous
Chickweed, mouse-eared		Ingenuous simplicity
Chicory		Frugality
Chrysanthemum		A desolate heart; cheerfulness; cheerfulness in old age; cheerfulness under adversity; optimism; with love; 'You're a wonderful friend'
Chrysanthemum, bronze		Friendship
Chrysanthemum, red		'I love'; reciprocated love
Chrysanthemum, white		Trust; truth
Chrysanthemum, yellow		Dejection; discouragement; slighted love
Cineraria		Delight; ever bright
Cinnamon tree		Forgiveness of injuries

Circaea		A spell
Cinquefoil		Beloved daughter; maternal affection; sisterly affection
Citron		Beauty with ill humour
Cistus		Favour; safety; approval; 'You are acclaimed the queen of beauty'
Clarkia	(godetia)	Farewell to spring; pleasing; pleasure; 'Will you dance with me?'
Cleavers		Tenacity
Clematis		Artifice; intellectuality; 'I pay tribute to your brilliance and cleverness'; mental beauty
Clematis, evergreen		Poverty
Clotburr		Rudeness
Clove		Dignity
Clover, four-leaved		'Be mine'; petition
Clover, pink		'Do not trifle with my affections'; injured dignity
Clover, purple		Provident
Clover, red		Entreaty; industry; petition; 'Will you be faithful to me though oceans part us?'
Clover, white		Happiness; 'I'll be true to you'; promise; 'Think of me'
Coboea		Gossip
Colchicum	(meadow saffron)	Middle age; 'My best days are past'
Coltsfoot		Justice shall be done; maternal care
Columbine		Disloyalty; folly; madness
Columbine, purple		Resolution; resolve to win
Columbine, red		Anxious and trembling
Columbine, white		Constancy; desertion; folly
Corchorus		Impatience in absence
Coreopsis		Always cheerful
Coreopsis, Arkansas		Love at first sight
Coriander		Concealed merit; hidden worth
Corn	(maize)	Holy communion; plenty; riches

Corn, broken		Quarrel
Corn, sraw		Agreement
Corncockle		Gentility
Cornel		Duration
Cornflower		'A dweller in heavenly places'; 'Be not over impetuous', blessedness; delicacy; hope in love; 'My heart cannot be stormed'; delicacy; sensitivity
Coronella		'Success crown your wishes'
Cosmos		Beautiful
Costmary		Impatience
Cowslip	(herb Peter)	Charm; comliness; early joys; happiness; pensiveness; rusticity; winning grace; winsome beauty; 'You are sweeter even than this charming spring flower'; 'Your grace and beauty have charmed me completely'
Cowslip, American		Divine beauty; pensiveness; 'You are my angel, my divinity'
Cranberry		Cure for heartache; hardiness
Cranesbill		Envy; steadfast piety
Crepis, bearded		Protection
Cresses		Stability
Cress, Indian		Eclat; resignation; warlike trophy
Crocus		'Abuse not'; cheerfulness; youthful gladness
Crocus, autumn		'My best days are over'
Crocus, saffron		Beware of excess; cheerfulness; 'I rejoice in you'; marriage; mirth; smiles
Crocus, spring		Joy of youth; 'My heart beats with yours'; pleasures of hope; youthful gladness
Cross of Jerusalem		Devotion
Crownbill		Envy
Crown imperial		Majesty; power
Crowsfoot		Ingratitude
Crowsfoot, aconite leaved		Lustre

Cucumber	Extent
Cucumber, squirting	Criticism
Cudweed	Never-ceasing remembrance
Cumin	Cupidity
Currants, branch	'You please all'
Currants, flower	Presumption
Cuscuta	Meanness
Cyclamen	Diffidence; distrust; goodbye; 'I choose not to hear your protestations'; indifference; resignation
Cypress	Death; despair; eternal sorrow; mourning
Cypress and marigold	Despair
Daffodil	Deceitful hope; 'I do not return your affections'; 'My fond hopes have been dashed by your behaviour'; regard; regret; refusal; unrequited love
Daffodil, great yellow	Chivalry
Dahlia	Dignity and elegance; instability; 'My gratitude exceeds your care'; pomp; 'Thine forever'
Dahlia, red	Rebuff
Dahlia, single	Good taste
Dahlia, group	Instability
Dahlia, white	Dismissal
Dahlia, yellow	Distaste
Daisy, double	'I partake of your sentiments'; participation
Daisy, field	'I will give you my answer presently'; temporization
Daisy, moon	Autumn love
Daisy, ox-eye (marguerite)	A token; hope; obstacle; temporization
Daisy, part coloured	Beauty
Daisy, red	Unconscious thoughts
Daisy, white	'I will'; 'You have as many virtues as this plant has petals'
Daisy, wild	Adoration; faithfulness; innocence; virginity

Daisies, wreath of wild	'I will think of it'
Dandelion	Absurdity; bitterness; depart; faithfulness; grief; rustic oracle
Daphne	Painting the lily
Darnel	Vice
Dates	The faithful
Datura	Deceitful charms
Day lily	Coquetry; reviving pleasure
Day lily, yellow	Coquetry
Dead leaves	Melancholy; sadness
Dead nettle, red	Coolness
Dead nettle, white	Coldness
Devil's bit	Unfortunate love
Dew plant	Serenade
Diosma	Inutility; uselessness
Dittany	Birth
Dittany, white	Passion
Dock	Patience
Dodder of thyme	Baseness; business
Dogwood	Durability; 'Our love will endure adversity'
Dogwood, blossom	'I am perfectly indifferent to you'
Dragon plant	Snare; 'The betrayer'
Dragonwort	Horror
Ebony	Blackness; hypocrisy; 'You are hard'
Elder	Compassion; death; 'My efforts will remain unremitting'; sorrow; zealousness
Elder, berries	Love
Elm	Dignity; 'Your queenly bearing and elegance delight me'
Elm, American	Patriotism
Endive	Frugality
Escholtzia	Petition; sweetness

Eupatorium		Delay
Evergreen		Humility; poverty; 'Thoughts of you are my only comfort'
Evergreen thorn		Solace in adversity
Everlasting	(cotton weed)	Never-ceasing remembrance
Everlasting flower		'At your wish I go away, to forget you never'; death of hope
Fennel		Force; praiseworthy; strength
Fern		Sincerity; stormy passions; 'You have deprived me of your heart and left mine in a wilderness'
Fern, hart's tongue		Gossip
Fern, flowering		Fascination; reverie
Fern, maidenhair		Virginity
Fern, royal		Contemplation
Feverfew		'Let me shield you'; protection
Fig		Argument; longevity; prolific
Fig marigold		Idleness
Fir		Boldness; elevation; pity; time
Flaming Katie	(kalanchoe)	Endurance; lasting affection
Flax		Domestic industry; fate; 'I feel your kindness'
Flax, blue		Gratitude
Flax, dried		Utility
Flax, leaf		Tardiness
Flax, red		Fervour
Fleur de Lis		Eloquence; flame of love; hope; light; 'My compliments'; power; royalty
Flora's bell		'You make no pretensions'
Flower of an hour		Delicate beauty
Flower		Death of hope; never forget
Flowers gathered		'We die together'
Fool's parsley		Silliness

Forget-me-not	Constancy; forget me not; 'Here is the key to my heart'; remembrance; true love
Forsythia	Anticipation; good nature; innocence
Foxglove	A wish; insincerity; shallowness; youth
Foxtail grass	Sporting
Frankincense	The incense of a faithful heart
Fraxinella	Ardour; fire
Freesia	Friendship; innocence; trust
Fritillary	Doubt; persecution
Fuchsia	Frugality; 'Take heed, your beloved is false'; warning
Fuchsia, scarlet	Amiability; taste
Fumitory	Anger; ill at ease; spleen
Gardenia	Concealed love; 'Like unto this virgin flower you are'; 'My love for you is secret'; purity; refinement; secret love; sweetness; sweet love; you're lovely
Genista	Neatness
Gentian	'I love you best when you are sad'; virgin pride; 'You are unjust'
Gentian, yellow	Ingratitude
Geranium	Comfort; gentility; 'I prefer you'; 'I miss you so, please comfort me'
Geranium, apple	Present preference
Geranium, dark (night-scented)	Melancholy
Geranium, fish	Disappointed expectation
Geranium, ivy-leaved	Bridal favour; 'I engage you for the next dance'
Geranium, lemon	Unexpected meeting
Geranium, nutmeg	Expected meeting
Geranium, oak-leaved	'Lady, deign to smile'; true friendship
Geranium, pencil-leaved	Ingenuity

Geranium, pink rose-scented		Doubt; 'I await your explanation'; preference
Geranium, scarlet		Folly; comforting; consolation; duplicity; 'I do not trust you'; preference; stupidity
Geranium, silver-leaved		Recall
Geranium, sorrowful		Melancholy mind; sorrowful remembrance
Geranium, white		'I must ponder yet a while'; indecision
Geranium, wild	(herb Robert)	'I am your slave forever'; steadfast devotion; steadfast piety
Gilly flower	(gilliflower, stock)	Bonds of affection; lasting beauty; 'You are very dear to me'; happy life; contented existence; 'You will always be beautiful to me'
Gladiolus		Pain; strength of character; 'Your words have wounded me'
Glasswort		Pretension
Globe flower		Welcome
Glory flower		Glorious beauty
Gloxinia		Love at first sight
Glycine		'Your friendship is agreeable and pleasing to me'
Goat's rue		Reason
Golden rod		Encouragement; indecision; 'Let us take care that our love remains undiscovered'; precaution; success
Goldilocks, flax-leaved		Tardiness
Good King Henry		Goodness; 'You are the soul of kindness'
Gooseberry		Anticipation
Goosefoot, grass-leaved		'I declare war against you'
Gorse		Anger; enduring affection; love for all seasons
Gourd		Bulk; extent
Grapevine, wild		Charity
Grapevine, cultivated		Christ, our life; drunkenness; intoxication; mirth; prosperity; 'This is my blood'

Grass		Submission; utility
Guelder Rose	(snowball tree)	Age; autumn love; good news; 'He who weds me must be a man in his first youth, not one already in his dotage'; winter
Gum cistus		'I shall die tomorrow'
Harebell		Aspiring; 'Delicate and lonely as this flower'; grief; 'I bow to your will, but hope and sigh for you still'; resignation; submission; thankfulness
Hawkweed		Quick-sightedness
Hawthorn	(quickthorn, may)	'Despite your answer I shall strive for your love'; hope; (to the Romans) marriage; contentment
Hazel		Peace; reconciliation
Heartsease		Happiness; 'You occupy my thoughts'
Heath (heather)		Solitude
Heather, white		Good luck; 'We shall be lucky in our life together'
Helenium		Tears
Heliotrope, Peruvian		Devotion; 'Devoted to you'; faithfulness; 'Intoxicated with pleasure'
Heliotrope, winter		'Justice shall be done to you'
Hellebore	(Christmas rose)	Anxiety; calumny; female inconstancy; lying tongues; 'Relieve my anxiety'; scandal; wit
Hellebore, stinking		Calumny; scandal
Hemlock		Perfidy; scandal; 'You will be my death'
Hemp		Fate
Henbane		Fault; imperfection
Hepatica		Confidence
Herb Paris		Betrothal
Herb patience		Patience
Hibiscus		Delicate beauty
Holly		'Am I forgiven?'; 'Am I forgotten?'; domestic happiness; foresight; 'I thank God that you are restored to health' ; recovery; resurrection

Hollyherb	Enchantment
Hollyhock	Ambition; fecundity; fruitfulness; truthfulness; 'You inspire me to achieve great things'
Hollyhock, white	Female ambition
Honesty	Frankness; honesty; 'I hide nothing from you'
Honey flower	'Love sweet and secret'
Honeysuckle	Affection; bonds of love; fidelity; generous and devoted; 'My affection for you is boundless'; plighted troth; sweetness of disposition; 'This is a token of my love'
Honeysuckle, coral	The colour of my fate
Honeysuckle, French	Rustic beauty
Honeysuckle, monthly (woodbine)	Bonds of love; domestic happiness; 'I will not answer hastily'; loves' ties
Honeysuckle, wild	Inconstancy in love
Hop	Apathy; injustice
Horehound	Fire; frozen kindness
Hornbeam	An ornament
Hornberry	Ornament
Horse chestnut	Luxuriance
Hortensia	'You are cold'
Houseleek	Domestic industry; vivacity
Houstonia	Content
Hoya	Sculpture
Humble plant	Despondency
Hungry grass	Sporting
Hyacinth	Cheerfulness; games; play; sport
Hyacinth, blue	Constancy; dedication; devotion; fidelity; 'I shall devote my life to you'
Hyacinth, purple	Sorrow
Hyacinth, white	Admiration; 'I esteem you highly'

Hydrangea	A boaster; aloofness; arrogance; changeableness; heartlessness; 'Thank you for understanding'; vanity; 'Why are you so fickle?'; 'You are cold'; 'You are too cold and proud for my taste, a coquette'
Hyssop	Cleanliness
Ice plant	Idleness; rejected addresses; 'You freeze me'
Imperial Montague	Power
Ipomoea, scarlet	Attachment; 'I attach myself to you'
Iris	A message; 'I have a message for you'; my compliments
Iris, Fleur de Luce	Eloquence; flame of love; hope; light; my compliments; power; royalty
Iris, German	Fame; 'I burn'
Iris, purple	Ardour; 'You have set my heart aglow'
Iris, yellow	Flame of love; 'I mourn with you'; passion; sorrow
Ivy	Absurdity; ambition; amiability; bonds; elegance; fidelity; friendship; 'I desire you above all else'; 'I die where I cling'; immortality; marriage; tenacity; the resurrection; wedded love; 'We shall cling together in the spirit of fidelity and lasting friendship – nothing will separate us'
Ivy sprig with tendrils	Assiduous to please; 'Be my bride'; bonds
Jacob's ladder	Come down; fidelity
Jasmine	Amiability; cheerful and graceful; elegance; 'How dainty and elegant you are'
Jasmine, Cape	Spring; transport of joy
Jasmine, Carolina	Separation
Jasmine, Indian	'I attach myself to you'
Jasmine, Spanish	Sensuality; sexy
Jasmine, white	Amiability
Jasmine, yellow	Grace with elegance
Jonquil	Appeal; desire; 'I desire a return of affection'; 'Please answer me soon, dare I hope you love me?'

Judas tree		Betrayal; unbelief
Juniper		Asylum; comfort; 'Please let me care for you for ever'; perfect loveliness; protection; succour
Justica		'The perfection of female loveliness'
Kalmia		Treachery
Kennedia		Mental beauty
King cup		'Bestow on me the incomparable treasure of your love'; desire for riches
Laburnum		Forsaken; neglect; pensive beauty; 'Why have you forsaken me?'
Lady's mantle		Fashion
Lady's slipper		Capricious beauty; fickleness; 'Win me and wear me'
Lady's smock	(cardamine; cuckoo flower)	Paternal error
Lagerstraemia		Eloquence
Lantana, various colours		Rigour; sharpness
Larch		Audacity; boldness; disguise; 'Only he who presses his suit with spirit shall win me'
Larkspur		Lightness; levity; 'Read my heart'; swiftness; trifling
Larkspur, double		Haughtiness
Larkspur, pink		Fickleness
Laurel		Glory; 'Your excellence is unsurpassed'
Laurel, cherry in flower		Perfidy
Laurel, mountain		Ambition; glory
Laurel, spurge		Coquetry; desire to please
Laurotinus		A token; 'I die if neglected'
Lavender		Acknowledgement of love; assiduity; diligence; distrust; 'I can only ever be your friend'; negation; sad refusal; sweets to the sweet
Lavender, sea		Dauntlessness

Lemon		Zest
Lemon, blossom		Discretion; fidelity in love
Lettuce		Cold-hearted; coldness
Liatris		Enthusiasm
Lichen		Dejection; solitude
Lilac		Brotherly love; first emotion of love; forsaken; memory
Lilac, field		Humility
Lilac, purple		First love
Lilac, white		'A tribute to your beauty and spirituality'; innocence; majesty; modesty; purity; virginity; youth; youthful innocence
Lily, eucharis	(pineapple lily)	Maiden charm
Lily, imperial		Majesty
Lily, Japanese		'You cannot deceive me'
Lily, pink		Acceptance; prosperity; wealth; youth
Lily, scarlet		High soul
Lily, tiger		Erotic love; 'My passion burns like a firebrand'; passion; pride; prosperity; wealth
Lily, white		Chastity; 'I kiss your fingertips'; modesty; purity; sweetness; virtue; youthful innocence; youthful love
Lily, yellow		Falsehood; gaiety; walking on air
Lily of the valley		Devotion; 'Friendship is too precious, talk to me not of love'; humility; let's forgive; maidenly modesty; return of happiness; sweetness
Linden	(lime tree)	Conjugal love
Linden, American		'Can we not wed soon, matrimony would be bliss'; matrimony
Lint		'I feel all my obligations'
Liquorice, wild		'I declare against you'
Liverwort		Confidence
Lobelia		Malevolence
Lobelia, blue		Dislike

Lobelia, white		Rebuff
Locust tree		Elegance
Locust tree, green		Affection beyond the grave
London pride		Flirtation; frivolity
Loosestrife, purple		Forgiveness; 'Take this flower as a peace offering'
Loosestrife, yellow		'I am sorry, accept this flower as a token of regret'; peace
Lords and Ladies	(arum; wake Robin)	Ardour; 'My heart is aflame with passion'; zeal
Lotus		Eloquence; 'Our present estrangement makes me sad, but I think we should remain apart'
Lotus, flower		Estranged love; silence
Lotus, leaf		Estranged love
Love in a mist		'I am perplexed by your behaviour, what is it you want?'; perplexity; uncertainty
Love in a puzzle		Embarrassment
Love in idleness		Love at first sight
Love lies bleeding		'All hope of love is lost for you and me'; broken heart; desertion; hopeless, not heartless
Lupin		Dejection; 'I conquer all'; over-assertiveness; voraciousness over boldness; 'Who goes softly goes far'
Lychnis		Religious enthusiasm
Lychnis, scarlet		Sun-beaming eyes
Lythrum		Pretension
Madder		Calumny
Madwort, rock		Tranquillity
Magnolia		'Although you have broken my heart I shall persevere with dignity'; 'Be not discouraged, better days are coming'; fortitude; love of nature; magnificence; nobility
Magnolia, laurel-leaved		Dignity
Magnolia, swamp		Perseverance

Maidenhair	Discretion
Mallow	Mildness; sweet disposition
Mallow, marsh	Beneficence; humanity
Mallow, Syrian	Consumed by love; persuasion
Mallow, Venetian	Delicate beauty
Manchineal tree	Duplicity; falsehood; hypocrisy
Mandrake	Horror; rarity
Maple	Reserve
Marigold	Anxiety; chagrin; despair; foreboding; grief; honesty; sacred affection; uneasiness
Marigold, African	Boorishness; constancy in love; cruelty; love; misery; pain; 'Refinement appeals to me'; vulgar minds; 'Your jealousy is without foundation'
Marigold, French	Jealousy
Marigold, small Cape	Presage; omen
Marigold and cypress	Despair
Marjoram	Blushes; maidenly innocence; maidenly modesty; 'Your passion brings blushes to my cheeks'
Marvel of Peru	Timidity
Meadow saffron	'My best days are past'
Meadowsweet	Adornment; 'I seek a lover who is something more than merely decorative'; uselessness
Mercury	Goodness
Mesembryanthemum	Idleness
Mexican orange blossom (choisya)	Eternal love; fruitfulness; innocence; loveliness; marriage; purity
Mezereum	Coquette; desire to please
Michaelmas daisy	Afterthought; cheerfulness with old age; daintiness; farewell; love; patience
Mignonette	Dull virtues; 'Your qualities surpass your charms'
Milk vetch	'Your presence softens my pain'
Milkwort	Hermitage

Mimosa		Concealed love; courtesy; sensitive love; sensitivity; 'You are too brusque with my tender feelings'
Mint		'Find someone your own age'; homeliness; virtue; 'You are over-reacting to a small thing'
Mistletoe		'I overcome all obstacles'; 'I am determined to overcome difficulties in this life, if it cannot be as we wish, we will meet in the next world'; 'I send you my kisses, as many as stars'; kisses; overcome
Mock orange		Cancelled wedding; counterfeit; fraternal love
Moneywort		Fidelity
Monk's hood	(aconite; wolf's bane)	Beware, a deadly foe is near; crime; misanthropy; knight errantry; 'Your attentions are unwelcome'
Moonwort		Forgetfulness
Morning glory		Affectation; obstinacy; repose at night; worth sustained
Moschatell		Insignificance; weakness
Moss		'How sweet is the bond between mother and child'; maternal move; recluse
Moss, Iceland		Health
Mosses		Ennui
Motherwort		Concealed love
Mournful widow		'I have lost all'; lucklessness; unfortunate attachments
Moving plant		Agitation
Mugwort		Happiness; tranquillity
Mulberry, black		'I shall not survive you'
Mulberry, red		Wisdom
Mulberry, white		Wisdom
Mullein		Comfort; take courage
Mullein, white		Friendship; good nature
Mushroom		Distrust; suspicion
Mushroom, on turf		An upstart
Musk		Over-adornment

Mustard		Indifference
Myrrh		Gladness; 'How bitter and precious is our love'; lessons hard won
Myrtle		'Be mine forever'; everlasting love; fertility; 'Jealously I do return your love'; love's fragrance; with joy; Hebrew emblem of marriage
Narcissus, poet's		Egotism; self esteem
Narcissus, yellow		Conceit; deceitful hope; disdain; regard; regret; refusal; self-obsession; 'You love none save yourself'
Nasturtium		Artifice; 'Beauty unadorned I seek'; conquest; optimism; patriotism; warlike trophy; victory in battle
Nasturtium, scarlet		Splendour
Nemophila		Prosperity
Nettle		Coolness; cruelty; pain; slander; 'You are cruel'; 'Your are spiteful'
Nettle, stinging		Slander
Nettle tree		Concert; plan
Nightshade		Dark thoughts; deception; falsehood; truth
Nightshade, deadly	(belladonna)	Deception; imagination; loneliness; silence; truth
Nightshade, enchanter's		Double dealing; fascination; scepticism; witchcraft; 'You have bewitched my heart with your charms'
Nightshade, woody	(bittersweet)	Truth
Nosegay		Gallantry
Oak		Hospitality; 'Your face and person will always be welcome at my door'; endurance and triumph
Oak, leaves		Bravery; courage and humanity; 'Take heart, love will find a way'
Oak, white		Independence
Oak and holly		Hospitality
Oats		The witching soul of music; 'Your voice is music to my ears'

Oleander		Beware; 'Beware others interfering in your affairs'; warning
Olive		Goodwill; peace
Ophrys, bee		Error
Ophrys, frog		Disgust
Ophrys, spider		Adroitness; skill
Orange flower, (blossom)		Bridal festivity; chastity; 'I greet you as your bride'; loveliness; purity; virginity; 'Your purity equals your loveliness'
Orange, tree		Generosity
Orchid		A belle; beauty; beautiful lady; flower of magnificence; love; luxury; pure affection; refinement; 'You are beautiful'
Orchid, bee		Industry; misunderstanding
Orchid, butterfly		Gaiety
Orchid, fly		Deceit; error
Orchid, lady's slipper		Pensive beauty
Osier		Frankness
Osmunda		Dreams; 'Never shall I escape from your enchantment'
Ox-eye		Obstacle; patience
Paeony	.	Bashfulness; bashful shame; 'Beauty is in the heart, not in the face'; contrition; happy marriage; 'Please forgive my brusqueness'; prosperity
Palm		Martyrdom; victory
Pansy	(heartsease; love in idleness)	'Kind thoughts of you'; merriment; tender thoughts; thoughtful reflection; virtue; 'You occupy my thoughts'
Pansy, purple		Memories; souvenirs; 'The thought of happy days spent together with you are my greatest treasure'
Pansy, white		'I cherish loving thoughts of you'; thoughts of love
Pansy, yellow		'Oceans part us but my heart stays with you'; remembrance
Parsley		Entertainment; feast or banquet; festivity; to win; useful knowledge

Pasque flower		Denial; expectation; sickness; 'You are without pretension'; 'You have no claims'
Passion flower		Belief; consecration; 'I am pledged to another'; religious fervour; religious superstition
Patience dock	(herb patience)	Patience
Pea, perennial		An appointed meeting; 'I long for a tryst'; lasting pleasure
Pea, pweet		Delicate pleasures; departure
Peach blossom		'I am your captive'
Pear		Affection; comfort; fairies' flower
Pelargonium		Acquaintance
Penny royal		Flee away; 'Leave me quickly'
Peppermint		Cordiality; warmth of feeling
Pepper plant		Satire
Periwinkle		First love; 'My heart was mine until we met'
Periwinkle, blue		Early friendship; sweet remembrances; tender recollections
Periwinkle, white		Pleasures of memory
Persicaria		Restoration
Persimmon		'Bury me amid nature's beauty'
Petunia		'Always stay by my side'; do not despair; proximity; 'We must bravely face the despair caused by our separation'
Phlox		Agreement; 'I think we could be friends'; our souls are united; unanimity
Phlox, pink		Friendship
Phlox, white		Interest
Pimpernel		Assignation; change; childhood; faithfulness; meeting; opportunity of a rendezvous
Pine, black		Hope; pity
Pine, pitch		Philosophy; time
Pine, Scots		Elevation; 'You are the sun of my life, which makes all things real'
Pineapple		'You are perfect'; welcome

Pink		Always lovely; boldness; 'No matter what the years may bring for me your beauty will never die'; perfection; woman's love
Pink, clove-scented		Dignity; fragrance; 'How sweet you are'
Pink, double red		Pure and ardent love
Pink, Indian		Always lovely; aversion
Pink, mountain		Aspiring
Pink, single red		Woman's love
Pink, striped or variegated		Refusal
Pink, white		Ingeniousness; talent
Plane tree		Genius; magnificence; shelter
Plum, myrobalan		Bereavement; privation
Plum tree		Fidelity; independence; promises to keep
Poinsettia	(Christmas star, Mexican flame leaf)	Good cheer; success; 'Wishing you mirth and celebration'
Polyanthus		'My spirit cleaves to yours'; pride of riches
Polyanthus, crimson		'The heart's mystery'
Polyanthus, lilac		Confidence
Pomegranate		Conceit; elegance; foolishness
Pomegranate, flower		Mature elegance
Poplar, black		Courage; funeral tree; 'Let not your heart be troubled, all will be well'; loss of hope
Poplar, white		Lamentation; 'Our love is timeless'; time
Poppy		Evanescent pleasure; eternal sleep; imagination; pleasure; wealth
Poppy, oriental		'My heart aspires in silence to thee'; silence
Poppy, pink		'May sweet sleep attend you, sweetest dreams beguile your slumbers'; sleep
Poppy, red		'Be cheered, you still have me'; consolation to the sick; moderation; oblivion
Poppy, scarlet		Fantastic extravagance
Poppy, white		Antidote; dreams and fantasies; 'I need time to consider'; my bane; sleep; sleep of the heart; temporization

Potato		Benevolence
Potentilla		'I claim, at least, your esteem'
Prickly pear		Satire
Pride of China		Dissension
Primrose		Believe me; dawning love; early youth; fears; grace; I cannot live without you; inconstancy; lovers' doubts; 'My heart is beginning to know you'; pensiveness; sadness; 'Your inconstancy saddens me'
Primrose, evening		Inconstancy; 'Humbly I adore you'; mute devotion; silent love; uncertainty
Primrose, red		Unpatronized merit
Primula		Diffidence; 'I accept your gift with thanks'
Privet		Defence; mildness; prohibition
Protea		Courage; diversity; don't despair
Pumpkin		Bulkiness
Pyracantha		Solace in adversity
Pyrus japonica	(faeries' fire)	'You are mysteriously lovely, as nature's wild heart'
Quaking grass		Agitation
Quamoclet		Busybody
Queen Anne's lace	(fool's parsley)	Festivity; folly; haven; sanctuary; self-reliance; silliness
Queen's rocket		Fashionable; 'You are the Queen of coquettes'
Quince		Temptation; 'Your charms are more than I can resist'; 'I will foreswear all others to be with you always'
Ragged Robin		Wit
Ranunculus		'I am dazzled by your charms'
Ranunculus, asiatis		'Your charms are resplendent'
Ranunculus, garden		'You are rich in attractions'
Ranunculus, wild		Ingratitude
Raspberry		Envy; remorse
Red bay		Love's memory

Reed, feathery		Indiscretion
Reed, flowering		Confidence in heaven
Reeds, bundle with panicles		Music; the spirit of music; 'You are the sweet inspiration which invests my life with harmony'
Reed, split		'Be careful, others are aware of our indiscretion'
Reedmace	(bullrush)	Docility; indiscretion; rashness
Restharrow		Hindrance
Rhododendron		Danger; forbearance; moderation; temperance
Rhubarb		Advice
Rocket		Deceit; rivalry
Rosa montiflora		Grace
Rosa mundi		Variety
Rose		Genteel; incorruption; love; pretty; silence; 'The loving heart of humanity'
Rose, Austrian		'Thou art all that is lovely'; very lovely
Rose, bridal		Happy love
Rose, burgundy		Pensive beauty; simplicity; unconscious beauty
Rose, cabbage		Ambassador of love
Rose, campion		'Only deserve my love'
Rose, Carolina		Love is dangerous
Rose, China	(monthly rose)	'Beauty always new'; departure; 'Everlasting loveliness' ; transient brilliance
Rose, daily		A smile; 'Thy smile I aspire to'
Rose, damask		Beauty ever new; brilliant complexion; maidenly blushes
Rose, deep red		Bashful love
Rose, dog		Maidenly beauty; pleasure mixed with pain; simplicity; 'You are as fair and innocent as this pure bloom'; 'You have enchanted me'

Rose eglantine	(sweet briar)	'I wound to heal'; fragrance; poetry; 'The fragrance of this flower brings sweet memories of you'
Rose eglantine, full-blown		Simplicity
Rose eglantine, yellow		Decrease of love
Rose, faded		Beauty is fleeting
Rose, full-blown		Beauty; 'I love you'
Rose, full-blown (placed over two buds)		Secrecy
Rose, hundred-leaved		Grace; pride
Rose, Japan		Beauty; 'Beauty is your only attraction'
Rose, lavender		'Our love is pure'; purity of heart
Rose, maiden's-blush		'If you love me you will find it out'
Rose, May		Precocity
Rose, moss		Ecstasy of enjoyment; pleasure without pain; shy love; voluptuous love
Rose, musk (single)		Affection; capricious beauty; capricious love; glad grace
Rose, musk (cluster)		'You are charming'
Rose, pink		Chaste love; friendship; 'Our love is perfect happiness'; 'To my friend'
Rose, Provence		'My heart is in flames'
Rose, pompon		Genteel; pretty
Rose, red		Desire; 'I love you'; love; passionate love
Rose, red (withered)		'Our love is over'
Rose, red-leaved		Beauty and prosperity; wishes for prosperity
Rose, single		Simplicity
Rose, thornless		Early attachment
Rose, unique		'Call me not beautiful'
Rose, white		Candour; ever beautiful; 'I am not worthy of you'; refusal; silence
Rose, white (dried)		'Death preferable to a loss of innocence'
Rose, white (withered)		'I would rather die'; transient impression

Rose, wild	Simplicity
Rose, yellow	Decrease of love; jealousy; infidelity; misplaced affection
Rose, Yorks and Lancs	War
Rose leaf	I am never importunate; you may hope
Rosebay willowherb	Celibacy; pretension
Rosebud	'I confess my love'; pure innocent love
Rosebud, with thorns and leaves	'I fear, but I hope'
Rosebud, stripped of leaves	'I am afraid'
Rosebud, stripped of thorns	'I am hopeful'
Rosebud, moss	Confession of love
Rosebud, red	Pure and lovely; 'You are young and beautiful'
Rosebud, white	Girlhood; charm and innocence; 'Heart ignorant of love'; 'Her heart knows naught of love'; 'Too young to love'
Roses, basket of	'Take the treasures of my heart'
Roses, crown of	Award of merit; reward of virtue
Roses, garland of	Life-long blessing
Roses, white and red together	Unity; warm of heart
Roses, wreath of	Beauty and virtue rewarded
Rosemary	Fidelity in love; gladness of spirit; remembrance; 'Your cherished memory will never fade from my heart'; 'Your presence revives me'
Rowan	'Love should bring us wisdom though our hearts are wild with ardour'; prudence; tree of life
Rudbeckia	Justice
Rue	Disdain; docility; domestic happiness; 'Do not annoy me with your unwelcome attentions'; grace; purification; repentance
Rush	Docility; meekness; 'Your sweet, childlike heart has utterly captivated me'

Safflower		'Do not abuse'
Sage, garden		Domestic virtue; esteem; household; long life; 'We have a wonderful family'; wisdom
Sainfoin		Agitation
St John's wort		Animosity; oblivion of life's troubles; superstition; superstitious sanctity
Sardony		Irony
Satin flower		Sincerity
Saxifrage		Humility; 'Only smile at me and my reward will be great'
Scabious		Unfortunate love
Scabious, sweet (or Indian)		'I have lost all'; 'Widowhood'
Scorpion grass, mouse-eared		Forget me not
Sea holly	(eryngo)	Austerity; independence; sternness
Sensitive plant		Bashfulness; sensitivity
Senvy		Indifference
Service tree		'From corruption, sweetness'; prudence
Shamrock		Light-heartedness
Shepherd's purse		'I offer you all'
Snakesfoot		Horror
Snake's lounge		Slander
Snapdragon		Deception; gracious lady; 'I cannot care for you'; presumption; refusal
Snowdrop		'Accept my consolation'; consolation; hope in sorrow; 'I make a fresh bid for your affections'; indiscretion; 'I wish you well for the future'; no!; purity; renewal
Solomon's seal		Concealment; discretion
Sorrel		Paternal affection; with affection
Sorrel, wild		Wit, ill-timed
Sorrel, wood		Joy; maternal tenderness
Southernwood		Bantering; jest
Spearmint		Warmth of sentiment

Speedwell		Female fidelity; 'I remain faithful'; resemblance; true love
Speedwell, germander		Facility
Speedwell, spiked		Semblance
Spider flower	(cleome)	Elope with me
Spiderwort		Esteem, but not love; transient happiness
Spindle		'Your charms are engraven on my heart'
Spruce		Boldness; fidelity; farewell; pity
Star of Bethlehem		Guidance; purity; reconciliation
Starwort, American		Cheerfulness in old age
Statice	(sea lavender)	Remembrance; sympathy
Stephanotis		Happiness in marriage; desire to travel; 'Will you accompany me to the Far East?'
Stitchwort		Hermitage
Stitchwort, greater		Afterthought
Stocks, ten-week		Promptness
Stonecrop		Tranquillity
Stramonium		Disguise
Strawberry		Be on the alert; foresight; perfection; 'I esteem you as a friend but not as a lover'
Strawberry tree		Esteemed love
Straw, broken		Dissension; rupture of contract
Sumach		Splendour
Sunflower, dwarf		Adoration
Sunflower, tall		False riches; haughtiness; lofty and pure thoughts; longevity; ostentation; pride; sunshine; 'That which glitters is not always precious'; warmth; your devoted admirer
Swallow wort		Cure for heartache; medicine
Sweet gale		Encouragement
Sweet pea		A meeting; blissful pleasure; delicate pleasure; departure; goodbye; 'Only in your departure will I find any pleasure'; 'Thank you for a lovely time'; tenderness; 'Your memory is a lingering fragrance'

Sweet sultan	Felicity; happiness; 'This is to wish you happiness'
Sweet Sultan, flower alone	Widowhood
Sweet vernal grass	Poor but happy
Sweet William	A smile; coquetry; craftiness; gallantry; 'Grant me one smile'; flirtation; 'I was only teasing you'
Sycamore	Curiosity; 'I wish to know more of you'; reserve; woodland beauty
Syringa, Carolina	Disappointment
Tamarisk	Crime
Tansy	'I declare war against you'; reflected addresses; refusal; resistance; 'Your feelings are not reciprocated'
Tarragon	Share
Teasel	Importunity; misanthropy
Tendrils of climbing plants	Ties
Thistle, common	Austerity; independence
Thistle, Scottish	Retaliation
Thornapple	Deceitful charms
Thorns, branch of	Severity or rigour
Thorns, flowering branch	Charms and enchantment; 'I worship you'
Thrift	Dauntlessness; interest; sympathy
Throatwort	Neglected beauty
Thyme	Activity; affection; courage; domestic virtue; energy; 'I need a wife as capable as you'; strength
Tiger flower	'For once may pride befriend me'
Toadflax	'Be more gentle in your wooing'; presumption; reluctant lips
Toothwort	Secret love
Tradescantia	Esteem not love; momentary happiness

Traveller's joy	Middle-aged love; safety; mature love; 'Though your youth has gone my love is still strong'
Tree of life	Old age
Trefoil	Revenge; unity
Tremella nestoe	Opposition; resistance
Truffle	Surprise
Trumpet flower	Flame of love; 'My heart is aflame for you'
Trumpet flower, ash-leaved	Separation
Tuberose	Dangerous pleasures; 'I have fluttered near the flame and singed my wings'; wounding
Tulip	Avowal; consuming love; eloquence; fame; romance; perfect lover; fame; elegance; grace
Tulip, purple	Undying love
Tulip, red	Declaration of love; believe me; passion
Tulip, variegated	Beautiful eyes
Tulip, white	Forgiveness
Tulip, yellow	Hopeless love; sunshine in your smile; cheerful thoughts
Tulip tree	Fame; rural happiness
Turnip	Charity
Tussilago, sweet-scented	'You shall have justice'
Valerian	Accommodating disposition; concealed merit; 'Conscious of my lowliness, I aspire none the less to wed you'; merit in disguise
Valerian, Greek blue-flowered	Rupture; warfare
Venus flytrap	Deceit; 'Fly with me'
Venus's looking glass	Flattery
Verbena	Enthralment; sensibility; 'You have cast a spell over me'; 'You have my confidence'; 'Will you get your wish?'
Verbena, pink	Family unity
Verbena, scarlet	Church unity; unite against evil

Verbena, white	Pray for me
Veronica (speedwell)	Female fidelity; 'Nothing shall part us'; true love
Vervain	Enchantment; purity; reconciliation; superstition; witchcraft; 'You have stolen away my soul'
Vetch bush	Shyness
Vine	Drunkenness; Christ our life; mirth; intoxication
Violet	Modesty; 'Pure and sweet art thou'
Violet, blue	Faithfulness
Violet, dame	Watchfulness; 'You are the queen of coquettes'
Violet, dog's	Lad's love; 'You are my first sweetheart'
Violet, purple	You occupy my thoughts
Violet, sweet	Constancy; faithfulness; modesty
Violet, small white	Candour; innocence
Violet, white	Purity of sentiment
Violet, yellow	Rare worth; rural happiness
Virginia creeper	Ever changing; 'I am spellbound by your infinite variety'; 'I will cling to you both in sunshine and in shade'
Virgin's bower	Artifice; filial love
Viscaria oculate	Will you dance with me?
Volkamenica japonica	'May you be happy'
Wallflower	Constancy; fidelity in adversity; misfortune; 'Through sunshine and storm I am true to you'
Wallflower, garden	Lasting beauty
Walnut	Intellect; 'Ours would be a marriage of true minds'; stratagem
Waterlily	Eloquence; purity of heart; silence; 'The light of your spirit ever shines through'
Watermelon	Bulkiness
Wax plant	Susceptibility
Wheat	Fruitfulness; 'I offer you all I have'; riches

Whin	(gorse)	Anger; enduring affections; love for all seasons
Whortleberry		Ingenuous simplicity; treason; treachery; 'You are faithless'
Willow		'Be mine again'; forsaken
Willow, creeping		Love forsaken
Willow, French		Bravery and humanity
Willow, water		Freedom
Willow, weeping		Melancholy; mourning
Willowherb		Pretension
Wisteria		Fair stranger; 'I cling to thee'; need; welcome; 'Meeting you means so much to me'
Witchhazel		A spell; spellbound
Woodbine		Fraternal love
Woodruff		Modest worth
Wormwood		Absence; 'Even the best of friends must say farewell'; heartache; sorrowful parting
Xanthium	(clot bur)	Rudeness
Xeranthermum		Cheerfulness in adversity; 'Never say die'
Yarrow		Elegance; friendship; war
Yew		Death; 'My heart feels with yours'; sorrow
Zephyr flower		Sadness
Zinnia		'I mourn your absence'; thoughts of absent friends; 'Where there is love there can be no separation'

Traditional Rules for Gathering Magical Plants

These are the traditional rules associated with gathering herbs for magical purposes. They were to be observed if the plant was to retain all of its potency of the spell or other usage to which it was to be put.

- Pick the plant with the left hand only, never the right.
- Never face into the wind.
- Never look behind you, or look back having gathered your plant.
- Never touch the plant with 'cold iron'. Consequently, knives, saws, scissors etc. must not be used in gathering plants for magical purposes.
- Do not let any blade touch the roots.
- Once cut do not let any part of the plant come into contact with the ground.
- Dress in clean white linen clothes, and no shoes, or simply wear no clothes at all.
- Herbs are best if gathered, in the appropriate phase of the moon, at night (at sunset or midnight), or at sunrise.
- More power will be obtained from the plant if you undergo a period of fasting and abstain from sexual intercourse for a day or two prior to going to collect the plant.
- Finally, you must speak to the herb and let it know of your intentions before cutting it.

There may be specific recommendations or requirement when gathering specific plants, or material for certain uses.

Plants in Dreams

To dream of		**Indicates**
Fir	*Abies*	Being in a fir forest is a sign of suffering.
Sycamore	*Acer*	Sycamore foretells jealousy to the married and marriage to the single.
Yarrow	*Achillea*	Picking yarrow for medicinal purposes means that you will soon receive good news.
Onions	*Allium*	Eating onions means that, within the domestic scene, secrets will be found out or else betrayed and many lies told. Fried onions indicate that a friend is ill but will recover. Onions signifies that you will have troubles and sorrow. Onions in any form are an ill omen indicating the loss of property and hard work combined with many worries.
Garlic	*Allium sativum*	Eating garlic (for a man) shows that he will discover hidden secrets, and meet some domestic problems. Having garlic in the home is a sign of good fortune.
Anemone	*Anemone*	Anemone suggests that a lover is unfaithful and should be spurned in favour of finding another.
Peanut	*Arachis*	Peanuts show that you will always be poor.
Wormwood	*Artemisia*	Wormwood is a good omen.
Arum Lily	*Arum*	Arum lily betokens an unhappy marriage.
Daisy	*Bellis*	Daisy in the spring and summer is a sign of good luck, but to dream of it in the autumn is an omen of misfortune.
Cabbage	*Brassica*	Cabbage growing is a sign of good fortune. Eating cabbage indicates the sickness of a loved one or the loss of money. Cutting cabbages indicates jealousy on the part of a husband/wife/partner. Cabbage indicates that friends are sick and there is little hope of their recovery. Eating cabbage shows that you will experience sorrow.

To dream of		Indicates
Turnip	*Brassica rapa*	Eating turnips means you will have family troubles. Turnips denote fruitless toil.
Box	*Buxus*	Box indicates a long life, prosperity and a happy marriage.
Marigold	*Calendula*	Marigolds means prosperity, success and a happy marriage. Failure in business, or coming wealth.
Thistle	*Carduus*	Being surrounded by thistles is a lucky omen, promising pleasant events shortly. May mean someone in whom you place implicit trust is disloyal to you.
Lemon	*Citrus*	Lemons indicates quarrels between man and wife, or the breaking off of an engagement.
Orange	*Citrus*	Oranges on a tree are an excellent omen. Picked oranges are always an ill omen. They indicate the loss of fidelity of a lover. Beware of placing implicit trust in a chance acquaintance.
Hazelnut	*Corylus*	Discovering hazelnuts predicts finding treasure. Cracking and eating hazelnuts means riches and contentment after toil.
Hawthorn	*Crataegus*	Hawthorn indicates a new lover.
Crocus	*Crocus*	Crocus indicates dangers in love. You should not trust the dark man who has attracted you.
Melon	*Cucumis melo*	For a woman, melons shows she will marry a foreigner and live in a foreign land. For a young man they indicate that he will wed a wealthy foreign woman and have a large family by her but that they will die young. For a sick person to dream of melons is a promise of recovery. A journey will bring much profit.
Cucumber	*Cucumis*	A green cucumber indicates someone will die. Cucumber indicates that an invalid will recover, and that you may then fall in love, or if in love already may wed. Cucumber promises success in trade, and to the sailor or traveller – a pleasant journey.
Cypress	*Cupressus*	Cypress foretells afflictions and obstruction in business.
Quince	*Cydonia*	Quince indicates a speedy release from troubles and sickness.

To dream of		Indicates
Carrot	*Daucus*	Carrot is a sign of profit and strength to those who are at law for an inheritance.
Carnation	*Dianthus*	Carnation indicates a passionate love affair or a secret admirer.
Wallflowers	*Erysimum*	Wallflowers tell a lover that his/her sweetheart is faithful. They tells an invalid that they will soon recover. For a woman, picking wallflowers for a bouquet means that the best of her admirers is yet to propose to her.
Fig	*Ficus*	Figs means health, prosperity and happiness, especially in old age. They are also said to indicate the realization of wishes and the inheritance of wealth through the form of a lottery or a will.
Flower		Flowers are bad luck, as they may indicate a death in the neighbourhood. Planting flowers is a sign of a death. Gathering flowers is a sign of a lasting friendship. Gathering flowers indicates a delightful surprise. A basket of flowers indicates a birth or a wedding. A wreath of flowers is a sign of a new love. Withered flowers is a sign of good fortune. Smelling flowers indicates that you must grasp an opportunity. Picking flowers is a sign of good luck. Bright flowers indicates a pleasant life. Casting flowers away signifies quarrels and despair. Flowers out of season mean bad luck.
Strawberry	*Fragaria*	They mean a new love and a happy marriage. They mean a visit to the country with someone who loves you.
Forest		Being lost in a forest is lucky, meaning success and prosperity. Viewing one from a distance indicates a loss of money.
Snowdrop	*Galanthus*	Snowdrops mean you should not conceal your secret. You will be happier if you confide in someone. Or simply that you have secrets to share.
Garden		A nice garden growing is a sign of a death. Garden growing well is an ill omen. Being led into a garden of flowers by someone you know who is dead portends a death.

To dream of		Indicates
		Being in a garden means that joy is coming to you. Garden means marriage to a beautiful woman or a handsome man. Gardens are a blessing for your spirit and indicate philanthropy.
Geranium	*Geranium*	Geranium indicates that there is no need to worry as your quarrel meant nothing and will soon be mended.
Grain		A large grain field foretells a happy marriage. Grain in abundance means respect and much honour coming to the dreamer; their household will be filled with joy and prosperity.
Grass		Green grass indicates that you will receive money. Fresh green grass is a good omen as it indicates long life, good luck and great wealth. Withered grass portends sickness and possible death of loved ones. Cutting grass portends great troubles. You will not succeed in the country and stand a better chance in the city.
Harvest		Harvesting shows prosperity for those engaged in business, and domestic concord. (One of the most favourable of dreams).
Ivy	*Hedera*	Ivy is lucky as it indicates that friendships will follow, happiness and good fortune, honour and success. It foretells the aid and comfort of a faithful friend.
Herb		Herbs means prosperity.
Bluebell	*Hyacinthoides*	Bluebell is an ill omen as it indicates a nagging spouse. It can indicate a stormy but passionate affair.
Hyssop	*Hyssopus*	Hyssop means that friend will bring you peace and happiness.
Iris	*Iris*	Iris means that you receive a letter bearing news.
Jasmine	*Jasminum*	Jasmine is lucky, especially for lovers.
Walnut	*Juglans*	Opening and eating walnuts signifies that you will receive money. Walnuts indicate worry, disappointments, rebukes and general upheavals.

To dream of		Indicates
Juniper	*Juniperus*	To dream of junipers is usually considered to be an indication of impending bad luck. However, a dream about juniper berries signifies good luck as it indicates coming success or the birth of an heir.
Lettuce	*Lactuca*	To dream of eating lettuce leaves is an omen of misfortunes ahead, indicating difficulties in the management of all your affairs.
Bay	*Laurus*	Bay laurel means achievement, great renown, fame.
Lily	*Lilium*	Lilies, in or out of season, are a sign of good luck. Lilies in flower foretell marriage, happiness and prosperity. Withered lilies indicate frustration of hopes and the illness or death of a loved one.
Honeysuckle	*Lonicera*	Honeysuckle indicates domestic troubles that will cause you sorrows, also tears and smiles.
Honesty	*Lunaria*	Honesty means luck in money.
Apple	*Malus*	An apple on the tree foretells prosperity. Seeing apples on a tree indicates prosperity and picking ripe apples off a tree and eating them is an omen of good fortune. Apples indicate a long life, success in trade and a faithful lover. Red apples means a gift of money coming within a few days. Green apples indicate a gift of money will be receives in several months time. Golden apples portend great wealth but also domestic strife. Apple blossom indicates a birth.
Mulberry	*Morus*	Mulberry shows that through weakness and indecision you will lose a friend.
Mushroom		Mushrooms prophesy fleeting happiness. Picking mushrooms indicates a lack of attachment on the part of a lover or spouse.
Forget-me-nots	*Myosotis*	Forget-me-nots suggest that you must be firm and break with your love as he or she is unsuitable, you must find a new lover.
Nutmeg	*Myristica*	Nutmeg means that you will rise to high position. A sign of many impending changes (for the better).
Myrtle	*Myrtus*	Myrtle promises many lovers and a legacy. In a married person it foretells a second marriage, especially if a similar dream is repeated.

To dream of		Indicates
Daffodil	*Narcissus*	Daffodils suggest that if you have been unjust to a friend you must seek a reconciliation. If you have neglected a friendship, you must make amends.
Nut		Eating nuts means that great happiness will come to you. It also indicates that you will become poorer by your own extravagance. Nuts growing show that you will have a rich marriage partner.
Olive	*Olea*	Gathering olives, or seeing them on the tree, is a sign of peace and happiness. It is especially favourable for sick people. Eating olives indicates a rise in position and the receipt of a favourable present.
Orchard		An orchard means that you will receive a large amount of money.
Paeony	*Paeonia*	Paeony means that your excessive modesty may cause you sorrows. You must grasp an opportunity.
Poppy	*Papaver*	Poppies indicate a message bringing you great disappointment.
Bean	*Phaseolus*	Picking green beans portends a death in the family. Beans, under any circumstances, indicate trouble of some kind.
Date palm	*Phoenix*	Palms are a good omen, particularly if in full bloom, when the dream predicts success and good fortune. It can betoken a journey to a foreign land, but if the tree is withered it will be a hazardous journey.
Pine	*Pinus*	Pine foretells dissolution, bad luck and possibly death.
Pea	*Pisum*	Eating peas is unlucky, especially if they are hard, as it denotes straitened circumstances and faithlessness in friends. If they are seen growing you may expect your enterprises to succeed.
Primrose	*Primula*	Primrose indicates that you will find happiness in a new friendship.

To dream of		Indicates
Cherry	*Prunus*	Picking and eating cherries means financial gain will follow. Cherries indicate inconstancy and disappointments in life, and unhappy circumstances in love affairs. Picking cherries and throwing them away indicates that you will suffer financial loss.
Plum	*Prunus*	Plums are a sign of a death in the family or sickness of a friend or relation. They are also a warning of ill health, losses, infidelity and much vexation in the married state.
Peach	*Prunus*	Peaches mean pleasure, wealth, contentment, reciprocated love, good health and many pleasant surprises.
Apricot	*Prunus*	Apricots denote health, a speedy marriage and every success in life. For a married person they indicate dutiful children.
Almond	*Prunus dulcis*	Eating almonds means a journey; if sweet almonds it will be prosperous, if bitter almonds it will be the opposite.
Pomegranate	*Punica*	Pomegranates are an omen of good fortune, happiness and success, to a lover a faithful and accomplished sweetheart, and to a married person, wealth, children and success in business.
Pear	*Pyrus*	Ripe pears promise riches and happiness. Unripe pears mean adversity. If baked, they indicate great success in business. To a woman it means that she will marry above her rank, have an abundance in business and a new friendship.
Oak	*Quercus*	Green oak signifies a long and happy life, calm and untroubled. Withered oak foretells poverty in old age. Thriving oaks promise male children who will win distinction. Oaks bearing acorns mean wealth. A blasted oak means death. If the tree falls you will never know prosperity.
Buttercup	*Ranunculus*	Buttercups signify that any business enterprise you are involved in will succeed.
Currant	*Ribes*	Currants promise happiness in life, success in all undertakings, constancy in loved ones and riches to a farmer or trade person.

To dream of		Indicates
Gooseberry	*Ribes*	For a sailor gooseberries indicate dangers on the next voyage. For a young girl they indicate an unfaithful husband. Beware a rival.
Rose	*Rosa*	Roses show that you will always have many friends. For a lover they indicate a happy marriage. For a farmer prosperity and ultimately independence. They foretell a wedding, perhaps your own, within a year. They also indicate that your heart's desire will come to you. Withered roses indicate a decay of fortunes and disappointments. Rosebuds mean many blessings.
Rosemary	*Rosmarinus*	Rosemary means that everlasting love will be yours.
Raspberry	*Rubus*	Raspberries indicate success and happiness in marriage, fidelity in a sweetheart and good news from abroad. They also mean great disappointment but consolation from an unexpected source.
Blackberry	*Rubus*	Blackberries presage a death or bad luck. They show that you will soon have troubles. Passing through places covered in brambles portends trouble; if they prick you then secret enemies will do you injury, but if you pass through unhurt you will triumph over your enemies. If they draw blood, expect heavy losses in trade.
Rushes		Rushes show that you must beware of a stranger; someone will try to win your confidence for their own ends; you must be discreet.
Willow	*Salix*	Mourning over some calamity beneath a willow tree is a happy omen and indicates good news.
Sage	*Salvia*	Sage indicates that you will marry.
Elder	Sambucus	Elder means sickness.
Potato	*Solanum*	Potatoes indicate that you have secret enemies of whom you are aware. Peeling potatoes and also seeing money indicates trouble in two days' time. A row of potatoes, with the potatoes protruding, is a sure sign of death. Digging for potatoes denotes gain. Eating them is an omen of heavy losses and dishonesty in business.

To dream of		Indicates
Straw		Drying straw is unlucky as it betokens a loss of money. Wet straw indicates a profitable journey.
Lilac	*Syringa*	Lilac means luck in love.
Dandelion	*Taraxacum*	Dandelions are an omen of misfortune, enemies, and deceit on the part of loved ones.
Yew	*Taxus*	Sitting beneath a yew tree means the loss of a friend through illness. Seeing a yew tree indicates a lucky escape from a serious accident.
Thorn		Thorns pricking you are an omen of failure leading to success. Only seeing thorns means the reverse.
Tree		Green trees mean that great wealth is coming to you. Trees in bud indicate a new love. Trees in luxuriant foliage mean a happy marriage and children. Bare of leaves they indicate matrimonial troubles. Fruit trees are always lucky, indicating great prosperity in business. Trees covered in green leaves are a good omen, particularly in affairs of the heart. Autumn leaves betoken sorrow, probably as a result of your own actions. Burning leaves tell of a great loss to a friend. Pulling up a tree by its roots indicates that a relative will die.
Clover	*Trifolium*	Clover is a happy augury indicating health, prosperity and great happiness. It indicates that someone who is very poor seeks your hand in marriage. It can also indicate a happy and prosperous marriage.
Wheat	*Triticum*	For a business person, fields of wheat mean much prosperity and happiness. For a lover, it indicates a happy marriage. For a sailor it indicates a safe voyage. For a married person it means a happy and comfortable life.
Nettle	*Urtica*	Being stung by nettles indicates vexation and disappointments. Gathering nettles means someone has formed a favourable impression of you. For a married person it indicates a harmonious family life.

To dream of		Indicates
Whortleberry	*Vaccinium*	Whortleberries indicate deception on the part of a friend.
Vegetable		Eating vegetables is a sign of good fortune. Generally, vegetables forebode misfortune.
Periwinkle	*Vinca*	Periwinkle indicates that a spirit watches over you.
Violet	*Viola*	Violets are a promise of advancement in life. They mean you will marry someone younger than yourself.
Vine	*Vitis*	Grapes are a sign of pleasure. They indicate success for a trader. They mean jealously for those in love. Vines denote health, prosperity and fertility.
Corn	*Zea*	Corn or a cornfield foretells an addition to the family. Corn is a sign of wealth. Gathering corn is very good luck in everything; success in all enterprises. Full ears of corn mean you can expect as many good years as there are ears. Thin ears of corn mean you expect as many bad years as there are ears. Blighted or mildewed corn indicate that you will be a considerable loser. Corn that suddenly becomes ripe means you will gain an unlooked for inheritance. A burning stack of corn indicates famine and death.

Plants with Gods and Goddesses

Plant	**Deities**
Acacia	Allat; Al Uzzah; Astarte; Diana; Ishtar; Osiris; Ra
Acanthus	Acantha
Adonis	Adonis
Agaric	Bacchus; Dionysus; Wotan
Agave	Agave
Agrimony	The Goddess
Alder	Bran; Cronus; Ellerkonig; the Goddess; Gwern; Helice; Phoroneus; Proteus; Saturnus
Almond	Artemis Caryatis; Attis; Carmenta; Hermes; Mercury; Metis; Thoth
Amaranth	Artemis
Anemone	Adonis; Aphrodite; Flora; Venus
Angelica	Sophia; Venus
Apple	Aphrodite; Apollo; Athene; Diana; Dionysus; Eve; the Goddess; Hera; Herakles Melon; Iduna; Olwen; Venus; Vertumnus; Zeus
Apricot	The Goddess
Arbutus	Cardea
Ash	Achilles; Andrasteia; Cerridwen; the Goddess; Gwydion; Mars; Nemesis; Neptune; Odin; Poseidon; Thor; Uranus; Wotan
Aspen	Hercules/Herakles; Leuce; Ua Ildak
Avens	The Goddess
Bael tree	Aegle; Shiva
Bamboo	Hina; Thoth
Banana	Kanaloa
Banyan tree	Maui; Shiva
Barley	Adonis; Alphito; Anna Perenna; Ariadne; Ceres; Cerridwen; Cronus; Damuzi; Dionysus; Isis; Osiris
Basil	Erzulie; Krishna; Vishnu
Bean	Cardea; Ceres; Demeter
Beech	Athena; Ammon; Apollo; Diana; Zeus

Plant	Deities
Belladonna	Atropos; Bellona; Circe; Hecate
Birch	Berkana; Thor
Bitterwort	The Goddess
Black hellebore	The Goddess
Black poplar	Hecate
Blackberry	Blodeuwedd; Brigid; the Goddess
Blackthorn	Eris
Bodhi tree	Buddha; Krishna; Vishnu
Box	Pluto
Bryony	The Goddess
Buckthorn	The Goddess
Buttercup	Hymen
Cannabis	Bast; Shiva
Cardamom	Erzulie
Carnation	Jupiter
Cat mint	Bast
Cedar	Arinna; Baalat; Osiris
Chaste tree	Asclepius
Cherry	The Goddess; Vertumnus
Cinnamon	Aphrodite; Chang-o; Venus
Cinquefoil	The Goddess
Clover	Artemis; Diana; Olwen; Rowen
Cocoa	Ek Chua
Corn	Abuk; Acca Laurentia; Adonis; Anath; Ashnan; Attis; Bhim Deo; Bhimsen; Ceres; Cerridwen; Chitariah Tubueriki; Cronus; Dagon; Damuzi; Danae; Demeter; Dionysus; Enlil; Ezinu; Fornax; Gauri; Heqet; Ino; Isis; Jehovah; Jesus; Kore; Llew/Lugh; Maneros; Metsik; Nebri; Neper Ninlil; Ninhursag; Nisaba; Osiris; Persephone; Robigus; Sita; Sud; Tailltiu; Tubueriki; Tammuz; Uma; Viribius; Volos
Coconut	Te Tuna
Cornel	Apollo
Costmary	Mary

Plant	Deities
Cow parsley	Mary
Cowslip	Freya
Crocus	Britomartis; Jove; Juno
Cupid's dart	Cupid
Cyclamen	Hecate
Cypress	Aphrodite; Apollo; Artemis; Ashtoreth; Astarte; Cranae; Cupid; Diana; Hebe; Heqet; Hercules/Herakles; Jupiter; Mithras; Osiris; Pluto; Zoroaster
Daisy	Artemis; Freya; Thor
Dandelion	Hecate
Date palm	Allah; Apollo; Artemis; Damuzi; Diana; Hecate; Isis; Latona; Tammuz
Dog rose	The Goddess
Dogwood	Circe; Cronus; Proteus; Saturn
Dropwort	The Goddess
Eglantine rose	The Goddess
Elder	Hulda; the Goddess; Venus
Elm	Dionysus; Hoenin; Lodr; Odin
Fennel	Dionysus; Prometheus
Fenugreek	Apollo
Fern	Laka; Puck
Fig	The Apsaras; Dionysus; the Goddess; Isis; Juno
Fir	Adonis; Artemis; Attis; Bacchus; the Birth Goddess; Diana; Dionysus; Druantia; Hathor; Io; Isis; Osiris; Pan
Flax	Apollo; the Goddess; Hulda; Isis; Linda
Flowers	Flora; Iduna
Frankincense	Baal; Bel; Ra
Galangal	Vulcan
Garlic	Hecate
Geranium	The Goddess
Gorse	Jupiter; Thor

Plant	Deities
Grain	Abuk; Acca Laurentia; Adonis; Anath; Ashnan; Athene Alea; Attis; Ceres; Cerridwen; Cronus; Dagon; Damuzi; Danae; Demeter; Dionysus; Enlil; Fornax; Heqet; Ino; Isis; Jehovah; Jesus; Llew/Lugh; Maneros Metsik; Neper; Osiris; Tammuz; Tubueriki; Volos
Hawthorn	Blodeuwedd; Cardea; Flora; Hymen; Maia; Olwen
Hazel	Aengus; Artemis; Diana; Hermes; MacColl; Mercury; Thor
Heather	Aphrodite Erycina; Cybele; Eryx; Isis
Heliotrope	Apollo
Hemlock	Hecate
Herb Robert	The Goddess
Hibiscus	The Goddess
Hindu lotus	Buddha
Holly	Tannus; Taranis; Thor
Hollyhock	Althea
Horehound	Horus
Houseleek	Jupiter; Thor
Hyacinth	Hyacinthus; Zeus
Iris	Hera; Iris; Isis; Juno
Ivy	Artemis Tridaria; Attis; Bacchus; Cissia; Dionysus; the Goddess; Hymen; Osiris; Rhea
Jasmine	Vishnu
Juniper	The Goddess
Kava kava	Kaneloa
Lady's bedstraw	Mary
Lady's mantle	Mary
Larkspur	Apollo
Laurel	Aesculapius; Apollo; Ceres; Daphne; Eros; Faunus
Lavender	Aradia
Lettuce	Min; Venus
Lily	Astarte; Britomartis; Hera; Juno; Kwan Yin; Nephthys; Venus
Lily of the valley	Apollo; Asclepius; Maia; Mercury
Linden	Lada; Venus

Plant	Deities
Loosestrife	The Goddess
Lotus	Hapy; Isis; Usnissa
Madonna lily	Mary
Maize	Chicomecohuatl; Cinteotl; Xipe Totec; Yum Caz
Mallow	The Goddess
Mandrake	Aphrodite; Circe; Hathor; Hecate; Selene; Venus
Manila palm	Adonis
Marigold	Artemis
Marigold (African)	Tages
Marjoram	Aphrodite; Venus
Marsh mallow	Althea
Meadow crowsfoot	Hymen
Meadowsweet	Blodeuwedd; the Goddess
Medlar	The Goddess
Milk thistle	Mary
Mint	Hecate; Minthe; Pluto
Mistletoe	Apollo; Asclepius; Baldur; Freya; Frigg; Ischys; Ixion; Jove; Manannan MacLir; Odin; Venus
Moneywort	The Goddess
Monkshood	Hecate; Hercules/Herakles; Minerva; Saturn
Mugwort	Artemis; Diana
Mulberry	Brahma; Diana; Minerva; San Ku Fu Jen
Mullein	Jupiter
Mustard	Asclepius
Myrrh	Adonis; Astarte; Isis; Mariamne; Marian; Mary; Myrrha; Ra
Myrtle	Aphrodite; Artemis; Ashtoreth; Astarte; Hathor; Mariamne Marian; Marina; Miriam; Myrrha; Myrto; Venus
Narcissus	Atropos; Clotho; The Fates; Hades; Lachesis; Narcissus; Persephone; Pluto
Nettle	Thor

Plant	Deities
Nutmeg	Myrrha
Oak	Allah; Ariadne; Artemis; Athena; Baldur; Belenos; Blodeuwedd; Cardea; Cernunnos; Circe; Cybele; the Dagda; Demeter; Dia; Diana; Dianus; Dione; Donar; Egeria; El; Erato; Esus; Hecate; Hercules/Herakles; Herne; Hou; Janicot; Janus; Jehovah; Jove; Jupiter; Mars; Mary; the Morrigan; Pan; Peirun; Perkunas; Picus; Tara; Taranis; Teutates; Thor; Thunor; Viribus; Zeus
Oats	Brigid
Olive	Allah; Apollo; Athene; Hercules/Herakles; Irene; Minerva; Ra; Zeus
Onion	Hecate; Isis
Opium poppy	Dionysus
Orange	Juno; Jupiter
Orris root	Aphrodite; Hera; Iris; Isis; Juno; Osiris
Ox-eye aisy	Artemis; Thor
Palm	Allah; Apollo; Artemis; Ashtoreth; Christ; Damuzi; Enlil; Hecate; Ishtar; Isis; Lat; Latona; Mariamne; Marian; Marina; Myrrha; Tamar; Tammuz; That
Pansy	The Goddess
Paeony	Apollo; Paeon
Paapyrus	Baalat; Hapy; Hathor
Parsley	Persephone
Pasque flower	Adonis; Venus
Peach	The Goddess; Harpocrates; Horus; Vertumnus
Pear	Athene Once; Hera; Priapus
Pennyroyal	Demeter
Peppermint	Pluto
Pine	Artemis; Astarte; Attis; Bacchus; Cybele; Diana; Dionysus
Pine	Neptune; Pan; Rhea; Sylvanus; Venus
Pirimrose	Freya
Plane tree	Apollo; the Goddess; Xerxes
Pomegranate	Adonis; Attis; Ceres; Damuzi; Hera; Judah; Ninib; Persephone; Rimmon; Tammuz

Plant	**Deities**
Poplar	Apollo; Leuce
Poplar (black)	Egeria; Hecate
Poplar (white)	Apollo; Juno; Leuce; Persephone; Zeus
Poppy	Aphrodite; Ceres; Demeter; Hypnos; Somnus
Primrose	The Goddess
Pumpkin	Chicomecohuatl
Purslane	The Goddess
Quince	Aphrodite
Reeds	Coventia
Rhubarb	Gayomart
Rice	Antaboda; Chang-o
Rose	Adonis; Aphrodite; Aurora; Cupid; Demeter; Eros; Harpocrates; Hathor; Horus; Hulda; Isis; Mary; Venus
Rosemary	Mary; Venus
Rowan	Akka/Mader; Akka/Rauni; Alys (Elis, Halys); Aphrodite; Brigantia; Brigid; Oeagrus; Ran; Thor
Rue	Aradia; Diana
Sacred fig	Krishna
Saffron	Ashtoreth; Eos; Thoth
Saint John's wort	Baldur; the Goddess
Sea Holly	The Goddess
Service tree	Oeagrus
Sesame	Ganesha
Shamrock	Artemis; Blodeuwedd; Diana
Southernwood	Artemis; Diana
Spurge	Munsa
Strawberry	Freya; Friggaa
Sweet William	The Goddess
Sycamore	Isis; Osiris
Tamarisk	Anu; Isis; Osiris

Plant	Deities
Tansy	Ganymede
Tarragon	Artemis; Diana
Thistle	Minerva; Thor
Ti	Kaneloa; Lono; Pele
Tormentil	Thor
Vervain	Aradia; Cerridwen; the Goddess; Isis; Juno; Jupiter; Mars; Mercury; Persephone; Thor; Venus
Vine	Bacchus; Dionysus; Hathor; Jesus; Osiris
Violet	Aphrodite; Attis; Io; Priapus; Venus; Zeus
Walnut	Artemis; Car; Carmenta; Carya; Jupiter; Metis
Waterlily	Coventia
Wheat	Adonis; Ceres; Demeter; Ishtar; Isis; Min; Osiris; Zeus Polleus
Willow	Anatha; Arawn; Artemis; Athena; Bel; Belenos; Belili; Cerridwen; Circe; Diana; Europe; Geshtinanna; Gwydion; Hecate; Helice; Hera; Ishtar; Jehovah; Luna; Mercury; Minerva; Orpheus; Osiris; Persephone; Zeus
Wormwood	Artemis; Diana; Iris
Yarrow	Achilles
Yew	Banba; Hecate; Saturn

Plant Planetary Rulers

Mercury Agaric; almond; aspen; bean; bittersweet; bracken; Brazil nut; caraway; celery; clover; dill; elecampane; fennel; fenugreek; fern; filbert; flax; goat's rue; hazel; honeysuckle; horehound; lavender; lemon grass; lemon verbena; lily of the valley; mace; maidenhair; male fern; mandrake; marjoram; mint; mulberry; orange bergamot; papyrus; parsley; peppermint; pimpernel; pistachio; pomegranate; senna; southernwood; speedwell; wax plant

Jupiter Agrimony; anise; avens; balm; banyan; betony; bilberry; bodhi; boneset; borage; chervil; chestnut; cinquefoil; clove; costmary; couch grass; dandelion; dock; dog rose; eglantine; endive; fig; hart's tongue fern; honeysuckle; horse chestnut; houseleek; hyssop; linden; liverwort (American); maple; meadowsweet; milk thistle; nutmeg; red rose; sage; samphire; sarsparilla; sassafras; star anis; ti; wood betony

Saturn Amaranth; *arbor vitae*; asphodel; beech; beet; belladonna; bistort; blackthorn; boneset; buckthorn; comfrey; cypress; darnel; datura; dodder; elm; euphorbia; fumitory; hellebore; hemlock; hemp; henbane; herb Christopher; horsetail; ivy; kavakava; knapweed; knotweed; lady's slipper; lobelia; mate; mimosa; monkshood; morning glory; moss; mullein; nightshade; pansy; patchouly; poplar (black and white); quince; royal fern; shepherd's purse; skullcap; skunk cabbage; slippery elm; Solomon's seal; tamarisk; thornapple; tutsan; willowherb; yew

Mars Agave; allspice; anemone; asafoetida; basil; berberis; black snakeroot; blood root; broom; bryony; buttercup; cactus; carrot; chilli pepper; coriander; cubeb; cumin; curry leaf; damiana; deerstongue; dragon plant; dragon's blood; galangal; garlic; gentian; ginger; gorse; grains of paradise; hawthorn; High John the Conqueror; holly; holy thistle; hops; horseradish; houndstongue; knot grass; leek; lesser celandine; lupin; maguey; masterwort; meadow crowsfoot; milk thistle; mustard; nettle; Norfolk Island pine; onion; pasque flower; pennyroyal; pepper; peppermint; pepper tree; pimento; pine; poke root; prickly ash; radish; reed; reed mace; shallot; sloe; snapdragon; squill; stonecrop; tarragon; thistle; toadflax; tobacco; Venus flytrap; white mustard; woodruff; wormwood; yucca

Venus Adam and Eve; African violet; alder; alfalfa; alkanet; apple; apricot; aster; avocado; bachelor's buttons; balm of Gilead; banana; barley; birch; blackberry; bleeding heart; blue flag iris; buckwheat; burdock; caper; cardamom; catmint; cherry; coltsfoot; columbine; corn; cowslip; creeping Jenny; crocus; cuckoo flower; cyclamen; daffodil; daisy; elder; feverfew; fleabane; foxglove; fragrant bedstraw; geranium; goldenrod; ground ivy; groundsel; heather; hibiscus; hyacinth; iris; lady's mantle; larkspur; life everlasting; lilac; liquorice; lucky hand; magnolia; maidenhair; marsh mallow; meadowsweet; mint; mugwort; myrtle; oats; orchid; orris; ox-eye daisy; passion flower; pea; peach; pear; periwinkle; plantain; ploughman's spikenard; plum; primrose; ragwort; raspberry; rhubarb; rose; rye; sanicle; sea holly; self heal; silverweed; sorrel; spearmint; spikenard; strawberry;

sweet pea; tansy; thyme; tomato; tonka; trillium; true love; tulip; valerian; vanilla; vervain; vertivert; violet; wheat; wild rose; willow; wood aloe; wood sorrel; yarrow

Sun Acacia; angelica; ash; bay; benzoin; bromeliad; burnet; carnation; cedar; celandine; centaury; chamomile; chicory; chrysanthemum; cinnamon; citron; copal; eyebright; frankincense; ginseng; goldenseal; gum Arabic; hazel; heliotrope; heartsease; juniper; lady's bedstraw; lime; liquidamber; lovage; marigold; mastic; mistletoe; oak; olive; orange; palm; peony; pimpernel; pineapple; rice; rosemary; rowan; rue; saffron; St John's wort; sandalwood; sesame; storax; sundew; sunflower; tea; tormentil; viper's bugloss; walnut; witch hazel

Moon Adder's tongue (American); aloe; bladderwrack; buchu; cabbage; calamus; camellia; camphor; cassia; chickweed; cleavers; costmary; club moss; coconut; cotton; cucumber; dulse; eucalyptus; gardenia; goose grass; gourd; honesty; jasmine; lemon; lemon balm; lesser duckweed; lettuce; Madonna lily; loosestrife; lotus; mallow; moonwort; myrrh; papaya; poppy; potato; privet; purslane; sandalwood; turnip; vine; wallflower; watercress; waterlily; white poppy; white rose; willow; wintergreen

Plants and Zodiac Signs

Aries	(21 March–20 April)	Chestnut; gorse; holly; nasturtium; thistle; thorn; tulip; wake Robin; wild rose; woodbine
Taurus	(21 April–21 May)	Almond; apple; ash; cherry; coltsfoot; lily; lily of the valley; lovage; myrtle; sycamore; violet; walnut; wild rose
Gemini	(22 May–21 June)	Dill; elder; fern; filbert; hazel; iris; parsley; rose; snapdragon
Cancer	(22 June–22 July)	Agrimony; delphinium; honesty; hyssop; moonwort; poppy; privet; sycamore; watercress; waterlily; white rose; willow
Leo	(23 July–22 August)	Angelica; borage; cowslip; forsythia; heliotrope; hops; laurel; marigold; oak; paeony; palm; pine; rue; sunflower
Virgo	(23 August–23 September)	Cornflower; daisy; elder; hazel; Madonna lily; rosemary; valerian
Libra	(24 September–23 October)	Almond; apple; hydrangea; love in a mist; myrtle; plum; violet; walnut; white rose
Scorpio	(24 October–22 November)	Blackthorn; chrysanthemum; holly; lesser celandine; paeony; purple heather; sweet basil; white thorn
Sagittarius	(23 November–21 December)	Carnation; chestnut; clove pink; mulberry; sage; vine; wallflower
Capricorn	(22 December–20 January)	Cypress; holly; nightshade; pine; rue; snowdrop; Solomon's seal; spruce; violet; yew
Aquarius	(21 January–19 February)	Foxglove; gentian; great valerian; mullein; orchid; pine; snowdrop
Pisces	(20 February–20 March)	Alstroemeria; carnation; heliotrope; opium poppy; violet

Plants and the Elements

Fire Agave; alder; allspice; amaranth; anemone; angelica; asafoetida; ash; avens; basil; bay; betony; black snakeroot; blackthorn; blood root; bryony; cactus; carnation; carrot; cedar; celandine; celery; centaury; chestnut; chrysanthemum; cinnamon; cinquefoil; clove; copal; coriander; cubeb; cumin; deerstongue; dill; dragon's blood; fennel; fig; flax; frankincense; galangal; garlic; gentian; ginger; ginseng; goldenseal; gorse; grains of paradise; hawthorn; heliotrope; High John the Conqueror; holly; horse chestnut; horseradish; houndstongue; hyssop; juniper; leek; lime; liquidamber; liverwort (American); lovage; mandrake; marigold; masterwort; milk thistle; mountain mahogany; mullein; mustard; nettle; Norfolk Island pine; nutmeg; oak; olive; onion; orange; peony; pasque flower; pennyroyal; pepper; peppermint; pepper tree; pine; pineapple; poke root; pomegranate; radish; reed mace; rosemary; rowan; rue; saffron; St John's wort; sarsparilla; sassafras; sesame; shallot; snapdragon; squill; storax; sundew; sunflower; tea; thistle; ti; toadflax; tobacco; tormentil; tutsan; Venus flytrap; walnut; white mustard; witch hazel; wood betony; woodruff; wormwood; yucca

Water Adam and Eve; adder's tongue (American); African violet; alkanet; aloe; apple; apricot; aster; avocado; bachelor's buttons; balm; balm of Gilead; banana; birch; blackberry; bladderwrack; bleeding heart; blue flag iris; boneset; buchu; buckthorn; burdock; cabbage; calamus; camellia; camphor; capers; cardamom; catmint; chamomile; cherry; chickweed; club moss; coconut; coltsfoot; columbine; comfrey; cowslip; crocus; cuckoo flower; cucumber; cyclamen; daffodil; daisy; datura; deadly nightshade; dodder; dulse; elder; elm; eucalyptus; euphorbia; feverfew; fleabane; foxglove; fragrant bedstraw; gardenia; gourd; groundsel; heartsease; heather; hellebore; hemlock; hemp; henbane; hibiscus; hollyhock; hyacinth; iris; ivy; jasmine; kava kava; knot grass; lady's mantle; lady's slipper; larkspur; lemon; lemon balm; lettuce; liquorice; lilac; Madonna lily; lobelia; lotus; lucky hand; maidenhair; mallow; mimosa; monkshood; moonwort; morning glory; myrrh; myrtle; orchid; orris; pansy; papaya; passion flower; peach; pear; periwinkle; ploughman's spikenard; plum; poplar; poppy; purslane; ragwort; raspberry; red rose; rose; sandalwood; sea holly; skullcap; skunk cabbage; Solomon's seal; spearmint; strawberry; sweet pea; tamarisk; tansy; thornapple; thyme; tomato; tonka; valerian; vanilla; vine; violet; watercress; waterlily; white poppy; white rose; willow; wintergreen; yarrow; yew

Earth Alfalfa; asphodel; barley; beet; bistort; buckwheat; corn; cotton; cypress; fern; fumitory; honesty; honeysuckle; horehound; horsetail; knotweed; loosestrife; magnolia; mugwort; oats; oleander; oregon grape; patchouly; pea; plantain; potato; primrose; quince; rhubarb; rye; sorrel; tulip; turnip; vervain; vertivert; wheat; wood sorrel

Air Acacia; agaric; agrimony; almond; anise; aspen; banyan; bean; benzoin; bittersweet; black poplar; bodhi; borage; bracken; brazil nut; bromeliad; broom; caraway; chicory; citron; clover; dandelion; dock; elecampane; endive; eyebright; fenugreek; fern; filbert; goat's rue; goldenrod; hazel; hops; houseleek; lavender; lemon grass; lemon verbena; lily of the valley; linden; mace; male fern; maple; marjoram; mastic; meadowsweet; mint; mistletoe; mulberry; orange bergamot; palm; papyrus; parsley; pimpernel; pine; pistachio; rice; sage; senna; slippery elm; southernwood; star anise; wax plant

The Oghams' Tree Alphabet

The Oghams were sacred glyphs or markings that were developed into an alphabet by the Celts. Each glyph is associated with a particular 'tree'.

B	Beth	Birch
L	Luis	Rowan
F	Fearn	Alder
S	Saille	Willow
N	Nuin	Ash
H	Huathe	Hawthorn
D	Duir	Oak
T	Tinne	Holly
C	Coll	Hazel
Q	Quert	Apple
M	Muin	Vine
G	Gort	Ivy
Ng	Ngetal	Reed
St	Straif	Blackthorn
R	Ruis	Elder
A	Ailim	Pine
O	Ohn	Gorse
U	Ur	Heather
E	Eadha	Poplar
I	Ioho	Yew

Flowers of the Scottish Clans

Buchanan	Birch
Cameron	Oak
Campbell	Myrtle
Chisholm	Alder
Colquhoun	Hazel
Cumming	Sallow
Drummond	Holly
Farquharson	Purple foxglove
Ferguson	Poplar
Forbes	Broom
Fraser	Yew
Gordon	Ivy
Graham	Spurge laurel
Grant	Cranberry heath
Gunn	Rosewood
Lamont	Crab apple tree
McAllister	Fine-leaved heath
McDonald	Heath bell
McDonnell	Mountain heath
McDougall	Cypress
McFarlane	Cloudberry heath
McGregor	Scots fir
McIntosh	Boxwood
McKay	Bulrush
McKenzie	Deer grass
McKinnon	St John's wort
McLachlan	Mountain ash
McLean	Blackberry heath
McLeod	Red whortleberry
McNab	Rose bush berry
McNeil	Seaweed bladder-fucus
McPherson	Variegated box
McQuarrie	Blackthorn
McRae	Clubmoss
Menzies	Ash
Murrey	Juniper
Ogilvie	Hawthorn
Oliphant	Great maple (sycamore)
Robertson	Fern
Rose	Briar rose
Ross	Bearberry
Sinclair	Clover
Stewart	Thistle
Sutherland	Cat's tail grass

Flowers of the Countries of the World

Some national flowers have cultural or religious roots that go back hundreds or even thousands of years, although they may not have been officially adopted.

The rose is the national flower of the United States, the United Kingdom and the Maldives as well as being the official flower of several states and the province of Alberta. Rose once served as Honduras' national flower as well.

Country	National Flower		Interesting Information
Antigua & Barbuda	Dagger's log	*Agave Karatto Miller*	The yellow flowers rise from the large rosette of the agave plant.
Argentina	Ceibo	*Erythrina crista-galli*	The flower was adopted on 2 December, 1942.
Australia	Golden wattle	*Acacia pycnantha*	1 September is National Wattle Day.
Austria	Edelweiss	*Leontopodium alpinum*	The star-like flowers are short-lived perennials.
Bahamas	Yellow elder or Yellow cedar	*Tecoma stans*	The flowers bloom in late summer/ early autumn.
Bangladesh	Waterlily	*Nymehaea nouchali*	Bangladesh adopted the flower in 1971.
Barbados	Pride of Barbados, also known as dwarf poinciana and flower fence	*Poinciana pulcherrima*	More common varieties of the flower are a fiery red and yellow.
Belarus	Flax	*Linum usitatissimum*	The flowers last only until the heat of the midday sun hits them.
Belgium	Red poppy	*Papaver rhoeas*	It is one of the easiest wildflowers to grow.
Belize	Black orchid	*Trichoglottis brachiata*	Black orchids acquired the name by virtue of their very dark intense colour, which tends to the dark brown and maroon.
Bermuda	Blue-eyed grass	*Sisyrinchium montanum*	Blue-eyed grass is a member of the iris family.
Bhutan	Blue poppy	*Meconopsis betonicifolia*	The flower is native to rocky mountain slopes of Tibet.

Country	National Flower		Interesting Information
Bolivia	Kantuta	*Cantua buxifolia*	The tubular flowers come in wild form, magenta, bicolour and subtle slightly bicoloured varieties.
Brazil	Tabebuia alba	*Ipè-amarelo-da-serra*/Golden trumpet tree	Its insect resistance and durability have made the timber from *Tabebuia* species popular for use in decking. In New York City the Department of Parks and Recreation has converted the timbers of the entire boardwalk to ipè, the common name of *Tabebuia*.
Bulgaria	Rose	*Rosa*	Roses are more fragrant on a sunny day.
Canada	Maple leaf	*Acer*	Maple syrup is made from the sap of sugar maple trees.
Cayman Islands	Wild banana orchid	*Schomburgkia thomsoniana*	This orchid species is found only in the Cayman Islands.
Chile	Copihue/Chilean bellflower	*Lapageria rosea*	The Chilean bellflower is best grown on a partially shady and sheltered wall.
China	Plum blossom	*Prunus mei*	Plum blossoms are the earliest blooms of the year, indicating the start of spring.
Colombia	Christmas orchid	*Cattleya trianae*	Christmas orchid has a foetid smell.
Costa Rica	Guaria morada (purple orchid)	*Cattleya skinneri*	The flower was adopted on 15 June 1939.
Croatia	Iris Croatica	*Iris perunika*	It grows only in northern and northwestern Croatia.
Cuba	Butterfly jasmine	*Mariposa*	The white butterfly jasmine is an endemic jasmine species.
Cyprus	Rose	*Rosa*	The more fragrant the rose, the shorter its vase life.
Czech Republic	Rose	*Rosa*	Miniature roses were first developed in China.
Denmark	Marguerite daisy	*Argyranthemum frutescens*	Marguerites produce large, single, daisy-like flowers most of the summer.

Country	National Flower		Interesting Information
Ecuador	Rose	*Rosa*	Rose-growing is a multi-million-pound industry in Ecuador, which employs mainly women. Roses given on Valentine's Day may well have been grown in Ecuador.
Egypt	Lotus	*Nymphaea lotus*	The pure white lotus flowers, are the only plant to fruit and flower simultaneously.
Estonia	Cornflower or Bachelor's button	*Centaurea (cyanus)*	The flower was adopted in Estonia on 23 June 1988.
Ethiopia	Calla lily		The flower is a solitary, showy, funnel-shaped unfurling spathe.
France	Iris	*Iris*	Iris flowers have three petals often called the 'standards', and three outer petal-like sepals called the 'falls'.
French Polynesia	Tiare	*Gardenia taitensis*	The flower is especially symbolic of Tahiti. The *tiare anei* is the emblem of the isle of Vavau. The *tiare apetahi* is the emblem of Raiatea.
Finland	Lily of the valley	*Convallaria majalis*	Lily of the valley is much used in bridal arrangements for its sweet perfume.
Germany	Knapweed	*Centaurea cyanus*	In Germany, it is a custom for an unmarried person to wear this flower in the buttonhole.
Greece	Bear's breech	*Acanthus mollis*	The fresh or dried flower spikes are used in floral arrangements.
Greenland	Willowherb	*Epilobium*	The name 'willowherb' refers to the willow-like form of the leaves.
Guam	Puti tai nobiu	*Bougainvillea spectabilis*	The flowers of the bougainvillea can be several different colours, from pink, to red, to orange, to white and yellow.
Guatemala	White nun orchid or monja blanca	*Lycaste skinnerialba*	The flower is rare. Found in the Verapaz distict of Guatemala, it symbolizes peace, beauty and art.
Guyana	Waterlily	*Victoria regia*	The largest flowers can measure 10 inches to one foot in diameter

Country	National Flower		Interesting Information
Honduras	Orchid	*Brassavola digbiana*	The rose was the national flower from 1946–1969.
Hong Kong	Orchid	*Bauhinia blakeana*	The flower is tubular with a corolla of five petals coloured a deep purple.
Hungary	Tulip	*Tulipa*	Tulip is the common name for between 50 and 150 species of the genus *Tulipa* in the lily family, Liliaceae.
Iceland	Mountain avens	*Dryas octopetala*	The flowers are produced on stalks up to four inches long, with eight creamy white petals.
India	Lotus	*Nelumbo nucifera*	The lotus is an aquatic perennial.
Indonesia	1. Melati (jasmine) 2. Moon orchid 3. Rafflesia	*Jasminum sambac* *Amabilis Phalaenopsis* *Rafflesia arnoldi Indonesia*	Indonesia adopted the three flowers on June 5, 1990 to mark the World Environment Day.
Iran	Red rose	*Rosa*	To make a dark red rose appear blacker, its stem can be put in water that has black ink in it.
Iraq	Rose	*Rosa*	The rose is said to be originally from Persia, and was introduced to the west by Alexander.
Ireland	Shamrock		Shamrock is the common name for several unrelated herbaceous plants with trifoliate leaves.
Italy	Daisy	*Bellis perennis*	The stylized lily is also used, being the symbol of Florence and the Medici family.
Jamaica	*Lignum vitae*, or wood of life	*Guaiacum sanctum*	The flower is indigenous to Jamaica and was found there by Christopher Columbus.
Japan	Chrysanthemum (imperial), cherry blossom sakura		The sakura trees are the subject of the annual National Cherry Blossom Festival in Japan.
Jordan	Black Iris	*Iris nigricans*	The dark purple coloured iris has six petals, three drooping and three upright.

Country	National Flower		Interesting Information
Laos	Champa flower, also known as plumeria	*Calophyllum Inophyllum*	The attractive white flowers are scented and waxy.
Kazakhstan	Lily serves as the unofficial national flower	*Lilium*	Lilies are believed to have been cultivated longer than any other ornamental flower, having existed in gardens 3,000 years ago.
Kuwait	Arfaj	*Rhanterum epapposum*	This woody perennial is a member of the daisy family and forms huge stands that cover vast areas of north-eastern Arabia.
Kyrgyzstan	Tulip	*Tulipa germanica*	The first Kyrgyz Revolution, which saw the overthrow of the government of President Askar Akayev in 2005, is known as the Tulip Revolution.
Latvia	Ox-eye daisy, or pipene	*Leucanthemum vulgare*	The flower was earlier known as *Chrysanthemum leucanthemum.*
Lebanon	Cedar of Lebanon	*Cedrus libanii*	There is no national flower but the Cedar of Lebanon is the national tree.
Liberia	Pepper		These are small, white, star-shaped flowers.
Libya	Pomegranate blossom	*Punica granata*	The flowers are fiery red.
Lithuania	Rue or herb of grace	*Ruta graveolens*	Rue's fragrance is strong, characteristically aromatic and sweet.
Luxembourg	Rose	*Rosa*	One of the most famous rose gardens was planted by Empress Josephine at the Chateau de la Malmaison in France in 1804.
Madagascar	Poinciana	*Delonix regia*	In early summer, the voluminous red blooms appear and hold for four to eight weeks.
Maldives	Pink rose	*Rosa*	The oldest painting in the world depicts a five-petalled pink rose.
Malta	Maltese centaury	*Paleocyanus crasifoleus*	The flower was adopted in the early 1970s.

Country	National Flower		Interesting Information
Netherlands	Tulip	*Tulipa*	Tulip bulbs are a good substitute for onions in cooking.
New Zealand	Kowhai	*Sophora microphylla*	Kowhai has a beautiful yellow or golden flower.
Paraguay	Jasmine-of-the-Paraguay		Jasmine flowers are white in most species.
Peru	Kantuta, Inca magic flower	*Cantua buxifolia*	Kantuta come in four varieties: wild, magenta, bicolour and subtile.
Philippines	Sampaguita	*Jasminum sambac*	The plant blooms all year and has white, small, dainty, star-shaped blossoms, which open at night and wilt in less than a day.
Poland	Corn poppy	*Papaver rhoeas*	Corn poppy is the wild poppy of agricultural cultivation.
Portugal	Lavender	*Lavendula angustifolium*	Used in cooking, the potency of the lavender flowers increases with drying.
Puerto Rico	Puerto Rican hibiscus, or *flor de maga*	*Thespesia grandiflora*	Closely related to the common hibiscus, but the *flor de maga* grows into a large tree.
Romania	Dog rose	*Rosa canina*	The white or pink five-petalled flowers are 1.5–2.5 in across and come in clusters of one to five.
Russia	Camomile	*Matricaria recutita*	The flower has an aromatic, fruity, floral fragrance.
San Marino	Cyclamen	*Cyclamen*	The flowers are produced in whorls of three to ten, each flower on a slender stem 2–5 in tall, with five united petals.
Scotland	Thistle	*Cirsium altissimum*	The thistle flower is a favourite among butterflies.
Seychelles	Tropicbird orchid	*Angraecum ebumeum*	These are sprays of white flowers with long spurs like the tails of tropical birds.
Sicily	Carnation	*Dianthus caryophyllus*	The carnation is native to Eurasia and has been cultivated for more than twenty centuries.
Singapore	Vanda Miss Joaquim orchid		The flower is a hybrid between *Vanda teres* and *Vanda hookeriana.*

Country	National Flower		Interesting Information
Slovakia	Rose	*Rosa*	The first historical reference to the rose is by the Sumerians from ancient Mesopotamia.
Slovenia	Carnation	*Dianthus caryophyllus*	Carnations can be propagated by planting young flowering shoots.
Spain	Red carnation		Wearing a red carnation on mother's day indicates your mother is still living.
Sri Lanka	Nil Mahanel waterlily	*Nympheae Stellata*	The flower, a blue water lily, was adopted on 26 February 1986.
South Africa	Protea	*Protea cynaroides*	The king protea is originally from the Cape Town area of South Africa.
South Korea	Rose of Sharon (Moogoonghwa)	*Hibiscus syriacus*	*Hibiscus syriacus* are pink-mauve single flowers, having a dark magenta eye. The flower is not a rose, but its large exotic blossoms attract hummingbirds and tiny insects.
Sweden	Linnea	*Linnea Borealis*	The flowers are pink, bell-like, very fragrant and grow in pairs.
Switzerland	Edelweiss	*Leontopodium alpinum*	The flowers are starfish-like white, woolly blooms.
Syria	Jasmine	*Jasminium sp.*	Jasmine flowers are generally white, although some species have yellow flowers.
Taiwan	Plum blossom	*Prunus mei*	Most plum blossoms have five petals and range in colour from white to dark pink.
Thailand	Ratchaphruek	*Cassia fistula*	Also called the golden shower tree, the yellow petals symbolizing Thai royalty.
Trinidad and Tobago	Chaconia	*Warszewiczia coccinea*	The flower is also known as 'pride of Trinidad and Tobago' or wild poinsettia.
Tonga	Red-blossomed heilala	*Garcinia pseudogutti fera*	The red-blossomed heilala festival in Tonga is celebrated during the Heilala Festival every 4 July.

Country	National Flower		Interesting Information
Turkey	Tulip	*Tulipa*	Tulips do not grow in the open in tropical climates as they need cold winters.
Ukraine	Sunflower	*Helianthus annuus*	Most flowerheads on a field of blooming sunflowers are turned towards the east, the direction of sunrise.
United States of America	Rose	*Rosa*	The rose was officially adopted on 20 November 1986.
United Kingdom (England)	Tudor rose	*Rosa*	The Tudor rose is a graphic design created by King Henry VII in 1485, with a red rose laid atop a white one.
United Kingdom (Wales)	Leek, Daffodil	*Babbingtons leek, Narcissus Amaryllidaceae*	The leek and the daffodil are both emblems of Wales. The national flower of Wales is usually considered to be the daffodil. However, the leek has even older associations as a traditional symbol of Wales – possibly because its colours, white over green, echo the ancient Welsh flag.
Uruguay	Ceibo	*Erythrina crista-galli*	Ceibo has bright red flowers.
Venezuela	Orchid	*Cattleya mossiae*	Orchids form the world's largest family of plants.
Virgin Islands	Yellow elder or yellow trumpet	*Tecoma stans*	The yellow flowers have a very sweet fragrance and attract hummingbirds and butterflies.
Yemen	Arabian coffee	*Coffea arabica*	Individual coffee flowers are white, fragrant, with waxy, linear petals.
Yugoslavia	Lily of the valley	*Convallaria majalis*	Lily of the valley are fragrant bell-shaped flowers.
Zimbabwe	Flame lily	*Gloriosa Rothschildiana*	The large, claw-like flowers open yellow and red and then change to a rich claret edged with gold.

Floral Symbols of American States

Each of the American states has had a flower designated as its floral symbol. The flower selected is usually chosen because of its prevalence in the state or due to a role it has played in its history.

Alabama	Camellia	*Camellia japonica L.*
Alaska	Forget-me-not	*Myosotis alpestris*
Arizona	Saguaro cactus blossom	*Carnegiea gigantea*
Arkansas	Apple blossom	*Malus*
California	Californian poppy	*Eschscholzia californica*
Carolina (North)	Dogwood	*Cornus florida*
Carolina (South)	Yellow jessamine	*Gelsemium sempervirens*
Colorado	Rocky mountain (blue) columbine	*Aquilegia caerulea*
Connecticut	Mountain laurel	*Kalmia latifolia*
Dakota (North)	Prairie rose	*Rosa blanda* or *arkansana*
Dakota (South)	American pasque flower	*Pulsatilla hirsutissima*
Delaware	Peach blossom	*Prunus persica*
Florida	Orange blossom	*Citrus sinensis*
Georgia	Cherokee rose	*Rosa laevigata*
Hawaii	Hibiscus	*Hibiscus brackenridgei*
Idaho	Lewis mock orange	*Philadelphus lewisii*
Illinois	Native wood violet	*Viola*
Indiana	Paeony	*Paeonia*
Iowa	Wild prairie rose	*Rosa arkansana*
Kansas	Sunflower	*Helianthus annuus*
Kentucky	Golden rod	*Soldiago gigantea*
Louisiana	Southern magnolia	*Magnolia grandiflora*
Maine	Pine cone	*Pinus strobus*
Maryland	Black-eyed susan	*Rudbeckia hirta*
Massachusetts	May flower	*Epigaea repens*

Michigan	Apple blossom	*Malus domestica*
Minnesota	Showy lady slipper	*Cypripedium reginae*
Mississippi	Southern magnolia	*Magnolia grandiflora*
Missouri	Downy hawthorn	*Crataegus mollis*
Montana	Bitter root	*Lewisia rediviva*
Nebraska	Golden rod	*Solidago gigantea*
Nevada	Sagebrush	*Artemisia tridentata*
New Hampshire	Purple lilac	*Syringa vulgaris*
New Jersey	Violet	*Viola sororia*
New Mexico	Yucca	*Yucca glauca*
New York	Rose	*Rosa*
Ohio	Scarlet carnation	*Dianthus caryophyllus*
Oklahoma	Mistletoe	*Phoradendron serotinum*
Oregon	Oregon grape	*Berberis aquifolium*
Pennsylvania	Mountain laurel	*Kalmia latifolia*
Rhode Island	Violet	*Viola palmate*
Tennessee	Iris	*Iris germanica*
Texas	Texas blue bonnet	*Lupinus texensis*
Utah	Sego lily	*Calochortus nuttallii*
Vermont	Red clover	*Trifolium pratense*
Virginia	Flowering dogwood	*Cornus florida*
Washington	Coast rhododendron	*Rhododendron macrophyllum*
Virginia, West	Great rhododendron	*Rhododendron maximum*
Wisconsin	Native violet	*Viola papilionacea*
Wyoming	Indian paintbrush	*Castilleja linariifolia*

Flowers of the Months

January	Carnation; snowdrop
February	Primrose, iris
March	Daffodil
April	Daisy
May	Lily; lily of the valley
June	Rose
July	Water lily; Sweet pea; delphinium
August	Gladiolus
September	Aster
October	Dahlia; marigold
November	Chrysanthemum
December	Holly; poinsettia

Anniversary Flowers

1st	Carnation
2nd	Lily of the valley
3rd	Sunflower
4th	Hydrangea
5th	Daisy
6th	Calla lily
7th	Freesia
8th	Lilac
9th	Bird of paradise
10th	Daffodil
11th	Tulip
12th	Peony
13th	Chrysanthemum
14th	Orchid
15th	Rose
20th	Aster
25th	Iris
30th	Lily
40th	Gladiolus
50th	Yellow roses and violets

APPENDIX 1: FLOWERS

Deities: Flora; Iduna
Meaning: Blood and bandages (red and white flowers)

When we give the gift of flowers we are maintaining a very ancient tradition, which dates back at least to the times of the ancient Egyptians. They gave flowers as tokens of good luck, believing that various deities dwelt in the flower or its scent. It is usually considered unlucky to give an even number of flowers in a bunch – not a superstition observed by florists, who tend to sell flowers in bunches of ten or twelve.

At religious festivals it was common for vast quantities of flowers to be placed onto the altars of their gods. Although baskets of flowers might be left it was also common for them to be woven into wreathes and garlands. The great sun god Ra is often depicted in papyri surrounded by flowers, and his statues, as well as those of other deities, were often crowned with circlets of blossoms. Some flowers species were particularly associated with certain gods. Helichrysum, for example, was worn by the virgins selected to serve in the temple of the sun god.

The practice of wearing flowers for personal adornment is equally ancient. It was originally the preserve of the gods and the aristocracy. When the common people started to wear flowers it was seen to be a challenge to the established order and in Rome new laws were issued to prevent the indiscriminate wearing of flower garlands. Punishments were severe. It was not unknown for someone who illegally wore a chaplet of roses to be sentenced to sixteen years in prison. One miscreant, who had crowned himself with flowers taken from a statue of Mersyas, was put in chains for the offence.

Flowers have been an important feature of funeral rites since the earliest times. Often, on such occasions, they and the greenery are the only sources of colour. In 1968 archaeologists excavating the Shanidar cave in north-eastern Iraq recovered soil samples from beneath a Neanderthal burial, some 60,000 years old. Traces of pollen, grouped in clusters, suggested that a wreath of some description had been woven onto a branch from a fir or similar plant. From the pollen samples it was possible to say that it contained cornflower, groundsel, grape hyacinth, hollyhock, St Barnaby's thistle and yarrow. The base of the wreath was of mare's tail (equisetum). The distribution of the species indicated that they were specifically gathered for their purpose.

In times gone by the sweet scent of the flowers used at funerals was also very important as it masked the smell of decaying flesh. The fear of being buried prematurely led to the dead being laid out for several days before being interred.

> Get six of my comrades to carry my coffin,
> Six girls of the city to bear me on high,
> And each of them carry a bunch of red roses,
> So that they don't smell me as they walk along.
> – *A Trooper Cut Down in His Prime* (folk song)

The practice of sending wreaths to funerals became popular throughout the eighteenth century, but in recent times it has become considerably less so as more people are cremated and money is donated to charities. But not all traditions use flowers for funerals; Orthodox Jews associate them with joy and so find them to be inappropriate for funerals.

One of the best known, and most often repeated, superstitions relating to flower colours is that red and white blooms should never be put together in the same vase, as it is supposed to cause misfortune or even death. This appears to be a widespread, though relatively recent belief. At one time it was not uncommon to see red and white flowers used together to decorate churches at Whitsuntide. They were said to represent the fire and wind of the Holy Spirit as it came upon the Apostles, now however, it seems that they are taken to represent blood and bandages. It is not only red and white flowers which might be taken as indicators of death. For a plant to bloom out of season or have two, or more, flowers on the same stem where it would normally only have one might also indicate someone's imminent demise.

One possible exception to the taboo of using red and white together appears to be in May Day decorations. Here they are taken to represent male and female put together in the fertility festivities.

Various colours and combinations have been said to have special significance. White is frequently a colour for misfortune, probably because of its connections with the White Goddess. Red represents blood, and therefore life, so is a bringer of good luck. Violet is usually said to show goodwill on the part of the giver. Yellow or golden flowers are said to please all, being the colour of the sun, although some writers say that yellow flowers in your room will bring bad luck. Red and yellow is supposed to be an ideal combination for decorating a house in readiness for a party as they are said to encourage a festive atmosphere.

Picking flowers is not without its risks; one must never pick up flowers that have been dropped on the ground.

> Pick up a flower,
> Pick up a fever.

This quite possibly has its origin at a time when most people would only have had flowers when there had been a death in the house. Flowers dropped in the street would have been deemed as unlucky as they may have fallen from the floral tributes placed on the coffin.

Picking wild flowers was seen as courting misfortune, as was taking wild flowers into the house. A Suffolk tradition has it that wild flowers, and therefore garden weeds, were planted by God and so should never be pulled up. A variation on this suggests that flowers should not be brought into the house before 1 May. This may date from a much earlier belief that the majority of white flowering plants were sacred to the White Goddess. To pick them and take them into the home would have been an affront to her and brought about her anger. The reference to 1 May, the date of her great festival, would seem to add weight to this.

A superstition on Guernsey recommends that you should swear at your young plants if you wish to encourage them to grow. Presumably the idea is that the plants will grow just to spite you. This assumes, of course, that your plants have feelings and so can feel grief. It is therefore recommended that if there has been a death in the household the houseplants should also be put into mourning.

Other superstitions say that flowers should never be touched by a menstruating woman, as it will kill them. Similar superstitions can be found relating to cropped farmland and fruiting trees. A second flowering by a plant that would normally only flower once is taken as a sign that winter will be harsh. To smell flowers in a house where there are no flowers is an omen of death.

Dreaming of flowers is also something of a mixed blessing. It is often taken as an omen of misfortune and possibly a death, particularly if you see red and white flowers in your dream. Dreaming of picking or gathering flowers indicates good fortune, a delightful surprise or a lasting friendship. Withered flowers are a sign of impending good fortune and bright flowers indicate a pleasant life. If you dream of casting flowers away it betokens quarrels and despair, whereas a basket of flowers indicates a wedding or a birth. Dreaming of smelling flowers suggests that you must grasp an opportunity that will soon be presented to you.

The giving and wearing of flowers has always borne great significance. In

Westphalia there was a belief that if a child under the age of twelve months old was allowed to wear a wreath of flowers it would soon die. In ancient Athens children who had reached the age of three at the spring equinox were crowned with flowers to mark their safe passage through infancy.

There are lucky flowers, based on the day of one's birth, much as some would talk about a birthstone.

Sunday:The day is, of course, ruled by the sun and is traditionally the day on which great leaders are born. Orange and gold are the colours of the sun and so are the lucky colours of the people born on this day. The bright flowers of these colours, such as Californian poppy, chrysanthemum, day lily, sunflower and marigolds are the appropriate flowers.

Monday: This day is ruled by the moon and those born on Monday are said to be thoughtful, home-loving people with close family ties. For them all the grey and silvery plants are considered to be lucky. Moon daisies, honesty, lily of the valley, white roses, snowdrops and scented white phlox are their flowers.

Tuesday: The day is ruled by Mars. In classical myth he was the god of war and those born under the influence of this planet are fighters, people with plenty of drive and determination. Mars is the red planet, and red flowers best suit the personality of those born on this day, such as scarlet nasturtiums, geraniums, dahlias and red-hot pokers. Red-berried plants will also prove lucky for those under the influence of Mars, so holly, rowan and glory vine would all be suitable.

Wednesday: Mercury rules Wednesday. In myth he is the messenger of the gods and those born on this day are usually seen as natural communicators. For them all colours are suitable, all at the same time. Irises are a good choice as are Russell lupins, columbines, salpiglossis and vervain. Plants that have variegated foliage will also prove lucky for these people.

Thursday: This day is under the influence of Jupiter. Those born on Thursday are happy, jovial people, but also those born to command. Lucky flowers for these people are those that are blue or mauve. Lilac, lavender, violet and stock would all be appropriate, as would cyclamen and Michaelmas daisy.

Friday: The influences of Venus fall on those born on a Friday. This will make them natural charmers and tactful diplomats. Their flowers are all clear blue or pink in colour. Plants such as forget-me-nots, carnations, nigella and pink roses will bring them luck.

Saturday: This is the day of Saturn, dark and brooding. Those born on a Saturday tend to take life very seriously and are often ambitious. Their flowers are the browns and dark reds. Heavily scented russet coloured wallflowers, bronze chrysanthemums, deep red fuchsias and dark-coloured dahlias will all be suitable. They could, of course, be backed by sprays of bronzed beech leaves.

Flowers have become associated with gods, saints, countries and cities. The individual entries identify any deities to whom each plant is particularly significant. Overleaf is a short list of the flowers associated with specific saints.

Flowers dedicated to saints

Bluebell	St George
Carnation	St Peter and St Paul
Canterbury bell	St Augustine of England
Crocus	St Valentine
Crown imperial	Edward the Confessor
Daisy	St Margaret
Daffodil	St David
Herb Christophe	St Christopher
Lady's smock	St Mary the Virgin
Lychnis	St John the Baptist
Madonna lily	St Mary the Virgin
Oak	St Cedd
Rhododendron	St Augustine
Rose	St Alban
Rose – blush	St Mary Magdalene
Rose – red or white	St George
Rose – three-leaved	St Boniface
Rose – yellow	St Nicomede
Rush	St Brigid
St John's wort	St John
St Barnaby's thistle	St Barnabas
Sycamore	St Fintan
Tulip	St Philip; St James
Valerian	St George
Yew	St Congar

Floral calendars

Floral calendars are quite ancient in their origins. The oldest are the Chinese, where each flower has a special meaning. The Japanese floral calendar is a little later but I suspect that the flowers are equally significant. The compiler of the English calendar (taken from 1866) can bee seen to have drawn some of their inspiration from the East in the choice of the plants.

Month	Chinese	Japanese	English
January	Plum blossom	Pine	Snowdrop and Carnation
February	Peach blossom	Plum blossom	Primrose
March	Tree peony	Cherry, Peach and Pear	Violet and Daffodil
April	Cherry blossom	Cherry blossom and Wisteria	Daisy
May	Magnolia	Azalea and Peony	Hawthorn and Lily of the valley
June	Pomegranate	Iris	Honeysuckle and Roses
July	Lotus flower	Morning glory and Mountain clover	Water lily and Sweet pea
August	Pear blossom	Lotus (and Hill crest†)	Poppy and Gladioli
September	Mallow blossom	Seven grasses of autumn*	Morning Glory and Aster
October	Chrysanthemum	Chrysanthemum and Maple	Hop and Dahlia
November	Gardenia	Maple and Willow	Chrysanthemum
December	Poppy	Camellia and Paulownia	Holly and Poinsettia

* The seven grasses of autumn are: hagi bush (Japanese clover); Susuki (pampas grass); kuzu (arrowroot); nadeshiko (wild carnation); ominaeschi (maiden flower); fujibakama (chinese agrimony); hirogao (convolvulus or wild morning glory)

† In some systems there is no flower given for the month of August. In these cases the month is symbolized by a hillcrest behind which the moon rises as a harbinger of good fortune.

APPENDIX 2: GRASS

Meanings: Submission; utility

Herodotus tells us that some of the classical peoples used grass as a sign of submission, and it is from this that the symbolic meaning is derived. If you write a wish on a stone using grass, or mark the stone with grass whilst visualizing the wish, and then throw it into running water, your wish will come true. If there is no stream or river nearby the stone may be buried. To protect your home from evil, and to exorcise any evil already there present, tie knots in grass, or form some into a ball, and hang it outside the front door.

According to a traditional country rhyme, grass can be used in weather forecasting. The rhyme says:

When the dew is on the grass,
Rain will never come to pass.
When grass is dry at morning light,
Look for rain before the night.

Another rhyme says:

If it rains on Good Friday and Easter Day,
There will be plenty of grass but little good hay.

A cat eating grass is supposed to indicate that rain is coming.

A superstition from Lincolnshire recommends that if a young girl wishes to discover her future she should go into a churchyard on St Mark's Eve. When the clock strikes midnight she must grab three tufts of grass from a grave on the south side of the church, the tougher and ranker the better. She should then put these tufts of grass beneath her pillow when she goes to bed, saying:

The eve of St Mark by prediction is blessed,
Set therefore my hopes and my fears all to rest.
Let me know my fate, whether weal or woe,
Whether my rank is to be high or low,
Whether to be single or to be a bride,
And the destiny my star doth provide.

If she does not dream that night it is an indication that she should expect misery and a single life. Dreams fills with storms and thunder mean nothing but difficulties ahead. A pleasant dream is a portent of a happy life in the future.

Dreaming of fresh, green grass is a very good omen and indicates that you may expect long life, good luck and great wealth. Some sources suggest that it simply means that you may be coming into money. Seeing withered grass in your dreams portends the sickness and possible death of loved ones. Dreaming of cutting grass suggests that you will encounter great troubles, that you will not succeed in your dealings in the countryside and would do better to try your luck in the towns.

Gardeners wishing for a lush green lawn ought to sow the grass seed on a windy day, as this is supposed to ensure better growth. It should also be sown in the light of the moon, that is to say in the period between new and full. Cutting the grass during the light of the moon will also encourage its growth:

Cut it in the light of the moon
And you'll cut it again soon.

Finally, if there is an abundance of grass at Christmas it is a sure indication that there will be lot of deaths in the area in the coming year.

APPENDIX 3: **HAY**

Hay, like many other valuable agricultural crops, has a number of odd beliefs associated with it. For example, to meet a load of hay is supposed to be the best of good luck, although many drivers who have been stuck behind one on a country lane might not agree. When you come upon a load of hay you should make a wish. If you do this without looking at the hay again your wish will come true. Not to make a wish will only bring you misfortune. If the hay has been baled, as is commonly the case now, the wish will not come true until the bales are undone.

In some counties, however, it is said to be *bad* luck to meet a load of hay and you must spit on the hay to cancel out any ill effects. Occasionally two of the scythes used to crop the hay would be placed on top of the load to act as a charm and to prevent the spontaneous combustion of the hay.

In some parts of England it used to be believed that feeding cattle with a little stolen hay on Christmas Eve would ensure that they would have good health in the coming year. On various other days in the calendar it was traditional that hay should be strewn on the floors of churches.

To see a whirl in the hay as the wind blew through it was said to be a good omen for the farmer as it was supposed to be the faeries making their way through it. A similar superstition can be found applied to corn and other cereal crops. Indeed it is probable that many such superstitions can be, and have been, transferred between various crops.

One old wives' cure for toothache is to add the flowers of hay grasses to bath water.

APPENDIX 4: **ROOTS**

Some superstitions are related to plants in general rather than to a specific species. Such is the situation with root talismans. According to one superstition if you are forced to sleep in the open at night, and have no form of protection, you should hang the root of any nearby plant about your neck. This will guard you from the attacks of wild beasts. If the root you select comes from a sacred site, such as a churchyard, so much the better as this will protect you from death.

Magicians can use roots as an indicator of their own level of skill, or that of their pupils. They must enter a field at night and pull a weed from the ground. The amount of soil that clings to the roots shows how much more power the magician will ultimately attain. They can do the same whilst thinking of their pupil, to see how successful their teaching will prove to be.

APPENDIX 5: TREES

Myth and Legend

A Siberian creation legend, which cannot be applied to any specific tree species, tells of the tree of life that sprang up with no branching. God caused nine limbs to grow from its trunk; and nine men were born at its base. These were the fathers of the nine races of mankind. Five of the tree's limbs, which grew from the eastern side of the tree, furnished man and beast with fruits but God forbade men from eating any of the fruits from those that grew on the western side. A serpent and a dog guarded them. One day, when the snake slept, Erlik, the tempter, climbed up into the branches on the western side and from there persuaded the woman, Edji, to take and eat the forbidden fruits. She took them and shared them with Torongoi, her husband. As in the biblical creation story in the book of Genesis, the pair saw their nudity and went and covered themselves in skins and hid beneath the tree.

Magic and Lore

Many trees were seen as deities in their own right or as synonymous with specific deities from various pantheons. Weapons were formed from the wood of specific trees, as it was believed that the characteristic associated with the tree would then pass into the weapon.

In many traditions, the planting of a tree was seen as an investment in life. There was a German custom that a tree would be planted at a wedding for the happy couple. Its health would show the character of their love. Similarly, as children arrived, trees would be planted: apples for boys and pears for girls.

The link between trees and people extends into the superstition that where a tree dies in the garden it indicates that there will be a death in the family. This can be extended so that the death of a tree in the street implies that there will be a death amongst the people living in the neighbourhood. When a tree planted on a grave dies it is an omen that the planter will also soon be dead. The blossoming of trees twice in the same year or out of season is another portent of a death in the family, and if you see a member of your family sitting in a tree when you know that they are really at home, it again portends their demise.

In ancient China there were four primary tree deities, the acacia, pine, hemlock and catalpa. In warfare Chinese conquerors would fell sacred forests rather than destroy temples. Marco Polo also reported that the great Khan had many trees planted in the belief that 'he who plants a tree will have a long life'.

The Sidhe, the faerie folk, protected certain trees, or trees on special sites. A lone thorn bush growing on uncultivated land, especially on the side of a hill, might be just such a faerie tree. Some writers are more specific regarding the sites where faerie trees grow, stating that they will be within a fort, within a faerie ring or amongst rough grass in a rocky field, especially if close to a boulder or running water. Hawthorn is a favourite faerie tree species but others, such as elder, blackthorn, rowan, willow, hazel, alder and ash might also come under the protection of the faeries. It is, of course, unwise to harm a tree protected by the faerie folk, as it will result in illness and misfortune.

The superstition of touching wood is thought by some to have been derived from the wood used to form the cross used at Christ's crucifixion. Touching any wood is, therefore, a remembrance of the crucifixion and so a sign of compassion and reverence.

Trees play their part as rural weather oracles. Rain is foretold when the leaves are seen withering on the tree, and when leaves turn upside down or inside out or curl up rain will follow within twenty-four hours. If, in the autumn, the leaves cling to the trees there will be a harsh winter and snow should be expected. Indeed it is claimed that if, throughout the year, there is a heavy crown of leaves on the trees it indicates that the winter will be particularly hard, whereas thin crowns in the summer suggest a mild winter. When frost clings to the leaves in the morning it indicates that snow is on the way and if, in the spring, thunderstorms are heard before the trees come back into leaf there will be more cold weather to come.

Not surprisingly there is also a certain amount of superstitious belief surrounding the cultivation of trees. There is a tradition in some parts that they ought be planted on 25 May, preferably with oats at their base to encourage them to thrive. In a twist to the idea that the wellbeing of a person could be linked to the health of a tree it has been recommended that newly planted trees should be named after prosperous people so that they will also prosper. Ideally trees should be planted 'when the moon is old' but clearly great care must be taken as it is claimed that trees will die if it cut into during the dark of the moon or at the new moon. Various traditions say that 13 August is a good day to kill trees and bushes. Or when a tree is cut on any date between 7 and 14 August, or on the 21st, it will die. Felling trees on any day between 1 October and the end of November will prevent the wood becoming wormy. It is, of course, always bad luck to burn the prunings of trees, bushes or vines. Finally, one old wives' tale claims that to prevent an ailing tree from dying you must knock a nail into the northern side of the trunk.

Dreaming of green trees means that great wealth is coming to you. If you dream of trees in bud it indicates a new love. Trees in luxuriant foliage means a happy marriage and children. Bare of leaves it indicates matrimonial troubles. Fruit trees in dreams are always lucky, indicating great prosperity in business. A dream of trees covered in green leaves is also said to be a good omen in affairs of the heart. Autumn leaves betoken sorrow, probably as a result of the dreamer's actions. Dreaming of burning leaves tells of a great loss to a friend. Dreaming of pulling up a tree by its roots indicates that a relative will die.

APPENDIX 6: **VEGETABLES**

Much of the information relating to specific species has been covered under separate headings; there are, however, some general points that may be applied to vegetables as a whole.

There has been a certain amount of interest in planting according to the phase of the moon. The basic premise being that 'above ground crops' should be planted in the light of the moon, i.e. on the days between the new moon and the full moon. 'Below ground crops' must be planted in the dark of the moon, that is between the full moon and the next new moon. Refinements on this require that leaf crops are planted at the new moon and fruit crops or flowers planted at the full moon.

If you find 'twin' vegetables when lifting your root crops, a forked carrot or parsnip for example, it indicates that your marriage is fast approaching.

To protect your crops from pests try scattering wood ash over them on Ash Wednesday.

BIBLIOGRAPHY

Websites

www.california-books.com

www.gardendigest.com/treeleg.htm

Mystical www – Plants, Flowers, Vegetables, Fruit and Herb Mystical Beliefs, http://www.mystical-www.co.uk

Miscellaneous books

Agnus Castus, 14th Century Middle English Manuscript

Bulletin of Miscellaneous Information, Royal Botanical Gardens, Kew, No. 37 HMSO, London, 1890

Celtic Tree Calendar http://www.dailyglobe.com/celtic.html

Complete Book of Fortune (Studio Editions, 1990)

Encyclopaedia of Magic and Superstition (Black Cat, 1988)

Encyclopaedia of World Mythology (Octopus, 1975)

Hallmark – Season and Reasons – Language of Flowers (http://www.hallmark.com.seasons)

Herb Lore (http://www.greycircle.demon.co.uk/grey/herbs.html)

The Language and Poetry of Flowers (Milner and Co., 1850)

Llewelyn's 1994 Organic Gardening Almanac (Llewelyn Publications, 1993)

Macer Floridus de Viribus Herbarum (Macer's Herbal) 12th Century Middle English Manuscript (http://www.ub.unibielefeld.de/diglib/kesmark/macerfloridus/)

Plant Folklore – Pocket Reference Digest (Geddes and Gosset, 1999)

The Readers Digest Complete Library of the Garden (Readers Digest Association, 1963)

Trees, and the Tree Alphabet (http://www.greycircle.demon.co.uk/grey/trees.html)

Books

Addison, Josephine & Hillhouse, Cherry, *Treasury of Flower Lore* (Bloomsbury Publications, 1977)

——, *Treasury of Tree Lore* (Andre Deutsch, 1999)

Alexander, H.B., *The World's Rim* (University of Nebraska Press, 1953)

Allardice, Pamela, *Aphrodisiacs and Love Magic* (Prism Press, 1990)

Allen, Andrew, *A Dictionary of Sussex Folk Medicine* (Countryside Books, 1995)

Ashe, Geoffery, *Mythology of the British Isles* (Methuen, 1993)

Askham, Anthony, *A Lytel Herball* (1550) (http://voynichcentral.com/users/gc/Ascham/herball_1550.html)

Baker, Margaret, *Discovering the Folklore of Plants* (Shire Publications, 1999)

——, *Folklore and Customs of Rural England* (David and Charles, 1974)

——, *The Gardener's Folklore* (David and Charles, 1977)

Barham, Richard H., *The Ingoldsby Legends* (William Heinemann, 1909)

Barrett, Francis, *The Magus* (University Books Inc., 1967)

Berboe, Edward, *The Browning Cyclopedia* (Swan Sonnenschein & Co., 1897)

Biles, Roy E., *The Complete Book of Garden Magic* (J.G. Ferguson Publishing Company, 1961)

Boland, Maureen & Bridget, *Old Wive's Lore for Gardeners* (The Bodley Head, 1977)

Boorde, Andrew (ed. by F.J. Furnivall) *Introduction of Knowledge & Dyetary of Health* (1542)

Branstone, Brian, *The Lost Gods of England* (Thames and Hudson, 1974)

Brewer, E. Cobham, *The Dictionary of Phrase and Fable* (Avenel Books, 1978)

Brickell, Chris (ed.) *A–Z Encyclopaedia of Garden Plants* (Dorling Kindersley, 1999)

Briggs, Katherine, *Abbey Lubbers, Banshees and Boggarts* (Viking, 1979)

Buchner, Greet and Hoogvelt, Fieke, *Nature on Your Side* (Pan Books, 1978)

Burke, L., *The Language of Flowers* (Hugh Evelyn, 1968)

Coder, Dr Kim D., *Cultural Aspects of Trees: Traditions and Myths* (http://stl20.ces.uga.edu/forestry/docs/for96-47.html)

Coles, William, *Adam in Eden* (Printed by J. Streater for Nathaniel Brooke, 1657)

——, *The Art of Simpling: An Introduction to the Knowledge and Gathering of Plants* (Kessinger Publishing Co., 2004)

Coghan, Thomas, *The Haven of Health* (Richard Field, 1596)

Coombes, Allen J., *A–Z of Plant Names* (Chancellor Press, 1994)

Cooper, J.C., *Brewer's Myth and Legend* (Cassell Publishers, 1992)

Crowley, Aleister, *Magick* (Routledge & Kegan Paul, 1973)

Culpeper, Nicholas, *The Complete Herbal And English Physician* (Foulsham and Co., 1997)

Cummingham, Scott, *Cummingham's Encyclopedia of Magical Herbs,* 16th edition (Llewelyn Publications, 1993)

Dalriada Celtic Heritage Society, *Herbs and Plants in Celtic Folklore* (gopher://gopher.almac.co.uk/11/scotland/dalriada, 1997)

Davies, B. & Knapp B., *Know Your Common Plant Names* (MDA Publications, 1992)

Day, Alice, *Gothic Gardening* (http://www.gothic.net/~malice/)

De Bray, Lys, *Midsummer Silver* (J.M. Dent and Sons, 1980)

De Gex, Jenny, *Shakespeare's Flora and Fauna* (Pavilion Books, 1999)

De Givry, Emile Grillot, *Illustrated Anthology of Sorcery, Magic, and Alchemy* (Zachary Kwinter Books, 1991)

Drayton, Michael, *Nymphidia* (1627) (from *A Sixteenth Century Anthology,* ed. by Arthur Symons, Blackie & Son, 1905)

Evans, George Ewart, *The Pattern Under The Plough* (Faber & Faber, 1966)

Evelyn, John, *Acetaria: A Discourse of Sallets* (1699, www.gutenberg.net)

Folkard, Richard, *Plant lore, legends, and lyrics* (Sampson, Low, Marston, Searle and Rivington, 1884)

Fraser, James G., *The Golden Bough* (Papermac, 1995)

Froud, Brian and Lee, Alan, *Faeries* (Souvenir Press, 1978)

Genders, Roy, *A Book of Aromatics* (Darton, Longman and Todd, 1977)

——, *Flowers and Herbs of Love* (Darton, Longman and Todd, 1978)

——, *Growing Herbs* (David McKay & Co., 1980)

Gerard, John, *The Herball* or *General Historie of Plants* (Bracken Books, 1985)

Gordon, Lesley, *Green Magic (Flowers, Plants and Herbs in Lore and Legend)* (Webb and Bower, 1977)

——, *The Mystery and Magic of Trees and Flowers* (Webb and Bower, 1985)

——, *Trees* (Grange Books, 1993)

Graves, Robert, *The White Goddess* (Faber & Faber, 1960)
Grieve, M., *A Modern Herbal* (Tiger Books International, 1992)
Griffiths, Mark, *Index of Garden Plants* (Macmillan, 1994)
Grigson, Geoffrey, *A Dictionary of English Plant Names* (Allen Lane, 1974)
——, *The Englishman's Flora* (Helicon Publishing, 1996)
——, *The Shell Country Book* (J.M. Dent & Son, 1973)
Gunther, Robert T., *The Greek Herbal Of Dioscorides* (Hafner Publishing Company, 1968)
Halliwell, James Orchard, *Popular Rhymes and Nursery Tales,* 4th edition (John Russell Smith, 1849)
Harrington, Sir John, *The Englishman's Doctor* (Printed for John Helme and John Busby Junior in Fleet Street, 1608) (http://user.icx.net/~richmond/rsr/ajax/harington.html, 2000)
Hatfield, Audrey Wynne, *Pleasure of Herbs* (Museum Press, 1964)
Hessayon, Dr D.G., *The Armchair Book of the Garden* (PBI Publications, 1986)
Hill, Thomas, *A Most Briefe and Pleausant Treastyse* (1563)
——, *The Proffitable Arte of Gardening* (Thomas Marshe, 1572)
——, *The Gardener's Labyrinth* (Oxford University Press, 1988)
Hohn, Reinhardt, *Curiosities of the Plant Kingdom (*Cassell Publishers, 1980)
Holland, Philemon, (trans.) *The Historie Of The World* (Commonly called *The Naturall Historie Of C. Plinius Secundus)* (1601) (http://Penelope.Uchicago.Edu/Holland/Index.Html)
Hopman, Ellen Evert, *Tree Medicine, Tree Magic* (Phoenix Publishing, 1992)
Hort, Sir Arthur, *Theophrastus, Enquiry Into Plants* (William Heinemann, 1916) (http://www.archive.org/stream/enquiryintoplant00theo/enquiryintoplant00theo_djvu.txt)
Howard, Elizabeth Jane, *Greenshades* (Aurum Press, 1991)
Howes, Michael, *Amulets* (Robert Hale, 1975)
Huxley, Anthony (ed.) *The New R.H.S. Dictionary of Gardening* (Macmillan, 1992)
Inwards, Richard, *Weather Lore* (Senate, 1994)
Jacob, Dorothy, *A Witch's Guide to Gardening* (Taplinger, 1965)
Jones, Evan John and Valiente, Doreen, *Witchcraft: A Tradition Renewed* (Robert Hale, 1990)
King, Francis X., *Mind and Magic* (Dorling Kindersley, 1991)
Launert, Edmund, *The Hamlyn Guide to Edible and Medicinal Plants of Britain and Northern Europe* (Hamlyn, 1989)
Lawson, William, *A New Orchard And Garden With The Country Housewife's Garden* (Prospect Books, 2003)
Lehane, Brendan, *The Power of Plants* (McGraw Hill, 1977)
Leyel, C.F., *Elixirs of Life* (Faber & Faber, 1948)
——, *Herbal Delights* (Faber & Faber, 1947)
Lightfoot, John, *Flora Scotica* (Gale ECCO, 2010)
Lyte, Henry, *Niewe Herball* (1578 – a translation of Dodoens, Rembert, *Cruydeboeck*, 1564)
Maundeville Sir John (ed. by A.W. Pollard) *The Travels of Sir John Maundeville* (Macmillan, 1900)
Mayhew, Ann, *The Rose* (New English Library, 1979)

Nahmad, Claire, *Garden Spells (The Magic of Herbs, Trees and Flowers)* (Parkgate Books, 1994)
Parkinson, John, *Paradisis in Sole Terrestris* (Aberdeen Press, 1904)
——, *Theatrum Botanicum* (Thomas Cotes, 1640)
Paterson, Jacqueline Memory, *Tree Wisdom* (Thorsons, 1996)
Pechey, John, *The Compleat Herbal of Physical Plants* (Gale Ecco, 2010)
Perry, Leonard, *Fall Asters* (Vermont Weathervane, http://www.vtweb.com/vermontweather vane/96.9september/seasonal.html)
Pettit, E. (ed. & trans.), *Anglo-Saxon Remedies, Charms, and Prayers from British Library MS 585: The 'Lacnunga'*, 2 vols. (Mellen Press, 2001)
Pickles, Sheila, *The Complete Language of Flowers* (Pavilion Books, 1998)
Porteous, Alexander, *The Lore of the Forest* (Senate, 1996)
Powell, Claire, *The Meaning of Flowers* (Jupiter Books, 1977)
Prior, R.C.A., *On the Popular Names of British Plants being an Explanation of the Origin and Meaning* (Williams & Norgate, 1863)
Roddins, R.H., *The Encyclopaedia of Witchcraft and Demonology* (Peter Nevill, 1964)
Rohde, Eleanour Sinclaire, *A Garden Of Herbs* (The Medici Society, 1922)
Rose, Graham, King, Peter and Squire, David, *The Love of Roses – From Myth to Modern Culture* (Quiller Press, 1990)
Salmon, William, *Botanologia: The English Herbal* (Printed by I. Dawks for H. Rhodes and J. Taylor, London, 1710)
Seymour, John, *Gardener's Delight* (Michael Joseph, 1978)
Schofield, Bernard, *A Miscellany of Garden Wisdom* (Harper Collins, 1995)
Singleton, Ester, *The Shakespeare Garden* (The Century Co., 1922)
Skinner, Charles M., *Myths and Legends of Flowers, Trees, Fruits and Plants* (J.B. Lippincott & Co., 1911)
Slade, Paddy, *Natural Magic* (Hamlyn, 1990)
Smith, A.W. (revised by W.T. Stearn), *A Gardener's Dictionary of Plant Names* (Cassell Publishers, 1972)
Smith, Keith Vincent, *The Illustrated Earth Garden Herbal* (Thomas C. Lothian Pty, 1994)
Spencer, Lewis, *The Encyclopaedia of the Occult* (Bracken Books, 1988)
Squire, Charles, *Celtic Myths and Legends* (Paragon, 1998)
Stace, Clive A., *New Flora of the British Isles* (Press Syndicate of the University of Cambridge, 1995)
Strabo, Walafrid, *Hortulus* or *The Little Garden* (840)
Surflet, Richard (trans.) *The Countrie Farme* (Printed by Edm. Bollifant for Bonham Norton, London, 1600)
Taylor, Brian, *Gardens of the Gods* (Clun Valley Publications, 1996)
Thistleton-Dyer, Thomas Firminger, *Folk-lore of Plants* (Llanerch, 1994)
Todd, Pamela, *(Flora's Gems) The Little Book of Daffodils* (Little, Brown, 1994)
——, *(Flora's Gems) The Little Book of Tulips* (Little, Brown, 1994)
Tomkins, Peter and Bird, Christopher, *Secret Life of Plants* (Penguin, 1971)
Tongue, Ruth L., *Forgotten Folk-Tales of the English Counties* (Routledge & Kegan Paul, 1970)

Turner, William, *A New Herbal* (1551)
——, *A New Herbal*, ed. by George T.L. Chapman and Marilyn N. Twiddle (Cambridge University Press, 1989)
Tusser, Thomas, *Five Hundred Points of Good Husbandry* (Oxford University Press, 1984)
——, *July's Husbandry* (1577)
Vickery, Roy, *A Dictionary of Plant-Lore* (Oxford University Press, 1995)
Ward, Bobby J., *A Contemplation Upon Flowers* (Timber Press, 1999)
Waring, Philippa, *Dictionary of Omens and Superstitions* (Treasury Press, 1984)
Wheatley, Dennis, *The Devil and all his Works* (Peerage Books, 1983)
Wilkinson, Gerald, *Epitaph for the Elm* (Arrow Books, 1979)
Withan Fogg, H.G., *A History of Popular Garden Plants From A–Z* (Kaye & Ward, 1976)
Wren, R.W., *Potter's New Cyclopaedia of Botanical Drugs and Preparations* (The C.W. Daniel Company, 1985)
Young, Norman, *The Complete Rosarian* (The Gardener's Book Club, 1972)
Zalewski, C.L., *Herbs in Magic and Alchemy* (Prism Press, 1999)
Zolar, *The Encyclopaedia of Ancient and Forbidden Knowledge* (Peerage Books, 1992)

INDEX